Geometry in Partial Differential Equations

GEOMETRY IN PARTIAL DIFFERENTIAL EQUATIONS

A. Pràstaro
Rome, Italy

Th. M. Rassias
Athens, Greece

World Scientific
Singapore • New Jersey • London • Hong Kong

Published by

World Scientific Publishing Co. Pte. Ltd.
P O Box 128, Farrer Road, Singapore 9128
USA office: Suite 1B, 1060 Main Street, River Edge, NJ 07661
UK office: 73 Lynton Mead, Totteridge, London N20 8DH

GEOMETRY IN PARTIAL DIFFERENTIAL EQUATIONS

ISBN 981-02-1407-3

Printed in Singapore by JBW Printers & Binders Pte. Ltd.

PREFACE

Geometric tools have been of particular importance to the understanding of the structure of Partial Differential Equations (PDEs). This has been shown in the pioneering works of eminent mathematicians such as L. Bianchi (1856–1928), C. Carathéodory (1873–1950), E. Cartan (1869–1951), L. Euler (1707–1783), C. F. Gauss (1777–1855), T. Levi-Cirita (1873–1941), S. Lie (1842–1899), E. Noether (1882–1935), H. Poincaré (1854–1912), G. Ricci-Curbastro (1853–1925), B. Riemann (1826–1866), H. Weyl (1885–1955), and, during recent times, in the works of C. Ehresmann, H. Goldschmidt, M. Kuranishi, B. Malgrange, D. Spencer, S. Steinberg, A. M. Vinogradov, among others. Differential Geometry and Algebraic Topology play a fundamental role in the development of a geometrical framework for the understanding of PDEs. In fact the first important step that has been taken in this direction has been the introduction of manifolds with some particular structure, which represent the natural environment for a PDE. This consists of the jet derivative spaces, first strictly formalized by C. Ehresmann. This way, PDEs can be regarded as submanifolds of these types of spaces and thus the machinery of Differential (Algebraic and Topological) Geometry is naturally introduced. In this context the Analysis of the space of solutions can be regarded as a sub-product of a more general geometric theory. In fact, what is important in the characterization of PDEs is not so much their space of solutions but rather the manifold that defines them.

The big advantage in the study of PDEs from the geometric point of view is the possibility of investigating the general behavior of their structure, without considering the set of solutions and without going into many particular cases. Moreover, when one uses for example modern Cartan theory one can obtain general results about the characterization of singular solutions of PDEs. This aspect is particularly interesting also from the applications point of view, where singular solutions can play a significant role in order to explain several fundamental phenomena.

It is important also to underline the current developments in the geometric approach. A few such approaches include: general methods to obtain conservation laws for PDEs, structural investigation for the understanding of the meaning of quantum geometry in PDEs, extensions to super PDEs (formulated in the category of supermanifolds) of the geometrical methods introduced for PDEs. The harmonic theory proved to be very important after the appearance of the Atiyah–Singer index theorem, which provides a link between geometry and topology.

The Yang–Mills equations are of interest to both physicists and mathematicians. M. Atiyah and R. Bott found applications of the Yang–Mills theory in the study of vector bundles over Riemann surfaces. After S.-T. Yau's solution of the Calabi conjecture several surprising results followed in Geometry and Topology. These developments include the solution of the positive mass conjecture and other applications of minimal surfaces, S. Donaldson's construction of four manifold invariants using the moduli spaces of connections, among other results. The most important quasilinear equation in Geometry is the minimal surface equation. Minimal surfaces have, over a long period of time, been a powerful tool in understanding the topology and the geometry of manifolds. Minimal surfaces also play a very important role in the study of the curvature of a manifold (works of J. Sacks and K. Uhlenbeck, R. Schoen and S.-T. Yau are very representative in this direction). The first problem in minimal surface theory is the Plateau problem. The works of L. Ljusternik and L. Schnirelman and of Marston Morse provide two important approaches towards the foundation of global analysis. In the 60's Morse theory was generalized by R. Palais and S. Smale to infinite-dimensional Hilbert manifolds. The Morse–Palais–Smale theory has proved to be useful (and usable) in the study of PDEs.

This volume emphasizes the interdisciplinary interaction in problems involving Geometry and Partial Differential Equations. It provides an attempt to follow certain threads that interconnect various approaches in the applications of geometry and their influence to PDEs.

We wish to express our wholehearted thanks to the scientists who contributed their works in this book and, last but not least, the World Scientific Publishing Company for their excellent cooperation.

March 1993

Agostino Prástaro
Themistocles M. Rassias

CONTENTS

Geometry in Partial Differential Equations (pp. 1–17)
eds. A. Pràstaro and Th. M. Rassias

SOME APPLICATIONS OF THE COAREA FORMULA TO PARTIAL DIFFERENTIAL EQUATIONS

FABRICE BETHUEL

Centre de Mathématiques et Leurs Applications
Ecole Normale Supérieure de Cachan et CNRS
61 Avenue du Président Wilson, 94235 Cachan Cedex, France
and
CERMA - ENPC, La Courtine, 93167 Noisy le Grand Cedex, France

and

JEAN-MICHEL GHIDAGLIA

Centre de Mathématiques et Leurs Applications
Ecole Normale Supérieure de Cachan et CNRS
61 Avenue du Président Wilson, 94235 Cachan Cedex, France

ABSTRACT *

In this article we give two applications of the coarea formula of Federer and Fleming to problems in P.D.E's. The first question we address concerns a functional inequality involving a Jacobian (know as Wente's estimate) and the use of the coarea formula allows us to show that the constant appearing in this inequality can be chosen independently of the domain. In the second part we address a totaly different question which is related to the stationary Euler equations for inviscid incompressible fluids. We propose a new proof of the fact that, in some circonstances, weak limits of approximate solutions are still weak solutions. This result is then applied to an existence result for weak solutions to a damped Euler equation.

1 Introduction

Let f be a C^2 function from $\mathbb{R}^2$ into $\mathbb{R}$ and h be a continous function from $\mathbb{R}^2$ into $\mathbb{R}_+ = [0, \infty[$. For every mesurable set $X \subset \mathbb{R}^2$, the coarea formula of Federer and Fleming[12] reads

$$\int_X h|\nabla f|dx = \int_{\mathbb{R}} \left(\int_{X_\lambda} h\, ds \right) d\lambda \qquad (1)$$

*Work partially supported by a Grant from the Ministère de la Recherche et de la Technologie (Contrat N^0 90S0315).

where $X_\lambda \equiv \{x \in X; f(x) = \lambda\}$ are the level sets of f and s denotes the curviligne abscissa on X_λ (recall that according to Sard's Lemma, for almost every λ the set X_λ is finite union of C^1 curves with regular points).

Formula (1), which is a particular case of the general coarea formula valid in $I\!R^n$ Federer[12], turns out to be a versatile tool in analysis. Our aim in this work is to present two totally different applications of (1) to problems in the theory of partial differential equations. The paper is organized as follows. In the Section 2, we present our first application (Theorem A) of the coarea formula to generalizations and complements about Wente's estimate. Section 3 is then devoted to some applications of this result to problems arising from geometry. We prove Theorem A in Section 4 and give some remarks and complements in Section 5. Then Section 6 deals with our second main application, related to the Euler equation for inviscid incompressible fluids in two dimensions (Theorem B). Thanks to this result, we construct in Section 7 a weak solution to the Charney's model for the Gulf-Stream. Finally Section 8 deals with the proof of Theorem B and its relation to the coarea formula.

2 Generalizations of Wente's estimates

Let Ω be a smooth and bounded domain in $I\!R^2$. We are given a second order self-adjoint elliptic operator in divergence form with variable coefficients : $-\frac{\partial}{\partial x_i}\left(a_{ij}(x)\frac{\partial}{\partial x_j}\right)$; and we consider the following problem on Ω.

Given u and v in $H^1(\Omega)$, find $\varphi \in W_0^{1,1}(\Omega)$ such that

$$-\frac{\partial}{\partial x_i}(a_{ij}(x)\frac{\partial\varphi(x)}{\partial x_j}) = g(u,v)\{u,v\}(x), \text{ in } \Omega, \tag{2}$$

$$\varphi = 0 \text{ on } \partial\Omega, \tag{3}$$

where we have denoted

$$\{u,v\} = \frac{\partial u}{\partial x_1}\frac{\partial v}{\partial x_2} - \frac{\partial u}{\partial x_2}\frac{\partial v}{\partial x_1}, \tag{4}$$

and $g \in C^0(I\!R^2) \cap L^\infty(I\!R^2)$.

The hypotheses on the a_{ij}'s are "minimal" i.e. we assume that there exists $\alpha > 0$ and $M < \infty$ such that for almost every $x \in \Omega$

$$\alpha|\xi|^2 \leqslant a_{ij}(x)\xi_i\xi_j \leqslant M|\xi|^2, \forall \xi \in I\!R^2; \tag{5}$$

that is : the a_{ij} are essentially bounded functions on Ω and the elliptic operator is coercive. We these notations we have the following result.

Theorem A Bethuel and Ghidaglia[4] *The problem (2)-(3) possesses a unique solution* $\varphi \in H^1(\Omega) \cap L^\infty(\Omega)$ *and there exists a universal constant* C_1 *such that*

$$|\varphi|_{L^\infty(\Omega)} + |\nabla\varphi|_{L^2(\Omega)} \leqslant \frac{C_1}{\alpha}|g|_{L^\infty(\mathbb{R}^2)}|\nabla u|_{L^2(\Omega)}|\nabla v|_{L^2(\Omega)}. \tag{6}$$

By taking $a_{ij}(x) \equiv \delta_{ij}$ and $g \equiv 1$ we deduce from Theorem A the following corollary.

Corollary 2.1 *There exists a universal constant* C_1, *such that for every* u *and* v *in* $H^1(\Omega)$, *the solution* φ *to*

$$-\Delta\varphi = \{u, v\}, \text{ in } \Omega, \tag{7}$$

$$\varphi = 0 \text{ on } \partial\Omega, \tag{8}$$

belongs to $C^0(\bar{\Omega}) \cap H^1(\Omega)$ *and satisfies*

$$|\varphi|_{L^\infty(\Omega)} + |\nabla\varphi|_{L^2(\Omega)} \leqslant C_1|\nabla u|_{L^2(\Omega)}|\nabla v|_{L^2(\Omega)}. \tag{9}$$

Comments 2.1

1. Since u and v belong to $H^1(\Omega), \{u, v\}$ belongs to $L^1(\Omega)$ and it is standard to construct a (unique) solution, φ, to (7)-(8) in $W_0^{1,1}(\Omega)$. This solution satisfies $-\Delta\varphi \in L^1(\Omega)$ but one cannot deduce from this fact that $\varphi \in W^{2,1}(\Omega)(\subset C^0(\bar{\Omega}))$.

2. The inequality (9) was first proved in Wente[17], in the case $\Omega = \mathbb{R}^2$. By using the maximum principle it is then standard to extend (9) to an arbitrary smooth domain in $\mathbb{R}^2$ but in general the constant obtained by this method will depend on Ω, see Brezis and Coron[6]. Note however that if the domain Ω is *simply connected*, by using a conformal transformation from Ω into the unit ball in $\mathbb{R}^2$, it is possible to show that the constant will not depend on Ω. We would like to mention that in some applications of (9)-(or (6)), see Bethuel and Ghidaglia[4], Ω is the level set of a function, hence Ω is not simply connected in general.

3 Applications to problems in geometry of Theorem A

3.1 *Surfaces of prescribed mean curvature*

We consider on a bounded domain $\Omega \in \mathbb{R}^2$ a function $u : \Omega \to \mathbb{R}^3$ that satisfies the equation of prescribed mean curvature H :

$$\Delta u = H(u)\frac{\partial u}{\partial x} \times \frac{\partial u}{\partial y}, \text{ in } \mathcal{D}'(\Omega), \tag{10}$$

where $H : \mathbb{R}^3 \to \mathbb{R}$ is smooth and bounded. This equations comes from geometry : if u is a conformal representation of a surface in $\mathbb{R}^3$, then the mean curvature of this surface at the point u is $H(u)$ when u satisfies (10).

We label two different types of hypotheses on the function H :

$$\text{there exists } C \text{ such that for every } \xi \in \mathbb{R}^3, |\nabla H(\xi)| \leqslant \frac{C}{1+|\xi|}, \tag{11}$$

$$\text{there exists } n \in \mathbb{R}^3 - \{0\} \text{ such that for every } \xi \in \mathbb{R}^3\, n.\nabla H(\xi) = 0. \tag{12}$$

Theorem 3.1 *Assume that* $H \in \mathcal{C}^\infty(\mathbb{R}^3;\mathbb{R}) \cap L^\infty(\mathbb{R}^3;\mathbb{R})$ *and satisfies either* (11) or (12). *If* $u \in H^1(\Omega;\mathbb{R}^3)$ *solves* (10) *then* $u \in \mathcal{C}^\infty(\Omega;\mathbb{R}^3)$.

Theorem 3.1, with hypothesis (11) is due to Heinz[13]. In Bethuel and Ghidaglia[4] we propose a totally different proof using Corollary 2.1 (we believe that our proof is simpler than that of Heinz[13]). The proof of Theorem 3.1 relies on Theorem A with $a_{ij} = \delta_{ij}$ and $g(u_1, u_2) = H(u_1, u_2, u_3)$ where $(u_1, u_2, u_3) \in \mathbb{R}^3$ represents the coordinates of u in a basis where $n = (0, 0, n_3)$. We refer to Bethuel and Ghidaglia[4] concerning the details of the proof.

3.2 *Uniqueness of harmonic maps into spheres*

For $\Omega \subset \mathbb{R}^2$ and $g \in H^{\frac{1}{2}}(\partial\Omega;\mathbb{S}^N), N \geqslant 2$, we study the set of solutions to the system : find $u \in H^1(\Omega;\mathbb{S}^N)$ such that

$$-\Delta u = u|\nabla u|^2, \text{ in } \Omega, \tag{13}$$

$$u = g \text{ on } \partial\Omega. \tag{14}$$

Theorem 3.2 *Let* Ω *be a bounded domain in* $\mathbb{R}^2$. *There exists a constant* $e_0(\Omega)$, *depending only on* Ω, *such that (13)-(14) has at most one solution satisfying*

$$E(u) \equiv \int_\Omega |\nabla u|^2 dx < e_0(\Omega).$$

The problem of finding a solution to (13)-(14) has always a solution obtained by minimizing the energy $E(u)$. In the case $\Omega = \mathbb{R}^2$, it is easy to see that $e_0 = 8\pi$.

Moreover, in this special case, Theorem 3.2 has already been proved by Sacks and Uhlenbeck[15] in the more general case of an arbitrary compact manifold instead of $\mathbb{S}^N$.

The constant in Theorem 3.2 must depend on Ω (see Bethuel, Brezis, Coleman and Helein[3] or [4]). We refer to Bethuel and Ghidaglia[4] concerning the proof of this Theorem.

4 Proof of Theorem A

Our aim is to prove inequality (6). We first observe that it is sufficient to show this inequality in the case where a_{ij}, g, u, v (hence φ) are smooth functions on $\bar{\Omega}$. Then (6) follows by a standard limiting procedure. Hence we assume in the sequel that $a_{ij}, g, u, v,$ and φ belong to $\mathcal{C}^\infty(\bar{\Omega})$.

4.1 *Preliminaries*

We introduce the Green function $G : \Omega \times \Omega \to I\!R$ associated to the elliptic operator $-\frac{\partial}{\partial x_i}(a_{ij}\frac{\partial}{\partial x_j})$.

For every $m \in \Omega, G(.,m)$ solves

$$-\frac{\partial}{\partial x_i}(a_{ij}(x)\frac{\partial G(x,m)}{\partial x_j}) = \delta_m(x), \text{ in } \mathcal{D}'(\Omega), \tag{15}$$

$$G(x,m) = 0 \, \textit{for} \; x \in \partial\Omega, \tag{16}$$

where δ_m denotes the Dirac mass at m. As it is well known, $G(.,.)$ is smooth except on the diagonal $x = m$ where it is infinite. Hence $G(.,m)$ is smooth on its level sets :

$$W(\gamma_1, \gamma_2; m) = \{x \in \Omega; \gamma_1 \leqslant G(x,m) \leqslant \gamma_2\}. \tag{17}$$

Moreover the function $G(.,m)$ is positive on Ω, a consequence of the maximum principle.

Althought $\nabla_x G(.,m) \notin L^2(\Omega)$ (since $\delta_m \notin H^{-1}(\Omega)$), $\nabla_x G(.,m) \in L^2_{loc}(\Omega\backslash\{m\})$ and we have the following remarkable identity.

Proposition 4.1 (Bethuel and Ghidaglia[4]) *With the notations above we have for every* γ_1, γ_2 with $0 \leqslant \gamma_1 \leqslant \gamma_2 < \infty$,

$$\int_{W(\gamma_1,\gamma_2;m)} a_{ij}(x)\frac{\partial G}{\partial x_i}(x,m)\frac{\partial G}{\partial x_j}(x,m)dx = \gamma_2 - \gamma_1. \tag{18}$$

We shall also need the following observation.

Proposition 4.2 *Let u and v be in $H^1(\Omega)$ and let A be an open and smooth subset of $\bar{\Omega}$. We have*

$$\int_A \{u,v\}dx = \int_{\partial A} u\frac{\partial v}{\partial \tau}ds = -\int_{\partial A} v\frac{\partial u}{\partial \tau}ds, \tag{19}$$

where (τ,ν) is the oriented Frenet Frame on ∂A and the curviligne abscissa, s, is chosen in order that ν is the exterior unit normal to A on ∂A.

4.2 *The estimate (6).*
Since φ solves (2)-(3), we have

$$\varphi(m) = \int_\Omega g(u,v)\{u,v\}G(.,m)dx. \tag{20}$$

Setting $w(x) = \int_0^{u(x)} g(s,v(x))ds$, we observe that

$$\{w,v\} = g(u,v)\{u,v\}. \tag{21}$$

Note that the norm of w in $H^1(\Omega)$ cannot be bounded in terms of those of u and v in $H^1(\Omega)$. Hence the proof of (6) with $g \equiv 1$ does not follow from that with $g \equiv 1$.

Given $\gamma > 0$, we set $W_n = W(\gamma_n,\gamma_{n+1};m)$ where γ_n will be chosen later and $\gamma_n \in]n\gamma,(n+1)\gamma[$ for $n \geqslant 0, \gamma_{-1} = 0$. Then, introducing

$$I_n = \int_{W_n} \{w,v\}G(.,m)dx, n = -1,0,1,...., \tag{22}$$

we rewrite (20) as

$$\varphi(m) = \sum_{n=-1}^{+\infty} I_n. \tag{23}$$

We decompose I_n into two parts :

$$I_n = J_n + \bar{\gamma}_n K_n \tag{24}$$

where

$$J_n = \int_{W_n} \{w,v\}(G(.,m) - \bar{\gamma}_n))dx, \tag{25}$$

$$K_n = \int_{W_n} \{w,v\}dx, \tag{26}$$

and $\bar{\gamma}_n = (\gamma_n + \gamma_{n+1})/2$.

By definition of W_n, on this set : $|G(x,m) - \tilde{\gamma}_n| \leqslant \frac{\gamma_{n+1}-\gamma_n}{2} \leqslant \frac{\gamma}{2}$ and therefore

$$|J_n| \leqslant \frac{\gamma}{2}\int_{W_n} |\{w,v\}|dx = \frac{\gamma}{2}\int_{W_n} |g(u,v)\{u,v\}|dx.$$

Hence finally

$$|J_n| \leqslant \frac{\gamma}{2}|g|_{L^\infty} \int_{W_n} |\nabla u||\nabla v|dx, n \geqslant 0. \tag{27}$$

Similarily,

$$|I_{-1}| \leqslant \frac{\gamma}{2}|g|_{L^\infty} \int_{W_{-1}} |\nabla u||\nabla v|dx. \tag{28}$$

On the other hand, K_n can be expressed thanks to (19) as

$$K_n = \int_{\partial W_n} w\frac{\partial v}{\partial \tau}ds, \tag{29}$$

and since $\partial W_n = V(\gamma_n) \cup V(\gamma_{n+1})$, where $V(\gamma) \equiv \{x \in \Omega; G(x,m) = \gamma\}$, we can write

$$S \equiv \sum_{n=0}^{\infty} \bar{\gamma}_n K_n = \sum_{n=0}^{\infty} \bar{\gamma}_n \left(\int_{V(\gamma_n)} w\frac{\partial v}{\partial \tau}ds - \int_{V(\gamma_{n+1})} w\frac{\partial v}{\partial \tau}ds \right). \tag{30}$$

Note that the (union of) curves $V(\gamma_{n+1})$ appears two times in S : as parts of the boundary of ∂W_n and ∂W_{n+1}; but the orientation of the normal changes from one set to the other. By chosing once for all, on each $V(\gamma_n)$, an orientation i.e. that induced by ∂W_n we obtain formula (30). We can also write it as

$$S = \bar{\gamma}_0 \int_{V(\gamma_0)} w\frac{\partial v}{\partial \tau}ds + \sum_{n=1}^{\infty}(\bar{\gamma}_n - \bar{\gamma}_{n-1}) \int_{V(\gamma_n)} w\frac{\partial v}{\partial \tau}ds, \tag{31}$$

from which it follows readily that

$$|S| \leqslant \gamma \sum_{n=0}^{\infty} | \int_{V(\gamma_n)} w\frac{\partial v}{\partial \tau}ds|. \tag{32}$$

We decompose $V(\gamma_n)$ into connected components ω_i and notice that

$$\int_{\omega_i} w\frac{\partial v}{\partial \tau}ds = \int_{\omega_i}\left(\int_{\lambda}^{u(s)} g(\sigma, v(\sigma))d\sigma\right)\frac{\partial v}{\partial \tau}ds, \forall \lambda \in I\!R. \tag{33}$$

Hence

$$|\int_{\omega_i} \omega\frac{\partial v}{\partial \tau}ds| \leqslant |g|_{L^\infty}\int_{\omega_i}|u(s)-\lambda||\frac{\partial v}{\partial \tau}|ds, \forall\lambda\in I\!R,$$

and taking $\lambda = \frac{1}{|\omega_i|}\int_{\omega_i} u(s)ds$ leads to

$$|\int_{V(\gamma_n)} w\frac{\partial v}{\partial \tau}ds| \leqslant |g|_{L^\infty}\int_{V(\gamma_n)}|\frac{\partial u}{\partial \tau}|ds\int_{V(\gamma_n)}|\frac{\partial v}{\partial \tau}|ds. \tag{34}$$

It remains to choose the γ_n. We apply the coarea formula (1) whith $h = \max(|\nabla u|, |\nabla v|), f = G(., m)$ and $X = W(n\gamma, (n+1)\gamma; m)$. Then

$$\int_{n\gamma}^{(n+1)}\int_{V(\lambda)} max(|\nabla u|, |\nabla v|)ds\, d\lambda = \int_{\tilde{W}_h} max(|\nabla u|, |\nabla v|)|\nabla G|dx, \tag{35}$$

where $\tilde{W}_n \equiv W(n\gamma, (n+1)\gamma; m)$.

It follows that for every $\varepsilon > 0$, we can find $\gamma_n \in]n\gamma, (n+1)\gamma[$ such that

$$\int_{V(\gamma_n)} max(|\nabla u|, |\nabla v|)ds \leqslant \frac{1+\varepsilon}{\gamma}\int_{\tilde{W}_h} h|\nabla G|dx. \tag{36}$$

Hence, returning to (32) and using (34) together with (36) we end up with the following estimate

$$|S| \leqslant \frac{(1+\varepsilon)^2}{\gamma}\sum_{n=0}^{+\infty}\left(\int_{\tilde{W}_n} h|\nabla G|dx\right)^2 |g|_{L^\infty}. \tag{37}$$

We have $\left(\int_{\tilde{W}_n} h|\nabla G|dx\right)^2 \leqslant \int_{\tilde{W}_n} h^2dx\int_{\tilde{W}_n}|\nabla G|^2dx$, therefore using (5) and (19) we deduce (recall that $\tilde{W}_n = W(h\gamma(n+1)\gamma; m)$)

$$\alpha\int_{\tilde{W}_h}|\nabla G|^2dx \leqslant \int_{\tilde{W}_h} a_{ij}\frac{\partial G}{\partial x_i}\frac{\partial G}{\partial x_j}dx = \gamma.$$

Hence

$$|S| \leqslant |g|_{L^\infty}\frac{(1+\varepsilon)^2}{\alpha}\sum_{n=0}^{+\infty}\int_{\tilde{W}_n} h^2dx = (1+\varepsilon)^2\int_\Omega max(|\nabla u|^2, |\nabla v|^2)dx. \tag{38}$$

Returning to (24)) and using (28))-(29)) we find finally

$$|\varphi(m)| \leqslant \frac{\gamma}{2}|g|_{L^\infty}\int_\Omega |\nabla u||\nabla v|dx + \frac{(1+\varepsilon)^2}{\alpha}|g|_{L^\infty}\int_\Omega max(|\nabla u|^2, |\nabla v|^2)dx. \quad (39)$$

In fact γ and ε where arbitrary and $\varphi(m)$ is independent of these numbers, hence

$$|\varphi(m)| \leqslant \frac{1}{\alpha}|g|_{L^\infty}\int_\Omega max(|\nabla u|^2, |\nabla v|^2)dx. \quad (40)$$

Finally, $max(|\nabla u|^2, |\nabla v|^2) \leqslant |\nabla u|^2 + |\nabla v|^2$ and we observe that if we change (u, v) into $(\lambda u, \lambda^{-1}v), \lambda > 0, \varphi$ remains unchanged. Hence (40)) implies by chosing the optimal λ that

$$|\varphi(m)| \leqslant \frac{2}{\alpha}|g|_{L^\infty}(\int_\Omega |\nabla u|^2 dx)^{1/2}(\int_\Omega |\nabla v|^2 dx)^{1/2} \quad (41)$$

which is the desired result (the estimate on $|\nabla\varphi|_{L^2(\Omega)}$ is obvious now).

Remark 4.1 The constant appearing in (41) is certainly not the optimal one since for e.g. $g \equiv 1, a_{ij} = \delta_{ij}$, Brezis-Coron[6] have shown in the case $\Omega = I\!R^2$ that

$$|\varphi(m)| \leqslant \frac{1}{2\pi}(\int_\Omega |\nabla u|^2 dx)^{1/2}(\int_\Omega |\nabla v|^2 dx)^{1/2}.$$

In fact $\frac{1}{2\pi}$ is the best constant in this case as it is easily seen by choosing

$$u(x) = x_1\, g((x_1^2 + x_2^2)^{1/2}), v(x) = x_2\, g((x_1^2 + x_2^2)^{1/2})$$

(then $\varphi(x) = \frac{1}{2}\int_{(x_1^2+x_2^2)^{1/2}}^\infty \sigma g^2(\sigma)d\sigma), g_\varepsilon(r) = r^{-1+\varepsilon}e^{-r}, \varepsilon > 0$ and then let ε tends to zero.

To our knowledge, finding the best constant in (6) is an open problem.

5 Remarks and complements

It has been observed recently by Coifman *et al* Coifman, Lions, Meyer and Semmes[9] that for u and v in $H^1(I\!R^2)$, the jacobian $\{u, v\}$ belongs to the Hardy space $\mathcal{H}^1(I\!R^2)$. Recall that $f \in L^1(I\!R^2)$ belongs to $\mathcal{H}^1(I\!R^2)$ if and only if $\int_{I\!R^2} f(x)dx = 0$ and $f^* \in L^1(I\!R^2)$ where

$$f^*(x) = \sup_{r>0} |\frac{1}{r^2}\int_{I\!R^2} f(y)\rho(\frac{x-y}{r})dy|$$

and $\rho \in \mathcal{D}(I\!R^2)$ satisfies supp $\rho \subset B(0,1), \rho \geqslant 0, \int_S dx = 1$. More precisely, there exists C such that

$$|\{u, v\}|_{\mathcal{H}^1(I\!R^2)} \leqslant C|\nabla u|_{L^2(I\!R^2)}|\nabla v|_{L^2(I\!R^2)}.$$

Then since the fundamental solution to Laplace's equation in $I\!R^2(-\frac{1}{2\pi}\log|x|)$ belongs to the dual of $\mathcal{H}^1(I\!R^2)$, one recovers Wente's estimate.
Actually, as shown in Coifman, Lions, Meyer and Semmes[9], one has a more general result namely : for every $p \in]1,\infty[, p' = \frac{p}{p-1}$, there exists C_p such that $|\{u,v\}|_{\mathcal{H}^1(I\!R^2)} \leqslant C_p|\nabla u|_{L^p(I\!R^2)}|\nabla v|_{L^{p'}(I\!R^2)}$.

More recently, Chanillo and Li[7] have extended the properties of the fundamental solution to general second order elliptic equation in $I\!R^2$. These extensions lead in particular to variants of Theorem A.

We have considered in Bethuel and Ghidaglia[4] the Neumann problem associated with (2) (with $g \equiv 1, a_{ij} = \delta_{ij}$) namely

$$-\Delta\varphi = \{u,v\} \text{ in } \Omega, \tag{42}$$

$$\frac{\partial\varphi}{\partial\nu} = 0 \text{ on } \partial\Omega. \tag{43}$$

We have the following estimate.

Theorem 5.1 (Bethuel and Ghidaglia[4]) There exists a constant $C_1(\Omega)$ such that for every u and v in $H^1(\Omega)$ such that $\int_{\partial\Omega} u\frac{\partial v}{\partial\tau}ds = 0$; the solution to (42)-(43) in $H^1(\Omega)$ belongs to $L^\infty(\Omega)$ and satisfies

$$|\nabla\varphi|_{L^2(\Omega)} + \inf_{c\in I\!R}|\varphi - c|_{L^\infty(\Omega)} \leqslant C_1(\Omega)|\nabla u|_{L^2(\Omega)}|\nabla v|_{L^2(\Omega)}. \tag{44}$$

In this result, we allow the constant to depend on the domain Ω. Actually Choné[8] has shown the following theorem.

Theorem 5.2 Choné[8]. (i) The constant in (44) cannot be chosen independently of Ω.
(ii) If u (or v) is constant on $\partial\Omega$ then the constant in (44) can be chosen independently of Ω.
(iii) If u and v are constant on each of the connected components of $\partial\Omega$ then the constant in (44) can be chosen independently of Ω.

6 Weak compactness of the set of solutions to Euler equation

We are going to construct in the next section a family of smooth vector valued functions $u^\varepsilon : \bar{\Omega} \to I\!R^2$ such that

(i) u^ε satisfies weakly the steady Euler equation whith force f^ε i.e. $(\nabla^\perp\eta \equiv (\frac{\partial\eta}{\partial x_2}, -\frac{\partial\eta}{\partial x_1}))$

$$\int_\Omega u^\varepsilon \otimes u^\varepsilon : \nabla\nabla^\perp \eta \, dx + \int_\Omega f^\varepsilon . \nabla^\perp \eta \, dx = 0, \forall \eta \in \mathcal{D}(\Omega), \tag{45}$$

$$div\, u^\varepsilon = 0, \text{ in } \Omega, \tag{46}$$

$$u^\varepsilon . \nu = 0 \text{ on } \partial\Omega \tag{47}$$

(ii) The family $(u^\varepsilon)_{\varepsilon>0}$ is bounded in $L^2(\Omega)$ and (curl $u^\varepsilon)_{\varepsilon>0}$ is bounded in $L^1(\Omega)$.

$$\sup_{\varepsilon>0} \int_\Omega (|u^\varepsilon|^2 + |\frac{\partial u_2^\varepsilon}{\partial x_1} - \frac{\partial u_1^\varepsilon}{\partial x_2}|)dx < \infty. \tag{48}$$

(iii) The family $(f^\varepsilon)_{\varepsilon>0}$ converges weakly to some limit f in $L^1(\Omega)$.

Our main result is then

Theorem B Bethuel and Ghidaglia[5] : *Let u be in the weak closure of the family $(u^\varepsilon)_{\varepsilon>0}$ in $L^2(\Omega)$. Then u satisfies weakly the steady Euler equation with force f.*

Comments 6.1 (i) This result was proved by Di Perna and Majda[10] and S. Alinhac[1] in the case where $\Omega = \mathbb{R}^2$, see also Evans[11]. To our knowledge, Theorem B is new. We would like to mention that our method for proving this result (which also works when $\Omega = \mathbb{R}^2$) is of a slightly different nature than that of the previous Authors. It relies in fact on the coarea formula (1) as we will see in Section 8.
(ii) We have stated Theorem B before actually constructing the family $(u^\varepsilon)_{\varepsilon>0}$ because this result is interesting for its own sake and does not depend on the particular family of (u^ε) which is considered.
(iii) In general Theorem B cannot be used to construct solutions to the steady Euler equation since this equation might have no solution (Minea[14]).

7 Weak solutions to the damped steady Euler equation

Let Ω be in this section a *simply connected* smooth open set. We are given $\gamma > 0$ and

$$g \in L^2(\Omega)^2 \text{ with } curl g \in L^1(\Omega). \tag{49}$$

A triplet (u, p, φ) is said to be solution to the Charney-Stommel model of the Gulf-Stream (see Barcilon *et al* Barcilon, Constantin and Titi[2] and references therein) if

$$\gamma u + (u.\nabla)u + \varphi + \nabla p = g, \text{ in } \Omega, \tag{50}$$

$$divu = 0, \text{ in } \Omega, \tag{51}$$

$$curl\varphi = u_2, div\varphi = 0, \text{ in } \Omega, \tag{52}$$

$$u.\nu = 0, \varphi.\nu = 0 \text{ on } \partial\Omega. \tag{53}$$

Using Theorem B we are going to prove the following result.

Theorem 7.1 *Under hypothesis (49), there exists $u \in L^2(\Omega)^2, \varphi \in H^1(\Omega)^2$ and $p \in \mathcal{D}'(\Omega)$ solution to the Charney-Stommel model of the Gulf-Stream.*

Comment 7.1 A solution to (50)-(53) was constructed by Barcilon *et al* Barcilon, Constantin and Titi[2] for $g \in W^{1,\infty}(\Omega)^2$. Later J.C. Saut[16] extended this result to the case where $g \in H^1(\Omega)^2$ (and then $u \in H^1(\Omega)$). He also proved in particular that solutions are smooth when the datum (g) is smooth. We shall rely on his result. We would like to thank him for bringing our attention on this model.

Proof of Theorem 7. In a first step we construct a family of functions $g^\varepsilon \in \mathcal{C}^\infty(\bar{\Omega})$ such that

$$\lim_{\varepsilon \to 0} \int_\Omega (|g^\varepsilon - g|^2 + |curlg^\varepsilon - curlg|)dx = 0. \tag{54}$$

This step is technical but standard : we can assume that g is supported in a small ball since $\theta g, \theta \in \mathcal{D}(\mathbb{R}^2)$, will satisfy (49) too. Then, the support of g being star-shaped (recall that Ω is smooth), the family g^ε can be obtained using standard arguments based on convolution with C^∞ mollifiers.

In a second step, we use Saut[16] to construct $u^\varepsilon, \varphi^\varepsilon$ and p^ε, smooth solutions to (50)-(53). In a third step, we derive estimates on (u^ε). We multiply (50) by u (in fact u^ε) and integrate over Ω. Thanks to Green's formula we end up with

$$\gamma \int_\Omega |u^\varepsilon|^2 dx + \int_\Omega \varphi^\varepsilon.u^\varepsilon dx = \int g^\varepsilon.u^\varepsilon dx. \tag{55}$$

At this stage we use the fact that Ω is simply connected to write

$$u^\varepsilon = \nabla^\perp \zeta^\varepsilon \text{ with } \zeta^\varepsilon = 0 \text{ on } \partial\Omega. \tag{56}$$

Hence

$$\int_\Omega \varphi^\varepsilon . u^\varepsilon dx = \int_\Omega \varphi^\varepsilon . \nabla^\perp \zeta^\varepsilon dx = -\int_\Omega \zeta^\varepsilon curl \varphi^\varepsilon dx =$$

$$= \int_\Omega \zeta^\varepsilon \frac{\partial \zeta^\varepsilon}{\partial x_1} dx = \frac{1}{2} \int_{\partial\Omega} (\zeta^\varepsilon)^2 d\sigma = 0,$$

and (55) reads

$$\gamma \int_\Omega |u^\varepsilon|^2 dx = \int g^\varepsilon . u^\varepsilon dx. \tag{57}$$

On the other hand taking the curl of (50) leads to

$$\gamma \omega^\varepsilon + (u^\varepsilon . \nabla) \omega^\varepsilon + u_2^\varepsilon = curl g^\varepsilon, \tag{58}$$

with

$$\omega^\varepsilon = curl u^\varepsilon. \tag{59}$$

Multiplying (58) by $sgn(\omega^\varepsilon)$ and integrating, we obtain

$$\gamma \int_\Omega |\omega^\varepsilon| dx \leqslant \int_\Omega |curl g^\varepsilon| dx + \int_\Omega |u_2^\varepsilon| dx. \tag{60}$$

Hence combining (57) and (60) with (54) we see that (48) holds true. Then we set $f^\varepsilon = g^\varepsilon - \gamma u^\varepsilon - \varphi^\varepsilon$, and we observe that according to (50) (with u replaced by u^ε) (45) holds true too.

Since $(u^\varepsilon)_{\varepsilon>0}$ is bounded in $L^2(\Omega)^2$, we can extract a subfamily (still denoted by (u^ε)), such that u^ε weakly converges to some u in $L^2(\Omega)^2$. It is straightforward to see that φ^ε weakly converges to φ (satisfying (52)-(53)) in $H^1(\Omega)^2$. Hence by (53) we know that f^ε weakly converges to $f = g - \gamma u - \varphi$ in $L^1(\Omega)$: point (iii) in Section 6 is therefore also satisfied. The assumptions of Theorem B are fulfilled and u satisfies

$$\int_\Omega u \otimes u : \nabla \nabla^\perp \eta dx + \int_\Omega (g - \gamma u - \varphi) . \nabla^\perp \eta dx = 0$$

for every $\eta \in \mathcal{D}(\Omega)$. This completes the proof of Theorem 7.1.

8 On the proof of Theorem B

The proof of Theorem B is given in full details in Bethuel and Ghidaglia[5]. Here we shall only show how the coarea formula (1) enters into this proof. We introduce the following notations.

The family ζ^ε, given by (56), is bounded in $H_0^1(\Omega)$. Hence ζ^ε converges weakly to ζ in $H_0^1(\Omega)$ where $u = \nabla^\perp \zeta$. Since ζ may not be smooth, we also construct a family $(\zeta_\delta)_{\delta>0}$ of smooth functions which vanish on $\partial\Omega$, and such that ζ_δ converges *strongly* to ζ in $H_0^1(\Omega)$ and $\Delta\zeta_\delta$ is bounded in $L^1(\Omega)$.

If ζ^ε would converge strongly to ζ in $H_0^1(\Omega)$, Theorem B would be obvious since then $u^\varepsilon = \nabla^\perp\zeta^\varepsilon$ would converge strongly to $u = \nabla^\perp\zeta$ in $L^2(\Omega)$. In order to estimate $|u^\varepsilon - u|^2$, we introduce the level sets of $\zeta^\varepsilon - \zeta_\delta$: for $\gamma \geqslant 0$, we set

$$W_\delta^\varepsilon(\gamma) = \{x \in \Omega, |\zeta^\varepsilon(x) - \zeta_\delta(x)| \geqslant \gamma\}. \tag{61}$$

Obviously the area of this set goes to zero as ε and δ go to zero. Indead

$$\text{Area } (W_\delta^\varepsilon(\gamma)) \leqslant \frac{1}{\gamma^2}|\zeta^\varepsilon - \zeta_\delta|^2_{L^2(\Omega)}, \tag{62}$$

and we know from Rellich's Theorem that ζ^ε converges strongly to ζ in $L^2(\Omega)$ as ε goes to zero. Hence with respect to Lebesgue's measure, the complement of $W_\delta^\varepsilon(\gamma)$ i.e. $W_\delta^\varepsilon(\gamma)^c = \Omega\backslash W_\delta^\varepsilon(\gamma)$ is a "large set" and we are going to estimate $|u^\varepsilon - u|$ on this set (see(65) where $u_\delta \equiv \nabla^\perp\zeta_\delta$).

Theorem 8.1 : *Let C_0 be such that*

$$\sup_{\varepsilon>0.\delta>0}\int_\Omega(|u^\varepsilon|^2 + |u_\delta|^2 + |\Delta\zeta^\varepsilon| + |\Delta\zeta_\delta|)dx \leqslant C_0 < \infty. \tag{63}$$

There exists a constant C_1, which only depends on C_0, such that for every $\varepsilon > 0, \delta > 0$ and $\gamma > 0$, there exists $\gamma_\delta^\varepsilon \in [\gamma, 2\gamma]$ such that the boundary of $W_\delta^\varepsilon(\gamma_\delta^\varepsilon)$ is the union of smooth curves whose total length is denoted by ℓ_γ^ε, and

$$l_\delta^\varepsilon \leqslant \frac{C_1}{\gamma^2}|\zeta^\varepsilon - \zeta_\delta|_{L^2(\Omega)}, \tag{64}$$

$$\int_{w_\gamma^\varepsilon(\delta_\delta^\varepsilon)^c}|u^\varepsilon - u_\delta|^2dx \leqslant C_1\gamma, \tag{65}$$

$$\text{Area } (w_\delta^\varepsilon(\delta_\delta^\varepsilon)) \leqslant \frac{1}{\gamma^2}|\zeta^\varepsilon - \zeta_\delta|^2_{L^2(\Omega)}. \tag{66}$$

Proof of Theorem 8.1 : We use (1) with $h = 1, f = |\zeta^\varepsilon - \zeta_\delta|$ and $X = W_\delta^\varepsilon(2\gamma)\backslash W_\delta^\varepsilon(\gamma)$. Then

$$\int_X |\nabla|\zeta^\varepsilon - \zeta_\delta||dx = \int_\gamma^{2\gamma} l_\delta^\varepsilon(\sigma)d\sigma,$$

where $\ell_\delta^\varepsilon(\sigma)$ denotes the length of the boundary of $W_\delta^\varepsilon(\sigma)$.

Hence,

$$\int_{\gamma}^{2\gamma} \ell_{\delta}^{\varepsilon}(\sigma) d\sigma \leqslant \int_X |\nabla(\zeta^{\varepsilon} - \zeta_{\delta})| dx. \tag{67}$$

Thanks to (62), (63), (65) and the Cauchy-Schwarz inequality we deduce that

$$\int_{\gamma}^{2\gamma} \ell_{\delta}^{\varepsilon}(\sigma) d\sigma \leqslant \frac{2C_0^{1/2}}{\gamma} |\zeta^{\varepsilon} - \zeta_{\delta}|_{L^2(\Omega)}. \tag{68}$$

According to Sard's Lemma the sets $W_{\delta}^{\varepsilon}(\gamma)$ are smooth for almost every $\gamma \geqslant 0$. It follows from (68) that we may find $\gamma_{\delta}^{\varepsilon} \in [\gamma, 2\gamma]$ such that $W_{\delta}^{\varepsilon}(\gamma_{\delta}^{\varepsilon})$ is smooth and

$$\ell_{\delta}^{\varepsilon} \equiv \ell_{\delta}^{\varepsilon}(\gamma_{\delta}^{\varepsilon}) \leqslant \frac{3C_0^{1/2}}{\gamma^2} |\zeta^{\varepsilon} - \zeta_{\delta}|_{L^2(\Omega)}. \tag{69}$$

Clearly, (66) follows from (62) and in order to complete the proof of Theorem 7 we only have to prove inequality (65). This will be a consequence of the fact that $\Delta\zeta^{\varepsilon}$ and $\Delta\zeta_{\delta}$ are bounded in $L^1(\Omega)$. Indead, we have

$$-\Delta(\zeta^{\varepsilon} - \zeta_{\delta}) = \omega^{\varepsilon} - \omega_{\delta}, \text{ in } \Omega, \tag{70}$$

and setting (ε, δ and γ are fixed)

$$W_+ = \{x \in \Omega, \zeta^{\varepsilon}(x) - \zeta_{\delta}(x) \geqslant \gamma\},$$

$$W_- = \{x \in \Omega, \zeta^{\varepsilon}(x) - \zeta_{\delta}(x) \leqslant -\gamma\},$$

$V_+ = \partial W_+$ and $V_- = \partial W_-$; we multiply (70) by $\zeta^{\varepsilon} - \zeta_{\delta}$ and integrate on $W_{\delta}^{\varepsilon}(\gamma)^c$. We find, using Green's formula :

$$\int_{W_{\delta}^{\varepsilon}(\gamma)^c} (\omega^{\varepsilon} - \omega_{\delta})(\zeta^{\varepsilon} - \zeta_{\delta}) dx = \int_{W_{\delta}^{\varepsilon}(\gamma)^c} |\nabla(\zeta^{\varepsilon} - \zeta_{\delta})|^2 dx - \int_{\partial(W_{\delta}^{\varepsilon}(\gamma)^c)} (\zeta^{\varepsilon} - \zeta_{\delta}) \frac{\partial(\zeta^{\varepsilon} - \zeta_{\delta}}{\partial\gamma}) ds. \tag{71}$$

On the other hand, integrating (70) on W_+(resp. W_-) we see that

$$\int_{V\pm} \frac{\partial(\zeta^{\varepsilon} - \zeta_{\delta})}{\partial\gamma} ds = \int_{W\pm} (\omega^{\varepsilon} - \omega_{\delta}) dx,$$

from which we deduce (using (63)) that

$$\left| \int_{V\pm} \frac{\partial(\zeta^{\varepsilon} - \zeta_{\delta})}{\partial\gamma} ds \right| \leqslant 2C_0.$$

Returning to (71) we deduce from the previous estimate (recall that $\zeta^{\varepsilon} = \zeta_{\delta} = 0$ on $\partial\Omega$) that

$$\int_{W_{\delta}^{\varepsilon}(\gamma)^c} |\nabla(\zeta^{\varepsilon} - \zeta_{\delta})|^2 dx \geqslant 4C_0\gamma,$$

notice that $\zeta^\varepsilon - \zeta_\delta$ is constant on V_+ on V_-. Whence (65) follows, which completes the proof of Theorem 7.

As already mentionned, this result is the first step in the proof of Theorem B. In the second step is of a different nature and is related to the effect of the boundary (altought Theorem B may seem a local result, it is not actually the case due to the pressure terms which are hiden in the formulation (45)). The third step then uses the two previous ones and leads to the result. It is based on the use of the Fourier transform like in previous works on the subject in the whole space (Di Perna and Majda[10] and Alinhac[1]).

We refer to Bethuel and Ghidaglia[5] concerning details.

References

1. S. ALINHAC, Un phénomène de concentration évanescente pour des flots non-stationnaires incompressibles en dimension deux, *Com. Math. Phys.* , **127** (1990) 585-596.

2. V. BARCILON, P. CONSTANTIN and E. TITI, Existence of solutions to the Charney model of the Gulf stream, *S.I.A.M.J. Math. Anal,* **19** (1988) 1355-1364.

3. F. BETHUEL, H. BREZIS, B.D. COLEMAN and F. HELEIN, Bifurcation analysis of minimizing harmonic maps describing the equilibrium of nematic phases between cylinders, *Arch. Rational Mech. Anal.*, to appear in (1992).

4. F. BETHUEL and J.M. GHIDAGLIA, Improved regularity of solutions to elliptic equations involving jacobians and applications, Publication du CMLA N° 9202, Cachan, France, (1992).

5. F. BETHUEL and J.M. GHIDAGLIA, Weak limits of solutions to the imcompressible 2D-Euler equations in a bounded domain, Publication du CMLA N^0 9210, Cachan, France, (1992).

6. H. BREZIS and J.M. CORON, Multiple solutions of H-systems and Rellich's conjecture, *Comm. Pure Appl. Math,***37** (1984) 149-187.

7. S. CHANILLO and Y.Y. LI, Continuity of solutions of uniformly elliptic equations in $\mathbb{R}^2$, Preprint (1992).

8. P. CHONE, Problèmes elliptiques avec second membre jacobien et condition de Neumann au bord, to appear.

9. R. COIFMAN, P.L. LIONS, Y. MEYER and S. SEMMES, Compacité par compensation et espaces de Hardy, *C.R. Acad. Sci. Paris,* **311**(1990) 519-524, and Compensated compactness and Hardy spaces, Prépublication CEREMADE 9123, U. Paris IX, (1991).

10. R. DI PERNA and A. MAJDA, Reduced Hausdorff dimension and concentration-cancellation for two dimensional incompressible flow, *J. of Amer. Math. Soc.,* **1** (1988) 59-95.

11. L.C. EVANS, Weak convergence methods for nonlinear partial differential equations, CBMS # 74, A.M.S. Providence, (1990).

12. H. FEDERER, *Geometric measure theory*, Springer, Berlin, 1969.

13. HEINZ, Uber die Regularität schwarcher Lösungen nicht linear elliptischer Systeme, *Nachr. Akad. Wiss. Göttingen II. Mathematisch Physikalische Klasse 1,* (1975) 1-13.

14. MINEA, Remarque sur les équations d'Euler dans un domaine possédant une symétrie de révolution, *C.R. Acad. Sci. Paris*, I, **284**(1977) 477-479.

15. J. SACKS and K. UHLENBECK, The existence of minimal immersions of 2 spheres, *Ann. of Math., 113,* (1981) 1-24.

16. J.C. SAUT, Remarks on the damped stationaly Euler equations, *Diff. Int. Eqn.,* **3**(1990) 801-812.

17. H. WENTE, An existence theorem for surfaces of constant mean curvature, *J. Math. Anal. Appl.,***26** (1969) 318-344.

Geometry in Partial Differential Equations (pp. 18–31)
eds. A. Pràstaro and Th. M. Rassias

LARGE SOLUTIONS FOR THE EQUATION OF SURFACES OF PRESCRIBED MEAN CURVATURE

FABRICE BETHUEL(1)(2) and OLIVIER REY(3)
(1) ENPC-CERMA, La Courtine, 93167 Noisy-le-Grand Cedex, France
(2) CMLA, Cachan, 61, av du Président Wilson, 94235 Cachan Cedex, France
(3) Centre de Mathématques, Ecole Polytechnique, 91128 Palaiseau Cedex, France

ABSTRACT

In this paper, we present recent developments concerning the multiplicity of solutions to the equation of surfaces of prescribed mean curvature.

1 *Introduction*

The questions we are going to consider originate from the classical Plateau problem. Γ being a Jordan curve in $\mathbf{R}^n$, the Plateau problem consists in finding surfaces of minimal area spanning Γ. Although this question has been largely investigated in the 19-th century, the first effective approach has been proposed by J. Douglas and T. Rado in 1930/31. Instead of considering the area functional, they turned their attention to the energy integral. More precisely, if $u : D^2 = \{(x,y) \in \mathbf{R}^2 \;\; s.t. \;\; x^2+y^2 \leq 1\} \to \mathbf{R}^n$ is a parametrization of a surface spanning Γ the area functional is given by

$$\mathcal{A}(u) = \int_{D^2} |u_x \wedge u_y|^2$$

and the Dirichlet energy integral reads

$$E(u) = \frac{1}{2} \int_{D^2} |\nabla u|^2$$

The main point is that $\mathcal{A}(u)$ is invariant with respect to every diffeomorphism of D^2, whereas $E(u)$ is invariant only for conformal transformations. Clearly $\mathcal{A}(u) \leq E(u)$, with equality if and only if u is conformal. On the other hand if u is critical for E

and is conformal, u turns out to be a solution to Plateau's problem (for more details, see [16]). Consequently, we are interested in finding solutions to the system

$$(1) \qquad \Delta u = 0 \quad \text{in} \quad D^2$$

$$(2) \qquad |u_x|^2 - |u_y|^2 = u_x.u_y = 0 \quad \text{in} \quad D^2$$

(3) u is a continuous monotone parametrization of Γ

However, in order to have a well defined variational formulation, one often prefers to consider the related Dirichlet-type problem :

$$(I) \qquad \begin{cases} \Delta u = 0 & \text{in} \quad D^2 \\ u = \gamma & \text{on} \quad \partial D^2 \end{cases}$$

where γ is a given function from ∂D^2 to $\mathbf{R}^n$. Then, one can take advantage of the freedom we have in the choice of γ as a parametrization of Γ to have condition (2) satisfied - see for instance [8] [3] and references therein.

Note that a solution to the classical Plateau problem has mean curvature equal to zero : a natural generalization is to seek for surfaces spanning Γ whose mean curvature is a given number $H \in \mathbf{R}$. Then, equation (1) is replaced by

$$(4) \qquad \Delta u = 2H u_x \wedge u_y \quad \text{in} \quad D^2$$

and we denote by (II) the corresponding Dirichlet problem. A further generalization is to look for surfaces whose mean curvature is prescribed, but not necessarily a constant. Namely, H being a given function from $\mathbf{R}^3$ to $\mathbf{R}$, we impose the mean curvature of the surface at the point $u(x,y)$ to be $H(u(x,y))$. This leads to consider instead of (1) the equation.

$$(5) \qquad \Delta u = 2H(u) u_x \wedge u_y \quad \text{in} \quad D^2$$

and the Dirichlet problem

$$(III) \qquad \begin{cases} \Delta u = 2H(u) u_x \wedge u_y & \text{in} \quad D^2 \\ u = \gamma & \text{on} \quad \partial D^2 \end{cases}$$

(III) has a variational structure : solutions may be obtained as critical points of the functional J defined by

$$(6) \qquad J(u) = E(u) + \frac{2}{3} \int_{D^2} Q(u).u_x \wedge u_y$$

for any $u \in H^1_\gamma(D^2, \mathbf{R}^3) = \{u \in H^1(D^2, \mathbf{R}^3)\ \ s.t.\ \ u|_{\partial D^2} = \gamma\}$. $Q(u)$ denotes the following map from $\mathbf{R}^3$ to $\mathbf{R}^3$

$$(7) \qquad Q(u) = \left(\int_0^{u_1} H(t, u_2, u_3)dt, \int_0^{u_2} H(u_1, t, u_3)dt, \int_0^{u_3} H(u_1, u_2, t)dt\right)$$

In case H is a constant, $Q(u) \equiv Hu$.

In this survey, we will focus on the existence and multiplicity of solutions to (III) ; note that similar results are available concerning the Plateau-type problem ((5)(2)(3)) as already mentionned, or are expected. Section 2 is devoted to the existence result by S.Hildebrandt. These solutions are obtained by minimizing the functional J on a suitable subset of H^1_γ. (Other existence results may be found in [7][20][18][12]). From the observation of some special cases, Rellich conjectured that there should exist pairs of solutions. Such a conjecture was settled by the result of Brezis and Coron [3] that we present in section 3. Indeed, they prove under the same assumptions as Hildebrandt the existence of a second solution, in the case where H is a constant. (For other statements concerning a "large" solution, see [20][13][14][15]). However, the proof cannot be easily extended to the case of variable H. The main obstacle is to analyse the behaviour of so called Palais-Smale sequences. To overcome this difficulty, Struwe introduced, as Sachs and Uhlenbeck in [19], a perturbed functional for which compactness properties hold. He was then able to prove in [17] a partial multiplicity result for variable curvatures close (in a special norm) to a constant. This result was slightly improved by Guofang in [6].

In [2] we propose a direct approach to the problem, relying on both a careful analysis of Palais-Smale sequences and a priori bounds on solutions. In section 4 we sketch the main arguments of [2], which allow to prove the existence of a second solution to (III) under much weaker assumptions, More precisely, we only need to know that H is close to a constant in L^∞-norm.

2 *The "small" solution*

Improving earlier results of Werner [21] and Heinz [7], S. Hildebrandt proved in [8] the following result :

Theorem 1.— *Let us assume that* $\gamma \in H^{1/2}(\partial D^2, \mathbf{R}^3) \cap C^0(\partial D^2, \mathbf{R}^3)$, $H \in C^1(\mathbf{R}^3, \mathbf{R}$ *with*

$$||\gamma||_{L^\infty(\partial D^2, \mathbf{R}^3)} \cdot \ ||H||_{L^\infty(D^3, \mathbf{R}^3)} \leq 1$$

(D^n denotes the unit ball of $\mathbf{R}^n$). Then Problem (III) has a solution $\underline{u}$, which satisfies $||\underline{u}||_{L^\infty(\partial D^2, \mathbf{R}^3)} \leq ||\gamma||_{L^\infty(\partial D^2, \mathbf{R}^3)}$

Remarks :

1 - Similar results have been obtained by Wente [18][19] and Steffen [12] in case of constant H, by Steffen [12] for variable H. The result of this theorem may be found again by mean of the heat flow associated to the equation - see [10].

2 - Using a minimizing procedure on the energy of the solutions of (III) with respect to the parametrization γ of Γ, one obtains a solution of the Plateau problem (2) (3)(5) - see [8].

We sketch the proof in the case $h = ||H||_{L^\infty(D^3,\mathbf{R}^3)} < 1, ||\gamma||_{L^\infty(\partial D^2,\mathbf{R}^3)} \leq 1$ and H smooth (the general case may be deduced using homogeneity and approximation arguments).

Let h' be such that

$$h < h' < 1$$

and $\tilde{H} : \mathbf{R}^3 \to \mathbf{R}^3$ a smooth function satisfying

$$H \equiv \tilde{H} \text{ on } D^3; \tilde{H}(|u|) = 0 \text{ for } |u| > \frac{1}{h'}; ||\tilde{H}||_{L^\infty(\mathbf{R}^3,\mathbf{R}^3)} < h'$$

We denote by $\widetilde{J}$ and $\widetilde{Q}$ the functionals corresponding to $\tilde{H}$ through (6) and (7). Note that we have

$$\sup_{u\in\mathbf{R}^3} |\widetilde{Q}(u)| \leq 1$$

so that

$$\frac{1}{3}E(u) \leq \widetilde{J}(u) \ \leq \ \frac{5}{3}E(u) \quad \forall u \in H^1_\gamma(D^2\mathbf{R}^3)$$

Then, it follows from a result of Morrey [9] that the infimum of $\widetilde{J}$ on $H^1_\gamma(D^2;\mathbf{R}^3)$ is achieved by a function $\underline{u}$ which is a solution to (III), with $\widetilde{H}$ instead of H. As we have also

$$-\Delta|\underline{u}|^2 = -2(|\nabla\underline{u}|^2 + \underline{u}.\Delta\underline{u}) \leq -2|\nabla\underline{u}|^2(1 - |\underline{u}||\widetilde{H}(\underline{u})|) \leq 0$$

$\underline{u}$ is subharmonic in D^2, and the maximum principle yields

$$|\underline{u}| \leq 1 \ \text{ in } \ D^2$$

$\widetilde{H}$ being equal to H on the unit ball of $\mathbf{R}^3$, $\underline{u}$ is a solution to (III).

Before ending this section, let us state a result which will be of fundamental importance in the sequel :

Lemma 1.— *Suppose that H is a constant, and that $|H|.||\gamma||_{L^\infty(\partial D^2,\mathbf{R}^3)} < 1$. Then, the second derivative of J at $\underline{u}$ is coercive, namely there exists $\delta > 0$ such that*

$$J''(\underline{u})(\varphi,\varphi) = \int_{D^2} |\nabla\varphi|^2 + 4H\underline{u}.\varphi_x \wedge \varphi_y \geq \delta \int_{D^2} |\nabla\varphi|^2$$

for any $\varphi \in H^1_0(D^2,\mathbf{R}^3)$

For a proof of this lemma, see [3].

3 *The "large" solution for constant H*

In order to prove the existence of a second solution to (III), a natural idea is to take advantage of the characterization of the first solution as a minimum, to use some mountain pass-type procedure. This program was performed by Brezis and Coron in [3] for constant H, and Struwe gave another presentation of the result in [15]. Namely, it is proved in [3] :

Theorem 2.— *Let $\gamma \in H^{1/2}(\partial D^2, \mathbf{R}^3) \cap L^\infty(\partial D^2, \mathbf{R}^3)$*

Assume that γ is not a constant on $\partial D^2, H > 0$, and that

$$H.||\gamma||_{L^\infty(\partial D^2, \mathbf{R}^3)} < 1$$

Then, there exist at least two distinct solutions of (III).

Remark : If γ is a constant, it follows from a result of Wente [20] that $u \equiv \gamma$ is the only solution.

3.1 - *The behaviour of Palais-Smale sequences*

The main difficulty which one has to deal with concerning functional J is that the Palais-Smale condition is not satisfied. A Palais-Smale sequence for J is a sequence (u_n) such that

$$J(u_n) \quad \text{is bounded}$$

$$J'(u_n) \quad \text{goes to zero in} \quad H^{-1}$$

and (P.S.) condition would imply that any sequence of this type is relatively compact. Note that we have the following property :

Lemma 2.— *Any Palais-Smale sequence is bounded in H^1.*

Proof. $J'(u_n) \to 0$ in H^{-1}writes

$$\Delta u_n = 2H u_{n_x} \wedge u_{n_y} + f_n, \quad f_n \to 0 \text{ in } H^{-1}$$

Substracting to this equation the equality

$$\Delta \underline{u} = 2H \underline{u}_x \wedge \underline{u}_y$$

multiplying by $\varphi_n = u_n - \underline{u}$ and integrating on D^2, we obtain

$$-\int_{D^2} |\nabla \varphi_n|^2 = 2H \int_{D^2} (u_{n_x} \wedge u_{n_y} - \underline{u}_x \wedge \underline{u}_y)\varphi_n + o(|\varphi_n|_{H_0^1})$$

Integration by parts yields

$$(8)\qquad -\int_{D^2}|\nabla\varphi_n|^2 = 4H\int_{D^2}\underline{u}.\varphi_{n_x}\wedge\varphi_{n_y} + 2H\int_{D^2}\varphi_n.\varphi_{n_x}\wedge\varphi_{n_y} + o(|\varphi_n|_{H^1_0})$$

On the other hand, expanding J at $\underline{u}$, we have

$$(9)\quad J(u_n) = J(\underline{u}) + \frac{1}{2}\int_{D^2}|\nabla\varphi_n|^2 + 2H\int_{D^2}\underline{u}.\varphi_{n_x}\wedge\varphi_{n_y} + \frac{2H}{3}\int_{D^2}\varphi_n.\varphi_{n_x}\wedge\varphi_{n_y}$$

It follows from (8) (9) that

$$|\int_{D^2}\varphi_n.\varphi_{n_x}\wedge\varphi_{n_y}| \leq C + o(|\varphi_n|_{H^1_0})$$

Then, Lemma 1 shows that (φ_n) is bounded in H^1_0, and Lemma 2 follows.

From Lemma 2 we deduce that any Palais-smale sequence is weakly relatively compact. It is no difficult to see that any weak limit is a solution to (III). Indeed, observe that

$$u_x\wedge u_y = \frac{1}{2}\left[\frac{\partial}{\partial x}(u\wedge u_y) + \frac{\partial}{\partial y}(u_x\wedge u)\right]$$

Therefore, for any $\varphi\in C_c^\infty(D^2,\mathbf{R}^3)$

$$\int_{D^2}u_x\wedge u_y.\varphi = -\frac{1}{2}\int_{D^2}u\wedge u_y.\varphi_x + u_x\wedge u.\varphi_y$$

and this last expression is continuous under weak convergence of u in H^1.

However, the Palais sequence are not necessarily strongly relatively compact in H^1. Brezis and Coron have precisely analysed this defect of convergence. Namely, it follows from [4] that :

Theorem 3.— *Suppose that (u_n) is a Palais-Smale sequence for J. Then, there exist*

(i) $u_0\in H^1$, solution to (III)

(ii) a finite number of solution $\omega^1,\ldots\omega^p$ of (4) on whole $\mathbf{R}^2$

(iii) sequences $(a_n^1)\ \cdots\ (a_n^p)$ in D^2

(iv) sequences $(\varepsilon_n^1)\ \cdots\ (\varepsilon_n^p)$ with $\varepsilon_n^i>0,\forall i,n$ and $\lim_{n\to+\infty}\varepsilon_n^i = 0,\forall i$

such that, up to a subsequence, we have

$$||u_n - u_0 - \sum_{i=1}^{p}\omega^i\,(\frac{\cdot - a_n^i}{\varepsilon_n^i})||_{H^1}\underset{n\to+\infty}{\to} 0$$

and

$$\int_{D^2} |\nabla u_n|^2 = \int_{D^2} |\nabla u_0|^2 + \sum_{i=1}^{p} \int_{\mathbf{R}^2} |\nabla \omega^i|^2 + o(1)$$

$$J(u_n) = J(u_0) + \sum_{i=1}^{p} J(\omega^i) + o(1)$$

(with the straight forward definition of J for maps from $\mathbf{R}^2$ to $\mathbf{R}^3$)

Theorem 3 emphazises the role of solutions to (4) on whole $\mathbf{R}^2$. These solutions are completely classified, as stated in the following lemma [4] :

Lemma 3.— *Let $\omega \in L^1_{\text{loc}}(\mathbf{R}^2, \mathbf{R}^3)$ be such that*

$$\Delta\omega = 2H\omega_x \wedge \omega_y \quad \text{on} \quad \mathbf{R}^2, \quad \int_{\mathbf{R}^2} |\nabla\omega|^2 < +\infty \ , \ H \neq 0$$

Then ω writes

$$\omega(z) = \frac{1}{H}\Pi(\frac{P(z)}{Q(z)}) + C, \ z = (x, y) = x + iy$$

where $\Pi : \mathbf{C} \to S^2$ denotes the stereographic projection, P, Q are polynomials and C is a constant. In addition

$$\int_{\mathbf{R}^2} |\nabla\omega|^2 = \frac{8\pi}{H^2} \max(degP, degQ)$$

and

$$J(\omega) = \frac{4\pi}{3H^2} \max(degP, degQ)$$

3.2. *The moutain-pass lemma*

Let us recall the classical mountain-pass lemma :

Lemma.— *Let F be a C^1 - functional on a Hilbert space V, which satisfies the following conditions :*

(i) $F(0) = 0$

(ii) $\exists r > 0, \rho > 0 \ \ s.t. \ \ F(v) \geq \rho, \forall v \in V, ||v|| = r$

(iii) $\exists v_0 \in V, ||v_0|| > r, s.t. \ F(v_0) < 0$

Setting

$$c = \inf_{p \in P} \max_{t \in [0,1]} F(p(t))$$

with $P = \{p \in C^0([0,1], V)$ *s.t.* $p(0) = 0, p(1) = v_0\}$, *there exists a sequence* (v_n) *of* V *such that*

$$F\,(v_n) \;\to\; c$$
$$F'(v_n) \;\to\; 0 \text{ in } V'$$

Up to the translation $u \mapsto u + \underline{u}$, we apply this lemma to the functional $J - J(\underline{u})$. Lemma 1 ensures that condition (ii) is satisfied. Moreover, condition (iii) is clearly satisfied, and the following lemma provides us with an upper bound on c :

Lemma 4.—

$$c < \frac{4\pi}{3H^2}$$

The proof of this lemma, which relies on careful computations along an explicit path, is given in [3].

3.3. *Construction of the large solution*

The mountain-pass lemma provides us with a Palais-Smale sequence (u_n) for J. Up to a subsequence, u_n converges weakly to a solution u_0 of (III). Moreover, using Theorem 3, we know that either

$$u_n \to u_0 \text{ in } H^1, \text{ and } J(u_0) = J(\underline{u}) + c$$

or

$$J(u_0) \leq J(\underline{u}) + c - \inf\, \{J(\omega), \omega \text{solution to (4) on } \mathbf{R}^2\}$$

This implies that either

$$J(u_0) \;>\; J(\underline{u})$$

or

$$J(u_0) \;<\; J(\underline{u})$$

since, according to Lemma 3, inf $\{J(\omega), \omega$ solution to (4) on $\mathbf{R}^2\} = \frac{4\pi}{3H^2}$

Then, in any case, $u_0 \neq \underline{u}$, and Theorem 2 follows.

Remark : In fact, it is possible to prove that we are in the first case, that is (u_n) converges strongly to u_0

4 *The second solution for variable H*

In this section, we no longer assume that H is a constant. In this context one might reasonably conjecture that Theorem 2 is still valid for variable H "close" enough to a constant $H_0 \neq 0$, which satisfies

$$|H_0|.\ ||\gamma||_{L^\infty(\partial D^2,\mathbf{R}^3)} \quad < 1 \tag{10}$$

A natural approach to prove this conjecture is to use a perturbation argument. However, the analysis of the Palais-Smale sequences which was exposed in the previous section does not carry over easily to the case $H \not\equiv$ *constant*, even if H is very close to H_0, whatever norm we consider. For instance, the argument that was used to prove that any weak limit of a Palais-Smale sequence is a solution to the equation does not hold any more as soon as H is not exactly a constant. Indeed, the nonlinear term is not weakly continuous. In order to overcome this difficulty, Struwe, following an argument of Sacks and Uhlenbeck [11], introduced a perturbed functional, for which the Palais-Smale condition is satisfied. For measuring the deviation of H from a constant H_0 he considers the affine space of Lipschitz maps H with

$$\begin{aligned}[H - H_0] = \operatorname{ess\,sup}[(1+|\xi|)(|H(\xi) - H_0| + |\nabla H(\xi)|) \\ + |Q(\xi) - H_0\xi| + |dQ(\xi) - H_0 id|; \xi \in \mathbf{R}^3] < +\infty\end{aligned} \tag{11}$$

He was then able to prove [17] :

Theorem 4.— *Assume that* $H_0 \in \mathbf{R} - \{0\}$ *satisfies* (10). *There exists a number* $\alpha > 0$ *such that for a dense set of curvature functions* H *in the* α*- neighbourhood of* H_0, *(III) admits at least two distinct solutions.*

This theorem was improved by Guofang, who showed in [6] that the same result holds for a whole neighbourhood of the constant H_0 in the []-norm.

In [2], we propose a direct approach to the problem, which allows us to release the assumptions on H. Namely, we prove :

Theorem 5.— *Assume that* $H_0 \in \mathbf{R} - \{0\}$ *satisfies* (10), *with* $\gamma \not\equiv$ *constant. There exists a number* $\alpha > 0$ *such that for any curvature function* H *which satisfies*

$$||H - H_0||_{L^\infty(\mathbf{R}^3,\mathbf{R})} \ < \ \alpha$$

(III) admits at least two distinct solutions.

We closely follow the strategy which was developped in Section 3. Namely, we begin with a careful analysis of the behaviour of the Palais-Smale sequences. First, we verify that :

Lemma 2'.— *Any Palais-Smale sequence is bounded in H^1, provided that*

(12) $$\sup_{\xi \in \mathbf{R}^3} [(1+|\xi|)\,|H(\xi)-H_0| + |Q(\xi)-H_0\xi|] < \beta$$

where $\beta > 0$ is small enough.

The proof of this lemma follows the argument of the proof of Lemma 2, the assumption on H ensuring that the perturbation terms are small.

Since the nonlinear term of the equation is not continuous any more under weak convergence, new arguments have to be developped to prove that the weak limit of a Palais-Smale sequence is a solution to the problem. This result was established in [1], i.e.

Theorem 6.— *Assume that $H : \mathbf{R}^3 \to \mathbf{R}$ satisfies*

$$\sup_{\xi \in \mathbf{R}^3} |\nabla H(\xi)| < +\infty$$

Let (u_n) be a bounded sequence in $H^1_\gamma(D^2, \mathbf{R}^3)$ such that

$$\Delta u_n = 2H(u_n) u_{n_x} \wedge u_{n_y} + f_n \quad \text{in} \quad D^2$$

with

$$f_n \to 0 \text{ in } H^{-1}$$

Any weak limit u of a subsequence of (u_n) verifies

$$\Delta u = 2H(u) u_x \wedge u_y \quad \text{in} \quad D^2$$

Once this convergence result is obtained, one can prove an equivalent to Theorem 3, characterizing the lack of compactness for Palais-Smale sequences. The statement reads as :

Theorem 3'.— *Assume that $H : \mathbf{R}^3 \to \mathbf{R}$ satisfies*

(13) $$\sup_{\xi \in \mathbf{R}^3} (1+|\xi|)|\nabla H(\xi)| < +\infty$$

Suppose that (u_n) is a Palais-Smale sequence for J. Then there exist

(i) $u_0 \in H^1$, *solution to (III)*

(ii) a finite number of solutions $\omega^1, \cdots, \omega^p$ of $\Delta\omega = 2H(\omega)\omega_x \wedge \omega_y$ in $\mathbf{R}^2$

(iii) sequences $(a_n^1), \cdots, (a_n^p)$ in D^2

(iv) sequences $(\varepsilon_n^1), \cdots, (\varepsilon_n^p)$ *with* $\varepsilon_n^i > 0, \forall i, n$ *and* $\lim_{n\to+\infty} \varepsilon_n^i = 0, \forall, i$ *such that, up to a subsequence, we have*

$$||u_n - u_0 - \sum_{i=1}^{p} \omega^i \left(\frac{\cdot - a_n^i}{\varepsilon_n^i} \right) ||_{H^1} \underset{n\to+\infty}{\longrightarrow} 0$$

and

$$\int_{D^2} |\nabla u_n|^2 = \int_{D^2} |\nabla u_0|^2 + \sum_{i=1}^{p} \int_{\mathbf{R}^2} |\nabla \omega^i|^2 + o(1)$$

$$J(u_n) = J(u_0) + \sum_{i=1}^{p} J(\omega^i) + o(1)$$

Let us denote by J_H, J_{H_0} the functionals corresponding respectively to H and H_0 and by $\underline{u}, \underline{u}_0$ the associated "small" solutions, given by Theorem 1.

Up to the translation $u \mapsto u + \underline{u}_0$ we apply the mountain pass lemma to the functional $J_H - J_H\ (\underline{u}_0)$. (Even if J_H is not C^1 on H^1_γ, the existence of continuous derivatives in any direction $\varphi \in H_0^1 \cap L^\infty$ is sufficient to construct a pseudo-gradient vector field and get a deformation lemma.) Comparing J_H to J_{H_0}, it is clear that conditions (i) (ii) (iii) are still satisfied, provided that

$$\sup_{\xi \in \mathbf{R}^3} |Q(\xi) - H_0 \xi|$$

is small enough ; hence the existence of a sequence (u_n) in H^1_γ such that

$$J_H(u) - J_H(\underline{u}_0) \to c_H = \inf_{p\in P} \max_{t\in[0,1]} (J_H(p(t) - J(\underline{u}_0))$$

$$J'_H(u_n) \to 0 \text{ in } H^{-1}$$

Moreover, assuming (13), from Lemma 2' and Theorem 3' we deduce that up to a subsequence, (u_n) converges weakly in H^1_γ to some limit $\overline{u}$, and that either

$$J_H(\overline{u}) = J_H(\underline{u}_0) + c_H$$

or

$$J_H(\overline{u}) \leq J_H(\underline{u}_0) + c_H - \inf\{J_H(\omega), \omega \quad \text{solution to (5) in} \quad \mathbf{R}^2\}$$

Note that, by a pertubation argument, it is easy to show the existence of some $\eta > 0$ such that

$$\eta \leq c_H \leq \frac{4\pi}{3H_0^2} - 2\eta$$

and

$$\inf\{J_H(\omega), \omega \quad \text{solution to (5) in} \quad \mathbf{R}^2\} > \frac{4\pi}{3H_0^2} - \eta$$

provided that (12) is satisfied, with β small enough. On the other hand, it follows directly from the way $\underline{u}$ is obtained - see the proof of Theorem 1 - that $\underline{u}$ goes to $\underline{u}_0$ in H^1 as H goes to H_0 in L^∞, and $J_H(\underline{u}) \to J_H(\underline{u}_0)$. Combining these facts, we see that either

$$J_H(\overline{u}) > J_H(\underline{u})$$

or

$$J_H(\overline{u}) < J_H(\underline{u})$$

as soon as H satisfies (13), and (12) with β small enough.

Remark : It is an open question to know whether in this case the convergence of (u_n) to $\overline{u}$ is strong, as for constant H.

We remark now that Lemma 2' provides us with a H^1- bound for $\overline{u}$- see the proof of Lemma 2 - which is independent of H satisfying (12), with β small enough. Thus, the last ingredient that we need to make the proof of Theorem 5 complete is an *a priori* bound for the L^∞- norm of the solutions in term of the H^1-norm :

Theorem 7.— *Assume that $H : \mathbf{R}^3 \to \mathbf{R}$ is smooth and bounded on $\mathbf{R}^3$. Let u be a smooth solution to (III).*

There exists $C > 0$, depending only on the L^∞ -norm of H, such that

$$||u||_{L^\infty} \leq ||\gamma||_{L^\infty} + C(\int_{D^2} |\nabla u|^2 + 1)$$

The proof of this Theorem, based upon an idea of Grüter [5], may be found in [2].

Let $\beta > 0$ be fixed, such that for H satisfying (12) and (13) there exists a second solution $\overline{u}$ to (III), with

$$||\overline{u}||_{H^1} < R$$

R independent of H. Theorem 7 implies that

$$||\overline{u}||_{L^\infty} \leq ||\gamma||_{L\infty} + C(R^2 + 1) = R'$$

There exists $\alpha > 0$ such that for any $H \in C^1(\mathbf{R}^3, \mathbf{R})$ which satisfies

$$||H - H_0||_{L^\infty} < \alpha$$

one can find some $\tilde{H} \in C^1(\mathbf{R}^3, \mathbf{R})$ which satisfies (12) and (13), such that

$$\tilde{H}(\xi) = H(\xi) \quad \forall \xi \in \mathbf{R}^3, |\xi| \leq R'$$

The two solutions $\underline{u}, \overline{u}$ corresponding to $\tilde{H}$ being such

$$||\underline{u}||_{L^\infty},\ ||\overline{u}||_{L^\infty} \ \leq \ R'$$

they are as well solutions to (III) for the initial curvature function H, and the proof of Theorem 5 is complete.

Acknowledgements

We are indebted to J.M. Coron for bringing this problem to our attention, and for numerous references in the literature. We also thank F. Helein for stimulating discussions.

References

1 - F. BETHUEL - Weak convergence of Palais-Smale sequences for some critical functionals, *to appear.*

2 - F. BETHUEL, O. REY - Multiple solutions to the Plateau problem for non-constant mean curvature *to appear.*

3 - H. BREZIS, J.M. CORON - Multiple solutions of H-Systems and Rellich's conjecture, *Comm. Pure Applied Math.* **37**, (1984), pp. 149-187.

4 - H. BREZIS, J.M. CORON - Convergence of solutions of H-Systems or how to blow bubbles, *Arch. Rat. Mech. Anal.* **89**, (1985), pp. 21-56.

5 - M.GRÜTER - Regularity of weak H-Surfaces, *J. Reine Angew. Math.* **329**, pp. 1-15.

6 - G. WANG -The Dirichlet problem for the equation of prescribed mean curvature, *Ann. Inst. Henri Poincaré, Anal. non linéaire* **IX**, (1992), pp. 643-655.

7 - E. HEINZ -Über die Existenz einer Fläche konstanter mittlerer Krümmung bei vorgegebener Berandung, *Math. Ann.* **127**, (1954), pp. 258-287.

8 - S. HILDEBRANDT - Randwertprobleme für Flächen unit vorgeschriebener mittlerer Krümmung und Anwendungen auf die Kapillaritätstheorie, *Math. Z.* **112**, (1969), pp. 205-213.

9 - C. MORREY - Multiple integrals in the calculus of variations *Springer, Berlin* (1966).

10 - O. REY - Heat flow for the equation of surfaces with prescribed mean curvature, *Math. Ann.* **291**, (1991), pp. 123-146.

11 - P. SACKS, K. UHLENBECK - The existence of minimal immersions of spheres, *Ann. Math.* **113**, (1981), pp. 1-24.

12 - K. STEFFEN - On the existence of surfaces with prescribed mean curvature and boundary, *Math. Z.* **146**, (1976), pp. 113-135.

13 - K. STEFFEN - On the uniqueness of surfaces with prescribed mean curvature spanning a given contour, *Arch. Rat. Mech. Anal.* **94**, (1986), pp. 101-122.

14 - M. STRUWE - Nonuniqueness in the Plateau problem for surfaces of constant mean curvature, *Arch. Rat. Mech. Anal.* **93**, (1986), pp. 135-157.

15 - M. STRUWE - Large H-surfaces via the mountain-pass Lemma, *Math. Ann.* **270**, (1985), pp. 441-459.

16 - M. STRUWE - Plateau's problem and the Calculus of Variations, *Mathematical Notes* **35**, Princeton University Press (1988)

17 - M. STRUWE - Multiple solutions to the Dirichlet problem for the equation of prescribed mean curvature, *Analysis, et cetera*, Academic Press, Boston (1990).

18 - H. WENTE - An existence theorem for surfaces of constant mean curvature, *J. Math. Analysis Appl.* **26**, (1969), pp. 318-344.

19 - H. WENTE - A general existence theorem for surfaces of constant mean curvature, *Math.Z.* **120**,(1971), pp. 277-278.

20 - H. WENTE - The differential equation $\Delta x = 2H(x_u \wedge x_v)$ with vanishing boundary values, *Proc. A.M.S.* **50**, (1975), pp. 113-137.

21 - H. WERNER - Das Problem von Douglas für Flächen konstanter mittlerer Krümmung, *Math. Ann.* **133**, (1957), pp. 303-319.

Geometry in Partial Differential Equations (pp. 32–50)
eds. A. Pràstaro and Th. M. Rassias

OPTICAL HAMILTONIAN FUNCTIONS

MISHA BIALY and LEONID POLTEROVICH*

Raymond and Beverly Sackler Faculty of Exact Sciences
School of Mathematical Sciences
Tel Aviv University, Tel Aviv 69978, Israel

ABSTRACT

Optical Hamiltonian functions can be defined (at least locally) on a symplectic manifold endowed with a Lagrangian distribution. They can be considered as a far-going generalization of fiberwise convex Hamiltonians on a cotangent bundle. It turns out that critical points of these functions have very specific properties which lead to global restrictions on the topology of underlying symplectic manifolds. We apply these results to a geometrical study of the group of Hamiltonian diffeomorphisms.

1. Introduction and Main Results

1.1. *From Legendre Condition to Twisting*

Recent progress in the qualitative theory of classical mechanical systems is closely related to realizing the role played by so-called *Legendre condition.*[15–17] In the canonical variables (p, q) on the symplectic phase space this condition means just the convexity of a Hamiltonian function with respect to momenta p. This property is satisfied in most of physically important examples, for instance in *natural systems* where the kinetic energy is a positive definite quadratic form. Geometrically, the Legendre condition can be reformulated as follows. Consider a distribution of tangent Lagrangian subspaces given by the equation $\{dq = 0\}$. Then the Legendre condition for a Hamiltonian function is equivalent to the fact that *the corresponding Hamiltonian flow twists the Lagrangian distribution in the positive direction.*

* Partially supported by United States-Israel Binational Science Foundation, Grant No. 90-00150/2.

In contrast with the Legendre condition, the last one still makes sense in much more general situation, namely on a symplectic manifold endowed with a Lagrangian distribution. In this setting Hamiltonian functions which have the twisting property as well as Hamiltonian diffeomorphisms generated by them are called *optical.* These objects were introduced by the authors[5] where it was demonstrated that they are very natural ones to which certain classical results by Birkhoff on area-preserving twist maps of the annulus can be extended. The starting point of the present paper is the following

Question 1.1.A. Given a symplectic manifold with a Lagrangian distribution, does it admit an optical Hamiltonian function?

As we shall see below in 1.5.A, optical functions always exist *locally.* The global situation, however, is much less understood. To be more precise, restrict for a moment our attention to manifolds without boundary. Such a manifold is called ***optical*** if it admits an exhausting (that is proper and bounded from below) function which is optical with respect to some symplectic structure and a Lagrangian distribution. The only examples of optical manifolds we know are cotangent bundles (see 1.4.A, 1.4.B for the discussion). Therefore naturally arises the following

Question 1.1.B. Is it true that every optical manifold without boundary is diffeomorphic to a cotangent bundle?

As we show (see 1.6.I) this is valid if $\dim M = 2$.

Though in the general the answer is not clear, it turns out that the topology of optical manifolds is very specific. Interestingly enough, exactly the same topological restrictions arise in the theory of Stein complex manifolds (see 1.6.D below).

Finally, we present one geometric application of our results on optical Hamiltonian functions. Namely we discuss the following remark by V.I. Arnold[3] which is presented below in our translation from Russian:

> "Symplectic diffeomorphisms which lie far away from the identity map are those to which one is led by paths with positively twisted Hamiltonians[1,2] ".

We put this assertion into the framework of Hofer's (pseudo) metric on the group of Hamiltonian diffeomorphisms, and confirm it in various cases using an idea by Eliashberg and Ratiu[12] (see §3).

1.2. *Hamiltonian Diffeomorphisms of Symplectic Manifolds*

A *symplectic structure* on a smooth even-dimensional manifold M is a closed differential 2-form, say ω which satisfies the following non-degeneracy condition: the homomorphism

$$I : TM \to T^*M ,$$
$$I(\xi) = -i_\xi \omega$$

is an isomorphism (here and below i denotes the interior product).

Symplectic manifolds naturally arise as phase spaces of classical mechanical systems. Given a (time-dependent) energy function $H(x,t)$ on $M \times \mathbf{R}$, the evolution of the system is described by the flow g_H^t corresponding to the (time-dependent) vector field

$$\xi_H^t = I^{-1}(dH(\cdot,t))$$

on M. Traditionally, H, g_H^t and ξ_H^t are called a *Hamiltonian* function, flow and vector field respectively.

We say that a diffeomorphism $f : M \to M$ is *Hamiltonian* if it can be represented as a time-one map g_H^1 of a Hamiltonian flow.

A fundamental property of Hamiltonian diffeomorphisms is that they preserve the symplectic structure.

1.3. *Lagrangian Grassmann Manifold*

Each tangent space $T_{(\cdot)}M$ of a symplectic manifold carries a structure of a symplectic vector space, that is a vector space endowed with a bilinear skew-symmetric non-degenerate form.

Let (E,ω) be a symplectic vector space. A subspace $\ell \subset E$ is called *Lagrangian* if $\dim \ell = \dim E/2$ and $\omega\big|_\ell$ vanishes. It turns out that the set Λ of all Lagrangian subspaces has a natural structure of a smooth manifold, which is called Lagrangian Grassmann manifold.[2,9] Let us describe a useful representation of tangent vectors to Λ. Given $\lambda \in \Lambda$, we identify $T_\lambda\Lambda$ with the space $S^2(\lambda)$ of all symmetric bilinear forms on λ in the following way. Every curve $\lambda(t) \subset \Lambda$ with $\lambda(0) = \lambda$ can be written as

$$\lambda(t) = \Phi(t)\lambda ,$$

where $\Phi(t)$ is a path of linear symplectic transformations with $\Phi(0) = \mathbb{1}$. Then the symmetric bilinear form corresponding to the vector

$$\frac{d}{dt}\bigg|_{t=0} \lambda(t)$$

is given by

$$(\xi,\eta) \mapsto \omega\Big(\xi, \frac{d}{dt}\Big|_{t=0} \Phi(t)\eta\Big) \quad \text{for all} \quad \xi,\eta \in \lambda \ .$$

One can easily check that this correspondence is well defined.

1.4. *Optical Hamiltonian Functions* (see M. Bialy and L. Polterovich[5])

Consider now a symplectic manifold (M,ω) endowed with a distribution $\mathcal{F}$ of tangent Lagrangian subspaces. We write $\Lambda(z)$ for the Lagrangian Grassmann manifold associated to the symplectic vector space $T_z M$.

For every time-independent Hamiltonian function $H \in C^\infty(M)$ define a tangent vector

$$b_H(z) = \frac{d}{dt}\Big|_{t=0} (g_H^t)_* \mathcal{F}(g_H^{-t} z) \in T_{\mathcal{F}(z)}\Lambda(z) \ .$$

Let $B_H(z) \in S^2(\mathcal{F}(z))$ be the bilinear form corresponding to $b_H(z)$ according to 1.3. By definition, a function $H \in C^\infty(M)$ is called $\mathcal{F}$*-optical* if $B_H(z)$ is positive definite for all $z \in M$.

Example 1.4.A. Consider the cotangent bundle T^*X endowed with the standard symplectic structure. Recall that in local canonical coordinates (p,q) this structure is written as $dp \wedge dq$ (see e.g. V.I. Arnold and A.B. Givental[4]). Let $H : T^*X \to \mathbb{R}$ be the energy function associated to a Riemannian metric on X. Then H is optical with respect to the distribution $\mathcal{F}$ of Lagrangian subspaces tangent to the fibers of the bundle (see M. Bialy and L. Polterovich[5] for the discussion). More generally, a function of T^*X is $\mathcal{F}$-optical if and only if its restriction on every fiber is strictly convex.

Example 1.4.B. Let (X,g) be a Riemannian manifold. The metric allows to identify its tangent and cotangent bundles. In particular we can assume that TX is endowed with a symplectic structure which comes from the standard one on T^*X. Denote by $\pi : TX \to X$ the natural projection and by M the complement to the zero section.

Let us define a Lagrangian distribution $\mathcal{F}$ on M in the following way. Let $L(x,v)$ be the set of all horizontal (in the sense of the Riemannian connection) tangent vectors $Y \in T_{(x,v)}M$ such that $\pi_* Y$ is orthogonal to v. Let e be the vector field on M tangent to the foliation of M by straight lines which lie in fibers and pass through the origin. Finally, define $\mathcal{F}$ as the span of L and e. One can easily check that it is a Lagrangian distribution.

Consider the energy function H on M given by $H(x,v) = |v|^2/2$. Recall that the Hamiltonian flow generated by this function is just the geodesic flow of the metric g.

Now we have enough data in order to define the form $B_H \in S^2(\mathcal{F})$. It turns out that this form has a nice geometric meaning. Namely, a direct computation (which can be held with the use of 2.1.A) shows that for every tangent vector $Y \in L(x,v)$ holds

$$2B_H(Y,Y) = (R(\pi_* Y, v)v, \pi_* Y),$$

where R is the curvature tensor of the Riemannian metric. Moreover, the function H is $\mathcal{F}$-optical if and only if the sectional curvature of g is positive.

Remark 1.4.C. Note, that the property of a function to be optical is *stable* in the following sense. Let $H : M \to \mathbb{R}$ be a function which is optical with respect to a symplectic structure ω and a Lagrangian distribution $\mathcal{F}$. Then this remains true for every sufficiently small C^2-perturbation H' of H, C^2-perturbation ω' of ω and C^1-perturbation $\mathcal{F}'$ of $\mathcal{F}$. For instance, every fiberwise strictly convex function on T^*X is optical with respect to a sufficiently small C^1-perturbation $\mathcal{F}'$ of the standard Lagrangian distribution $\mathcal{F}$. Let us emphasize that one can choose such $\mathcal{F}'$ to be *non-integrable*.

Remark 1.4.D. Recall that if a Lagrangian distribution $\mathcal{F}$ is integrable then its leafs carry natural affine structure.[4] In this case the form B_H associated to an $\mathcal{F}$-optical function H determines *a Hessian structure* on every leaf (that is a Riemannian metric which coincides in local affine coordinates with a Hessian of some function.[19] It would be interesting to understand the role of this structure in analysis of optical Hamiltonian functions. We return to affine properties of leafs in 1.6 below.

1.5. *Critical Points of Optical Hamiltonian Functions*

It follows immediately from 1.4.A that there are specific restrictions on critical points of optical Hamiltonian functions on cotangent bundles. We show now that this remains valid in a much more general situation. Let (M, ω) be an arbitrary symplectic manifold endowed with a Lagrangian distribution $\mathcal{F}$.

Theorem 1.5.A. *A function is $\mathcal{F}$-optical in a neighborhood of its critical point, say z if and only if its second differential is positive definite on the subspace $\mathcal{F}(z)$ of the Lagrangian distribution passing through this point.*

The theorem is proved below in §2. Let us formulate two immediate consequences.

Corollary 1.5.B. *The index of a non-degenerate critical point of an $\mathcal{F}$-optical Hamiltonian function does not exceed the half of the dimension of the manifold M.*

Corollary 1.5.C. *(The maximum principle)*
An $\mathcal{F}$-optical Hamiltonian function on M cannot achieve a (weak) local maximum at an interior point of M.

Return for a moment to optical Hamiltonian functions on cotangent bundles. As we saw in 1.4.A each such function is fiberwise strictly convex, and therefore every critical point of its restriction to a fiber is a non-degenerate local minimum. As a generalization of this phenomenon we present the following

Theorem 1.5.D. *Let (M,ω) be a symplectic manifold endowed with a Lagrangian distribution $\mathcal{F}$. Let $L \subset M$ be a Lagrangian leaf of $\mathcal{F}$. Then each critical point of the restriction to L of an $\mathcal{F}$-optical Hamiltonian function is a non-degenerate local minimum.*

Theorem 1.5.D is proved below in §2.

1.6. *Optical Symplectic Manifolds*

It follows immediately form 1.5.C that a compact manifold without boundary admits no optical Hamiltonian functions.

Definition 1.6.A. Assume that a connected manifold M admits a symplectic structure ω, a Lagrangian distribution $\mathcal{F}$ and an $\mathcal{F}$-optical Hamiltonian function $H : M \to \mathbb{R}$ which is proper and bounded from below. Suppose that one of the following conditions holds:

(i) M is open and without boundary.

(ii) M is compact and H is constant on the boundary ∂M.

Then M is called *an optical* manifold and $\mathcal{F}$ is called *an optical* distribution of M.

Theorem 1.6.B. *Every optical manifold admits a proper Morse function which is bounded from below and is constant on the boundary and such that the indexes of all its critical points do not exceed the half of the dimension of the manifold.*

Proof: Note that a sufficiently small smooth perturbation of an $\mathcal{F}$-optical function is an $\mathcal{F}$ optical function (see 1.4.C). Therefore one can choose H in 1.6.A to be a Morse function. The assertion follows now from 1.5.A. ■

Using the Morse theory one obtains the following

Corollary 1.6.C. *Every optical manifold has homotopy type of a CW-complex of the dimension less or equal to dim $M/2$.*

Corollary 1.6.D. *Assume that M is an optical manifold of the dimension greater than 4. Then M admits a structure of a Stein complex manifold (with pseudo-convex boundary), and hence of a Weinstein symplectic manifold.*[11]

This corollary follows immediately from a result of Y. Eliashberg.[10] Note that a Weinstein symplectic structure from 1.6.D is in general different from the original one on M.

Example 1.6.E. If M is an optical manifold then ∂M is connected provided dim $M \geq 4$. Indeed otherwise $H_{\dim M-1}(M)$ has positive dimension but this contradicts 1.6.C. Note that ∂M can be disconnected if dim $M = 2$. For instance, the two-dimensional annulus is an optical manifold with the disconnected boundary.

It turns out that the existence of $\mathcal{F}$-optical Hamiltonian functions leads to essential restrictions on the topology of leafs of the distribution $\mathcal{F}$ (if such leafs exist).

Theorem 1.6.F. *Let $\mathcal{F}$ be an optical distribution of M, and let L be a properly embedded leaf of $\mathcal{F}$ such that $\partial L \subset \partial M$. Then L is diffeomorphic to an open (respectively closed) ball of dimension dim $M/2$ provided M is open (respectively compact).*

The theorem follows immediately from 1.5.D and standard arguments of the Morse theory.

Consider now the case when M is open and the distribution $\mathcal{F}$ is *integrable*, i.e. $\mathcal{F}$ defines a Lagrangian foliation on M. Then each leaf of the foliation carries a natural affine structure. We shall say that an integrable distribution $\mathcal{F}$ is *convex* if the universal cover of each fiber is affine equivalent to a convex domain in $\mathbb{R}^n$.

Example 1.6.G. Assume that dim $M = 2$. Then every distribution on M is Lagrangian, integrable, and convex.

We call an open subset of a cotangent bundle T^*B convex, if its intersection with every fiber is convex with respect to the canonical affine structure. Recall that a twisted symplectic structure on T^*B is one of the form $\Omega + \pi^*\sigma$, where $\pi : T^*B \to B$ the canonical projection, Ω is the standard symplectic structure on T^*B and σ is a closed 2-form on B.

Theorem 1.6.H. *Let M be an open manifold endowed with an integrable convex optical Lagrangian distribution. Then there exists a fiber-preserving symplectomorphism between M and an open convex neighborhood of the zero-section of a cotangent bundle endowed with a twisted symplectic structure.*

Corollary 1.6.I. *Every optical surface without boundary is diffeomorphic to either $\mathbb{R}^2$ or $\mathbb{R}^1 \times S^1$.*

Proof of Theorem 1.6.H.

Set $2n = \dim M$. Let $H : M \to \mathbb{R}$ be a proper, $\mathcal{F}$-optical function bounded from below. We divide the proof into several steps.

Step 1. Define a subset $B \subset M$ as follows:

$$B = \left\{ x \in M : \text{ such that } dH \text{ vanishes on } \mathcal{F}(x) \right\} .$$

Note that B is not empty since it contains critical points of H. We claim that B is properly embedded n-dimensional submanifold of M transversal to the fibres. Indeed, introduce local coordinates p, q in a neighborhood of a point of B such that with these coordinates $\omega = dp \wedge dq$ and $\mathcal{F} = \{dq = 0\}$. Then B is locally given by the equation $\{\frac{\partial H}{\partial p} = 0\}$. Note that $\frac{\partial^2 H}{\partial p^2} > 0$ since H is optical. The claim follows from the implicit function theorem.

Step 2. We claim that the set B intersects each fiber L of $\mathcal{F}$. Indeed, the function H is proper and hence attains an absolute minimum on the Closure (L) at some point z. Note that the fiber L' passing through z is entirely contained in Closure (L). Moreover, $z \in B$ and L' intersects B transversally by Step 1. Therefore, L intersects B. Our claim follows.

Step 3. Let L be a fiber of $\mathcal{F}$. then $H|_L$ is a convex function with a critical point. Therefore our assumption on convexity of $\mathcal{F}$ implies that this critical point is unique, L is simply connected and hence affine equivalent to a convex open subset of $\mathbb{R}^n$.

Step 4. It follows from the previous steps that B intersects each fiber at unique point. Define the map $\Theta : M \to B$ which corresponds to each point $x \in M$ the intersection point between B and the fiber passing through x. Note that Θ is a submersion since B is transversal to the fibers. Summing up the results of the previous steps we obtain that M admits a submersion onto some manifold B with Lagrangian simply-connected fibers.

Now one can use a well-known construction[4] (see 4.2 of Chapter 2) in order to find a diffeomorphism between M and a convex neighborhood of the zero-section in

T^*B. Moreover, one can check that this diffeomorphism is symplectic with respect to the original form ω on M and the twisted symplectic form $\Omega + \pi^*(\omega|_{TB})$ on T^*B. This completes the proof. ■

1.7. *Hofer's (Pseudo) Metric*

Let (M, ω) be a compact symplectic manifold with the boundary Σ. Let $\mathfrak{d}$ be the set of all smooth functions on M which vanish on Σ. Note that $\mathfrak{d}$ is a Lie algebra with respect to the Poisson bracket on M. Set

$$\mathcal{H} = \left\{ H \in C^\infty(M \times [0;1]) | H(\cdot, t) \in \mathfrak{d} \quad \text{for all } t \in [0;1] \right\} .$$

All Hamiltonian diffeomorphisms of M generated by Hamiltonian functions from $\mathcal{H}$ form a group which we denote by $\mathcal{D}$. We consider $\mathcal{D}$ as an infinite-dimensional Lie group with the Lie algebra $\mathfrak{d}$.

Define a norm on $\mathfrak{d}$ as follows:

$$\|H\| = \max_M H - \min_M H$$

for all $H \in \mathfrak{d}$.

Obviously this norm is invariant under the adjoint action of $\mathcal{D}$ on $\mathfrak{d}$:

$$\|\varphi^* H\| = \|H\|$$

for all $\varphi \in \mathcal{D}$, $H \in \mathfrak{d}$. Therefore one can define a biinvariant (pseudo) metric ρ on $\mathcal{D}$ in the following way:

$$\rho(\varphi, \psi) = \inf \int_0^1 \|\dot{\gamma}(t)\| dt ,$$

where the infinum is taken over all smooth paths $\gamma : [0;1] \to \mathcal{D}$ with $\gamma(0) = \varphi$ and $\gamma(1) = \psi$.

Note that such a pseudometric was originally introduced by H. Hofer[13,14] who studied it for compactly supported Hamiltonian diffeomorphisms of $\mathbb{R}^{2n}$. In particular, Hofer showed that in this case his pseudometric is a genuine *metric*. In other words, the pseudodistance between any pair of distinct points is strictly positive. Hofer's result was extended by L. Polterovich[18] to a wider class of so-called *tame rational* symplectic manifolds without boundary, which includes for instance cotangent bundles of closed manifolds.

It turns out that this remains true also for certain symplectic manifolds with boundary. For example, in our setting ρ is a genuine metric provided M lies in $\mathbb{R}^{2n}$ or T^*X.

1.8. *Metrical Properties of Optical Hamiltonian Diffeomorphisms*

Below we continue the discussion started in 1.7. Assume in addition that M is endowed with a Lagrangian distribution $\mathcal{F}$. A Hamiltonian diffeomorphism from $\mathcal{D}$ is called *$\mathcal{F}$-optical* if it can be represented as a time one map of a Hamiltonian flow generated by a Hamiltonian function $H(x,t) \in \mathcal{H}$ which is $\mathcal{F}$-optical (see 1.4) for every $t \in [0;1]$.

Let us pose the following question:

What is the geometry (in the sense of Hofer's (pseudo) metric) of the subgroup $\mathcal{D}_f \subset \mathcal{D}$ generated by a given optical Hamiltonian diffeomorphism $f \in \mathcal{D}$?

As we shall see below the answer crucially depends on symplectic properties of the boundary ∂M.

Example 1.8.A. Consider the linear complex space $\mathbb{C}^n$ endowed with the standard symplectic form. Let $M \subset \mathbb{C}^n$ be the unit Eucledian ball $\{|z| \leq 1\}$, and let $\mathcal{F}$ be a *translation invariant* Lagrangian distribution on $\mathbb{C}^n$. For every $\alpha \in \mathbb{R}$ set $f_\alpha(z) = e^{i\alpha}z$. We claim that f_α is an $\mathcal{F}$-optical Hamiltonian diffeomorphism from $\mathcal{D}$. Indeed, denote by $\{\alpha\}$ the minimal *positive* element of the subset $\alpha + 2\pi\mathbb{Z} \subset \mathbb{R}$. Then f_α is the time one map of the Hamiltonian flow generated by the function $H_\alpha(z,t) = \frac{1}{2}\{\alpha\}(|z|^2 - 1)$. Obviously, $H_\alpha \in \mathcal{H}$. Moreover, H_α is *convex* and therefore $\mathcal{F}$-optical. Our claim follows.

Note that $\max_M H_\alpha - \min_M H_\alpha = \frac{1}{2}\{\alpha\}$. Thus

$$\rho(\mathbb{1}, f_\alpha) \leq \frac{1}{2}\{\alpha\} \leq \pi\ .$$

We conclude that *the subgroup $\mathcal{D}_{f_\alpha}$ is a ρ-bounded subset of $\mathcal{D}$.*

It turns out that the situation described in 1.8.A is *very untypical* at least for *exact* symplectic manifolds. We show below that for a *generic* exact symplectic form on M and for an arbitrary optical Hamiltonian diffeomorphism $f \in \mathcal{D}$ the sequence

$$\frac{1}{n}\rho(\mathbb{1}, f^n)$$

is bounded away from zero.

In order to formulate our result let us recall that the *characteristic foliation* of a hypersurface of a symplectic manifold is a 1-dimensional foliation whose tangent lines are symplectically orthogonal to the tangent spaces to the hypersurface.

Theorem 1.8.B. *Let (M,ω) be an exact compact symplectic manifold endowed with a Lagrangian distribution $\mathcal{F}$. Assume that the characteristic foliation of the boundary ∂M has at least one non-closed leaf. Then for every F-optical Hamiltonian diffeomorphism $f \in \mathcal{D}$ the sequence*

$$\frac{1}{n}\rho(\mathbb{1}, f^n)$$

is bounded away from zero when $n \to \infty$.

The proof of this theorem, which essentially uses the maximum principle 1.5.C, as well as some generalizations one can find in §3.

The following result is an immediate consequence of 1.8.B.

Corollary 1.8.C. *Assume that (M,ω) satisfies the conditions of 1.8.B. Then the identity map is not $\mathcal{F}$-optical with respect to any Lagrangian distribution $\mathcal{F}$.*

(We refer the reader to M. Bialy and L. Polterovich[5] (Theorem 2.6) for another approach to this problem.)

Remark 1.8.D. The assertions of 1.8.B and 1.8.C do not remain true without an assumption on the characteristic foliation of the boundary. Indeed, in the setting of 1.8.A the hypersurface ∂M is the usual Eucledian sphere and hence all the leafs are closed. As we have seen in this case there exist optical Hamiltonian diffeomorphisms for which $\frac{1}{n}\rho(\mathbb{1}, f^n) \to 0$ for $n \to \infty$. Moreover, the identity map coincides with f_0 and therefore is $\mathcal{F}$-optical.

2. Differential Geometry of Optical Hamiltonian Functions

2.1. *Bennequin's Identity*

Let (M,ω) be a symplectic manifold endowed with a Lagrangian distribution $\mathcal{F}$. Recall that according to 1.4 every Hamiltonian function $H \in C^\infty(M)$ defines a symmetric bilinear form $B_H(z) \in S^2(\mathcal{F}(z))$ for all $z \in M$. The purpose of this section is to describe this form from differential geometric point of view. This description is used below in the proof of 1.5.A and 1.5.D. Note that in a particular case the identity 2.1.A below was obtained by D. Bennequin.[7]

Fix a point $z \in M$ and consider a germ of a vector field X at z such that $X(w) \in \mathcal{F}(w)$ for all points w in a neighborhood of the point z. Recall that we write ξ_H for the Hamiltonian vector field of a function H.

Lemma 2.1.A. *For every function $H \in C^\infty(M)$ and for every $Y \in \mathcal{F}(z)$ the following equality holds*

$$B_H(X(z), Y) = \omega([\xi_H, X](z), Y)$$

Proof: Choose a path $\Phi(t)$ of symplectic transformations of the linear symplectic space $T_z M$ satisfying the following conditions:

- $\Phi(0) = \mathbb{1}$;
- $(g_H^t)_* \mathcal{F}(g_H^{-t} z) = \Phi(t) \mathcal{F}(z)$;
- $(g_H^t)_* X(g_H^{-t} z) = \Phi(t) X(z)$

for all sufficiently small t. By the definition of the form B_H we have

$$B_H(X(z), Y) = \omega\Big(Y, \frac{d}{dt}\Big|_{t=0} \Phi(t) X(z)\Big) =$$
$$= \omega(Y, -\mathcal{L}_{\xi_H} X) = \omega(Y, -[\xi_H, X]) = \omega([\xi_H, X], Y)\ .$$

(here and below $\mathcal{L}$ denotes the Lie derivative).

This yields the proof of the lemma. ■

Example 2.1.B. Consider the cotangent bundle T^*X endowed with the standard symplectic structure $\omega = \Sigma dp_i \wedge dq_i$, where $p_i, q_i, i = 1 \dots n$ are local coordinates. Consider the distribution $\mathcal{F}$ of Lagrangian subspaces tangent to the fibers of the cotangent bundle. Let $H \in C^\infty(T^*X)$ and set $v_i = \partial/\partial p_i$ be vector fields tangent to the fibers of T^*X. Lemma 2.1.A gives

$$B_H(v_i, v_k) = \omega([\xi_H, v_i], v_k)$$

where with our notations

$$\xi_H = -\sum_j \partial H/\partial q_j\ \partial/\partial p_j + \sum_j \partial H/\partial p_j\ \partial/\partial q_j\ .$$

Then

$$[\xi_H, v_i] = \sum_j \partial^2 H/\partial q_j \partial p_i\ \partial/\partial p_j - \sum_j \partial^2 H/\partial p_j \partial p_i\ \partial/\partial q_j\ ,$$

and therefore

$$\omega([\xi_H, v_i], v_k) = \partial^2 H/\partial p_i \partial p_k\ .$$

Thus one concludes that the function H is $\mathcal{F}$-optical in this case if and only if it is fiberwise strictly convex.

2.2. *Proof of Theorem 1.5.A*

Let H be a function on M and let z be a critical point of H. Let X be a germ of a non-vanishing vector field at z which is tangent to $\mathcal{F}$.

In view of 1.2.A it is sufficient to check that

$$d^2H(X(z)) = \omega([\xi_H, X](z), X(z)) .$$

Indeed,

$$d^2H(X(z)) = \mathcal{L}_X\mathcal{L}_X H(z) = \mathcal{L}_X(\omega(X, \xi_H))(z)$$

according to the definition of the Hamiltonian vector field. But the last expression is equal to

$$(\mathcal{L}_X\omega)(X(z), \xi_H(z)) + \omega([\xi_H, X](z), X(z)) .$$

Note that $\xi_H(z) = 0$ since z is a critical point of H, and hence we get the desired identity. ■

2.3. *Proof of Theorem 1.5.D*

Let H be an $\mathcal{F}$-optical function on M, and let z be a critical point of its restriction h to a leaf L of the foliation $\mathcal{F}$. Fix a germ of a vector field X at z which is tangent to $\mathcal{F}$. In view of 2.1.A it is sufficient to show that

$$d^2h(X(z)) = \omega([\xi_H, X](z), X(z)) .$$

As in 2.2 we easily derive that

$$d^2h(X(z)) = (\mathcal{L}_X\omega)(X(z), \xi_H(z)) + \omega([\xi_H, X](z), X(z)) .$$

Note that $dH = -i_{\xi_H}\omega$ vanishes on $\mathcal{F}(z)$ since z is critical for h, and hence $\xi_H(z) \in \mathcal{F}(z)$ since $\mathcal{F}(z)$ is a Lagrangian subspace. Moreover, $\mathcal{L}_X\omega$ vanishes on $\mathcal{F}(z)$ since ω vanishes identically on TL. Thus $(\mathcal{L}_X\omega)(X(z), \xi_H(z)) = 0$, and we get the desired identity. This completes the proof. ■

3. Hofer's (Pseudo) Metric and Calabi Invariant

In the present section we prove Theorem 1.8.B. Let us fix some notations. We identify each tangent space to the group $\mathcal{D}$ with its Lie algebra $\mathfrak{d}$ using the left action of $\mathcal{D}$ on itself. We write $\mathfrak{d}_{\mathcal{F}}$ for the subset of $\mathfrak{d}$ of all $\mathcal{F}$-optical Hamiltonian functions. With this language, a diffeomorphism from $\mathcal{D}$ is $\mathcal{F}$-optical if it can be joined with the identity map by a smooth path $\gamma(t)$ whose tangent vector $\dot{\gamma}(t)$ belongs to $\mathfrak{d}_{\mathcal{F}}$ for all t.

Following Eliashberg and Ratiu[12] we *calibrate* the cone $\mathfrak{d}_{\mathcal{F}}$ by a suitable linear functional on $\mathfrak{d}$ which generates an exact 1-form on $\mathcal{D}$ and whose primitive is the *Calabi homomorphism* .

3.1. *Calabi Invariant*

Define a functional $c : \mathfrak{d} \to \mathbb{R}$, $c(H) = \int H\omega^n$. Note that c is invariant under the adjoint action of the group: $c(\varphi^* H) = c(H)$ for all $\varphi \in \mathcal{D}$, $H \in \mathfrak{d}$. Therefore c defines a biinvariant 1-form, say θ on $\mathcal{D}$.

Proposition 3.1.A. *The form θ is closed, that is for every smooth map $\psi : D^2 \to \mathcal{D}$ holds $\int_{S^1} \psi^*\theta = 0$.*

Proof: Since θ is biinvariant 1-form, it is sufficient to check that c vanishes on the commutant of the Lie algebra: $c|_{[\mathfrak{d},\mathfrak{d}]} = 0$. In other words we have to check that

$$\int_M \{H, F\}\omega^n = 0$$

for all $H, F \in \mathfrak{d}$ (here and below $\{\ ,\ \}$ denote the Poisson bracket). Note that

$$\int_M \{H, F\}\omega^n = \int_M dH \wedge dF \wedge \omega^{n-2} = \int_{\partial M} H(dF \wedge \omega^{n-2}) = 0$$

since $H|_{\partial M} = 0$. Therefore θ is closed. ■

Assume now for a moment that the 1-form θ is *exact*, that is for every diffeomorphism $f \in \mathcal{D}$ the integral

$$\mathrm{Cal}(f) = \int_\gamma \theta$$

does not depend on the choice of a path γ which joins $\mathbb{1}$ with f. In this case the map Cal:$\mathcal{D} \to \mathbb{R}$ is a homomorphism which is called *the Calabi homomorphism.* The value Cal(f) is called *the Calabi invariant* of f. This homomorphism was introduced by E. Calabi.[8] A detailed exposition of its properties on open symplectic manifolds can be found in A. Banyaga.[6]

3.2. *An Estimate for the Calabi Invariant*

It turns out that the functional c is bounded with respect to the norm $\|\quad\|$ on $\mathfrak{d}$. More precisely, the following estimate is true.

Proposition 3.2.A. *For every $H \in \mathfrak{d}$*

$$|c(H)| \leq \text{Volume}(M) \cdot \|H\| \; .$$

The proof is straightforward. As an obvious consequence we obtain the following

Corollary 3.2.B. *Assume that the form θ is exact. Then for every $f \in \mathcal{D}$*

$$|\operatorname{Cal}(f)| \leq \text{Volume}(M) \cdot \rho(\mathbb{1}, f) \; .$$

Remark 3.2.C For every $f \in \mathcal{D}$ the sequence $\rho(\mathbb{1}, f^n)$ is *sublinear*, that is $\rho(\mathbb{1}, f^{n+m}) \leq \rho(\mathbb{1}, f^n) + \rho(\mathbb{1}, f^m)$. Therefore there exists the limit $\sigma(f) = \lim_{n\to\infty} \frac{1}{n}\rho(\mathbb{1}, f^n)$. With this language 3.2.B implies that $\sigma(f) \geq |\operatorname{Cal}(f)|/\text{Volume}(M)$ provided the form θ on $\mathcal{D}$ is exact. It is not clear however how to compute $\sigma(f)$ even in simple examples.

3.3. *Calibration*

Proposition 3.3.A. *For every $\mathcal{F}$-optical function $H \in \mathfrak{d}$ the value of $c(H)$ is strictly negative.*

Proof: Since H vanishes on the boundary ∂M then the maximum principle 1.5.C implies that H is strictly negative on the interior of M. Thus $c(H) < 0$. ■

As an obvious consequence we obtain the following

Corollary 3.3.B. *Assume that the form θ is exact. Then for every $\mathcal{F}$-optical diffeomorphism $f \in \mathcal{D}$*

$$\operatorname{Cal}(f) < 0 \; .$$

Corollary 3.3.C. *Assume that the form θ is exact. Then for every $\mathcal{F}$-optical diffeomorphism $f \in \mathcal{D}$ the sequence $\frac{1}{n}\rho(\mathbb{1}, f^n)$ is bounded away from zero.*

Proof: From 3.2.B we have

$$\rho(\mathbb{1}, f^n) \geq |\operatorname{Cal}(f^n)|/\text{Volume}(M) = n|\operatorname{Cal}(f)|/\text{Volume}(M) \; .$$

But $|\operatorname{Cal}(f)| \neq 0$ due to 3.3.B. The needed assertion follows. ■

3.4. *Exactness of the Form θ*

Proposition 3.4.A. *Assume that the form ω is exact and the characteristic foliation of the boundary ∂M has at least one non-closed leaf. Then the form θ is exact.*

Proof of Theorem 1.8.B.

The Theorem follows immediately from 3.3.C and 3.4.A. ■

Proof of Proposition 3.4.A.

In fact we have to prove that for an arbitrary smooth loop $\gamma : [0;1] \to \mathcal{D}$ with $\gamma(0) = \gamma(1) = \mathbb{1}$ the integral $\int_\gamma \theta$ vanishes. Let $H(x,t) \in \mathcal{H}$ be a Hamiltonian function which generates such a loop, that is $g_H^t = \gamma(t)$. We have to show that

$$\int_0^1 dt \int_M H(x,t)\omega^n = 0 .$$

Denote by Γ_x the trajectory of a point $x \in M$ under the flow g_H^t. Let λ be a 1-form on M such that $\omega = d\lambda$. Define a function $S : M \to \mathbb{R}$ as follows:

$$S(x) = \int_{\Gamma_x} \lambda - \int_0^1 H(g_H^t x, t)dt .$$

(Usually this function is called *the action* along the trajectory Γ_x.)

The rest of the proof is divided into 3 steps.

Step 1. One can easily check that

$$\int_M S(x)\omega^n = -(n+1)\int_0^1 dt \int_M H(x,t)\omega^n .$$

(We refer the reader to A. Banyaga[6] (II. 4.3) for a similar computation.)

Step 2. We claim that S is a *constant* function. Indeed,

$$dS = \int_0^1 (g_H^t)^*(di_{\xi^t}\lambda - dH)dt ,$$

where ξ^t denotes the Hamiltonian field. Recall that $dH = -i_{\xi^t}\omega$, and therefore $di_{\xi^t}\lambda - dH$ is just the Lie derivative of λ along the field ξ^t. Thus

$$dS = \int_0^1 (g_H^t)^* \mathcal{L}_{\xi^t} \lambda dt = (g_H^1)^*\lambda - \lambda = 0$$

since $g_H^1 = \mathbb{1}$. Our claim is proved.

Step 3. In view of conclusions of the previous steps, it remains to show that S vanishes at least at one point. Choose x to be a point on a non-closed leaf K of the characteristic foliation of ∂M. We claim that $S(x) = 0$. Indeed,

$$S(x) = \int_{\Gamma_x} \lambda - \int_0^1 H(g_H^t x, t) dt \ .$$

Note that the whole trajectory of x belongs to K, and therefore $H(g_H^t x, t) = 0$ for all t. Moreover, the trajectory Γ_x is closed since $g_H^1 = \mathbb{1}$. Thus

$$\int_{\Gamma_x} \lambda = 0$$

since Γ_x is homologous to zero in K provided the leaf K is non-closed. Our claim, and therefore the Proposition follow. ■

We complete this section with a discussion on possible generalizations of 3.4.A.

Let (M, ω) be a compact connected symplectic manifold with boundary, and let θ be the closed 1-form on the group $\mathcal{D}$ defined in 3.1. By definition, ***the Calabi spectrum*** $C(M)$ is the image of $H_1(\mathcal{D}; \mathbb{Z})$ under the cohomology class of θ, that is

$$C(M) = \left\{ \int_\gamma \theta \;\middle|\; \text{over all smooth loops } \gamma : [0; 1] \to \mathcal{D} \text{ with } \gamma(0) = \gamma(1) = \mathbb{1} \right\} .$$

Question 3.4.B. How to compute the Calabi spectrum of a given symplectic manifold?

Note that due to 3.4.A $C(M) = \{0\}$ provided ω is exact and the characteristic foliation of the boundary has at least one non-closed leaf. In fact, our proof of 3.4.A allows to obtain some information about $C(M)$ also in the case when all leafs are closed.

Indeed assume that ω is exact and all leafs of the characteristic foliation of ∂M are closed. Define a set

$$\Gamma(M) = \Big\{\mu \in \mathbb{R} \mid \text{ for every 1-form } \lambda \text{ on } M \text{ with } d\lambda = \omega$$

$$\text{and for every closed leaf } K \text{ holds } \mu \in \Big(\int_K \lambda\Big) \cdot \mathbb{Z}\Big\}.$$

Proposition 3.4.C. $C(M) \subset \Gamma(M)$.

In certain cases this Proposition allows computation of the spectrum $C(M)$. Denote by $B^{2n}(R)$ the ball $\{z \in \mathbb{C}^n \mid |z| \leq R\}$ and by $A^{2n}(r,R)$ the spherical annulus $\{z \in \mathbb{C}^n \mid r \leq |z| \leq R\}$.

Corollary 3.4.D. $C(B^{2n}(R)) = \pi R^2 \mathbb{Z}$.

Corollary 3.4.E. *Suppose that r^2 and R^2 are independent over rational numbers. Then $C(A^{2n}(r,R)) = \{0\}$.*

Question 3.4.F. Is it true that $C(A^{2n}(\frac{1}{\sqrt{2}}, 1)) = \{0\}$?

Note that the answer is positive in the case $n = 1$ in view of 4.1.B. Indeed, each of the boundary circles of ∂A^2 is not homologous to zero, and therefore one can choose a primitive λ of ω in such a way that its integrals over these circles are independent over rationals.

Acknowledgments

We are deeply grateful to Y. Eliashberg for useful discussions.

References

1. V.I. Arnold, Characteristic class entering in quantization conditions, *Funct. Anal. Appl.* **1** (1967) 1-13.
2. V.I. Arnold, The Sturm theorems and symplectic geometry, *Funct. Anal. Appl.* **19** (1985) 251-259.
3. V.I. Arnold, First steps in symplectic topology, *Russ. Math. Surv.* **41** (1986) 1-21.

4. V.I. Arnold and A.B. Givental, Symplectic geometry, in *Dynamical Systems 4, Encyclopedia of Math. Sciences*, (Springer, 1990), pp. 1-136.
5. M. Bialy and L. Polterovich, Hamiltonian diffeomorphisms and Lagrangian distributions, *Geometric and Functional Analysis* **2** (1992) 173-210.
6. A. Banyaga, Sur la structure du group des diffeomorphismes preservent une forme symplectique, *Comm. Math. Helv.* **53** (1978) 174-227.
7. D. Bennequin, Quelques remarques simples sur la rigidité symplectique, in *Géométrie Symplectique et de Contact: Autour de Théorème de Poincare-Birkhoff*, eds. P. Dazord, N. Desolneux-Moulis, (Herman, Paris, 1984), pp. 1-50.
8. E. Calabi, On the group of automorphisms of a symplectic manifold, in *Problems in Analysis* (Symposium in honour of S. Bochner), (University Press, Princeton, 1970), pp. 1-26.
9. J.J. Duistermaat, On the Morse index in variational calculus, *Adv. Math.* **21** (1976) 173-195.
10. Y. Eliashberg, Topological characterisation of Stein manifolds of dimension > 2, *International J. of Mathematics* **1** (1990) 29-46.
11. Y. Eliashberg and M. Gromov, Convex symplectic manifolds, *Proc. of Symposia in Pure Math.* **52** (1991) 135-162.
12. Y. Eliashberg and T. Ratiu, The diameter of the symplectomorphism group is infinite, *Invent. Math.* **103** (1991) 327-340.
13. H. Hofer, On the topological properties of symplectic maps, *Proc. Royal Soc. Edinburgh* **115A** (1990) 25-38.
14. H. Hofer, Estimates for the energy of a symplectic map, preprint (1991).
15. M. Herman, Inequalites "a priori" pour des tores Lagrangiens invariants par des diffeomorphismes symplectiques, *Publications Mathematiques* n**70** (1989) 47-101.
16. J. Mather, Action-minimizing invariant measures for positive definite Lagrangian systems, *Math. Z.* **207** (1991) 169-207.
17. J. Moser, Monotone twist mappings and the calculus of variations, *Ergod. Theory & Dynam. Syst.* **6** (1986) 401-413.
18. L. Polterovich, Symplectic displacement energy for Lagrangian submanifolds, *Ergod. Theory & Dynam. Syst.*, to appear.
19. H. Shima, Homogeneous Hessian manifolds, *Ann. Inst. Fourier* **30** (1980) 91-128.

Geometry in Partial Differential Equations (pp. 51–67)
eds. A. Pràstaro and Th. M. Rassias

ON THE GEOMETRY OF THE HODGE–DE RHAM LAPLACE OPERATORS

M. CRAIOVEANU AND M. PUTA
Department of Mathematics, University of Timisoara, Blvd V Parvan 4, 1900 Timisoara, Romania

and

THEMISTOCLES M. RASSIAS
Department of Mathematics, University of LaVerne, P.O. Box 51105, Kifissia, Athens, Greece 14510

1. Introduction

The usual Laplace operator acting on smooth real functions has a natural extension to differential forms. In the theory of this new operator, also called the Hodge-De Rham Laplacian, the interplay between analysis, topology and geometry is fascinating. The aim of this paper is to make a survey of this theory and to give the readers the possibility to understand the flavour of this interesting subject.

2. The Hodge–De Rham Laplace Operators

Let us start with some elementary considerations concerning spectral properties of the Hodge-De Rham Laplace operators.

Let (M, g) be a closed, connected, oriented, n-dimensional smooth Riemannian manifold and $A^p(M)$ the space of smooth differential p-forms on M. Recall that for each p, $0 \leq p \leq n$, there is a linear operator $*$ called the Hodge star operator which assigns to each $\omega \in A^p(M)$ a smooth differential $(n-p)$-form $*\omega \in A^{n-p}(M)$ which satisfies $** = (-1)^{p(n-p)}$. For each $p, 0 \leq p \leq n$, let us consider the operator δ: $A^p(M) \to A^{p-1}(M)$ by setting $\delta = (-1)^{n(p+1)+1} * d*$. On smooth differential 0-forms i.e. on smooth functions, δ is simply the zero linear functional. The Hodge-De Rham Laplace operator (Hodge-De Rham Laplacian for short) is defined for each $p, 0 \leq p \leq n$, by

$$\Delta_g^{(p)} = \delta d + d\delta \ . \tag{2.1}$$

It is a linear operator on $A^p(M)$ for each $p, 0 \leq p \leq n$. It is easy to see that on smooth functions on the Euclidean space $\mathbb{R}^n$, the Hodge-De Rham Laplacian is just $-\sum_{i=1}^n \frac{\partial^2}{\partial x_i^2}$. Also, it is a straightforward computation to check that the Hodge-De Rham Laplacian $\Delta_g^{(p)}$ commutes with $*$ for each p, that is $*\Delta_g^{(p)} = \Delta_g^{(p)} * (0 \leq p \leq n)$.

Let us consider an inner product on the vector space $A^p(M)(0 \leq p \leq n)$ defined by

$$\langle \alpha, \beta \rangle = \int_M \alpha \wedge *\beta \tag{2.2}$$

for $\alpha, \beta \in A^p(M)$, and we denote the corresponding norm by $\| \ \ \|$. We extend the inner products (2.2) for $0 \leq p \leq n$ to an inner product on the direct sum $\bigoplus_{p=0}^n A^p(M)$ simply by requiring the different $A^p(M)$ to be orthogonal.

Proposition 2.1. *δ is the adjoint of d on $\bigoplus_{p=0}^n A^p(M)$, that is $\langle d\alpha, \beta \rangle = \langle \alpha, \delta\beta \rangle$.*

Proof. By linearity and the orthogonality of the $A^p(M)$, the proof reduces to the consideration of the case when $\alpha \in A^{p-1}(M)$ and $\beta \in A^p(M)$.

In this case we have:

$$\begin{aligned} d(\alpha \wedge *\beta) &= d\alpha \wedge *\beta + (-1)^{p-1}\alpha \wedge d * \beta \\ &= d\alpha \wedge *\beta - \alpha \wedge *\delta\beta \ . \end{aligned}$$

By integrating both sides over M and applying the Stokes theorem we obtain:

$$\begin{aligned} 0 &= \int_M (d\alpha \wedge *\beta - \alpha \wedge *\delta\beta) \\ &= \langle d\alpha, \beta \rangle - \langle \alpha, \delta\beta \rangle \ . \end{aligned}$$

Thus

$$\langle d\alpha, \beta \rangle = \langle \alpha, \delta\beta \rangle \ . \qquad \square$$

Remark 2.1. As a consequence of the above Proposition 2.1 we can deduce easily that $\Delta_g^{(p)}$ $(0 \le p \le n)$ is self-adjoint, that is

$$\langle \Delta_g^{(p)}\alpha, \beta \rangle = \langle \alpha, \Delta_g^{(p)}\beta \rangle$$

for each $\alpha, \beta \in A^p(M)$.

Proposition 2.2. (i) $\Delta_g^{(p)}$ $(0 \le p \le n)$ *is positive, that is*

$$\langle \Delta_g^{(p)}\alpha, \alpha \rangle \ge 0$$

for each $\alpha \in A^p(M)$.

(ii) $\Delta_g^{(p)}\alpha = 0$ *if and only if* $d\alpha = 0$ *and* $\delta\alpha = 0$.

Proof. (i) $\langle \Delta_g^{(p)}\alpha, \alpha \rangle = \langle (d\delta + \delta d)\alpha, \alpha \rangle = \langle \delta\alpha, \delta\alpha \rangle + \langle d\alpha, d\alpha \rangle \ge 0$.

(ii) Clearly $\Delta_g^{(p)}\alpha = 0$ if $d\alpha = 0$ and $\delta\alpha = 0$. Now

$$\langle \Delta_g^{(p)}\alpha, \alpha \rangle = \langle \delta\alpha, \delta\alpha \rangle + \langle d\alpha, d\alpha \rangle \ .$$

Thus, if $\Delta_g^{(p)}\alpha = 0$, it follows that $d\alpha = 0$ and $\delta\alpha = 0$. $\square$

Proposition 2.3. *The Hodge-De Rham Laplacian is an elliptic operator.*

Proof. It is easy to see that for each $\xi \in T_m^* M$ there is a well-defined linear transformation $\sigma_\Delta(\xi)$ on $\wedge^p T_m^*(M)$ defined by

$$\sigma_\Delta(\xi)(v) = \Delta_g^{(p)}(\varphi^2\alpha)(m) \ ,$$

where $v \in \wedge^p(T_m^* M)$, α is any smooth differential p-form such that $\alpha(m) = v$, and where φ is any smooth real function such that $\varphi(m) = 0$ and $d\varphi(m) = \xi$. This linear transformation $\sigma_\Delta(\xi)$ is known as the symbol of $\Delta_g^{(p)}$. Now, the ellipticity of $\Delta_g^{(p)}$ is equivalent to the property that the symbol $\sigma_\Delta(\xi)$ is an isomorphism at each point $m \in M$ for each non-zero $\xi \in T_m^* M$, and this follows by a straightforward computation (see for details [17]). $\square$

Definition 2.1. Consider the Hodge-De Rham Laplacian $\Delta_g^{(p)}$ acting on the real vector space $A^p(M)$ for some fixed $p, 0 \leq p \leq n$. A real number $\lambda^{(p)}$ corresponding to which there exists a not identically zero smooth differential p-form ω such that $\Delta_g^{(p)}\omega = \lambda^{(p)}\omega$ is called an eigenvalue of $\Delta_g^{(p)}$. If $\lambda^{(p)}$ is an eigenvalue, then any smooth differential p-form ω as before such that $\Delta_g^{(p)}\omega = \lambda^{(p)}\omega$ is called an eigenform of $\Delta_g^{(p)}$ corresponding to the eigenvalue $\lambda^{(p)}$. The zero p-form and the eigenforms corresponding to a fixed eigenvalue $\lambda^{(p)}$ form a vector subspace of $A^p(M)$ denoted by $V_{\lambda^{(p)}}(M,g)$ and is called the eigenspace corresponding to the eigenvalue $\lambda^{(p)}$.

Proposition 2.4. (i) *The eigenvalues of* $\Delta_g^{(p)}$ $(0 \leq p \leq n)$ *are non-negative.*

(ii) *The eigenspaces of* $\Delta_g^{(p)}$ $(0 \leq p \leq n)$ *are finite dimensional.*

(iii) *The eigenforms corresponding to distinct eigenvalues are orthogonal.*

Proof. (i) Let $\omega \in V_{\lambda^{(p)}}(M,g)$ be an eigenform corresponding to the eigenvalue $\lambda^{(p)}$, that is $\Delta_g^{(p)}\omega = \lambda^{(p)}\omega$ and then $\lambda^{(p)} = \frac{\langle \Delta_g^{(p)}\omega, \omega\rangle}{\|\omega\|^2} \geq 0$.

(ii) This follows via Proposition 2.3.

(iii) Let $\lambda_1^{(p)} \neq \lambda_2^{(p)}$ and ω_1, ω_2 be the corresponding eigenforms. Then

$$\langle \Delta_g^{(p)}\omega_1, \omega_2\rangle = \lambda_1^{(p)}\langle \omega_1, \omega_2\rangle$$

and

$$\langle \Delta_g^{(p)}\omega_1, \omega_2\rangle = \langle \omega_1, \lambda_2^{(p)}\omega_2\rangle = \lambda_2^{(p)}\langle \omega_1, \omega_2\rangle \ .$$

Therefore

$$0 = (\lambda_1^{(p)} - \lambda_2^{(p)})\langle \omega_1, \omega_2\rangle$$

and since $\lambda_1^{(p)} \neq \lambda_2^{(p)}$ it follows that $\langle \omega_1, \omega_2\rangle = 0$, that is, ω_1 and ω_2 are orthogonal. □

Using the general theory of elliptic operators it can be proved that the p-spectrum of $\Delta_g^{(p)}$ denoted by $\mathrm{Spec}^{(p)}(M,g)$ lies on the non-negative real half line and is pure discrete, i.e., given by a sequence

$$\mathrm{Spec}^{(p)}(M,g) = \{0 \leq \lambda_0^{(p)} \leq \lambda_1^{(p)} \leq \cdots \to +\infty\} \ ,$$

all eigenvalues being of finite multiplicity. Moreover,

$$\mathrm{Spec}^{(p)}(M,g) = \mathrm{Spec}^{(n-p)}(M,g), \quad \text{for each } 0 \leq p \leq n \ .$$

It is an interesting problem to know which are the sequences of eigenvalues that are possible for the Hodge-De Rham Laplacian of an arbitrary smooth Riemannian metric on a given closed smooth manifold. Let us point out a partial answer to this question for the particular case of the Laplacian ($p = 0$): If M is a closed smooth manifold of dimension ≥ 3 and $s_N^{(0)} = \{\lambda_1^{(0)} = 0 < \lambda_2^{(0)} \leq \cdots \leq \lambda_N^{(0)}\}$ an arbitrary finite sequence, then there exists a Riemannian metric on M such that the corresponding Laplacian admits $s_N^{(0)}$ as the sequence of the first N eigenvalues with multiplicities for any N. In the case of dimension 2 the same result is true for any sequence of the form $\lambda_1^{(0)} = 0 < \lambda_2^{(0)} < \cdots < \lambda_N^{(0)}$ [8].

3. Explicit Examples of p-Spectra

The explicit calculation of spectra even of the standard manifolds is generally a very difficult problem; many representation theorems of classical analysis must be considered in order to determine the spectra of the Laplace operator on a certain space. The calculation of the spectrum of a standard space usually requires the translation of the problem into the language of representation theory, number theory and harmonic analysis on classical groups.

The $\mathrm{Spec}^{(p)}(M,g)$ are known for the spheres [17], [30], [31], flat tori (T_Γ^n, g_Γ) [4], [10], complex projective spaces with Fubini-Study metric [30], [35] and SU(2), SU(3), SO(3), SO(4) and SO(5) [1], [16]. In general their expressions are very complicated. For instance:

(i) $\lambda \in \mathrm{Spec}^{(p)}(T_\Gamma, g_\Gamma)$ if and only if there exists $x \in \Gamma^*$ such that $\|x\|^2 = \lambda/4\pi^2$.

(ii) $\mathrm{Spec}^{(1)}(S^n, g_0) = \left\{\frac{k^2+k(n-2)}{2(n-2)}; \frac{k^2+k(n-2)+n-3}{2(n-2)}\right\}$, where $k \in \mathbb{N}\backslash\{0\}$ and $n \geq 6$.

(iii) $\mathrm{Spec}^{(1)}(\mathbb{C}P^{m-1}, g_{FS}) = \left\{\frac{n-1}{2}(k^2+k); \frac{n-1}{2}(k^2+k) - k; \frac{n-1}{2}(k^2 - k) + k; \frac{n-1}{2}(k^2-k) + 2k\right\}$ where $k \in \mathbb{N}\backslash\{0\}$ and $n \geq 3$.

Also, $\mathrm{Spec}^{(p)}(M,g)$ can be completely described for a Riemannian Heisenberg manifold M, which is the quotient of the $(2n+1)$-dimensional Heisenberg group, endowed with a left invariant metric, by a discrete co-compact subgroup [25], [28].

4. Asymptotic Expansions

Let (M,g) be a closed, connected, n-dimensional, oriented Riemannian manifold. There are important relations between the geometry of (M,g) and the global structure of the spectrum $\mathrm{Spec}^{(p)}(M,g) = \{0 \le \lambda_1^{(p)}(g) \le \lambda_2^{(p)}(g) \le \cdots \to +\infty\}$. For instance, global metric invariants like the volume, dimension and certain integrals of higher order local invariants of the metric g are contained in the asymptotic distribution behaviour of the eigenvalues of $\Delta_g^{(p)}$. There exists an asymptotic expansion of the series

$$Z_g^{(p)}(t) = \sum_{k=1}^{\infty} \exp(-t\lambda_k^{(p)}(g)) ,$$

which is the trace of the heat operator $\exp(-t\Delta_g^{(p)})$. The function $Z_g^{(p)}$ is C^∞ on $(0,+\infty)$ and it admits, when $t \searrow 0$, the Minakshisundaram-Pleijel asymptotic expansion

$$Z_g^{(p)}(t) \sim (4\pi t)^{-n/2} \sum_{k\ge 0} a_k^{(p)}(g) t^k \, [39] .$$

The giving of the function $Z_g^{(p)}(t)$ is equivalent to the giving of the spectrum $\mathrm{Spec}^{(p)}(M,g)$. On the other hand, this asymptotic expansion (i.e. the giving of the coefficients $a_k^{(p)}, k \in \mathbb{N}$) is not sufficient for the determination of the spectrum $\mathrm{Spec}^{(p)}(M,g)$.

The coefficients $a_k^{(p)}(g)$ are universal expressions of the curvature of (M,g) and its covariant derivatives.

Proposition 4.1. ([38]) $a_0^{(p)}(g) = \binom{n}{p}\mathrm{Vol}(M,g)$,

$$a_1^{(p)}(g) = -\frac{1}{6}\left(\binom{n}{p} - 6\binom{n-2}{p-1}\right)\int_M \tau v_g ,$$

$$\begin{aligned} a_2^{(p)}(g) = \frac{1}{360}\int_M \Bigg\{ & \left[2\binom{n}{p} - 30\binom{n-2}{p-1} + 180\binom{n-4}{p-2}\right]|R|^2 \\ & + \left[-2\binom{n}{p} + 180\binom{n-2}{p-1} - 720\binom{n-4}{p-2}\right]|c_2 R|^2 \\ & + \left[5\binom{n}{p} - 60\binom{n-2}{p-1} + 180\binom{n-4}{p-2}\right]\tau^2 \Bigg\} v_g , \end{aligned}$$

where v_g denotes the Riemannian measure, τ the scalar curvature of (M,g), R the Riemann tensor field of (M,g) and c_2R the Ricci tensor field of (M,g) (see P. Gilkey [19] *for the expression of $a_3^{(p)}$).*

These properties allow us to derive geometric and topological information from the knowledge of $\mathrm{Spec}^{(p)}(M,g)$. For example, the following version of H. Weyl's formula is valid:

$$\lim_{j\to\infty}\frac{(\lambda_j^{(p)})^{n/2}}{j}=\frac{(4\pi)^{n/2}\Gamma\left(\frac{n}{2}+1\right)}{\binom{n}{p}\mathrm{Vol}(M,g)},\quad 0\le p\le n\ [10]\ .$$

On the other hand, the real p-th cohomology vector space $H^p(M)$ of M is isomorphic to $\mathrm{Ker}\Delta_g^{(p)}$ and the Euler-Poincaré characteristic $\chi(M)=\sum_{p=0}^{m}(-1)^p\dim H^p(M)$ of M can be expressed as follows

$$\sum_{p=0}^{m}(-1)^p Z_g^{(p)}(t)=\chi(M)$$

for any $t\in(0,+\infty)$ (McKean-Singer "téléscopaje" formula), which implies the following

Corollary 4.2.

$$\sum_{p=0}^{n}(-1)^p a_k^{(p)}(g)=\begin{cases}0 & \text{if } k\neq n/2\ ,\\ (4\pi)^{n/2}\chi(M) & \text{if } k=\frac{n}{2}\ .\end{cases}$$

The above considerations and Corollary 4.2 show the possibility to obtain, in principle, a formula for the Euler-Poincaré characteristic of M in terms of the curvature R of (M,g), namely

$$\chi(M)=\int_M P(R,\dots,\nabla^k R,\dots)v_g\ ,$$

where P is a polynomial in R and its covariant derivatives. On the other hand, in the alternating sum $\sum_{p=0}^{n}(-1)^p\ \mathrm{Trace}[\exp(-t\Delta_g^{(p)})]$, associated to the family of Hodge-De Rham Laplacians of (M,g), some remarkable cancellations take place, the final result being that P depends only on R

and not on its covariant derivatives. If n is odd, Corollary 4.2 implies that the polynomial P vanishes. If $n = 2m$ is even, then using Gilkey's algebraic theory of invariants [18], [21] one can prove

Theorem 4.3. (Chern-Gauss-Bonnet)

$$\chi(M) = C(n) \int_M \mathrm{Pff}(R) v_g$$

where $C(n)$ is a constant depending only on the dimension n and $\mathrm{Pff}(R)$ is the so-called Pfaffian associated to R and which is essentially defined by

$$\mathrm{Pff}(R) dx^1 \wedge \cdots \wedge dx^n = C(n) \sum \varepsilon(i)\varepsilon(j) R_{i_1 i_2 j_1 j_2} \cdots R_{i_{2m-1} i_{2m} j_{2m-1} j_{2m}} \cdot dx^{j_1} \wedge dx^{j_2} \wedge \cdots \wedge dx^{j_{2m-1}} \wedge dx^{j_{2m}} ,$$

where $\varepsilon(i) = \varepsilon(i_1 \cdots i_{2m})$ [resp. $\varepsilon(j) = \varepsilon(j_1 \cdots j_{2m})$] denotes the sign of the permutation $(i_1 \cdots i_{2m})$ [resp. $(j_1 \cdots j_{2m})$] of the set $\{1, \ldots, 2m\}$, the sum being extended to all the permutations $(i_1 \cdots i_{2m}), (j_1 \cdots j_{2m})$ of this set.

Also, one should point out the importance of the asymptotic expansions for the Atiyah-Singer index theorem [21]. On the other hand, Gilkey's theory of invariants has also been successfully applied [20], [21] to prove the Lefschetz fixed point formulas of Atiyah-Bott and Atiyah-Singer.

Following the classical paper of Mark Kac [32] and thinking of the eigenvalues as the frequencies of the normal modes of vibration of an idealized elastic medium, the "drum", we say that a geometric property of (M, g) can be "heard" if it is determined by the $\mathrm{Spec}^{(p)}(M, g)$. Thus the above results, in particular Proposition 4.1 and Corollary 4.2, show that $\dim M$, $\mathrm{Vol}(M, g)$, the coefficients of the asymptotic expansion of the function $Z_g^{(p)}$, the integral of curvatures of (M, g), $\chi(M), \ldots$ can be heard. In fact, $\mathrm{Spec}^{(p)}(M, g)$, $0 \leq p \leq \dim M$, can determine the aspects of dimension one and $\dim M$ of the geometry of (M, g), but not aspects of intermediary dimensions [11]. On the other hand, D. M. DeTurck, H. Gluck, C. Gordon, D. Webb [12] have exhibited specific geometric invariants on a particular 6-dimensional closed smooth manifold M that cannot be "heard". Essentially, these invariants are functions defined by using the volumes of higher than one-dimensional minimizing cycles of homology classes of M. They allow us to show that certain isospectral metrics (see Sec. 5) on M are not isometric and in turn help us to answer the question "How can a drum change shape, while sounding the same?".

Also, let us mention another function that can be associated to $Spec^{(p)}(M,g)$, namely the zeta function defined by

$$\zeta_g^{(p)}(z) = \sum_{\lambda_k^{(p)} \neq 0} \frac{1}{[\lambda_k^{(p)}]^z}, \quad z \in \mathbb{C} ,$$

whose convergence and analytic prolongation is controlled by the operator $\exp(-t\Delta_g^{(p)})$. Certain special values of these functions $(0 \leq p \leq \dim M)$ have a special geometrical significance for M. Thus, $\sum_{p=0}^{\dim M}(-1)^p \zeta_g^{(p)}(0)$ gives, essentially, the Euler-Poincaré characteristic of M and another linear combination of the values $\{(\zeta_g^{(p)})'(0) | 0 \leq p \leq \dim M\}$ determines the Reidemeister torsion. One can show that the function $\zeta_g^{(p)}$ incorporates all the information on $\mathrm{Spec}^{(p)}(M,g)$, except the Betti number of order p of M [44]. For interesting results concerning the function $\zeta_g^{(p)}$ and their applications, see [44].

Finally, let us point out that the existence and uniqueness of the fundamental solution $E_g^{(p)}\colon M \times M \times (0, +\infty) \to \wedge^{(p)}T^*M \otimes \wedge^p T^*M$ of the heat operator associated to $\Delta_g^{(p)}$, which can be used in deriving the asymptotic expansion of $Z_g^{(p)}$, gives the possibility to prove the Hodge-De Rham decomposition: for any $\omega \in A^p(M)$, $\omega = H_g^{(p)}(\omega) + \Delta_g^{(p)} \circ G_g^{(p)}(\omega)$, where the harmonic projection $H_g^{(p)}\colon A^p(M) \to \mathrm{Ker}\Delta_g^{(p)}$ and the Green operator $G_g^{(p)}\colon A^p(M) \to A^p(M)$ are respectively defined by

$$[H_g^{(p)}(\omega)](x) = \lim_{t\to\infty} \int_{y\in M} E_g^{(p)}(x,y,t) \wedge *\omega(y) ,$$

$$[G_g^{(p)}(\omega)](x) = \int_0^\infty \left(\int_{y\in M} E_g^{(p)}(x,y,t) \wedge *[\omega - H_g^{(p)}(\omega)](y) \right) dt, \quad x \in M .$$

Moreover, $||G_g^{(p)}(\omega)|| \leq \frac{1}{\lambda^{(p)}} ||\omega||$, where $\lambda^{(p)}$ is the first nonvanishing eigenvalue of $\Delta_g^{(p)}$ [15].

5. p-Isospectral Closed Riemannian Manifolds

Definition 5.1. We shall say that two closed n-dimensional Riemannian manifolds (M,g) and (M',g') are p-isospectral if their p-spectra (where $0 \leq p \leq n$) are the same, that is

$$\mathrm{Spec}^{(p)}(M,g) = \mathrm{Spec}^{(p)}(M',g') .$$

Two isometric closed n-dimensional Riemannian manifolds are p-isospectral for any $p, 0 \leq p \leq n$.

Now it is natural to ask: Let (M,g) and (M',g') be two p-isospectral closed n-dimensional Riemannian manifolds for some $p \in [0,n]$. Then are (M,g) and (M',g') isometric? The answer is *no* and the first counterexample was given by J. Milnor [37] in 1964: there exists two 16-dimensional flat tori that are p-isospectral for each $p \in [0,16]$ but which are not isometric. Later, M. Kneser [34] (resp. Y. Kitaoka [33]) has constructed p-isospectral flat tori in dimension 12 (resp. 8) for any p, which are not isometric.

Starting from these examples, it is possible to construct non-isometric isospectral n-dimensional flat tori for any $n \geq 8$. On the other hand, two 2-dimensional flat tori are p-isospectral $(0 \leq p \leq 2)$ if and only if they are isometric [4], [10], [40]. M. Kneser (unpublished) and S. Wolpert [51] have established that there exists only a finite number of isospectral flat tori which are two by two isometric.

Let us note that the isometry classes of Riemannian Heisenberg manifolds are also completely described [24], [21]. Recently, H. Pesce [40] gives in any dimension an explicit bound of the number of isometry classes of flat tori isospectral to a given flat torus. In dimension three, he obtains a bound which only depends on the dimension. The problem of the existence of a universal bound for this number of isometry classes in dimension $n \geq 4$ is open. In the case when the number of isometry classes of closed Riemannian manifolds p-isospectral to a given closed Riemannian manifold is finite, it would be interesting to find bounds for this number in terms of the geometry of the given manifold and to establish under what assumptions these bounds depend only on the dimension.

In 1978, M. F. Vignéras [50] has given examples of non-isometric Riemann surfaces that are p-isospectral for any p. The construction of all these examples is based essentially on the possibility of describing their p-spectra. T. Sunada [46] gave a method that allows the generalizations of these examples. This method reduces partially the construction of p-isospectral closed Riemannian manifolds to a well-understood problem of finite groups. Indeed, let M be a closed smooth manifold and assume that the finite group G acts without fixed points on M. Let H_1, H_2 be subgroups of G and take the corresponding quotient manifolds $M_1 = M/H_1$ and $M_2 = M/H_2$. Choosing a Riemannian metric on M/G, it induces Riemannian metrics g (resp. g_1, resp. g_2) on M (resp. M_1, resp. M_2) such that the corresponding canonical projections became Riemannian coverings.

Proposition 5.2. ([46]). *If H_1 and H_2 are not conjugate subgroups of G and there exists a bijection $F : H_1 \to H_2$ such that for any $h_1 \in H_1$, there exists $g = g_{h_1} \in G$ such that $F(h_1) = gh_1g^{-1}$, then the closed Riemannian manifolds (M_1, g_1) and (M_2, g_2) are p-isospectral for any p, but non-isometric.*

The p-isospectrality of these two closed Riemannian manifolds follows immediately from a direct computation showing that the operators $\exp(-t\Delta_{g_1}^{(p)})$ and $\exp(-t\Delta_{g_2}^{(p)})$ have the same trace.

For examples of finite groups G and subgroups H_1, H_2 of G that satisfy the hypothesis of Proposition 5.2 see [5], [6], [46].

Let us consider another interesting example given by D. De Turck and C. Gordon [14]. Let G be a simply-connected Lie group, which is nilpotent of rank m and Γ a cocompact discrete subgroup of G. Let (M, g) be a Riemannian manifold without boundary and assume that one has a left G-action on M by isometries, which is faithful and proper. Also assume that Γ acts freely on M and that the (left) quotient manifold $\Gamma\backslash M$ is closed.

Proposition 5.3. ([2], [14]) *Let $\phi : G \to G$ be an automorphism, but not an interior automorphism of G, with the following property: for any $\gamma \in \Gamma$, there exists $g = g_\gamma \in G$ such that $\phi(\gamma) = g\gamma g^{-1}$. Then $\phi(\Gamma)$ acts freely on M, the quotient manifold $\phi(\Gamma)\backslash M$ is closed and the closed Riemannian manifolds $(\phi(\Gamma)\backslash M, g)$ and $(\Gamma\backslash M, g)$ are p-isospectral for any p and non-isometric.*

For examples of Lie groups G and cocompact discret subgroups Γ of G satisfying the above assumptions, see [25], [27], [13].

In fact, D. DeTurck and C. Gordon [14] have generalized Sunada's construction in such a way that Propositions 5.2 and 5.3 became consequences of a much more general result.

The above non-isometric examples of closed Riemannian manifolds are p-isospectral for any p. It is natural to ask if this is always the case. C. Gordon [25], [26] has given examples of pairs of non-isometric, 0-isospectral, Riemannian Heisenberg manifolds which are not 1-isospectral. On the other hand, for a given p, A. Ikeda [29] exhibits examples of k-isospectral lenticular spaces for $0 \leq k \leq p$, that are not $(p+1)$-isospectral.

6. p-Spectra and Geometry

Various geometric and topological properties of (M, g) have been found to be determined by $\mathrm{Spec}^{(p)}(M,g)$ (see also Sec. 4). A great deal is known about such properties.

Let us review some of them:

Theorem 6.1. ([38]) *Let (M,g) and (M',g') be two closed, connected, oriented, Riemannian manifolds of the same dimension.*

(i) *If $a_2^{(p)} = a'^{(p)}_2$ for $p = 0,1$, then M is flat if and only if M' is flat.*

(ii) *If $a_2^{(p)} = a'^{(p)}_2$ for $p = 0,1$ and $a_i^{(p)} = a'^{(p)}_i$ for $i = 0,1$ and $p = 0$, then (M,g) is a space of constant sectional curvature K if and only if (M',g') is so.*

Theorem 6.2. ([42]) *Let $p \in \mathbb{N}^*$ and $(M,g),(M',g')$ be closed, connected, oriented Riemannian manifolds with $\mathrm{Spec}^{(p)}(M,g) = \mathrm{Spec}^{(p)}(M',g')$ (hence $\dim M = \dim M' = n$). Assume that M' has constant sectional curvature K. Then:*

(i) *M has constant sectional curvature K too, whenever $p \geq 24$ and*

$$n \in [1.02p;\ 1.2p] \cup [1.32p;\ 1.6p] \cup [2.6p;\ 4.4p] \cup [5p;\ 84.79p]\ .$$

(ii) *If M is already an Einstein manifold, then it has constant sectional curvature K too, whenever*

$$(n,p) \notin \{(15,2),(15,3),(16,2),(16,14)\}$$

and

$$n(n-1) - 6p(n-p) \neq 0\ .$$

Remark 6.1. For the values of p not mentioned in (i), the values of n can be determined by numerical methods. This was done for $p = 1,2,3,4,5$ by Tanno [47], Tsagas and Kockinos [49], Stresse [45] and Puta [41].

For the particular case of the Euclidean sphere $S^n \subset \mathbb{R}^{n+1}$ with its induced canonical Riemannian metric g_0, one can improve substantially the previous results. More precisely we have:

Theorem 6.3. ([38]) *The sphere (S^n, g_0) is completely characterized up to an isometry by its spectrum on functions, 1-forms and 2-forms, i.e. by $\mathrm{Spec}^{(p)}(M,g)$, $0 \leq p \leq 2$.*

Theorem 6.4. ([48]) *For a given n there exists an integer* $p \in [0, n]$ *such that* $\text{Spec}^{(p)}(M, g) = \text{Spec}^{(p)}(S^n, g_0)$, *then g is a metric of the same constant sectional curvature as* g_0.

Theorem 6.5. ([24]) *The Euclidean sphere* S^{2n+1} *is characterized by its 2-spectrum in the class of connected, oriented, compact normal contact Riemannian manifolds.*

Theorem 6.6. ([36]) *The sphere* $S^n, n > 2$, *with its canonical Riemannian metric is completely characterized in the class of compact, connected, oriented, boundaryless Einstein manifolds by the p-spectrum, whenever its dimension and the value of p satisfy the condition (ii) from Theorem 6.2.*

For a Kähler manifold (M, g, J) Gilkey and Sacks [22] showed that it is possible to determine whether g is of constant holomorphic sectional curvature from the spectrum of its Laplacian on (p, q)-forms, $0 \leq p + q \leq \dim_R M$. This problem was originally proposed by M. Berger [3]. Since the only manifold with a metric of constant positive holomorphic sectional curvature is $\mathbb{C}P^n$ with the Fubini-Study metric we have:

Theorem 6.7. ([22]) *Let M be a compact analytic manifold of complex dimension n. Suppose that the spectrum of the Laplacians*

$$\Delta^{(p,q)} : A^{p,q}(M) \to A^{p,q}(M)$$

$0 \leq p + q \leq 2n$, *on M agrees with the corresponding spectrum on* $\mathbb{C}P^n (n = \dim_{\mathbb{C}} M)$. *Then* $\mathbb{C}P^n$ *and M are isometric holomorphically.*

Their result was substantially improved. More precisely we have:

Theorem 6.8. ([23]) *Let* (M, g, J) *be a compact Kähler manifold with* $\text{Spec}^{(2)}(M, g) = \text{Spec}^{(2)}(\mathbb{C}P^n, g_{FS})$. *Then* (M, g) *is holomorphically isometric to* $(\mathbb{C}P^n, g_{FS})$ *for all n.*

For the particular cases $p = 1, 8 \leq n \leq 51$ and $p = 2$, $n \neq 2, 8$ we refined the results of Tanno [47] and respectively Chen and Vanhecke [7].

Now the natural question is: Is it possible to extend Goldberg's result for a fixed but arbitrary $p, p > 2$? The answer is partially affirmative. More precisely we have:

Theorem 6.9. ([43]) *Let* $p \in \mathbb{N}^*$ *and* $(M, J, g), (M', g', J')$ *be compact Kähler manifolds with* $\mathrm{Spec}^{(p)}(M, g) = \mathrm{Spec}^{(p)}(M', g')$ (*hence* $\dim_{\mathbb{C}} M = \dim_{\mathbb{C}} M' = n$). *Assume that* M' *has constant holomorphic sectional curvature* h. *Then*

(i) M *has constant holomorphic sectional curvature* h *too, whenever* $p \geq 100$ *and*

$$n \in [0.51p;\ 0.6198p] \cup [0.6454p;\ 0.82225p] \cup [1.2755p;\ 2.2799p] \cup [2.4669p;\ 42.396p]\ .$$

(ii) *If* M *is already an Einstein manifold then it has constant holomorphic sectional curvature equal to* h *whenever*

$$(n, p) \notin \{(8, 2); (8, 14)\}$$

and

$$n(2n - 1) - 3p(2n - p) \neq 0\ .$$

As a particular case we obtain.

Theorem 6.10. ([43]) *The complex projective space* $(\mathbb{C}P^n, g_{FS})$ *with the Fubini-Study metric is completely characterized in the class of compact Kähler-Einstein manifolds by the p-spectrum whenever its complex dimension n and the value p satisfy the conditions (ii) from the above theorem.*

References

1. B. L. Beers, R. Millman, *The spectra of the Laplace-Beltrami operator on compact semisimple Lie groups*, Amer. J. of Math. **99** (1977), 801–807.
2. P. Bérard, *Variétés riemanniennes isospectrales non isométriques*, Séminaire Bourbaki, 41ème année, Vol. 1988–89, n° 705, Astérisque 177–178 (1989), 127–154.
3. M. Berger, *Eigenvalues of the Laplacian*, Proc. Symp. Pure Math. **16** (1970), 121–125.
4. M. Berger, P. Gauduchon, E. Mazet, *Le spectre d'une variété riemannienne*, Lecture Notes in Mathematics **194**, Springer-Verlag, Berlin, 1971.
5. R. Brooks, *Constructing isospectral manifolds*, Amer. Math. Monthly **95** (1988), 823–839.
6. P. Buser, *Cayley graphs and planar isospectral domains*, Lecture Notes in Mathematics, **1339**, Springer, 1988, pp. 64–77.
7. B. Y. Chen and L. Vanhecke, *The spectrum of the Laplacian of Kähler manifolds*, Proc. Amer. Math. Soc. **79** (1980), 82–86.

8. Y. Colin de Verdière, *Spectres de variétés riemanniennes et spectres de graphes*, Proc. of the International Congress of Mathematicians, Berkeley, California, USA, 1986, pp. 522–530.
9. M. Craioveanu, *Geometrie spectrală cu forme diferentiale*, Lucrările Conferinţei Naţionale de Geometrie şi Topologie, P. Neamţ, 1983, Tipografia Univ. "Al. I. Cuza" Iaşi, 1984, pp. 47–61.
10. M. Craioveanu and M. Puta, *Introducere în Geometria Spectrală*, Editura Academiei R. S. România, Bucureşti, 1988.
11. D. M. DeTurck, *Audible and inaudible geometric properties*, preprint Univ. Pennsylvania, 1988.
12. D. M. DeTurck, H. Gluck, C. S. Gordon, and D. Webb, *You cannot hear the size of a homology class*, Comment. Math. Helv. **64** (1989), 589–617.
13. D. M. DeTurck and C. S. Gordon, *Isospectral deformations I: Riemannian structures on two-step nilspaces*, Comm. Pure Appl. Math. **40** (1987), 367–387.
14. D. M. DeTurck and C. S. Gordon, *Isospectral Riemannian metrics and potentials*, preprint Univ. Pennsylvania, 1987.
15. J. Dodziuk and V. K. Patodi, *Riemannian structures and triangulations of manifolds*, J. Indian Math. Soc. **40** (1976), 1–52.
16. H. D. Fegan, *The spectrum of the Laplacian for forms over a Lie group*, Pacific Math. J. **90** (1980), 373–387.
17. S. Gallot and D. Meyer, *Opérateur de courbure et Laplacien des formes differentielles d'une variété riemannienne*, J. Math. Pure Appl. **54** (1975), 259–284.
18. P. B. Gilkey, *Curvature and the eigenvalues of the Laplacian*, Adv. in Math. **10** (1973), 344–382.
19. P. B. Gilkey, *The spectral geometry of a Riemannian manifold*, J. Diff. Geometry **10** (1975), 601–618.
20. P. B. Gilkey, *Lefschetz fixed point formulas and the heat equation*, Proc. Conf. Park City, Utah, 1977, Lecture Notes in Pure and Appl. Math. **48**, Marcel Dekker, New York, 1979, pp. 91–147.
21. P. B. Gilkey, *Invariance Theory, the Heat Equation and the Atiyah-Singer Index Theorem*, Publish or Perish, Washington, 1984.
22. P. B. Gilkey and J. Sacks, *Spectral geometry and manifolds of constant holomorphic sectional curvature*, Proc. Sympos. Pure Math. **27** (1975), pp. 281–285.
23. S. I. Goldberg, *A characterization of complex projective space*, C. R. Math. Rep. Acad. Sci. Canada **6** (1984), 193–198.
24. S. I. Goldberg and H. Gauchman, *Characterizing S^m by the spectrum of the Laplacian on 2-forms*, Proc. Amer. Math. Soc. (4) **99** (1987), 750–756.
25. C. Gordon, *Riemannian manifolds isospectral on functions but not on 1-forms*, J. Diff. Geometry (1) **24** (1986), 79–96.
26. C. S. Gordon, *The Laplace spectrum versus the length spectra of Riemannian manifolds*, Contemporary Math. **51** (1986), 63–80.

27. C. S. Gordon and E. N. Wilson, *Isospectral deformations of compact solvmanifolds*, J. Diff. Geometry **19** (1984), 241–256.
28. C. S. Gordon and E. N. Wilson, *The spectrum of the Laplacian on Riemannian Heisenberg manifolds*, Michigan Math. J. **33** (1986), 253–271.
29. A. Ikeda, *Riemannian manifolds p-isospectral but not $(p+1)$-isospectral*, preprint Hiroshima Univ. 1989.
30. A. Ikeda and Y. Taniguchi, *Spectra and eigenforms of the Laplacian on S^n and $P^n(\mathbb{C})$*, Osaka J. Math. **15** (1978), 515–546.
31. I. Iwasaki and K. Katase, *On the spectra of Laplace operator on $\wedge^*(S^n)$*, Proc. Japan Acad. **55** (1979), 141–145.
32. M. Kac, *Can one hear the shape of a drum?* Amer. Math. Monthly **73** (1966), 1–23.
33. Y. Kitaoka, *Positive definite quadratic forms with the same representation numbers*, Arch. Math. **28** (1977), 495–497.
34. M. Kneser, *Lineare Relationen zwischen Darstellungszahlen quadratischer Formen*, Math. Ann. **168** (1967), 31–39.
35. A. Lévy-Bruhl-Laperriere, *Spectre du de Rham-Hodge sur l'espace projectif complexe*, C. R. Acad. Sci. Paris **284** (1977), 1265–1267.
36. S. Marchiafava, M. Puta, and A. Török, *On the spectrum of the p-Laplacian*, Proc. 20th NCGT, Timişoara October 5–7, 1989, pp. 125–129.
37. J. Milnor, *Eigenvalues of the Laplace operator on certain manifolds*, Proc. Nat. Acad. Sci. **51** (1964), 542.
38. V. K. Patodi, *Curvature and the fundamental solution of the heat operator*, J. Indian Math. Soc. **34** (1979), 269–285.
39. V. K. Patodi, *Curvature and the eigenforms of the Laplace operator*, J. Differential Geom. **5** (1971), 233–249.
40. H. Pesce, *Borne explicite du nombre de tores plats isospectraux à un tore donné*, Manuscripta Math. **75** (1992), 211–223.
41. M. Puta, *The spectrum of the Laplacian for 5-forms*, Exposition. Math. **3** (1985), 75–79.
42. M. Puta and A. Török, *On the spectrum of the Laplacian on p-forms*, Ann. Univ. Timişoara, Ser. Stiinţ. Mat. **25** (1987), 67–73.
43. M. Puta and A. Török, *The spectrum of the p-Laplacian on Kähler manifolds*, Rendiconti di Matematica Ser. VII, vol. 11, Roma (1991), 257–271.
44. S. Rosenberg, *The variation of the De Rham Zeta function*, preprint Brandeis Univ., 1985.
45. H. Strese, *Zum spectrum des Laplaces Operators auf p-formen*, Math. Nachr. **106** (1982), 35–40.
46. T. Sunada, *Riemannian coverings and isospectral manifolds*, Ann. of Math. **121** (1985), 169–186.
47. S. Tanno, *The spectrum of the Laplacian for 1-forms*, Proc. Amer. Math. Soc. **45** (1974), 125–129.
48. Gr. Tsagas, *The spectrum of the Laplace operator for a special Riemannian manifold*, Kodai Math. J. **4** (1981), 377–382.

49. Gr. Tsagas and C. Kockinos, *The geometry and the Laplace operator on the exterior 2-forms on a compact Riemannian manifold*, Proc. Amer. Math. Soc. **73** (1979), 109–116.
50. M. F. Vignéras, *Exemples de sous-groupes discrets non conjugués de PSL(2,ℝ) qui ont même fonction zêta de Selberg*, C. R. Acad. Sci. Paris **287** (1978), 47–49.
51. S. Wolpert, *The eigenvalue spectrum as moduli for flat tori*, Trans. Amer. Math. Soc. **244** (1978), 312–321.

Geometry in Partial Differential Equations (pp. 68–90)
eds. A. Pràstaro and Th. M. Rassias

MINIMAL SURFACES IN ECONOMIC THEORY

JERRY DONATO
Institute for Mathematics and its Applications
University of Minnesota
206 Church Street S.E.
Minneapolis, Minnesota 55455-0436
U.S.A.

ABSTRACT

A minimal surface is reflected in a constant elasticity of substitution (CES) function used in economic theory and in specifying extremum conditions. The paper gives an account of some of the main developments connected with minimal surfaces in economic theory.

1. Introduction

Section 2 briefly reviews the algebra of the composition of functions and the inverses of functions.

Section 3 reviews the general equation of the second degree found in analytic geometry. Some of its geometric properties are noted along with the operation of rotating coordinates to eliminate cross terms. The fundamental theorem for space curves is given along with the Serret-Frenet equations. Its skew-symmetric form is noted along with the tangent, normal and binormal vectors with their corresponding orthonormal planes. Section 4 presents the general integrability conditions of differential equations. Gaspard Monge's approach to solving partial differential equations of order two with variable coefficients is briefly outlined. The second derivative of an implicit function is presented and its similarity with Monge's approach is noted along with the extremum conditions.

Section 5 describes the Monge coordinate patch. The first and second fundamental forms of surfaces are presented and their properties are outlined.

The mean curvature and Gaussian curvature are defined. The mean curvature relationship $EN-2MF+GL$ and Gaspard Monge's relationship $p^2t-2s\,pq+q^2r$ to solving partial differential equation of order two with variable coefficients is noted along with extremum conditions. The above equation may be displayed in the following bordered determinant form

$$-\begin{vmatrix} 0 & p & q \\ p & r & s \\ q & s & t \end{vmatrix}$$

where the corresponding first and second fundamental form coefficients p, q, r, s, t are noted. Gauss's Theorema Egregium is stated. The Gauss-Weingarten equations for a Monge patch is presented and its relationship to E, F, G, L, M and N are listed. Using differential forms and Cartan's approach, the mean and Gaussian curvature is presented and applied to Monge's approach.

Section 6 defines minimal surfaces as those surfaces that have zero mean curvature. The relations in Section 4 and Section 5 are used to describe minimal surfaces. Th.M. Rassias's investigations into minimal surfaces using an eigenvalue formulation of the Morse-Smale index is outlined and his comments are noted.

Section 7 outlines the conditions for an extremal using the calculus of variational approach. Again Monge's equations are presented in this different format. A very basic linear constrained economic maximization example is given. The example suggest analogously that the second fundamental form (a type of say production function) be maximized subject to the first fundamental form (a type of say economic constraints with prices and quantities). Hence the intrinsic, covariant nonlinear aspects of the budget constraint are suggested. The non-commutability of multiplication of prices and quantities in economic theory is also noted. The extrinsic properties (that is, the visualization of various geometric objects) of the second fundamental form possess a relative ease in communicating aspects of the problem when compared to the complexity of communicating the fundamentally important intrinsic invariant properties of the first fundamental form are noted.

Section 8 presents the notion of a hyperplane and the corresponding theorem of the separating hyperplanes. The algebraic properties of a hyperplane are noted as being reflected in the construction of the second fundamental form which uses differentials. Gerard Debreu's geometric approach to using the hyperplane concept to demonstrate the first and second fundamental propositions of welfare economics is outlined. Debreu's investigations of applying differential topology to economics is noted along with his notion of a non-zero Gaussian curvature. Stephen Smale's investigations into economic analysis are also noted. The non-zero Gaussian curvature Debreu considers appears to be extrinsic, hence, non-invariant and appears to have a non-zero mean curvature, hence, a non-minimal surface.

A brief review of a production function and its neoclassical properties is given. The elasticity of substitution concept in production theory is defined and its properties are listed. Murray Brown's derivation of the constant elasticity of substitution (CES) production function using an ordinary non-linear second order differential equation is noted and the equation is presented. The infinite value of the elasticity of substitution is selected for the subsequent analysis; this infinite value makes the isoquants a straight line and indicates that the factors of production are perfect substitutions for each other. But these are the same geometric conditions that determine a zero mean curvature. This suggests that a minimal surface reflects a large elasticity of substitution value, that is, a minimal surface reflects the CES function. A bordered determinant generated from the CES function is displayed and compared with a bordered determinant generated from a minimal surface condition. The corresponding elements are reviewed. A rotation concept also appears.

2. Algebra of Functions

A function maps the set of first coordinates (the domain) to the set of second coordinates (the range). Functions can be combined in a way called composition of functions, denoted by the symbol *o*. This is a type of multiplication of functions. For example, consider the functions $f : A \to B$ and $g : B \to C$. The normal composition of these functions is denoted as $g \circ f : A \to C$. But in general $g \circ f \neq f \circ g$. This suggests non-commutability of multiplication. The identity function is denoted as $1 : A \to A$. This suggests a mathematical group property. The inverse function for $f : A \to B$ is $f^{-1} : B \to A$ with $f \circ f^{-1} = 1$ where f^{-1} is the inverse and 1 is the identity function. This suggests another group property. Also, this type of function considers a push forward operation and a pull back operation where the function assumes its original starting position; this would be a path independent notion. If these operations are not fulfilled, the notion of path dependence appears ([6,7]).

3. Tangents, Normals and Binormals

All the curves that the following equation $Ax^2 + Cy^2 + Dx + Ey + F = 0$ can represent are presented in the following theorem.

Theorem 3.1. *If A and C are both positive or both negative, then the locus is an ellipse, a circle (if $A = C$), a point or nothing; if A and C are of opposite signs, then the locus is a hyperbola or two intersecting lines and if either A or C is zero, then the locus is a parabola, two parallel lines, one line or nothing.*

Theorem 3.2. *Consider the following general equation of the second degree $Ax^2 + Bxy + Cy^2 + Dx + Ey + F = 0$ where A, B, C are not all zero. It is always possible to rotate the coordinates in such a way that there are no cross terms in the new coordinate system.*

Theorem 3.3. *For general equations of the second degree, the following properties* *exist: If* $B^2 - 4AC < 0$, *then the curve is an ellipse, a circle, a point or there is no curve; if* $B^2 - 4AC > 0$, *then the curve is a hyperbola or two intersecting straight lines and if* $B^2 - 4AC = 0$, *then the curve is a parabola, two parallel lines, one line or there is no curve.*

Remark 3.4. The circle, ellipse, parabola and hyperbola are often called conic sections because all of them can be obtained as sections cut from a right circular cone by planes, flat surfaces. For example, when the plane is perpendicular to the axis of the cone, a circle appears which is a special case of an ellipse. This visualization of a geometric object suggests the notions of a tangent, denoted as T, and a normal, denoted as N where they are perpendicular to each other. When the plane is parallel to one of the generators of the cone, the intersection results in another geometric figure called a parabola and when the plane intersects, both nappes of the cone, the geometric figure of a hyperbola appears.

Remark 3.5. The subject of descriptive geometry presents the graphical solution of a point, line and plane problems in space. These solutions are accomplished by means of the same principles of orthographic drawing which are involved in making a simple three-view drawing of an object.

Theorem 3.6. The fundamental existence and uniqueness theorem for space curves states: *Let* $k(s)$ *and* $\tau(s)$ *be arbitrary continuous functions on* $a \le s \le b$, *then there exists, except for position in space, one and only one space curve* C, *for which* $k(s)$ *is the curvature,* $\tau(s)$ *is the torsion and* s *is the natural parameter along* C.

The Serret-Frenet equations of a curve are as follows:

$$\dot{T} = +kN,\ \dot{N} = -kN + \tau B,\ \dot{B} = -\tau B$$

where T, N and B represent the tangent, normal and binormal vectors respectively, k and τ represent the curvature and torsion parameters respectively, and $\dot{T}, \dot{N}$ and $\dot{B}$ represent the derivatives with respect to the natural parameters. The three vectors, T, N and B form a right-handed orthonormal triplet.

For each point on a curve there exists (a) the following characteristic lines: a tangent line; a principal normal line and a binormal line and (b) the following characteristic planes: a normal plane, a rectifying plane and an osculating plane.

4. Partial Differential Equations

Theorem 4.1. *Any Pfaffian differential equation in* $\mathbb{R}^2$

$$\omega = Pdx + Qdy = 0,\ P = P(x,y);\ Q = Q(x,y)$$

admits an integrating factor.

Theorem 4.2. *The Frobenius condition $d\omega \wedge \omega = 0$ is called integrability condition for Pfaffian equation $\omega = 0$.*

Example 4.3. Given the following Pfaffian equation in $\mathbf{R}^3$ (which can be extended to $\mathbf{R}^n$) where $\omega = P(x,y,z)dx + Q(x,y,z)dy + R(x,y,z)dx = 0$ then

$$d\omega \wedge \omega = \left[P\left(\frac{\partial R}{\partial y} - \frac{\partial Q}{\partial z}\right) + Q\left(\frac{\partial P}{\partial z} - \frac{\partial R}{\partial x}\right) + R\left(\frac{\partial Q}{\partial x} - \frac{\partial P}{\partial y}\right)\right] dx \wedge dy \wedge dz\ .$$

Hence $\omega = 0$ is integrable if and only if the term in the square brackets vanishes (cf. [1]).

Remark 4.4. Observe the cycle properties of P, Q and R and x, y and z. Also observe the symmetry properties.

The following is a brief review of Gaspard Monge's approach to solving partial differential equations of order two with variable coefficients (cf. [3], [13]).

The most general linear partial differential equation of order two in two independent variables has the following form $Rr + Ss + Tt + Pp + Qq + Zz = F$ where R, S, T, P, Q, Z, F are functions of x and y and not all R, S, T are zero.

There are a number of special types of these equations; each type suggests a corresponding procedure for its solvability. The Laplace transformation also suggests a procedure for solving the general form.

Many examples of solving linear differential equations suggest another possible method for solving the following given non-linear partial differential equation of order two $F(x, y, z, p, q, r, s, t) = 0$ where $p = \frac{\partial z}{\partial x}$, $q = \frac{\partial z}{\partial y}$, $r = \frac{\partial^2 z}{\partial x^2}$, $s = \frac{\partial^2 z}{\partial x\, \partial y}$, and $t = \frac{\partial^2 z}{\partial y^2}$.

In many of those examples, the first step consisted in finding a relation of the form

$$u = \psi(v), \quad \psi \text{ arbitrary}$$

where $u = u(x, y, z, p, q)$ and $v = v(x, y, z, p, q)$ from which the given differential equation could be derived by eliminating the arbitrary function; such a relation is called an intermediate integral.

The most general partial differential equation having

$$u = \psi(v), \quad \psi \text{ arbitrary}$$

where $u = u(x, y, z, p, q)$ and $v = v(x, y, z, p, q)$, as an intermediate integral has the following form $Rr + Ss + Tt + U(rt - s^2) = V$ where R, S, T, U, V are functions of x, y, z, p, q. However, from the definitions of $R, S, \ldots, V$, not every equation of the above form has an intermediate integral. The following discussion of Monge's method for determining an intermediate integral of the above equation assumes that one exists. Now consider the following equation $Rr + Ss + Tt = V$ where U from above is zero. Consider z as a function of x and y, hence, the following equations appear

$$dz = \frac{\partial z}{\partial x}dx + \frac{\partial z}{\partial y}dy = pdx + qdy$$
$$dp = \frac{\partial p}{\partial x}dx + \frac{\partial p}{\partial y}dy = rdx + sdy$$
$$dq = \frac{\partial q}{\partial x}dx + \frac{\partial q}{\partial y}dy = sdx + tdy\ .$$

Solving the latter two equations for

$$r = \frac{dp - sdy}{dx},\ t = \frac{dt - sdx}{dy}$$

and substituting one gets

$$R\frac{dp - sdy}{dx} + Ss + T\frac{dq - sdx}{dy} = V\ .$$

Therefore

$$s[R(dy)^2 - Sdxdy + T(dx)^2] = Rdydp + Tdxdq - Vdxdy\ .$$

The following are Monge's equations:

$$R(dy)^2 - Sdxdy + T(dx)^2 = 0,\ Rdydp + Tdxdq - Vdxdy = 0\ .$$

Suppose $R(dy)^2 - Sdxdy + T(dx)^2 = (Ady + Bdx)^2 = 0$. If $u = u(x, y, z, p, q) = a$, $v = v(x, y, z, p, q) = b$ satisfy the following system $Ady + Bdx = 0$, $Rdydp + Tdxdq - Vdxdy = 0$ then $u = \psi(v)$ is an intermediate integral since $u = a, v = b$.

Suppose

$$R(dy)^2 - Sdxdy + T(dx)^2 = (A_1dy + B_1dx)(A_2dy + B_2dx) = 0$$

where $A_1B_2 - A_2B_1 \neq 0$. Then these are two systems

$$\left\{\begin{matrix} A_1dy + B_1dx = 0 \\ Rdydp + Tdxdq - Vdxdy = 0 \end{matrix}\right\} \quad \text{and} \quad \left\{\begin{matrix} A_2dy + B_2dx = 0 \\ Rdydp + Tdxdq - Vdxdy = 0 \end{matrix}\right\}\ .$$

If either system is integrable, then there is an intermediate integral. If both are integrable, then there are two intermediate integrals.

Example 4.5. ([25],[2]) If $dz = pdx + qdy = 0$ then

$$\frac{dy}{dx} = -\frac{p}{q}, \text{ where } p = \frac{\partial z}{\partial x}, \ q = \frac{\partial z}{\partial y}.$$

The second derivative is described as

$$\frac{d^2y}{dx^2} = \frac{d}{dx}\left(\frac{dy}{dx}\right) = \frac{d}{dx}\left(-\frac{p}{q}\right) = -\frac{qdp - pdq}{q^2}.$$

Apply the total differentials to get

$$dp = \frac{\partial p}{\partial x}dx + \frac{\partial p}{\partial y}dy = rdx + sdy$$

$$dq = \frac{\partial q}{\partial x}dx + \frac{\partial q}{\partial y}dy = sdx + tdy$$

$$\text{where } r = \frac{\partial^2 z}{\partial x^2}, \ s = \frac{\partial^2 z}{\partial x \partial y}, \ t = \frac{\partial^2 z}{\partial y^2}.$$

Substitute into the formula to get

$$\frac{d^2y}{dx^2} = -\frac{1}{q^2}(qrdx + qsdy - psdx - ptdy).$$

Divide numerator by dx and substituting for dy/dx, one obtains

$$\frac{d^2y}{dx^2} = -\frac{qr + qs\frac{dy}{dx} - ps - pt\frac{dy}{dx}}{q^2}$$

$$\frac{d^2y}{dx^2} = -\frac{1}{q^3}(q^2r - 2spq + p^2t).$$

The above equation may also be rewritten as

$$-\frac{d^2y}{dx^2}q^3 = q^2r - 2spq + p^2t.$$

The sign of $\frac{d^2y}{dx^2}$ will depend on what value is contained in the brackets. If the value is positive, then $\frac{d^2y}{dx^2} < 0$. This means that the simple curvature of the curve is concave upwards. If the value is negative, then $\frac{d^2y}{dx^2} > 0$. This means the simple curvature of the curve is concave downwards. If the value is zero, then $\frac{d^2y}{dx^2} = 0$. This means the curve has no curvature, that is, the curve is a straight line. This Example results in a second order partial differential equation similar to that equation when Monge's approach and symbols were used. Example 4.5 clearly reveals a simple geometric curvature concept whereas Monge's approach presents the analysis without highlighting the geometry in the equations.

5. Mean Curvature and Gaussian Curvature

Following [14] one has:

Definition 5.1. *A regular parametric representation of class $C^m (m \geq 1)$ of a set of points S in E^3 is a mapping $x = f(u, v)$ of an open set U in the uv plane onto S such that: (i) f is of class C^m in U and (ii) if (e_1, e_2, e_3) is a basis in E^3 and*

$$f(u,v) = f_1(u,v)e_1 + f_2(u,v)e_2 + f_3(u,v)e_3 \ ,$$

then for all (u, v) in U

$$\text{rank} \begin{pmatrix} \frac{\partial f_1}{\partial u} & \frac{\partial f_1}{\partial v} \\ \frac{\partial f_2}{\partial u} & \frac{\partial f_2}{\partial v} \\ \frac{\partial f_3}{\partial u} & \frac{\partial f_3}{\partial v} \end{pmatrix} = 2$$

Definition 5.2. *The three components of the vector product are given by*

$$x_u \times x_v = \det \begin{pmatrix} e_1 & \frac{\partial x_1}{\partial u} & \frac{\partial x_1}{\partial v} \\ e_2 & \frac{\partial x_2}{\partial u} & \frac{\partial x_2}{\partial v} \\ e_3 & \frac{\partial x_3}{\partial u} & \frac{\partial x_3}{\partial v} \end{pmatrix}$$

where $x = x(u, v)$, $x_u = \frac{\partial x}{\partial u}$, $x_v = \frac{\partial x}{\partial v}$, $x_{uu} = \frac{\partial^2 x}{\partial u^2}$, $x_{uv} = \frac{\partial^2 x}{\partial u \partial v}$, etc.

Definition 5.3. *A coordinate patch of class $C^m (m \geq 1)$ in S is a mapping $x = x(u, v)$ of an open set U into S such that (i) x is of class C^m on U, (ii) $x_u \times x_v \neq 0$ for all (u, v) in U and (iii) x is $1-1$ and bicontinuous on U.*

Definition 5.4. *The functions of the form $x = ue_1 + ve_2 + f(u,v)e_3$ or $x = ue_1 + f(u,v)e_2 + ve_3$ or $x = f(u,v)e_1 + ue_2 + ve_3$ define a coordinate patch of class C^m simply if $f(u, v)$ is a function of class C^m.*

Theorem 5.5. The following theorem describes a Monge patch: *If a set S in E^3 has a regular parametric representation of class C^m, then for every point P in S there exists a Monge patch of C^m in S containing P.*

Definition 5.6. *Let $x = x(u, v)$ be a coordinate patch on a surface of class ≥ 1. The differential of the mapping $x = x(u, v)$ at (u, v) is a $1-1$ linear mapping $dx = x_u du + x_v dv$ of the vectors (du, dv) in the uv plane onto the vectors $x_u du + x_v dv$ parallel to the tangent plane. Now consider the following function called the first fundamental form (FFF) of $x = x(u, v)$:*

$$I = dx \cdot dx = Edu^2 + 2Fdudv + Gdv^2 \ .$$

It is a homogeneous function of the second degree in du and dv with coefficients E, F and G called the first fundamental coefficients which are functions of u and v and vary from point to point on the coordinate patch. The form is quadratic. Also I is positive definite and its coefficients must satisfy $E > 0$, $G > 0$ and $EG - F^2 > 0$. Using a vector approach

$$EG - F^2 = |x_u \times x_v|^2$$

where at every point $x_u \times x_v \neq 0$, hence $EG - F^2 > 0$.

Definition 5.7. *The following equation describes the arc length on the surface as the integral of the square root of the first fundamental form*

$$s = \int_a^b \left[E\left(\frac{du}{dt}\right)^2 + 2F\frac{du}{dt}\frac{dv}{dt} + G\left(\frac{dv}{dt}\right)^2 \right]^{\frac{1}{2}} dt$$

where $x = x(u(t), v(t))$, $a \le t \le b$ is a regular arc on a patch $x = x(u,v)$.

Definition 5.8. *Consider $dx = x_u du + x_v dv$ and $\delta x = x_u \delta u + x_v \delta v$ to be two vectors parallel to the tangent plane at a point x. If α is the angle between dx and δx then*

$$\cos\alpha = \frac{dx \cdot \delta x}{|dx|\,|\delta x|}\,.$$

Definition 5.9. *If β is the angle between the u- and the v-parameter curves at x, that is, the angle between x_u and x_v, then*

$$\cos\beta = \frac{x_u \cdot x_v}{|x_u| \cdot |x_v|} = \frac{F}{\sqrt{EG}}\,.$$

Theorem 5.10. *(a) The tangent vectors $dx = x_u du + x_v dv$ and $\delta x = x_u \delta u + x_v \delta v$ are perpendicular if and only if*

$$E\, du\delta u + F(du\delta v + dv\delta u) + G\, dv\delta v = 0$$

and (b) the u and v parameter curves at a point are perpendicular if and only if $F = 0$.

Definition 5.11. *Let $x = x(u,v)$ be a coordinate patch on a surface of class ≥ 2 (compare to Definition 5.6). Then at each point on the patch there is a unit normal vector*

$$N = \frac{x_u \times x_v}{|x_u \times x_v|}$$

which is a function of u and v of class C' with differential $dN = N_u du + N_v dv$. Now consider the following function called the second fundamental form (SFF) of $x = x(u,v)$ $II = -dx \cdot dN = Ldu^2 + 2Mdudv + Ndv^2$. It is homogeneous of the second degree in du and dv with coefficients L, M and N called the second fundamental coefficients which are continuous functions of u and v.

Note that II is invariant in the same sense that I is invariant under a parameter transformation which preserves the direction of N; otherwise II changes its sign, (compare with Definition 5.6). Note that dN is orthogonal to N since it is parallel to the tangent plane of the spherical image of N. This observation also follows from the following, $0 = d(1) = d(N \cdot N) = 2dN \cdot N$. Hence dN is a vector parallel to the tangent plane. (Again compare with Definition 5.6.)

Definition 5.12. *Depending upon the value of the discriminant $LN - M^2$, the following four cases appear.*

(1) *A point is called an elliptic point if $LN - M^2 > 0$.*
(2) *A point is called a hyperbolic point if $LN - M^2 < 0$.*
(3) *A point is called a parabolic point is $LN - M^2 = 0$ and $L^2 + M^2 + N^2 \neq 0$ that is $LN - M^2 = 0$ and the coefficients L, M and N are not all zero.*
(4) *A point is called a planar point is $L = M = N = 0$.*

Remark 5.13. When discussing the SFF which has a built-in normal, the Remark 3.4 made regarding conic sections should be kept in mind. Also, note that the observer is looking at these geometric figures extrinsically as is done in descriptive geometry. See Remark 3.5.

Remark 5.14. The nature of the surface at a planar point is described by the next higher order terms of the expansion of $x(u, v)$. The geometric figure of a monkey saddle uses a cubic equation. The geometric saddle concept appears also in nonlinear programming analysis.

Theorem 5.15. *All curves through a point P on a patch which are tangent to the same line through P have the same normal curvature at P.*

Theorem 5.16. *All curves on a surface through a point P which have the same osculating plane through P have the same curvature k at P provided that the osculating plane is not tangent to the surface.*

Theorem 5.17. *The curvature of a normal section of a patch at a point P is equal to the normal curvature of the section at P.*

Definition 5.18. *The following equation called a Dupin's indicatrix determines a conic section in the x_1x_2 plane: $\pm 1 = Lx_1^1 + 2Mx_1x_2 + Nx_2^2$ where the distance r from a point (x_1, x_2) to the origin is the reciprocal of the square root of $|k_n|$ (k_n is the normal curvature) in the direction $\cos\theta : \sin\theta$.*

If P is an elliptic point ($LN - M^2 > 0$), the indicatrix is an ellipse. If P is a hyperbolic point ($LN - M^2 < 0$), the indicatrix consists of a pair of conjugate hyperbolas. Along one of the hyperbola k_n is positive and along the other k_n is negative. The common asymptotes correspond to the directions for which $k_n = 0$. In the parabolic case ($LN - M^2 = 0$, $L^2 + N^2 + M^2 \neq 0$) the quantity $Lx_1^2 + 2Mx_1x_2 + Nx_2^2$ factors and the indicatrix is a pair of parallel lines whose directions are also in the direction in which $k_n = 0$. In the planar case ($L = M = N = 0$), the indicatrix does not exist.

Definition 5.19. *The two perpendicular directions for which the values of k_n take on maximum and minimum values are called principal directions and the corresponding normal curvatures k_1 and k_2 are called the principal curvatures.*

Remark 5.20. The Dupin indicatrix can be helpful when considering a Monge patch function of say the following form: $x = ue_1 + ve_2 + f(u,v)e_3$.

Theorem 5.21. (cf. [14]) *A real number k is a principal curvature at P in the direction $du : dv$ if and only if k, du and dv satisfy $(L - kE)du + (M - kF)dv = 0$, $(M - kF)du + (N - kG)dv = 0$ where $du^2 + dv^2 \neq 0$.* The above is a homogeneous system of equations and will have a nontrivial solution du, dv if and only if

$$\det \begin{pmatrix} L - kE & M - kF \\ M - kF & N - kG \end{pmatrix} = 0$$

or expanding

$$(EG - F^2)k^2 - (EN + GL - 2FM)k + (LN - M^2) = 0\,.$$

The discriminant of the above equation is greater than or equal to zero. Hence the equation has either two roots k_1 and k_2, the principal curvatures at a nonumbilical point or a single real root k with multiplicity two, the curvature of an umbilical point.

Theorem 5.22. *A number k is a principal curvature if and only if k is a solution of the equation*

$$(EG - F^2)k^2 - (EN + GL - 2FM)k + (LN - M^2) = 0\,.$$

Theorem 5.23. *A point is an umbilical point if and only if the fundamental coefficients are proportional in which case the normal curvature is*

$$k = \frac{L}{E} = \frac{M}{F} = \frac{N}{G}\,.$$

Definition 5.24. The mean curvature H, is

$$H = \frac{1}{2}(k_1 + k_2) = \frac{EN - 2FM + GL}{2(EG - F^2)}$$

which is the average of the roots k_1 and k_2.

Definition 5.25. *The Gaussian curvature K is*

$$K = k_1 k_2 = \frac{LN - M^2}{EG - F^2}$$

which is the product of the roots.

Remark 5.26. The following relationship in the numerator defining the mean curvature, that is $EN-2MF+GL$ corresponds to the following relationship developed in section 4 (wherein Gaspard Monge's approach to solving partial differential equations of order two with variable coefficients was outlined) $p^2t - 2spq + q^2r$.

The above equation may also be recasted into a bordered determinant form as follows

$$-\begin{vmatrix} o & p & q \\ p & r & s \\ q & s & t \end{vmatrix} .$$

The following correspondence between the various coefficients then appear as follows $E \sim p^2$, $N \sim t$, $M \sim s$, $F \sim pq$, $G \sim q^2$, $L \sim r$ where $\sim$ indicates a correspondence.

This suggests that the above bordered determinant be constructed as follows:

$$-\begin{vmatrix} o & p & q \\ p & L & M \\ q & M & N \end{vmatrix}$$

where the corresponding FFF and SFF coefficients are noted.

Theorem 5.27. *The parameter curves on a Monge patch $x = ue_1 + ve_2 + f(u,v)e_3$ form an orthogonal family of curves if and only if $f_u f_v = 0$.*

Theorem 5.28. *The surface area on a Monge patch $x = ue_1 + ve_2 + f(u,v)e_3$ is given by the integral*

$$A = \iint\limits_W \sqrt{1 + f_u^2 + f_v^2} du\, dv .$$

Theorem 5.29. *The second fundamental form on a Monge patch*

$$x = ue_1 + ve_2 + f(u,v)e_3$$

is

$$II = (f_u^2 + f_v^2 + 1)^{-\frac{1}{2}} [f_u f_u du^2 + 2f_{uv} du dv + f_{vv} dv^2] .$$

Example 5.30. The Gauss-Weingarten equations for a Monge patch $x = ue_1 + ve_2 + f(u_1 v)e_3$ are given by

$$\begin{aligned}
gx_{uu} &= prx_u + qrx_v + rg^{\frac{1}{2}} N \\
gx_{uv} &= psx_u + qsx_v + sg^{\frac{1}{2}} N \\
gx_{vv} &= ptx_{uv} + qtx_v + tg^{\frac{1}{2}} N \\
g^{\frac{3}{2}} N_u &= (spq - rq^2 - r)x_u + (rpq - sp^2 - s)x_v
\end{aligned}$$

$$g^{\frac{3}{2}} N_v = (tpq - sq^2 - s)x_u + (spq - tp^2 - t)x_v$$

where $p = f_u$, $q = f_v$, $r = f_{uu}$, $s = f_{uv}$, $t = f_{vv}$ $g = 1 + p^2 + q^2$, $x_u = e_1 + pe_3$, $x_v = e_2 + qe_3$, $x_{uv} = re_3$, $x_{uv} = se_3$, $x_{vv} = te_3$. $E = x_u \cdot x_u = 1 + p^2$, $F = x_u \cdot x_u = pq$, $G = x_v \cdot x_v = 1 + q^2$, $EG - F^2 = 1 + p^2 + q^2$, $N - q^{-\frac{1}{2}} - (pe_1 + qe_2 - e_3)$, $L = rg^{-\frac{1}{2}}$, $M = sg^{-\frac{1}{2}}$, $N = tg^{-\frac{1}{2}}$.

The Christoffel symbols of the second kind are defined by: $\{11,1\} = prg^{-1}$, $\{12,1\} = psg^{-1}$, $\{22,1\} = ptg^{-1}$, $\{11,2\} = qrg^{-1}$, $\{12,2\} = qsg^{-1}$, $\{22,2\} = qtg^{-1}$. The coefficients of the Weingarten equations are

$$\beta_1^1 = (spq - rq^2 - r)g^{-\frac{3}{2}}, \qquad \beta_2^1 = (tpq - sq^2 - s)g^{-\frac{3}{2}}$$
$$\beta_1^2 = (rpq - sp^2 - s)g^{-\frac{3}{2}}, \qquad \beta_2^2 = (spq - tp^2 - t)g^{-\frac{3}{2}}.$$

Example 5.31. Using differential forms and Cartan's approach, the mean curvature H, and the Gaussian curvature K can be described as follows (cf. [10]):

The mean curvature H is

$$H = -\frac{1}{nw}\left(\Delta u - \sum \frac{1}{w^2} p_i r_{ij} p_j\right)$$

where n is the space dimension (the mean curvature is related to the trace) and Δu is the Laplacian of u.

The Gaussian curvature K is

$$K = \frac{(-1)^n}{w^{n+2}}|R|.$$

For the special case $n = 2$ for surfaces Monge's approach would reveal the mean curvature as

$$H = -\frac{1}{2w^3}[(1+q^2)r - 2spq + (1+p^2)t],$$

and the Gaussian curvature as $K = \frac{(rt-s^2)}{w^4}$, where $w^2 = 1 + p^2 + q^2$.

Remark 5.32. The equation for mean curvature presented in Example 5.31 correspond with Remark 5.26 and Example 5.30. There are many different ways to look at mean curvature.

6. Minimal Surfaces

Theorem 6.1. *Minimal surfaces are surfaces of zero mean curvature.*

Remark 6.2. Theorem 6.1 suggests the following relations presented in Section 4 and Section 5 can be used to describe a minimal surface $p^2t - 2spq + q^2r = 0$, $EN - 2MF + GL = 0$, $w^3[(1+p^2)t - 2spq + (1+q^2)r = 0$.

The minimal surface equation leads to partial differential equations and conversely (Dirk Struik [23]).

Remark 6.3. Themistocles M. Rassias [19] considered an eigenvalue formulation of the Morse-Smale index to the study of minimal surface in $\mathbf{R}^3$ to get global results for complete minimal surfaces and to examine the eigenvalues of certain elliptic operators on manifolds. Such an approach leads to investigating the eigenvalues of the Laplacian. The following three examples of complete minimal surfaces in $\mathbf{R}^3$ were analyzed: (1) the plane, $\mathbf{R}^2 \subset \mathbf{R}^3$, (2) the catenoid and (3) Enneper's minimal surface. The following presents some of Th.M. Rassias's investigations and his comments.

The following eigenvalue problem was studied by Th.M. Rassias in a number of papers (cf. [19])

$$\begin{cases} -\Delta f + 2KWf = \lambda f, \ f = f(u,v) \text{ defined on } \overline{D}^2_{r_0} \\ f \equiv 0 \text{ on } \partial D^2_{r_0} \quad \text{(boundary condition)} \end{cases}$$

where f is a variation in the direction of the surface normal, Δ is the Laplacian of the surface, K is the Gaussian curvature of the minimal surface and W is the discriminant of the first fundamental form, that is, $W = (EG - F^2)^{\frac{1}{2}}$ where E, F, G are the coefficients of the first fundamental form of the minimal surface.

The eigenvalue problem for the plane is as follows

$$\begin{cases} \nabla^2 f + \lambda f = 0, \quad f = f(u,v) \text{ on } \overline{D}^2_{r_0} \\ f \equiv 0 \text{ on } \partial D^2_{r_0} \, . \end{cases}$$

Observe that $\nabla^2 f + \lambda f$ implies $\nabla^2 f = -\lambda f$ and $-\nabla^2$ is a positive definite operator, hence, it is impossible to find negative eigenvalues. This implies the Morse-Smale index of the plane is zero. The eigenvalue problem for the catenoid $q : D^2_{r_0} \to \mathbf{R}^3$ is given by

$$\begin{cases} \dfrac{\partial^2 f}{\partial u^2} + \dfrac{\partial^2 f}{\partial v^2} + 2f + \lambda(\cosh^2 u) f = 0, \ f = f(u,v) \text{ on } \overline{D}^2_{r_0} \\ \quad f \equiv 0 \text{ on } \partial D^2_{r_0} \, . \end{cases}$$

The catenoid is not an algebraic minimal surface.

The eigenvalue problem for Enneper's minimal surface $q : D^2_{r_0} \to \mathbf{R}^3$ is

$$\begin{cases} \dfrac{\partial^2 f}{\partial u^2} + \dfrac{\partial^2 f}{\partial v^2} + 8f + \lambda f = 0, \ f = f(u,v)\overline{D}^2_{r_0} \\ \quad f \equiv 0 \text{ on } \partial D^2_{r_0} \, . \end{cases}$$

From a topological point of view, Enneper's surface is simply connected. It is properly immersed in $\mathbf{R}^3$ but this is not an embedding because the surface has self-intersections. The surface is given in isothermal coordinates. The surface is algebraic — a polynomial minimal surface of degree 3.

7. Calculus of Variations

7.1. One of the main problems of the calculus of variations is to determine that curve connecting two given points which either minimizes or maximizes some given integral. Another type of problem in the calculus of variations is the one which involves a constraint condition; such problems can generally be solved by using the method of Lagrange multipliers. Following [22] one has:

Theorem 7.2. *Consider a surface S having equation $z = z(x,y)$ to be bounded by a closed curve C whose projection on the xy plane is C' which forms the boundary of a region $\mathcal{R}$. A necessary condition for $z = z(x,y)$ to be an extremal of*

$$\iint_R F(x,y,z,p,q)dx\,dy$$

is

$$\frac{\partial F}{\partial z} - \frac{\partial}{\partial x}\left(\frac{\partial F}{\partial p}\right) - \frac{\partial}{\partial y}\left(\frac{\partial F}{\partial q}\right) = 0$$

where $p = \frac{\partial z}{\partial x}$, $q = \frac{\partial z}{\partial y}$.

Theorem 7.3. *The surface area of S in Theorem 7.2 is given by*

$$\iint_R \sqrt{1+p^2+q^2}\; dx\,dy\ .$$

Theorem 7.4. *The surface with minimum surface area must be a solution of the equation*

$$\frac{\partial}{\partial x}\left(\frac{p}{\sqrt{1+p^2+q^2}}\right) + \frac{\partial}{\partial y}\left(\frac{q}{\sqrt{1+p^2+q^2}}\right) = 0$$

which can be written as $(1+p^2)t - 2spq + (1+q^2)r = 0$.

Theorem 7.5. *The expressions $pdx + qdy$ and $(pdy - qdx)(1+p^2+q^2)^{-\frac{1}{2}}$ are both complete differentials.*

Theorem 7.6. *A necessary condition for $z = z(x,y)$ to be an extremal of*

$$\iint_R F(x,y,z,p,q,r,s,t)dx\,dy$$

is

$$\frac{\partial F}{\partial z} - \frac{\partial}{\partial x}\left(\frac{\partial F}{\partial p}\right) - \frac{\partial}{\partial y}\left(\frac{\partial F}{\partial q}\right) + \frac{\partial^2}{\partial x^2}\left(\frac{\partial F}{\partial r}\right) + \frac{\partial^2}{\partial x \partial y}\left(\frac{\partial F}{\partial s}\right) + \frac{\partial}{\partial y}\left(\frac{\partial F}{\partial t}\right) = 0\,.$$

Example 7.7. Consider the output of a business firm as a quantity z of a certain product where z is a function $f(x,y)$ where x is the amount of labor used and y is the amount of capital equipment or investment (cf. [16]). If the price of labor is α and the price of capital is β and the firm can spend no more than B dollars, then the following question appears: How can the amount of labor and capital be found in order to maximize the output z? If the amount of labor and capital is increased then the output should be

$$\frac{\partial z}{\partial x} = p \geq 0 \text{ and } \frac{\partial z}{\partial y} = q \geq 0\,.$$

If more labor is added to a given amount of capital equipment, then the output should decrease, that is,

$$\frac{\partial^2 z}{\partial x^2} = r < 0, \qquad \frac{\partial^2 z}{\partial^2 y} = t < 0\,.$$

These ideas suggest a set of level curves of output (called isoquants) $z(x,y) = c$ to appear. The convexity of the isoquant can be viewed as follows. Movements to the right along a given isoquant takes more and more capital to replace a unit of labor and still produce the same output. The budget constraint means staying inside the triangle bounded by the axes and the following line $\alpha x + \beta y = B$. Geometrically this suggests than an isoquant should be selected in such a way as to just touch the budget constraint line. This approach suggests that the method of Lagrange multipliers be applied to find the maxima, that is, look for the critical points of the following auxiliary function $h(x,y,\lambda) = f(x,y) - \lambda(\alpha x + \beta y - B)$ where λ is the multiplier of Lagrange.

Performing the operators gives the following conditions to be met to maximize output

$$\frac{\partial z}{\partial x} = \lambda\alpha\,, \qquad \frac{\partial z}{\partial y} = \lambda\beta, \ \alpha x + \beta y = B$$

or rewrite as $\frac{1}{\alpha}\frac{\partial z}{\partial x} = \frac{1}{\beta}\frac{\partial z}{\partial y} = \lambda$.

At the optimum production point, the marginal change in output per dollar's worth of additional capital investment is equal to the marginal change of output per dollar's worth of additional labor and λ is this common value.

Remark 7.8. ([6–9]) The following comments on Example 7.7 are made: The algebraic operation of production factor prices α and β on the production factors, x and y, themselves should be evaluated in terms of their commutability, that is $\alpha x \neq x\alpha$ and $\beta y \neq y\beta$ prices and quantities may not be symmetric. This suggests the problem of integrating monetary theory into value theory in economic theory.

The constrainted maximization process suggests that the second fundamental form (SFF) be maximized subject to the first fundamental form (FFF). Now the budget constraint notion suggests an intrinsic covariant highly non-linear functional manifold or surface to exist incorporating interdependent prices and quantities. The production function to be maximized suggests using the extrinsic notions of the second fundamental form (SFF). The geometric figures visualized from the SFF reflects the constraints of the first fundamental form. Theorems 7.2, 7.3 and 7.4 give an indication as to how to approach the extremal problem geometrically taking into consideration both the intrinsic and extrinsic properties existing in the problem.

The formulation of the constraint where the intrinsic invariant geometric properties live greatly influences the visualized results which are reflected in the geometric coefficients of the second fundamental form (SFF) which is not invariant. The messages from the SFF are not necessarily the messages in the FFF but the notions of the SFF are easier to transmit and communicate. See Definition 5.18 on Dupin's indicatrix. When approaching extremal problems using the notions of FFF and SFF, the proportionality or nonproportionality of the fundamental coefficients should be kept in mind, that is, the umbilical point or nonumbilical point should be considered (see Theorem 5.23).

8. Economic Theory

Definition 8.1. *A hyperplane in E^n is defined to be the set of points $X = \{x | cx = z\}$ with $c \neq 0$ being a given n-component row vector and z a given scalar.*

The equation for a hyperplane is as follows:

$$cx = c_1x_1 + c_2x_2 + \cdots + c_nx_n = z$$

and any x satisfying the above equation lies on the hyperplane.

A hyperplane passes through the origin if and only if $z = 0$. Then the above equation becomes $cx = 0$. Hence c is orthogonal to every vector on the hyperplane and hence c is normal to the hyperplane. If $z \neq 0$, and x_1, x_2 are any two distinct points lying on the hyperplane, then

$$cx_1 - cx_2 = c(x_1 - x_2) = z - z = 0$$

hence c is orthogonal to every vector $x_1 - x_2$. Therefore even with $z \neq 0$, c is normal to the hyperplane.

If $cx = z$ is multiplied by the scalar $\lambda \neq 0$, then $(\lambda c)x = \lambda z$. The same hyperplane is defined by either λc and λz or c, z. Hence, if c is normal to a hyperplane, so is λc.

Theorem 8.2. (cf. [11]) *The following can be considered the theorem of the separating hyperplanes. Given any closed convex set X, a point y either belongs to the set X or there exists a hyperplane which contains y such that all of X is contained in one open half-space produced by that hyperplane.*

Definition 8.3. (cf. [11]) *Given a boundary point w of a convex set X; then $cx = z$ is called a supporting hyperplane at w if $cw = z$ and if all of X lies in one closed half-space produced by the hyperplane, that is, $cu \geq z$ for all $u \in X$ or $cu \leq z$ for all $u \in X$.*

Remark 8.4. The algebraic properties of a hyperplane may be reflected in the construction of the second fundamental form (SFF) of classical differential geometry of surfaces. The concept of the hyperplane is normally not developed in terms of differentials; the subject of classical differential geometry does this and the notion of the SFF appears.

Example 8.5. The economist V. Pareto, showed graphically (1896, 1897, 1906) that every competitive equilibrium is optimal. He observed that at equilibrium two indifference curves are tangent and this implies optimality. A local interpretation of Pareto's figure (1909) reveals that at an equilibrium point and as well as at an optimum, the marginal rate of substitution is equal for two micro units (or agents). A global interpretation of the figure reveals that the tangent (the straight line) through an equilibrium as well as through an optimum point, say c, separates the two indifference curves. Gerard Debreu [5] demonstrated that the original geometric argument of Pareto can be extended to a general proof if properly formulated as follows: consider an economic system defined by individual preferences $\geq_i$ $(i = 1, \ldots, m)$ production sets $Y_j (j = 1, \ldots, n)$ and total initial endowments z^0. Hence no externalities in consumption and production are allowed. If (x_i^0, y_j^0, p^0) denotes a competitive equilibrium where individual income is given by $p^0 \cdot x_i^0$, then the hyperplane $\{x \in \mathbf{R}^\ell | p^0 \cdot x = p^0 \cdot x_i^0\}$ with normal p^0 through x_i^0 supports the set $X_i(x_i^0) = \{x \in \mathbf{R}^\ell |_x \geq_i x_i^0\}$. This means that $p^0 \cdot x \geq p^0 \cdot x_i^0$ for every $x \in X_i(x_i^0)$ and that the hyperplane $\{y \in \mathbf{R}^\ell p^0 \cdot y = p^0 \cdot y_j^0\}$ with normal p^0 through y_j^0 supports the production set Y_j, that is, $p^0 \cdot y \geq p^0 \cdot y_j^0$ for every $y \in Y_j$. Consequently the first welfare proposition turns out to be a consequence of the definition of competitive equilibria and Pareto optima, provided no externalities in consumption or production prevail. In this case individual decisions, decentralized by the price system, leads to an optimum allocation of economic resources. Using further geometric insights, Debreu demonstrated the second welfare proposition as follows: Let (x_i^0, y_j^0) be a Pareto optimum of the economy defined by $(\leq_i, Y_j, z^0)$. Then the vector z^0 of initial total resources of the economy lies on the boundary of the set

$$Z = \sum_{i=1}^{m} X_i(x_i^0) + \sum_{j=1}^{n} Y_j \, .$$

Hence, if the set Z is convex, which will be the case if preferences and the total production set are convex, then there exists a hyperplane through z^0 supporting the set Z. The normal to this hyperplane defines a price system $p^0 \in \mathbf{R}^\ell$ which decentralizes the given optimum (x_i^0, y_j^0) in the following way: if every consumer receives the "income" $w = p^0 \cdot x_i^0$ and if this income is not at the minimum subsistence level, then the state (x_i^0, y_j^0) with the price vector p^0 forms a competitive equilibrium. In another paper Debreu [5] investigates the concept of cardinal utility in three different situations by means of the same mathematical result that gives a topological characterization of three families of parallel straight lines in a plane. In a latter paper, "The Application to Economics of Differential Topology and Global Analysis: Regular Differentiable Economies," Debreu presents a detailed discussion of the concept of equilibrium and in "Smooth Preferences," Debreu incorporates the notion of a Gaussian curvature of a hypersurface in $\mathbf{R}^\ell$ in his approach to economics.

Stephen Smale [21] also mentions Debreu's (1972) hypothesis of positive Gaussian curvature in his paper (1976) titled "Global Analysis and Economics VI - Geometric Analysis of Pareto Optima and price equilibria under classical hypotheses".

Theorem 8.6. (The Gauss-Bonnet Theorem) *If M is a compact orientable geometric surface, then the total Gaussian curvature of M is $2\pi\chi(M)$, where $\chi(M)$ is the Euler-Poincaré's characteristic of M.*

Remark 8.7. Debreu initially uses the concept of a normal (a hyperplane) without using a differential geometric structure. He then uses notions from topology without relating them to the intrinsic invariant Gaussian curvature that the above classical Gauss-Bonnet Theorem suggests. When Debreu uses a non-zero Gaussian curvature notion, the curvature is extrinsic, a type based implicitly on the second fundamental form (SFF); the invariant intrinsic concept of the Gaussian curvature based on the first fundamental form (FFF) is not explicitly considered, that is, Gauss's Theorema Egregium is not considered in the analysis. Debreu's use of the non-zero Gaussian curvature reflects a non-minimal surface, that is, a surface with mean curvature greater than zero or less than zero, $H \lessgtr 0$. The partial differential equations, that is, $p^2t - 2spq + q^2r$ and/or the coefficients E, F, G, L, M and N related to the second fundamental form (SFF), that is, $EN - 2MF + GL$ are not explicitly revealed in the analysis. His geometric insights are, however, very helpful in outlining the economic problems being investigated.

Definition 8.8. *A production function expresses the relation between the maximum quantity of output and the inputs required to produce it and the relation between the inputs themselves. A production function can be represented by a set of curves, called isoquants, each curve representing various combinations of inputs which produce a given output. Each isoquant is convex downwards which reflects the law of diminishing marginal rate of substitution between production factors. For any given technology, no two isoquants can intersect. An isoquant which represents a larger output is located further out from the origin than one representing a smaller output ([4]).*

The neoclassical production function has the following four characteristics: (1) efficiency of technology, (2) technologically determined economics of scale, (3) capital intensity of a technology and (4) the ease with which capital is substituted for labor (the elasticity of substitution). There are also two types of technological change—neutral and non-neutral. The above characteristics emphasize the mathematical properties of rescaling the axis, specifying the slopes of the isoquants, comparing the slopes in different reference systems and measuring the degree of curvature of the isoquants.

The elasticity of substitution (ES) indicates how rapidly diminishing returns sets in to one factor of production when its price falls relative to another factor price. The ES formula relates the proportional change in the relative factor inputs to a proportional change in the marginal rate of substitution between labor and capital. The ES can take on any value between zero and infinity always positive. If ES is infinite, the factors are for all purposes identical and the isoquants tend to be a straight line and the factors of production are perfect substitutes for each other. If ES is zero, the factors of production combine in fixed proportions and the substitution of one factor for another is impossible and the isoquants are right-angled curves which are concave upward and parallel to the axes. The reciprocal of the ES is the slope of the logarithmic relation between the marginal rate of substitution and relative factor inputs.

The elasticity of substitution (ES) denoted as η is defined as follows: $\eta = (d\xi/\xi)/(d\rho/\rho)$ where $\xi = x/y$ and $\rho = q/p$, $z = z(x,y)$ and z is the gross output, x represents services of labor and y represents services of capital and there is a functional relationship between the inputs and the outputs. The marginal products are considered to be positive, that is, $p > 0$ and $q > 0$. Each marginal product should decrease when labor and capital increase, that is,

$$\frac{\partial^2 z}{\partial x^2} = r < 0\,, \quad \frac{\partial^2 z}{\partial y^2} = t < 0\,.$$

Each marginal product of labor, for example, would increase when capital increases, that is

$$\frac{\partial^2 z}{\partial x \partial y} = s > 0\,.$$

The slope of the isoquants are represented by the marginal rate of substitution of labor for capital, that is,

$$\frac{dy}{dx} = \frac{\frac{\partial z}{\partial x}}{\frac{\partial z}{\partial y}} = \frac{p}{q}\,.$$

Remark 8.9. Consider the following ordinary non-linear second order differential equation presented by Murray Brown in his derivation of the constant elasticity of substitution (CES) production function using the above notation

$$\eta x y \frac{d^2 y}{dx^2} = x\left(\frac{p}{q}\right)^2 - y\left(\frac{p}{q}\right)\,.$$

Now multiply both sides of the equation by $-q^3$ to get

$$-\eta xy \frac{d^2y}{dx^2} q^3 = -q^3 \left(x\frac{p^2}{q^2} - y\frac{p}{q} \right) .$$

Thus

$$\eta xy \frac{d^2y}{dx^2} q^3 = -\left(xp^2q - ypq^2 \right) .$$

Therefore

$$-\frac{d^2y}{dx^2} q^3 = -\frac{pq(xp - yq)}{\eta xy} .$$

When η, the elasticity of substitution becomes infinite, that is, $\eta \to \infty$, the above equation equals zero, that is, $-\frac{d^2y}{dx^2}q^3 = -\frac{pq(xp-yq)}{\eta xy} = 0$. This suggests that the numerator equals zero, that is,

$$-pq(xp - yq) = 0 .$$

In Example 4.5, a similar equation was developed and then it was specified that $\frac{d^2y}{dx^2} = 0$ means that a curve has no curvature. This suggests that the equation in Example 4.5 be presented as

$$-\frac{d^2y}{dx^2} q^3 = p^2t - 2spq + q^2r = 0 .$$

The above equation reflects Monge's approach. As Remark 5.26 notes, the above equation corresponds to the numerator defining the mean curvature, and further, Remark 6.2 notes that the mean curvature for a minimal surface is zero, that is, $EN - 2MF + GL = 0$. The above observations suggest the following equation or relationship

$$-pq(xp - yq) = p^2t - 2spq + q^2r .$$

This suggests that when the elasticity of substitution, η, is very large, the mean curvature can be viewed as zero, that is, a minimal surface reflects a large elasticity of substitution.

Consider the above relationship to be displayed in the following bordered determinant form

$$\begin{vmatrix} 0 & p & q \\ p & -yp & 0 \\ q & 0 & xq \end{vmatrix} = - \begin{vmatrix} 0 & p & q \\ p & r & s \\ q & s & t \end{vmatrix} .$$

The above suggests that when η is very large, $s = 0$, that is the cross terms have been reduced to zero. This suggests that a rotation of the coordinates has been performed and in a sense the second fundamental form with coefficients r, s and

t has been diagonalized [8]. The diagonalization process uses angles, that is non-algebraic notions which may be reflected in a CES with minimal surface. Also, when comparing the following elements in the respective determinants, that is,

$$r \text{ compared to } -yp \text{ and } t \text{ compared to } xq$$

the signs are not equal and the marginal product conditions presented in Definition 8.8 are violated. See Murray Brown's ([4]) comments on these marginal product conditions. The above display also indicates that r and p are related through $-y$ and that t and q are related through x. This suggests that a type of relationship exists between the coefficients of the second fundamental form (SFF) r and t and those related to the first fundamental form (FFF) p and q. The minimal surface problem mixes FFF and SFF coefficients in a certain way based on the mean curvature which is fundamentally non-invariant only the Gaussian curvature is fundamentally invariant.

REFERENCES

1. R. Abraham, J.E. Marsden, and T. Ratiu, *Manifolds, Tensor Analysis, and Applications*, second edition, (Springer-Verlag, New York, 1988), 392–448.
2. R.G.D. Allen, *Mathematical Analysis for Economists*, (St. Martin's Press, New York, 1938), 326–350.
3. F. Ayers, Jr., *Differential Equations*, (McGraw-Hill Book Company, New York, 1952), 231–293.
4. M. Brown, *On the Theory and Measurement of Technological Change*, (Cambridge University Press, New York, 1966).
5. G. Debreu, *Mathematical Economics: Twenty Papers of Gerard Debreu* with introduction by Werner Hildenbrand, (Cambridge University Press, New York, 1983), 1–29, 120–131, 186–202, 232–241.
6. J. Donato, *Donato's Research on Economics, Social Choice, Statistics and Control Theory through 1986*, (MC Printing Company, Syracuse, New York, April 1987), copyright 1987, Jerry Donato.
7. J. Donato, "Path Dependent Analysis", in: *Topics in Mathematical Analysis, A Volume Dedicated to the Memory of A.L. Cauchy*, Ed. Th. M. Rassias, (World Scientific Publishing Company, Singapore, 1989), 210–230.
8. J. Donato, "Cartan's Method and Plateau's Problem", in: *The Problem of Plateau: A Tribute to Jesse Douglas and Tibor Radó*, Ed. Th.M. Rassias, to appear in this volume.

9. J. Donato, "Entropy and Curvature", in: *Constantin Carathéodory: An International Tribute*, Ed. Th.M. Rassias, to appear.
10. H. Flanders, *Differential Forms with Applications to the Physical Sciences*, (Academic Press, Inc., New York, 1963), 1-4, 112-149.
11. G. Hadley, *Linear Algebra*, (Addison-Wesley Publishing Company, Reading, Massachusetts, 1961), 188–235.
12. M.C. Hawk, *Descriptive Geometry*, (McGraw-Hill Book Company, New York, 1962), 1–20.
13. M. Kline, *Mathematical Thought from Ancient to Modern Times*, (Oxford Univesity Press, New York, 1972), 531–543, 573–591.
14. M.M. Lipschutz, *Differential Geometry*, (McGraw-Hill Book Company, New York, 1969), 150–226.
15. S. Lipschutz, *Finite Mathematics*, (McGraw-Hill Book Company, New York, 1966), 66–80.
16. J.E. Marsden, and A.J. Tromba, *Vector Calculus*, third edition, (W.H. Freeman and Company, New York, 1988), 241–302.
17. B. O'Neill, *Elementary Differential Geometry*, (Academic Press, New York, 1966), 304–390.
18. M.H. Protter and C.B. Morrey, Jr., *College Calculus with Analytic Geometry*, (Addison-Wesley Publishing Company, Inc., Reading, Massachusetts, 1964), 286–330.
19. Th. M. Rassias, "On the Morse-Smale Index Theorem for Minimal Surfaces", in: *Differential Geometry, Calculus of Variations and Their Applications*, (eds: G.M. Rassias and Th.M. Rassias), (Marcel Dekker, Inc., Vol. 100, 1985), 429–453.
20. P.A. Samuelson, *Foundations of Economic Analysis*, Enlarged Edition, (Harvard University Press, Cambridge, Massachusetts, 1983), based on first edition of 1947, 57–89.
21. S. Smale, *Global Analysis and Economics VI—Geometric Analysis of Pareto Optima and Price Equilibria Under Classical Hypotheses*, Journal of Mathematical Economics **3**, (1976), 1–14.
22. M.R. Spiegel, M.R., *Advanced Mathematics*, (McGraw-Hill Book Company, New York, 1971), 375–398.
23. D.J. Struik, *Lectures on Classical Differential Geometry*, second edition, (Dover Publications, Inc., New York, 1961), 162– 204.
24. C. Von Westenholz, *Differential Forms in Mathematical Physics*, revised edition, (North-Holland Publishing Company, 1981), 141– 256.
25. T. Yamane, *Mathematics for Economists*, (Prentice-Hall, Inc., Englewood Cliffs, New Jersey, 1962), 74–132.

Geometry in Partial Differential Equations (pp. 91–99)
eds. A. Pràstaro and Th. M. Rassias

THE MORIMOTO PROBLEM

B. DOUBROV
International Sophus Lie Center
ul. Kalinovskogo, 61–53, Minsk, 220103, Belarus
A. KUSHNER
International Sophus Lie Center
ul. Mechnikova, 2–2, Astrahan, 414000, Russia

ABSTRACT

This work is devoted to solving the problem of finding particular solutions of one non-linear partial differential equation of second order. This problem was posed by T. Morimoto. First, the symmetry algebra of this equation is found. The work also gives the classification (up to the group of automorphisms) of its one-dimensional subalgebras and their normalizers, which are then used for finding particular solutions of the equation. The methods employed in this work can also be applied to quadrature of other partial differential equations.

1°. While classifying Monge–Ampère equations T. Morimoto [1,2] found the following normal form for one class of equations:

$$u_{q_1q_1}u_{q_2q_2} - u_{q_1q_2}^2 = (u - q_1u_{q_1} - q_2u_{q_2})^4. \tag{1}$$

He also pointed out [1] that classical methods of integration are not applicable to finding any solutions of Eq. 1 and posed the problem of finding its particular solutions.

In this work, the algebra of contact symmetries of Eq. 1 is found. We also classify (up to the group of automorphisms) its one-dimensional subalgebras and their normalizers. Using pairs "subalgebra and its normalizer", we construct some k-parameter families of particular solutions of Eq. 1.

2°. In order to find the algebra of contact symmetries of Eq. 1 we use connection (established by V. V. Lychagin [3]) between Monge–Ampère equations and differential forms on the space of 1-jets of smooth functions. Let us recall some basic definitions.

Let M be a two-dimensional smooth manifold. By J^kM denote the manifold of k-jets of smooth functions defined on M. To any differential 2-form ω on J^1M one can assign the non-linear differential operator

$$\Delta_\omega : C^\infty(M) \to \Lambda^2(M)$$

defined by

$$\Delta_\omega(h) = \sigma^*_{j_1(h)}(\omega),$$

where $h \in C^\infty(M)$ and

$$\sigma_{j_1(h)} : M \ni x \mapsto j_1(h)|_x \in J^1M$$

is the section corresponding to the 1-jets of the function h (see [4]). Let us remark the same operator can, generally speaking, be assigned to different differential forms. In order to establish a one-to-one correspondence between forms and operators, we must restrict ourselves to effective forms.

Let U_1 be the Cartan form on J^1M (in specific coordinates, $U_1 = du - p_1dq_1 - p_2dq_2$), $a \in J^1M$, and let $C_a = \ker U_{1,a}$ be a Cartan subspace.

Consider the restriction $\Omega_a = dU_1|_{C_a}$ of the differential of the form U_1 to the subspace $C_a \subset T_a(T^*M)$. Ω_a is a symplectic structure on C_a.

Let X_1 be contact vector field such that its generating function is 1. A differential 2-form $\omega \in \Lambda^2(J^1M)$ is called *effective* (see [3]) if

(1) $X_1 \lrcorner \omega = 0$,
(2) $\omega \wedge \Omega = 0$.

Let ω be an effective differential 2-form and Δ_ω the corresponding differential operator. The operator Δ_ω is called a *Monge–Ampère operator*, and the corresponding equation $\mathcal{E}_\omega = \{\Delta_\omega = 0\} \subset J^2M$ is called a *Monge–Ampère equation*. A *generalized solution* of the equation $\mathcal{E}_\omega$ is a two-dimensional submanifold $L \subset J^1M$ such that $U_1|_L = 0$ and $\omega|_L = 0$.

By $\Lambda^s(C^*)$ denote the set of differential s-forms ω, such that $L_{X_1}(\omega) = 0$. Consider the projection

$$p : \Lambda^s(J^1M) \to \Lambda^s(C^*)$$

defined by

$$p(\omega) = \omega - U_1 \wedge (X_1 \lrcorner \omega) \text{ for } \omega \in \Lambda^s(J^1M).$$

A contact transformation $\varphi : J^1M \to J^1M$ is called a *symmetry* of the Monge–Ampère equation $\mathcal{E}_\omega$, if there exists a smooth function $\lambda \in C^\infty(J^1M)$ such that $p(\varphi^*(\omega)) = \lambda\omega$. A contact vector field X_f on the manifold J^1M is called an *infinitesimal symmetry* of the Monge–Ampère equation $\mathcal{E}_\omega$, if there exists a function $h \in C^\infty(J^1M)$ such that

$$p(L_{X_f}(\omega)) = h\omega. \tag{2}$$

Using Eq. 2, one can find a system of partial differential equations such that the generating function f is a solution of the system.

3°. The effective differential 2-form corresponding to Eq. 1 has the form

$$\omega = (u - q_1p_1 - q_2p_2)^4 dq_1 \wedge dq_2 - dp_1 \wedge dp_2. \tag{3}$$

The Legendre transformation

$$q_1 \mapsto p_1,\; q_2 \mapsto p_2,\; u \mapsto u - q_1p_1 - q_2p_2,\; p_1 \mapsto -q_1,\; p_2 \mapsto -q_2$$

takes form Eq. 3 into the form

$$u^4 dp_1 \wedge dp_2 - dq_1 \wedge dq_2. \tag{4}$$

The corresponding equation has the form

$$u^4(u_{q_1q_1}u_{q_2q_2} - u^2_{q_1q_2}) = 1. \tag{5}$$

Generating functions of infinitesimal symmetries of Eq. 5 can be found from the following system of differential equations:

$$\begin{aligned}
&u^4(f_{q_2q_2} + 2p_2 f_{uq_2} + p_2^2 f_{uu}) + f_{p_1p_1} = 0,\\
&u^4(f_{q_1q_2} + 2p_1 f_{uq_2} + p_2 f_{uq_1} + p_1p_2 f_{uu}) - f_{p_1p_2} = 0,\\
&u^4(f_{q_1q_1} + 2p_1 f_{uq_1} + p_1^2 f_{uu}) + f_{p_2p_2} = 0,\\
&u^4(f_{p_1q_1} + f_{p_2q_2} + p_1 f_{p_1u} + p_2 f_{p_2u} + f_u) -\\
&\quad - 2(p_1 f_{p_1} + p_2 f_{p_2} - f) = 0.
\end{aligned}$$

The functions

$$\begin{aligned}
g_1 &= q_1(p_1q_1 + p_2q_2 - u), & g_5 &= p_1q_2,\\
g_2 &= q_2(p_1q_1 + p_2q_2 - u), & g_6 &= p_2q_1,\\
g_3 &= 3p_1q_1 - u, & g_7 &= p_1,\\
g_4 &= 3p_2q_2 - u, & g_8 &= p_2
\end{aligned}$$

form a basis of the algebra of generating functions of symmetries for Eq. 5. Using the Legendre transformation, we obtain the following basis of the algebra of generating functions of symmetries for Eq. 1:

$$\begin{aligned}
f_1 &= up_1, & f_5 &= p_2q_1,\\
f_2 &= up_2, & f_6 &= p_1q_2,\\
f_3 &= u + 2p_1q_1 - p_2q_2, & f_7 &= q_1,\\
f_4 &= u + 2p_2q_2 - p_1q_1, & f_8 &= q_2.
\end{aligned}$$

Further, using the methods described in [4], we obtain the following basis of the symmetry algebra of Eq. 1:

$$\begin{aligned}
X_1 &= -u\frac{\partial}{\partial q_1} + p_1^2\frac{\partial}{\partial p_1} + p_1p_2\frac{\partial}{\partial p_2},\\
X_2 &= -u\frac{\partial}{\partial q_2} + p_1p_2\frac{\partial}{\partial p_1} + p_2^2\frac{\partial}{\partial p_2},\\
X_3 &= -2q_1\frac{\partial}{\partial q_1} + q_2\frac{\partial}{\partial q_2} + u\frac{\partial}{\partial u} + 3p_1\frac{\partial}{\partial p_1},\\
X_4 &= q_1\frac{\partial}{\partial q_1} - 2q_2\frac{\partial}{\partial q_2} + u\frac{\partial}{\partial u} + 3p_2\frac{\partial}{\partial p_2},\\
X_5 &= -q_1\frac{\partial}{\partial q_2} + p_2\frac{\partial}{\partial p_1},\\
X_6 &= -q_2\frac{\partial}{\partial q_1} + p_1\frac{\partial}{\partial p_2},\\
X_7 &= q_1\frac{\partial}{\partial u} + \frac{\partial}{\partial p_1},\\
X_8 &= q_2\frac{\partial}{\partial u} + \frac{\partial}{\partial p_2}.
\end{aligned}$$

The commutation table for these vector fields has the form:

	X_1	X_2	X_3	X_4	X_5	X_6	X_7	X_8
X_1	0	0	$3X_1$	0	X_2	0	$\frac{2X_3+X_4}{3}$	0
X_2	0	0	0	$3X_2$	0	X_1	X_5	$\frac{X_3+2X_4}{3}$
X_3	$-3X_1$	0	0	0	$3X_5$	$-3X_6$	$3X_7$	0
X_4	0	$-3X_2$	0	0	$-3X_5$	$3X_6$	0	$3X_8$
X_5	X_2	0	$-3X_5$	$3X_5$	0	$\frac{X_3-X_4}{3}$	0	X_7
X_6	0	$-X_1$	$3X_6$	$-3X_6$	$-\frac{X_3-X_4}{3}$	0	X_8	0
X_7	$-\frac{2X_3+X_4}{3}$	$-X_5$	$-X_7$	0	0	$-X_8$	0	0
X_8	0	$-\frac{X_3+2X_4}{3}$	0	$-3X_8$	$-X_7$	0	0	0

Let $\mathfrak{g}$ denote the symmetry algebra of Eq. 1. Note that the Lie algebra $\mathfrak{g}$ is isomorphic to the Lie algebra $\mathfrak{sl}(3,\mathbb{R})$ by the mapping $\pi : \mathfrak{g} \to \mathfrak{sl}(3,\mathbb{R})$:

$$\begin{array}{ll} X_1 \mapsto -E_{31}, & X_5 \mapsto E_{23}, \\ X_2 \mapsto E_{21}, & X_6 \mapsto E_{32}, \\ X_3 \mapsto E_{11} + E_{22} - 2E_{33}, & X_7 \mapsto E_{13}, \\ X_4 \mapsto E_{11} - 2E_{22} + E_{33}, & X_8 \mapsto -E_{12}, \end{array}$$

where E_{ij} is the square matrix of order 3 whose elements e_{kl} are defined by

$$e_{kl} = \begin{cases} 1, & \text{if } l = i \text{ and } k = j, \\ 0, & \text{if } l \neq i \text{ or } k \neq j. \end{cases}$$

Let us describe (up to the group $\mathrm{Int}(\mathfrak{g})$ of inner automorphism of $\mathfrak{g}$) one-dimensional subalgebras of $\mathfrak{g}$ and also their normalizers. This is equivalent to classifying (up to conjugation) one-dimensional subalgebras of the Lie algebra $\mathfrak{sl}(3,\mathbb{R})$.

Proposition. *All one-dimensional subalgebras of the Lie algebra $\mathfrak{g}$ (viewed up to $\mathrm{Int}(\mathfrak{g})$) and their normalizers are as follows:*

(1) $\mathfrak{g}_1 = \mathbb{R}(X_3 + X_4)$,
$$N(\mathfrak{g}_1) = \mathbb{R}X_3 \oplus \mathbb{R}X_4 \oplus \mathbb{R}X_5 \oplus \mathbb{R}X_6;$$
(2) $\mathfrak{g}_2 = \mathbb{R}(X_3 + \mu X_4)$ $(1 < \mu \leqslant 2)$,
$$N(\mathfrak{g}_2) = \mathbb{R}X_3 \oplus \mathbb{R}X_4;$$
(3) $\mathfrak{g}_3 = \mathbb{R}(\lambda X_3 + X_2 + X_8)$ $(\lambda \geqslant 0)$,
$$N(\mathfrak{g}_3) = \mathbb{R}X_3 \oplus \mathbb{R}(X_2 + X_8);$$
(4) $\mathfrak{g}_4 = \mathbb{R}(X_3 + X_8)$,
$$N(\mathfrak{g}_4) = \mathbb{R}X_3 \oplus \mathbb{R}X_8;$$
(5) $\mathfrak{g}_5 = \mathbb{R}(X_5 - X_8)$,
$$N(\mathfrak{g}_5) = \mathbb{R}(2X_3 + X_4) \oplus \mathbb{R}(X_5 - X_8) \oplus \mathbb{R}X_7;$$
(6) $\mathfrak{g}_6 = \mathbb{R}X_5$,
$$N(\mathfrak{g}_6) = \mathbb{R}X_3 \oplus \mathbb{R}X_4 \oplus \mathbb{R}X_5 \oplus \mathbb{R}X_8.$$

4°. Now let us describe the method we use in order to find particular solutions of Monge–Ampère equations. Let $\mathfrak{g}$ be the Lie algebra of infinitesimal symmetries of an equation $\mathcal{E}_\omega$. The set of generating functions of symmetries of $\mathcal{E}_\omega$ can be turned into the Lie algebra by introducing the Lagrange bracket

$$[f,g]=U_1([X_f,X_g]),$$

where X_f, $X_g \in \mathfrak{g}$ (see [4]). We denote the corresponding Lie algebra of generating functions by $\tilde{\mathfrak{g}}$. A generalized solution L of the equation $\mathcal{E}_\omega$ is called *f-automodel*, if L is invariant under shifts along field X_f. Let us describe how one can construct automodel solutions. First of all, note that an f-automodel solution L belongs to the manifold $E=\{f=0\}\subset J^1M$. This follows from the formula $U_1(X_f)=f$ (see [4]). Assume that there exists a submanifold $\Gamma\subset E$ of codimension 1 such that Γ is transversal to the field X_f and any trajectory of X_f meets Γ at only one point. Since X_f preserves the solution L, we see that, starting from $L_\Gamma=L\cap\Gamma$, we can uniquely reconstruct the solution itself. Note that $\dim L_\Gamma=1$. If Γ contains no singular points of the equation $E=\{f=0\}$, then the restriction of the Cartan form U_1 to Γ, $\omega_1=U_1|_\Gamma$, is a contact structure on Γ, and L_Γ is an integral manifold of the 1-form ω_1. In addition, since $\omega|_L=0$, L_Γ is an integral manifold of the 1-form $\omega_2=(X_f\lrcorner\,\omega)|_\Gamma$. Thus, the problem of constructing an f-automodel solution reduced to that of constructing an integral manifold of the completely integrable distribution $\mathcal{F}(\omega_1,\omega_2)$ corresponding to the 1-forms ω_1 and ω_2. The distribution $\mathcal{F}(\omega_1,\omega_2)$ is completely integrable, since it is one-dimensional. Not any completely integrable distribution $\mathcal{F}$ can be quadratured. However, if there exists k-dimensional solvable algebra of shuffling symmetries of $\mathcal{F}$ ($k=\operatorname{codim}\mathcal{F}$), it can be done (see [5]). In our case, it is sufficient to know a 2-dimensional algebra of shuffling symmetries of the distribution $\mathcal{F}(\omega_1,\omega_2)$.

Recall (see [5]) that a vector field X is called an *infinitesimal symmetry* of the distribution $\mathcal{F}(\omega_1,\omega_2)$ if $L_X(\omega_i)\wedge\omega_1\wedge\omega_2=0$ $(i=1,2)$. An infinitesimal symmetry X of the distribution $\mathcal{F}$ is called *characteristic* wherever X belongs to $\mathcal{F}$. Let $\operatorname{Char}\mathcal{F}$ denote the set of all characteristic symmetries of $\mathcal{F}$ and $\operatorname{Sym}\mathcal{F}$ the Lie algebra of infinitesimal symmetries of $\mathcal{F}$. Then $\operatorname{Char}\mathcal{F}$ is an ideal in $\operatorname{Sym}\mathcal{F}$. The quotient algebra $\operatorname{Sym}\mathcal{F}/\operatorname{Char}\mathcal{F}$ is called the *algebra of shuffling symmetries* of the distribution $\mathcal{F}$ and denoted by $\operatorname{Shuf}\mathcal{F}$.

Suppose $\mathfrak{h}$ is a one-dimensional subalgebra of the Lie algebra $\tilde{\mathfrak{g}}$ such that the dimension of its normalizer $N_\mathfrak{h}$ is equal to 3. Let $f\in\mathfrak{h}$, $g\in N_\mathfrak{h}$. Then $[f,g]=\lambda f$, where $\lambda\in C^\infty(J^1M)$. On the other hand, $[g,f]=X_g(f)-X_1(g)f$. Therefore $X_g(f)=\mathfrak{s}f$, where $\mathfrak{s}=X_1(g)-f$. Thus, the hypersurface $E=\{f=0\}$ is invariant under the action of those vector fields whose generating functions belong to $N_\mathfrak{h}$.

Let $\mathfrak{h}=\mathbb{R}X_f$, $N_\mathfrak{h}=\mathbb{R}X_f\oplus\mathbb{R}X_{g_1}\oplus\mathbb{R}X_{g_2}$. Suppose X_{g_1} and X_{g_2} are tangent vector fields of the manifold Γ. Then any element of $N_\mathfrak{h}/\mathfrak{h}$ is a symmetry of the distribution $\mathcal{F}(\omega_1,\omega_2)$.

5°. Consider the subalgebra $\mathfrak{h}=\mathbb{R}(f_5-f_8)$ in the Lie algebra $\tilde{\mathfrak{g}}$ of generating functions of symmetries of Eq. 1. The normalizer of $\mathfrak{h}$ has the form

$$N_\mathfrak{h}=\mathbb{R}(2f_3+f_4)\oplus\mathbb{R}(f_5-f_8)\oplus\mathbb{R}f_7.$$

The corresponding vector fields are:

$$X_f = X_{f_5-f_8} = -q_1\frac{\partial}{\partial q_2} - q_2\frac{\partial}{\partial u} + p_2\frac{\partial}{\partial p_1} - \frac{\partial}{\partial p_2},$$
$$X_{g_1} = \frac{1}{3}X_{2g_3+g_4} = -q_1\frac{\partial}{\partial q_1} + u\frac{\partial}{\partial u} + 2p_1\frac{\partial}{\partial p_1} + p_2\frac{\partial}{\partial p_2},$$
$$X_{g_2} = X_{f_7} = q_1\frac{\partial}{\partial u} + \frac{\partial}{\partial p_1}.$$

Let $\Gamma = \{p_2 = 0,\, q_2 = 0\}$. The manifold Γ lies on the hypersurface $E = \{f = p_2q_1 - q_2 = 0\}$. Then

$$\omega_0 = U_1|_\Gamma = du - p_1dq_1,$$
$$\omega_1 = (X_f \lrcorner \omega)|_\Gamma = -q_1(u_1 - p_1q_1)^4dq_1 + dp_1.$$

The restrictions of the vector fields X_{g_1} and X_{g_2} to Γ are

$$X_1 = X_{f_1}|_\Gamma = -q_1\frac{\partial}{\partial q_1} + u\frac{\partial}{\partial u} + 2p_1\frac{\partial}{\partial p_1},$$
$$X_2 = X_{f_2}|_\Gamma = q_1\frac{\partial}{\partial u} + \frac{\partial}{\partial p_1}.$$

Since now we shall use the algorithm described in [5]. Put $x = q_1$, $p = p_1$, $v = u$. Then

$$\omega_0 = dv - pdx, \qquad \omega_1 = dp - x(v - px)dx.$$

$$X_1 = -x\frac{\partial}{\partial x} + v\frac{\partial}{\partial v} + 2p\frac{\partial}{\partial p},$$
$$X_2 = x\frac{\partial}{\partial v} + \frac{\partial}{\partial p}.$$

Consider the matrix

$$\|\omega_i(X_j)\| = \begin{pmatrix} v + xp & x \\ 2p + x^2(v - xp)^4 & 1 \end{pmatrix}.$$

Its determinant has the form

$$T = \det\|\omega_i(X_j)\| = (v - xp)(1 - x^3(v - xp)^3).$$

Suppose $T \neq 0$. Then the matrix $\|\omega_i(X_j)\|$ is invertible and

$$\|\omega_i(X_j)\|^{-1} = T^{-1}\begin{pmatrix} 1 & -x \\ -2p - x^2(v - xp)^4 & v + xp \end{pmatrix}.$$

Define the form ω_0' and ω_1' by

$$\begin{pmatrix} \omega_0' \\ \omega_1' \end{pmatrix} = \|\omega_i(X_j)\|^{-1} \begin{pmatrix} \omega_0 \\ \omega_1 \end{pmatrix},$$

so that $\omega_i'(X_j) = \delta_{ij}$, δ_{ij} being the Kronecker delta. Then

$$\omega_0' = \alpha dx + \beta dv + \gamma dp,$$

where

$$\begin{aligned} \alpha =& -p(v-xp)^{-1}(1-x^3(v-xp)^3)^{-1} + \\ &+ x^2(v-xp)^3(1-x^3(v-xp)^3)^{-1}, \\ \beta =&(v-xp)^{-1}(1-x^3(v-xp)^3)^{-1}, \\ \gamma =& -x(v-xp)^{-1}(1-x^3(v-xp)^3)^{-1}; \end{aligned}$$

and

$$\omega_1' = \lambda dx + \mu dv + \nu dp,$$

where

$$\begin{aligned} \lambda &= (2p^2 - xv(v-xp)^4)(v-xp)^{-1}(1-x^3(v-xp)^3)^{-1}, \\ \mu &= -(2p + x^2(v-xp)^4)(v-xp)^{-1}(1-x^3(v-xp)^3)^{-1}, \\ \nu &= (v+xp)(v-xp)^{-1}(1-x^3(v-xp)^3)^{-1}. \end{aligned}$$

The form ω_0' is locally closed, and, under assumption that

$$1 - x(v-xp) \neq 0, \tag{6}$$

we have

$$\int \omega_0' = \ln \frac{C|v-xp|}{|1-x^3(v-xp)^3|^{1/3}}.$$

In our case, it is convenient to consider the solution $F = \exp \int \omega_0'$,

$$F = \frac{C(v-xp)}{(1-x^3(v-xp)^3)^{1/3}}.$$

Let H_1 be a level surface of the function F, i.e.

$$H_1 = \{(v-xp)(1-x^3(v-xp)^3)^{-1/3} = C_1\}. \tag{7}$$

Suppose $C_1 \neq 0$ and put $a = C_1^{-1}$. From Eq. 7 it follows that

$$p = x^{-1}(v - (x^3 + a^3)^{-1/3}).$$

The restriction of the 1-form ω_1' to H_1 has the form

$$\omega_1'|_{H_1} = \rho dx + \tau dv,$$

where

$$\rho = x^{-2}(x^3 + a^3)^{-1/3} - x^{-2}v,$$
$$\tau = x^{-1}.$$

The form $\omega_1'|_{H_1}$ is locally closed. Integrating this form, we obtain

$$x^{-1}v + \int x^{-2}(x^3 + a^3)^{-1/3}dx = b,$$

where b is an arbitrary constant. It follows that

$$v = -x\int x^{-2}(x^3 + a^3)^{-1/3}dx + bx.$$

The corresponding f-invariant solution of Eq. 1 has the form

$$u = \frac{1}{2}q_1^{-1}q_2^2 - q_1\int q_1^{-2}(q_1^3 + a^3)^{-1/3}dq_1 + bq_1.$$

Canceling restrictions Eq. 6 and $C_1 \neq 0$, we obtain two more solutions:

$$u = \frac{1}{2}q_1^{-1}q_2^2 + \frac{1}{2}q_1^{-1} + Cq_1,$$
$$u = \frac{1}{2}q_1^{-1}q_2^2 + Cq_1.$$

One can obtain other particular solutions of Eq. 1, using the subalgebras $\mathbb{R}f_5$ and $\mathbb{R}(f_3 + f_4)$ of the Lie algebra $\tilde{\mathfrak{g}}$. In the first case we obtain trivial solutions $u = Cq_2$. In the second case we obtain the following solutions:

$$u = \frac{1}{q_1^2 v(t)},$$

where $t = -q_2/2q_1$ and the function $v(t)$ can be determined from the functional equation

$$t = \int (\frac{c_1}{v} + 108)^{-1/2}dv.$$

5°. In our case it is possible to write out the global action of the symmetry group of equation (1): the group $SL(3,\mathbb{R})$ acts naturally on $\mathbb{R}^3$. In particular, this implies

Proposition. *Let $u(q_1, q_2)$ be a solution of equation (1) and*

$$\begin{pmatrix} a & b & c \\ d & e & f \\ g & h & k \end{pmatrix} \in SL(3,\mathbb{R}).$$

Consider the functions $\varphi : \mathbb{R}^2 \longrightarrow \mathbb{R}^2$ and $\psi : \mathbb{R}^2 \longrightarrow \mathbb{R}$ defined by

$$\varphi(x,y) = (ax + by + cu(x,y),\ dx + ey + fu(x,y)),$$
$$\psi(x,y) = gx + hy + ku(x,y).$$

Then the function $v(q_1, q_2) = \psi(\varphi^{-1}(q_1, q_2))$ is also a solution of equation (1).

Corollary. *If $u(q_1, q_2)$ is a solution of equation* (1) *and*

$$\begin{pmatrix} a & b \\ c & d \end{pmatrix} \in GL(2, \mathbb{R}), \quad e, f \in \mathbb{R},$$

then the function

$$v(q_1, q_2) = \frac{u(aq_1 + bq_2,\ cq_1 + dq_2)}{ad - bc} + eq_1 + fq_2$$

is also a solution of Eq. 1.

The authors are very grateful to A. V. Churyumov, S. V. Duzhin, B. P. Komrakov, V. V. Lychagin, T. Morimoto, and V. N. Rubtsov for their contributions. Thanks are also due to Z. Belyankina and L. Levin for the help in preparing the manuscript.

REFERENCES

1. T. Morimoto, *La dèomètrie des èquations de Monge–Ampère.* C.R. Acad. Sc. Paris **289** (2 juillet 1979), Sèrie A, 25–28..
2. T. Morimoto, *Open Problems in str. theory of Non-linear Integrable Differential and Difference Systems.* The 15 international symposium held at Katata (1984), 27–29.
3. V. V. Lychagin, *Contact Geometry and Non-linear Second Order Differential Equations.* Uspekhi Mat. Nauk **34** (1979); English transl., Russian Math. Surveys **34** (1979), 149–180.
4. V. V. Lychagin, *The Local Classification of Non-linear Differential Equations in First Order Partial Derivatives.* Uspekhi Math. Nauk **30** (1975), 101–171; Englisn transl., Russian Math. Surveys **30** (1975), 105–175.
5. S. V. Duzhin, V. V. Lychagin, *Symmetries of Distributions and Quadrature of Ordinary Differential Equations.* Acta. Appl. Math **24** (1991), 25–27.

Geometry in Partial Differential Equations (pp. 100–113)
eds. A. Pràstaro and Th. M. Rassias

ASYMPTOTIC EXPANSIONS IN SPECTRAL GEOMETRY

PETER B. GILKEY
Department of Mathematics, University of Oregon
Eugene Oregon 97403 USA

ABSTRACT

Let M be a compact Riemannian manifold with a smooth boundary. We examine various asymptotic expansions arising in the heat equation trace and the heat content in spectral geometry. MOS classification number 58G25.

§1 Introduction

In this brief note, we will summarize some of the asymptotic formulas which play a crucial role in spectral geometry. In §1, we establish notation. In §2, we discuss the asymptotics of the heat equation for an operator of Laplace type. In §3, we discuss heat content asymptotics; these are relatively new and their precise relationship to the heat equation asymptotics is not well understood. In §4, we discuss the asymptotics associated with an operator of Dirac type. The Mellin transform relates the invariants of §2 to the zeta function and the invariants of §4 to the eta function; both the zeta and eta functions play an important role in index theory. We conclude in §5 by discussing the heat equation asymptotics for second order operators which do not have scalar leading symbol and which are not of Laplace type. The literature on this subject is extensive; we present a short bibliography at the end of this paper and refer to [10, 11] for additional references.

Let M be a compact Riemannian manifold with smooth boundary ∂M. Let $x = (x^1, ..., x^m)$ be a system of local coordinates on M. Near ∂M, we choose coordinates $x = (y, x^m)$ so x^m is the geodesic distance to the boundary and so the curves $y(s) = (y, s)$ are unit speed geodesics perpendicular to ∂M. Greek indices ν, μ

(resp. Roman indices i, j) range from 1 through m and index coordinate (resp. orthonormal) frames for the tangent $T(M)$ and cotangent $T^*(M)$ bundles of M. Greek indices α, β (resp. Roman indices a, b) range from 1 through $m-1$ and index coordinate (resp. orthonormal) frames for $T(\partial M)$ and $T^*(\partial M)$. We adopt the Einstein convention and sum over repeated indices. Thus

$$ds^2 = g_{\nu\mu} dx^\nu \circ dx^\mu. \tag{1.1}$$

Let dx and dy be the Riemannian measures on M and ∂M.

Let R_{ijkl} be the curvature tensor of the Levi-Civita connection ∇^M of M; with our sign convention $R_{1212} = -1$ for the standard sphere in $\mathbf{R}^3$. Let $\tau = R_{ijji}$ and $\rho_{ij} = R_{ikkj}$ be the scalar curvature and the Ricci tensor. Let Γ be the Christoffel symbols of ∇^M;

$$\begin{aligned} \Gamma_{\nu\mu}{}^\sigma &= \tfrac{1}{2} g^{\sigma\epsilon}(\partial_\nu g_{\mu\epsilon} + \partial_\mu g_{\nu\epsilon} - \partial_\epsilon g_{\nu\mu}) \\ R_{\nu_1\nu_2\nu_3}{}^{\nu_4} &= \partial_{\nu_1}\Gamma_{\nu_2\nu_3}{}^{\nu_4} - \partial_{\nu_2}\Gamma_{\nu_1\nu_3}{}^{\nu_4} + \Gamma_{\nu_1\mu_2}{}^{\nu_4}\Gamma_{\nu_2\nu_3}{}^{\mu_2} - \Gamma_{\nu_2\mu_2}{}^{\nu_4}\Gamma_{\nu_1\nu_3}{}^{\mu}. \end{aligned} \tag{1.2}$$

Let V be a smooth vector bundle over M, let ∇^D be a connection on V, and let $E \in C^\infty(\mathrm{End}(V))$. Let

$$D = -(\Sigma_{\nu,\mu} g^{\nu\mu} \nabla^D_\nu \nabla^D_\mu + E) \tag{1.3}$$

be an operator of Laplace type on $C^\infty(V)$; every operator of Laplace type can be put in this form. Let ω be the connection 1-form and Ω the curvature tensor in some local system. If we decompose

$$\begin{aligned} D &= -(g^{\nu\mu}\partial_\nu\partial_\mu + p^\nu\partial_\nu + q), \text{ then :} \\ \omega_\nu &= \tfrac{1}{2} g_{\nu\mu}(P^\mu + g^{\sigma\epsilon}\Gamma_{\sigma\epsilon}{}^\mu), \\ \Omega_{\nu\mu} &= \partial_\nu\omega_\mu - \partial_\mu\omega_\nu + \omega_\nu\omega_\mu - \omega_\mu\omega_\nu, \\ E &= q - g^{\nu\mu}(\partial_\nu\omega_\mu + \omega_\nu\omega_\mu - \omega_\sigma\Gamma_{\nu\mu}{}^\sigma). \end{aligned} \tag{1.4}$$

Example 1: Let $\Delta = -g^{-1}\partial_\nu g g^{\nu\mu}\partial_\mu$ for $g = \det(g_{\nu\mu})^{1/2}$ be the scalar Laplacian. The associated connection is flat and the associated endomorphism vanishes. More generally, if

$$\Delta_p = \delta_p d_p + d_{p-1}\delta_{p-1} \tag{1.5}$$

is the Laplacian on the space of smooth p forms $C^\infty(\Lambda^p)$, then $\nabla^D = \nabla^M$ and E is given by the Weitzenböch formulas in terms of curvature; $E^{\Delta_1} = -\rho$.

Example 2: Let M be spin and let D be the spin Laplacian. Then ∇^D is the spin connection and $E = -\frac{1}{4}\tau$ by the Lichnerowicz formula.

Let ";" and ":" be multiple covariant differentiation with respect to ∇^M and

$\nabla^{\partial M}$. When sections of bundles built from V are involved, ";" and ":" will mean

$$\nabla^M \otimes 1 + 1 \otimes \nabla^D \text{ and } \nabla^{\partial M} \otimes 1 + 1 \otimes \nabla^D. \tag{1.6}$$

Thus $D\psi = -(\psi_{;ii} + E\psi)$. The second fundamental form L measures the difference between ∇^M and $\nabla^{\partial M}$;

$$L_{\alpha\beta} = (\partial_m, \nabla_{\partial_\alpha}\partial_\beta) = \Gamma_{\alpha\beta}{}^m = -\tfrac{1}{2}\partial_m g_{\alpha\beta}. \tag{1.7}$$

Thus, for example,

$$f_{;ab} = f_{:ab} - L_{ab} f_{;m}. \tag{1.8}$$

Since the boundary of M is non empty, we must impose suitable boundary conditions. Let $S \in C^\infty(\text{End}(V))$ and let e_m be the inward pointing unit normal. Let

$$\mathcal{B}_S^+ f = (\nabla_{e_m}^D f + Sf)|_{\partial M} \text{ and } \mathcal{B}^- f = f|_{\partial M} \tag{1.9}$$

be Neumann and Dirichlet boundary conditions. More generally, we can mix these boundary conditions. Near ∂M, we assume given a splitting $V = V^+ \oplus V^-$ as the direct sum of two vector bundles; the trivial splitting $V = V^+$ or $V = V^-$ is permitted. Let $\Pi^\pm$ be the associated projections on $V^\pm$ and let $\chi = \Pi^+ - \Pi^-$. Let

$$\mathcal{B} = \mathcal{B}(\chi, S, \nabla^D)f = \{\Pi^+(\nabla_m^D + S)\Pi^+ f\}|_{\partial M} \oplus \Pi^- f|_{\partial M}. \tag{1.10}$$

Absolute and relative boundary conditions for the DeRham complex are such boundary conditions; these are important in index theory and are boundary conditions of this type. Let

$$i^* : \Lambda M \to \Lambda \partial M. \tag{1.11}$$

be the pullback via the inclusion $i : \partial M \to M$. If $\omega \in C^\infty \Lambda M$, decompose

$$\omega|_{\partial M} = \omega^T + \omega^N \text{ where } \omega^T = i^*\omega \in C^\infty \Lambda \partial M. \tag{1.12}$$

Let π^T and π^N be the corresponding projection operators. Let

$$\mathcal{B}^a \omega = \pi^N \omega \oplus \pi^N d\omega. \tag{1.13}$$

Dually, let $\star$ be the Hodge operator and let $\mathcal{B}^r \omega = \mathcal{B}^a \star \omega$. We say ω satisfies absolute boundary conditions if $\mathcal{B}^a \omega = 0$; dually, ω satisfies relative boundary conditions if $\mathcal{B}^r \omega = 0$. By the Hodge decomposition theorem,

$$\begin{aligned} \text{kernel}(\Delta_{k,\mathcal{B}^a}) &= H^k(M), \\ \text{kernel}(\Delta_{k,\mathcal{B}^r}) &= H^k(M, \partial M). \end{aligned} \tag{1.14}$$

Let $\mathcal{B}$ be a suitable boundary condition and let $h(x,t) = e^{-tD_\mathcal{B}} f_1(x)$ be the solution for $t>0$ to the equations:

$$\partial_t h(x,t)=-Dh(x,t),\ \mathcal{B}h(x,t)=0,\ \text{and}\ \lim_{t\to 0+} h(x,t)=f_1(x). \tag{1.15}$$

§2 Heat Equation Asymptotics

Standard elliptic methods [12, 28, 29, 51] show there is an asymptotic series as $t\to 0^+$ of the form:

$$Tr_{L^2}(fe^{-tD_{\mathcal{B}}})\simeq \sum_{n=0}^{\infty} a_n(f,D,\mathcal{B})t^{(n-m)/2}; \tag{2.1}$$

the $a_n(f,D,\mathcal{B})$ are expressible in terms of local information and are defined for quite general elliptic boundary value problems. They have been studied by many authors [2, 3, 13, 14, 18, 21, 22, 23, 24, 25, 31, 33, 34, 35, 37, 38, 39, 40, 42, 41, 43, 44, 46, 47, 49, 50, 52, 53, 54].

Suppose for the moment $D_{\mathcal{B}}$ is self-adjoint and scalar. Let $\{\lambda_\nu,\phi_\nu\}$ be a spectral resolution of $D_{\mathcal{B}}$; the ϕ_ν are smooth sections to V which form a complete orthonormal basis for $L^2(V)$ satisfying:

$$D\phi_\nu=\lambda_\nu\phi_\nu \text{ and } \mathcal{B}\phi_\nu=0.$$

We recover the standard asymptotics of the heat equation by setting $f=1$;

$$Tr_{L^2}e^{-tD_{\mathcal{B}}}=\Sigma_\nu e^{-t\lambda_\nu}\simeq \sum_{n=0}^{\infty} a_n(1,D,\mathcal{B})t^{(n-m)/2}. \tag{2.2}$$

This shows the $a_n(1,D,\mathcal{B})$ are spectral invariants of $D_{\mathcal{B}}$ which are locally computable; they play a crucial role in spectral geometry.

The heat kernel is the fundamental solution of the heat equation:

$$\begin{aligned} K(t,x,y)&=\Sigma_\nu e^{-t\lambda_\nu}\phi_\nu(x)\phi_\nu(y)\\ e^{-tD_{\mathcal{B}}}\phi(x)&=\textstyle\int_M K(t,x,y)\phi(y)dy. \end{aligned} \tag{2.3}$$

The kernel K behaves like a distribution near the boundary on the diagonal. By studying

$$Tr_{L^2}(fe^{-tD_{\mathcal{B}}})=\textstyle\int_M f(x)K(t,x,x)dx \tag{2.4}$$

we recover this local behavior as $t\to 0^+$. This is the origin of the terms involving the normal derivatives of the function f in the formulas below. The interior asymptotics are independent of the boundary condition. We refer to [14] for the following results:

Theorem 2.1: *Let $\mathcal{B}^-$ define Dirichlet boundary conditions.*

(a) $a_0(f,D,\mathcal{B}^-)=(4\pi)^{-m/2}\int_M Tr(f)dx.$

(b) $a_1(f,D,\mathcal{B}^-)=-\frac{1}{4}(4\pi)^{-(m-1)/2}\int_{\partial M} Tr(f)dy.$

(c) $a_2(f,D,\mathcal{B}^-)=(4\pi)^{-m/2}\frac{1}{6}\{\int_M Tr(6fE+f\tau)dx+\int_{\partial M} Tr(2fL_{aa}-3f_{;m})\}dy.$

(d) $a_3(f,D,\mathcal{B}^-) = -(4\pi)^{-(m-1)/2}\frac{1}{384}\{\int_{\partial M}Tr\{f(96E+16\tau+8R_{amam}$
$+7L_{aa}L_{bb}-10L_{ab}L_{ab})-30f_{;m}L_{aa}+24f_{;mm}\}dy\}$.

(e) $a_4(f,D,\mathcal{B}^-) = (4\pi)^{-m/2}\frac{1}{360}\{\int_M Tr\{f(60E_{;kk}+60\tau E+180E^2+30\Omega_{ij}\Omega_{ij}$
$+12\tau_{;kk}+5\tau^2-2\rho^2+2R^2)\}dx+\int_{\partial M}Tr\{f(-120E_{;m}$
$-18\tau_{;m}+120EL_{aa}+20\tau L_{aa}+4R_{amam}L_{bb}-12R_{ambm}L_{ab}$
$+4R_{abcb}L_{ac}+24L_{aa:bb}+\frac{40}{21}L_{aa}L_{bb}L_{cc}$
$-\frac{88}{7}L_{ab}L_{ab}L_{cc}+\frac{320}{21}L_{ab}L_{bc}L_{ac})+f_{;m}(-180E$
$-30\tau-\frac{180}{7}L_{aa}L_{bb}+\frac{60}{7}L_{ab}L_{ab})+24f_{;mm}L_{aa}-30f_{;iim})\}dy\}$.

Theorem 2.2: *Let $\mathcal{B}_S^+$ define modified Neumann boundary conditions.*

(a) $a_0(f,\mathcal{B}_S^+) = (4\pi)^{-m/2}\int_M Tr(f)dx$.

(b) $a_1(f,D,\mathcal{B}_S^+) = \frac{1}{4}(4\pi)^{-(m-1)/2}\int_{\partial M}Tr(f)dy$.

(c) $a_2(f,D,\mathcal{B}_S^+) = (4\pi)^{-m/2}\frac{1}{6}\{\int_M Tr(6fE+f\tau)dx+\int_{\partial M}Tr(2fL_{aa}+3f_{;m}$
$+12fS)dy\}$.

(d) $a_3(f,D,\mathcal{B}_S^+) = (4\pi)^{-(m-1)/2}\frac{1}{384}\int_{\partial M}Tr\{f(96E+16\tau+8R_{amam}$
$+13L_{aa}L_{bb}+2L_{ab}L_{ab}+96SL_{aa}+192S^2)+f_{;m}(6L_{aa}+96S)+24f_{;mm}\}dy$.

(e) $a_4(f,D,\mathcal{B}_S^+) = (4\pi)^{-m/2}\frac{1}{360}\{\int_M Tr\{f(60E_{;kk}+60\tau E+180E^2+30\Omega_{ij}\Omega_{ij}$
$+12\tau_{;kk}+5\tau^2-2\rho^2+2R^2)\}dx+\int_{\partial M}Tr\{f(240E_{;m}+42\tau_{;m}$
$+24L_{aa:bb}+120EL_{aa}+20\tau L_{aa}+4R_{amam}L_{bb}$
$-12R_{ambm}L_{ab}+4R_{abcb}L_{ac}+\frac{40}{3}L_{aa}L_{bb}L_{cc}+8L_{ab}L_{ab}L_{cc}$
$+\frac{32}{3}L_{ab}L_{bc}L_{ac}+720SE+120S\tau$
$+144SL_{aa}L_{bb}+48SL_{ab}L_{ab}+480S^2L_{aa}+480S^3+120S_{:aa})$
$+f_{;m}(180E+30\tau+12L_{aa}L_{bb}+12L_{ab}L_{ab}+72SL_{aa}+240S^2)$
$+f_{;mm}(24L_{aa}+120S)+30f_{;iim}\}dy\}$.

Theorem 2.3: *Let $\mathcal{B}=\mathcal{B}(\chi,S,\nabla^D)$ define mixed boundary conditions.*

(a) $a_0(f,D,\mathcal{B}) = (4\pi)^{-m/2}\int_M fTr(1)dx$.

(b) $a_1(f,D,\mathcal{B}) = \frac{1}{4}(4\pi)^{-(m-1)/2}\int_{\partial M}fTr(\chi)dy$.

(c) $a_2(f,D,\mathcal{B}) = \frac{1}{6}(4\pi)^{-m/2}\{\int_M fTr(6E+\tau)dx+\int_{\partial M}\{fTr(2L_{aa}+12S)$
$+f_{;m}Tr(3\chi)\}dy\}$.

(d) $a_3(f,D,\mathcal{B}) = \frac{1}{384}(4\pi)^{-(m-1)/2}\int_{\partial M}\{fTr\{96\chi E+16\chi\tau+8f\chi R_{amam}$
$+(13\Pi^+-7\Pi^-)L_{aa}L_{bb}+(2\Pi^++10\Pi^-)L_{ab}L_{ab}+96SL_{aa}$
$+192S^2-12\chi_{:a}\chi_{:a}\}+f_{;m}Tr\{(6\Pi^++30\Pi^-)L_{aa}+96S\}+f_{;mm}Tr(24\chi)\}dy$.

(e) $a_4(f,D,\mathcal{B}) = \frac{1}{360}(4\pi)^{-m/2}\{\int_M fTr\{60E_{;kk}+60\tau E+180E^2+30\Omega_{ij}\Omega_{ij}+12\tau_{;kk}$

$$+5\tau^2-2\rho^2+2R^2\}dx+\int_{\partial M}\{fTr\{(240\Pi^+-120\Pi^-)E_{;m}+(42\Pi^+-18\Pi^-)\tau_{;m}$$
$$+24L_{aa:bb}+120EL_{aa}+20\tau L_{aa}+4R_{amam}L_{bb}-12R_{ambm}L_{ab}$$
$$+4R_{abcb}L_{ac}+\tfrac{1}{21}((280\Pi^++40\Pi^-)L_{aa}L_{bb}L_{cc}$$
$$+(168\Pi^+-264\Pi^-)L_{ab}L_{ab}L_{cc}+(224\Pi^++320\Pi^-)L_{ab}L_{bc}L_{ac})+720SE$$
$$+120S\tau+144SL_{aa}L_{bb}+48SL_{ab}L_{ab}+480S^2L_{aa}+480S^3$$
$$+120S_{:aa}+60\chi\chi_{:a}\Omega_{am}-42\chi_{:a}\chi_{:a}L_{bb}+6\chi_{:a}\chi_{:b}L_{ab}-120\chi_{:a}\chi_{:a}S\}$$
$$+f_{;m}Tr\{180\chi E+30\chi\tau+\tfrac{1}{7}(84\Pi^+-180\Pi^-)L_{aa}L_{bb}$$
$$\tfrac{1}{7}(84\Pi^++60\Pi^-)L_{ab}L_{ab}+72SL_{aa}+240S^2-18\chi_{:a}\chi_{:a}\}$$
$$+f_{;mm}Tr(24L_{aa}+120S)+f_{;iim}Tr(30\chi)\}dy\}.$$

Remark: We recover Theorems 2.1 and 2.2 by setting $\chi=\pm I_V$.

Partial information about all the coefficients exists. The following result is crucial to compactness results in spectral geometry; see [25, 46]:

Theorem 2.4: *Let the boundary of M be empty and let $n\geq 3$. Modulo cubic and higher degree terms which have fewer covariant derivatives,*

$$\begin{aligned}a_{2n}(1,D)=&(-1)^n\frac{1}{(4\pi)^{m/2}2^{n+1}\cdot 1\cdot 3\cdot\ldots\cdot(2n+1)}\int_M\\ &Tr\{(n^2-n-1)|\nabla^{n-2}\tau|^2+2|\nabla^{n-2}\rho|^2\\ &+4(2n+1)(n-1)\nabla^{n-2}\tau\cdot\nabla^{n-2}E\\ &+2(2n+1)\nabla^{n-2}\Omega\cdot\nabla^{n-2}\Omega\\ &+4(2n+1)(2n-1)\nabla^{n-2}E\cdot\nabla^{n-2}E+\ldots\}dx.\end{aligned}$$

§3 Heat Content Asymptotics

Denote the natural pairing between V and the dual bundle V^* by $\langle\cdot,\cdot\rangle$. Let $\tilde{D}$ be the formal adjoint of D on $C^\infty(V^*)$. and let $\tilde{\mathcal{B}}$ be the adjoint boundary condition. To study the short time behavior of the heat content $h=e^{-tD_{\mathcal{B}}}f_1$ for $f_1\in C^\infty(V)$, we introduce an auxiliary smooth test function $f_2\in C^\infty(V^*)$ and define:

$$\beta(f_1,f_2,D,\mathcal{B})(t)=\int_M\langle e^{-tD_{\mathcal{B}}}f_1,f_2\rangle dx. \tag{3.1}$$

If $D_{\mathcal{B}}$ is self-adjoint and if $\{\lambda_\nu,\phi_\nu\}$ is a spectral resolution, let

$$\gamma_\nu(f)=\int_M\langle f,\phi_\nu\rangle dx \tag{3.2}$$

be the Fourier coefficient. Then

$$\beta(f_1,f_2,D,\mathcal{B})(t)=\Sigma_\nu e^{-t\lambda_\nu}\gamma_\nu(f_1)\gamma_\nu(f_2). \tag{3.3}$$

For example, let $D = \Delta_0$ be the scalar Laplacian. Let $f_1 \in C^\infty(M)$ represent the initial temperature of M. We keep the boundary of M at temperature 0 for $t > 0$; this corresponds to Dirichlet boundary conditions $\mathcal{B}^-$. Then $\beta(f_1, 1, D, \mathcal{B}^-)$ is the total heat energy content of the manifold; we can also study the corresponding localized problem by choosing f_2 suitably.

Standard elliptic methods show there is an asymptotic series as $t \to 0^+$:

$$\beta(f_1, f_2, D, \mathcal{B})(t) \simeq \sum_{n=0}^{\infty} \beta_n(f_1, f_2, D, \mathcal{B}) t^{n/2}; \tag{3.4}$$

these are the heat content asymptotics. They are expressible in terms of local invariants. The study of heat content asymptotics is relatively recent; see [4, 5, 7, 9, 36, 48] for work by other authors on this subject. We refer to [6, 8, 20] for:

Theorem 3.1: *Let $\mathcal{B}^-$ define Dirichlet boundary conditions.*

(a) $\beta_0(f_1, f_2, D, \mathcal{B}^-) = \int_M \langle f_1, f_2 \rangle dx.$

(b) $\beta_1(f_1, f_2, D, \mathcal{B}^-) = -2\pi^{-1/2} \int_{\partial M} \langle f_1, f_2 \rangle dy.$

(c) $\beta_2(f_1, f_2, D, \mathcal{B}^-) = -\int_M \langle D f_1, f_2 \rangle dx + \int_{\partial M} \{ \langle \frac{1}{2} L_{aa} f_1, f_2 \rangle - \langle f_1, f_{2;m} \rangle \} dy.$

(d) $\beta_3(f_1, f_2, D, \mathcal{B}^-) = -2\pi^{-1/2} \int_{\partial M} \{ \frac{2}{3} \langle f_{1;mm}, f_2 \rangle + \frac{2}{3} \langle f_1, f_{2;mm} \rangle - \langle f_{1:a}, f_{2:a} \rangle$
$+ \langle E f_1, f_2 \rangle - \frac{2}{3} L_{aa} \langle f_{1;m}, f_2 \rangle - \frac{2}{3} L_{aa} \langle f_1, f_{2;m} \rangle + \langle (\frac{1}{12} L_{aa} L_{bb} - \frac{1}{6} L_{ab} L_{ab}$
$+ \frac{1}{6} R_{amam}) f_1, f_2 \rangle \} dy.$

(e) $\beta_4(f_1, f_2, D, \mathcal{B}^-) = \frac{1}{2} \int_M \langle D f_1, \tilde{D} f_2 \rangle + \int_{\partial M} \{ \frac{1}{2} \langle (D f_1)_{;m}, f_2) \rangle + \frac{1}{2} \langle f_1, (\tilde{D} f_2)_{;m} \rangle$
$- \frac{1}{4} \langle L_{aa} D f_1, f_2 \rangle - \frac{1}{4} \langle L_{aa} f_1, \tilde{D} f_2 \rangle + \langle (\frac{1}{8} E_{;m} - \frac{1}{16} L_{ab} L_{ab} L_{cc} + \frac{1}{8} L_{ab} L_{ac} L_{bc}$
$- \frac{1}{16} R_{ambm} L_{ab} + \frac{1}{16} R_{abcb} L_{ac} + \frac{1}{32} \tau_{;m} + \frac{1}{16} L_{ab:ab}) f_1, f_2 \rangle - \frac{1}{4} L_{ab} \langle f_{1:a}, f_{2:b} \rangle$
$- \frac{1}{8} \langle \Omega_{am} f_{1:a}, f_2 \rangle + \frac{1}{8} \langle \Omega_{am} f_1, f_{2:a} \rangle \} dy.$

Theorem 3.2: *Let $\mathcal{B}_S^+$ define modified Neumann boundary conditions.*

(a) $\beta_0(f_1, f_2, D, \mathcal{B}_S^+) = \int_M \langle f_1, f_2 \rangle dx.$

(b) $\beta_1(f_1, f_2, D, \mathcal{B}_S^+) = 0.$

(c) $\beta_2(f_1, f_2, D, \mathcal{B}_S^+) = -\int_M \langle D f_1, f_2 \rangle dx + \int_{\partial M} \langle \mathcal{B}_S^+ f_1, f_2 \rangle dy.$

(d) $\beta_3(f_1, f_2, D, \mathcal{B}_S^+) = \frac{2}{3} \cdot 2\pi^{-1/2} \int_{\partial M} \langle \mathcal{B}_S^+ f_1, \tilde{\mathcal{B}}_S^+ f_2 \rangle dy.$

(e) $\beta_4(f_1, f_2, D, \mathcal{B}_S^+) = \frac{1}{2} \int_M \langle D f_1, \tilde{D} f_2 \rangle dx + \int_{\partial M} \{ -\frac{1}{2} \langle \mathcal{B}_S^+ f_1, \tilde{D} f_2 \rangle$
$- \frac{1}{2} \langle D f_1, \tilde{\mathcal{B}}_S^+ f_2 \rangle + \langle (\frac{1}{2} S + \frac{1}{4} L_{aa}) \mathcal{B}_S^+ f_1, \tilde{\mathcal{B}}_S^+ f_2 \rangle dy.$

(f) $\beta_5(f_1, f_2, D, \mathcal{B}_S^+) = 2\pi^{-1/2} \int_{\partial M} \{ -\frac{4}{15} \langle \mathcal{B}_S^+ D f_1, \tilde{\mathcal{B}}_S^+ f_2 \rangle - \frac{4}{15} \langle \mathcal{B}_S^+ f_1, \tilde{\mathcal{B}}_S^+ \tilde{D} f_2 \rangle$
$- \frac{2}{15} \langle (\mathcal{B}_S^+ f_1)_{:a}, (\tilde{\mathcal{B}}_S^+ f_2)_{:a} \rangle + \langle (\frac{2}{15} E + \frac{4}{15} S^2 + \frac{4}{15} S L_{aa} + \frac{1}{30} L_{aa} L_{bb} + \frac{1}{15} L_{ab} L_{ab}$
$- \frac{1}{15} R_{amam}) \mathcal{B}_S^+ f_1, \tilde{\mathcal{B}}_S^+ f_2 \rangle \} dy.$

(g) $\beta_6(f_1, f_2, D, \mathcal{B}_S^+) = -\frac{1}{6} \int_M \langle D^2 f_1, \tilde{D} f_2 \rangle + \int_{\partial M} \{ \frac{1}{6} \langle \mathcal{B}_S^+ D f_1, \tilde{D} f_2 \rangle$

$$+\tfrac{1}{6}\langle D^2 f_1,\tilde{\mathcal{B}}_S^+ f_2\rangle + \tfrac{1}{6}\langle \mathcal{B}_S^+ f_1,\tilde{D}^2 f_2\rangle - \tfrac{1}{6}\langle S\mathcal{B}_S^+ D f_1,\tilde{\mathcal{B}}_S^+ f_2\rangle - \tfrac{1}{6}\langle S\mathcal{B}_S^+ f_1,\tilde{\mathcal{B}}_S^+\tilde{D} f_2\rangle$$
$$-\tfrac{1}{12}\langle L_{aa}\mathcal{B}_S^+ D f_1,\tilde{\mathcal{B}}_S^+ f_2\rangle - \tfrac{1}{12}\langle L_{aa}\mathcal{B}_S^+ f_1,\tilde{\mathcal{B}}_S^+\tilde{D} f_2\rangle + \langle(\tfrac{1}{24}E_{;m} + \tfrac{1}{12}EL_{aa}$$
$$+\tfrac{1}{48}L_{ab}L_{ab}L_{cc} + \tfrac{1}{24}L_{ab}L_{ac}L_{bc} - \tfrac{1}{48}R_{ambm}L_{ab} + \tfrac{1}{48}R_{abcb}L_{ac} - \tfrac{1}{24}R_{amam}L_{bb}$$
$$+\tfrac{1}{96}\tau_{;m} + \tfrac{1}{48}L_{ab:ab} + \tfrac{1}{12}SL_{aa}L_{bb} + \tfrac{1}{12}SL_{ab}L_{ab} - \tfrac{1}{12}SR_{amam}$$
$$+\tfrac{1}{12}(SE+ES) + \tfrac{1}{4}S^2L_{aa} + \tfrac{1}{6}S^3 + \tfrac{1}{6}S_{:aa})\mathcal{B}_S^+ f_1,\tilde{\mathcal{B}}_S^+ f_2\rangle$$
$$-\tfrac{1}{12}L_{aa}\langle(\mathcal{B}_S^+ f_1)_{:b},(\tilde{\mathcal{B}}_S^+ f_2)_{:b}\rangle - \tfrac{1}{12}L_{ab}\langle(\mathcal{B}_S^+ f_1)_{:a},(\tilde{\mathcal{B}}_S^+ f_2)_{:b}\rangle$$
$$-\tfrac{1}{6}\langle S(\mathcal{B}_S^+ f_1)_{:a},(\tilde{\mathcal{B}}_S^+ f_2)_{:a}\rangle - \tfrac{1}{24}\langle\Omega_{am}(\mathcal{B}_S^+ f_1)_{:a},\tilde{\mathcal{B}}_S^+ f_2\rangle$$
$$+\tfrac{1}{24}\langle\Omega_{am}\mathcal{B}_S^+ f_1,(\tilde{\mathcal{B}}_S^+ f_2)_{:a}\rangle\}dy.$$

Theorem 3.3: *Let $\mathcal{B} = \mathcal{B}(\chi, S, \nabla^D)$ define mixed boundary conditions.*

(a) $\beta_0(f_1,f_2,D,\mathcal{B}) = \int_M\langle f_1,f_2\rangle dx.$

(b) $\beta_1(f_1,f_2,D,\mathcal{B}) = -2\pi^{-1/2}\int_{\partial M}\langle f_1^-,f_2^-\rangle dy.$

(c) $\beta_2(f_1,f_2,D,\mathcal{B}) = -\int_M\langle Df_1,f_2\rangle dx + \int_{\partial M}\{\langle f_{1;m}^+ + Sf_1^+,f_2^+\rangle + \langle\tfrac{1}{2}L_{aa}f_1^-,f_2^-\rangle$
$-\langle f_1^-,f_{2;m}^-\rangle\}dy.$

(d) $\beta_3(f_1,f_2,D,\mathcal{B}) = 2\pi^{-1/2}\int_{\partial M}\{-\tfrac{2}{3}\langle f_{1;mm}^-,f_2^-\rangle - \tfrac{2}{3}\langle f_1^-,f_{2;mm}^-\rangle$
$+\tfrac{2}{3}L_{aa}\langle f_{1;m}^-,f_2^-\rangle + \tfrac{2}{3}L_{aa}\langle f_1^-,f_{2;m}^-\rangle + \langle(-\tfrac{1}{12}L_{aa}L_{bb} + \tfrac{1}{6}L_{ab}L_{ab}$
$-\tfrac{1}{6}R_{amam})f_1^-,f_2^-\rangle + \tfrac{2}{3}\langle f_{1;m}^+ + Sf_1^+,f_{2;m}^+ + Sf_2^+\rangle - \langle Ef_1^-,f_2^-\rangle$
$-\tfrac{2}{3}\langle Ef_1^-,f_2^+\rangle - \tfrac{2}{3}\langle Ef_1^+,f_2^-\rangle + \langle f_1^-{}_{:a},f_2^-{}_{:a}\rangle + \tfrac{2}{3}\langle f_1^+{}_{:a},f_2^-{}_{:a}\rangle$
$+\tfrac{2}{3}\langle f_1^-{}_{:a},f_2^+{}_{:a}\rangle\}dy.$

Remark: We recover Theorems 3.1 (a)-(d) and 3.2 (a)-(d) by setting $\chi = \pm I_V$ in Theorem 3.3. We refer to [36] for another proof of Theorem 3.3 using a modified DeWitt ansatz.

Remark: A detailed comparison of the formulas of §2 and §3 shows there is no obvious relationship between the heat equation asymptotics and the heat content asymptotics.

§4 Operators of Dirac type

In some cases, it is possible to take a square root of D within the category of differential operators; for example the total form valued Laplacian $d\delta+\delta d = (d+\delta)^2$ has this property. This leads to additional spectral invariants. The Clifford algebra bundle is the natural setting for this discussion; it is the universal complex unital algebra bundle generated by the tangent bundle subject to the Clifford commutation relations

$$\partial_\nu * \partial_\mu + \partial_\mu * \partial_\nu = -2g_{\nu\mu}. \tag{4.1}$$

A Clifford module structure on V is a unital algebra morphism γ from the Clifford algebra of M to the bundle of endomorphisms of V. If $\gamma^\nu = \gamma(dx^\nu)$, then $\gamma^\nu\gamma^\mu + \gamma^\mu\gamma^\nu = -2g^{\nu\mu}$. We may always choose a connection ∇ on V so $\nabla\gamma = 0$; fix such a connection henceforth. We refer to [26, 27] for further details.

Let ψ be a smooth endomorphism of V and let

$$P = P(\nabla, \psi) = \gamma^\nu\nabla_\nu - \psi : C^\infty(V) \to C^\infty(V) \tag{4.2}$$

be an operator of Dirac type; any operator of Dirac type can be expressed in this form. Let $D = P^2$ be the associated operator of Laplace type.

We suppose the boundary of M empty for the moment. Standard elliptic methods show there is an asymptotic series as $t \to 0^+$ of the form:

$$Tr_{L^2}(fPe^{-tD}) \simeq \sum_{n=0}^{\infty} a_n(f,P)t^{(n-m)/2}; \tag{4.3}$$

the $a_n(f,P)$ are expressible in terms of local information and vanish for n even. Let $\Psi = \gamma_i\psi\gamma_i$ and let $W_{ij} = \Omega_{ij} - \frac{1}{4}R_{ijkl}\gamma_k\gamma_l$. We refer to [15] for

Theorem 4.1: *If the boundary of M is empty,*

(a) $a_1(P)(x) = (4\pi)^{-m/2}(m-1)\,\mathrm{Tr}\{\psi\}$.

(b) $a_3(P)(x) = -\frac{1}{12}(4\pi)^{-m/2}Tr\{\{2(1-m)\psi_{:i} + 3(4-m)\psi\gamma_i\psi + 3\Psi\gamma_i\psi\}_{:i}$
$+(m-3)\{\tau\psi + 6\gamma_i\gamma_jW_{ij}\psi + 6\psi\psi_{:i}\gamma_i + (4-m)\psi\psi\psi + 3\psi\psi\Psi\}\}$.

If $\partial M \neq \emptyset$, we must impose boundary conditions analogous to those discussed in previously. Let χ be an endomorphism of $V|_{\partial M}$ with $\chi^2 = 1$ so that

$$\gamma_m\chi + \chi\gamma_m = \gamma_a\chi - \chi\gamma_a = 0; \tag{4.4}$$

χ always exists if m is even; if m is odd, χ need not exist. Let

$$\mathrm{domain}(P_\chi) = \{\phi \in C^\infty(V) : (1+\chi)(\phi|_{\partial M}) = 0\}. \tag{4.5}$$

As before, we may expand the trace in an asymptotic series:

$$Tr_{L^2}(fP_\chi e^{-tD_\chi}) \sim \Sigma_n a_n(f,P,\chi)t^{(n-m)/2}. \tag{4.6}$$

We refer to [16] for:

Theorem 4.2 :

(a) $a_0(f,P,\chi) = 0$.

(b) $a_1(f,P,\chi) = (4\pi)^{-m/2}\int_M (m-1)f\,\mathrm{Tr}(\psi)dx$.

(c) $a_2(f,P,\chi) = \frac{1}{4}(4\pi)^{-(m-1)/2}\int_{\partial M}(m-2)fTr(\psi\chi)dy$.

(d) $a_3(f,P,\chi) = -\frac{1}{12}(4\pi)^{-m/2}\{\int_M fTr\{2(1-m)\psi_{;i} + 3(4-m)\psi\gamma_i\psi + 3\Psi\gamma_i\psi\}_{;i}$
$+(m-3)fTr\{\tau\psi + 6\gamma_i\gamma_jW_{ij}\psi + 6\psi\psi_{;i}\gamma_i + (4-m)\psi\psi\psi + 3\psi\psi\Psi\}$

$+\int_{\partial M} 6(2-m) f_{;m} \operatorname{Tr}\{\chi\psi\} + f \operatorname{Tr}\{(18-6m)\chi\psi_{;m} + (2-2m)\psi_{;m}$
$-6\chi\gamma_m\gamma_a\psi_{;a} + 6(m-2)\chi\psi L_{aa} + 2(m-3)\psi L_{aa} + 6(3-m)\chi\gamma_m\psi\psi$
$+3\gamma_m\psi\Psi_T + 3(3-m)\chi\gamma_m\psi\chi\psi + 6\chi\gamma_a W_{am}\}\}.$

The eta invariant plays an important role in the index theorem for manifolds with boundary. Recent work by Grubb [30] permits one to examine the heat equation asymptotics with non local boundary conditions of Dirac type; this permits the generalization of Theorem 4.2 to situations where local boundary conditions do not exist. Suitable generalizations of the results of the theorems discussed §2, §3, and §4 to these boundary conditions should provide very promising and exciting new developments for the future.

§5 Operators with non scalar leading symbol

In previous sections, we have studied the asymptotics associated to operators of Laplace type. These operators have scalar leading symbol and we use the leading symbol to define the metric tensor. There are results for more general classes of operators, see [13, 32] for details. We shall restrict our attention here to the generalized Ahlfors Laplacian which appears in geometrical contexts; see [17, 45] for further details.

Let a and b be positive constants and let ϵ be an arbitrary constant. Let

$$D = a d_0 \delta_0 + b \delta_1 d_1 - \epsilon\rho \text{ on } C^\infty T^*M; \qquad (5.1)$$

The classical Ahlfors operator $(1-\frac{1}{m})d_0\delta_0 + \frac{1}{2}\delta_1 d_1 - \rho$ plays a fundamental role in the study of quasiconformal geometry; see Ahlfors [1].

Theorem 5.1: *Let $\mathcal{B} = \mathcal{B}^a$ define absolute boundary conditions.*

(a) $a_0(P,\mathcal{B}^a) = (4\pi)^{-m/2}(a^{-m/2} + (m-1)b^{-m/2})\int_M dx.$

(b) $a_1(P,\mathcal{B}^a) = \frac{1}{4}(4\pi)^{-(m-1)/2}(a^{(1-m)/2} + (m-3)b^{(1-m)/2})\int_{\partial M} dy.$

(c) $a_2(P,\mathcal{B}^a) = \frac{1}{6}(4\pi)^{-m/2}\{\{a^{(2-m)/2} + (m-7)b^{(2-m)/2}$
$+6\epsilon m^{-1}(a^{-m/2} + (m-1)b^{-m/2})\}\int_M \tau dx$
$+2\{a^{(2-m)/2} + (m-7)b^{(2-m)/2}\}\int_{\partial M} L_{aa} dy\}.$

(d) $a_3(P,\mathcal{B}^a) = \frac{1}{384}(4\pi)^{-(m-1)/2}\int_{\partial M}\{a^{(3-m)/2}(16\tau + 8R_{amam} + 13L_{aa}L_{bb}$
$+2L_{ab}L_{ab} + b^{(3-m)/2}\{(16m-144)\tau + (8m-216)R_{amam}$
$+(13m-129)L_{aa}L_{bb} + (2m+102)L_{ab}L_{ab}\}$
$+\epsilon(\beta_4\tau + \beta_5 R_{amam})\}dy.$
$\beta_4 = 96(m-1)^{-1}(a^{(1-m)/2} + (m-2)b^{(1-m)/2})$ and
$\beta_5 = 96(m-1)^{-1}\{a^{(1-m)/2} + (2m-3)b^{(1-m)/2}\}.$

Theorem 5.2: *Let $\mathcal{B} = \mathcal{B}^r$ define relative boundary conditions.*

(a) $a_0(P, \mathcal{B}^r) = (4\pi)^{-m/2}(a^{-m/2} + (m-1)b^{-m/2})\int_M dx.$

(b) $a_1(P, \mathcal{B}^r) = -\frac{1}{4}(4\pi)^{-(m-1)/2}(a^{(1-m)/2} + (m-3)b^{(1-m)/2})\int_{\partial M} dy.$

(c) $a_2(P, \mathcal{B}^r) = \frac{1}{6}(4\pi)^{-m/2}\{\{a^{(2-m)/2} + (m-7)b^{(2-m)/2}$
$+6\epsilon m^{-1}(a^{-m/2} + (m-1)b^{-m/2})\}\int_M \tau dx$
$+2\{a^{(2-m)/2} + (m-7)b^{(2-m)/2}\}\int_{\partial M} L_{aa} dy\}.$

(d) $a_3(P, \mathcal{B}^r) = \frac{1}{384}(4\pi)^{-(m-1)/2}\int_{\partial M}\{a^{(3-m)/2}(-16\tau - 8R_{amam} - 7L_{aa}L_{bb}$
$+10L_{ab}L_{ab} + b^{(3-m)/2}\{(144-16m)\tau + (216-8m)R_{amam}$
$+(123-7m)L_{aa}L_{bb} + (10m-114)L_{ab}L_{ab}\}$
$-\epsilon(\beta_4\tau + \beta_5 R_{amam})\}dy$ where the β_i are as above.

Acknowledgements: Research partially supported by the NSF and by Max Planck Institut fuï Mathematik

References

Any list of references in this field is bound to be incomplete. We list below a selection of some of the papers in the field. We refer to [10] and [11] for a more comprehensive listing.

[1] L. Ahlfors, *Conditions for quasiconformal deformations in several variables* in: **Contributions to analysis, a collection of papers dedicated to L. Bers**, Academic Press, New York (1974); *Quasiconformal deformations and mappings in* $\mathbf{R}^n$, J. Analyse Math. **30**, 74-97 (1976).

[2] P. Amsterdamski, A. Berkin, and D. O'Connor, b_8 *Hamidew coefficient for a scalar field*, Class. Quantum Grav. **6** (1989), 1981-1991.

[3] I. G. Avramidi, *The covariant technique for the calculation of the heat kernel asymptotic expansion*, Physics Letters B. **238** (1990) #1 92-97; see also articles in Theor. Math. Fiz. **79** (1989) 219 and also Yad. Fiz. **49** (1989) 1185. (Russian).

[4] M. van den Berg, *Heat equation on a hemisphere*, Proceedings of the Royal Soc. of Edinburgh **118A**, 5-12 (1991).

[5] M. van den Berg and E. B. Davies, *Heat flow out of regions in* R^m, Math. Z. **202**, 463-482 (1989).

[6] M. van den Berg, S. Desjardins, and P. Gilkey, *Functorality and heat content asymptotics for operators of Laplace type* (preprint).

[7] M. van den Berg and J-F Le Gall, *Mean curvature and the heat equation* (preprint).

[8] M. van den Berg and P. Gilkey, *Heat content asymptotics of a Riemannian manifold with boundary* (preprint).

[9] M. van den Berg and S. Srisatkunarajah, *Heat flow and Brownian motion for a region in* $\mathbf{R}^2$ *with a polygonal boundary*, Probab. Th. Rel. Fields **86**, 41-52 (1990).
[10] P. Bérard, *Variétés Riemanniennes isospectrales non isométriques,* Séminaire Bourbaki **41** ème année 1989 # 705.
[11] P. Bérard and M. Berger, *Le spectre d'une variété Riemannienne en 1981*, Lecture notes in Math. Springer **1207** (1986).
[12] L. Boutet de Monvel, *Boundary problems for pseudo-differential operators*, Acta. Mat. **126** (1971).
[13] T. Branson, S. Fulling, and P. Gilkey (1991), *Heat equation asymptotics of nonminimal operators on differential forms*, J. Math. Phys. **32**, 2089-2091.
[14] T. P. Branson and P. B. Gilkey, *The asymptotics of the Laplacian on a manifold with boundary*, Comm. in PDE **15**, 245-272 (1990).
[15] T. P. Branson and P. B. Gilkey, *Residues of the eta function for an operator of Dirac type*, J. Funct. Anal. (to appear).
[16] T. P. Branson and P. B. Gilkey, *Residues of the eta function for an operator of Dirac type with local boundary conditions*, Differential Geometry and its Applications (to appear).
[17] T. P. Branson, P. B. Gilkey, B. Ørsted, and A. Pierzchalski, *Heat equation asymptotics of a generalized Ahlfors Laplacian on a manifold with boundary*, Proceedings of the Lambrecht Conference, Birkhäuser (to appear).
[18] T. Carleman, *Propriétés asymptotiques des fonctions fondamentales des membranes vibrantes*, Åttonde Skand. Matematikerkongressen i Stockholm, 1934, 34-44.
[19] H. S. Carslaw and J. C. Jaeger, **Conduction of heat in solids**, Clarendon Press, Oxford (1986).
[20] S. Desjardins and P. Gilkey, *Heat content asymptotics for operators of Laplace type with Neumann boundary conditions* (preprint).
[21] J. S. Dowker and J. P. Schofield, J. Math. Phys. **31** (1990), 808.
[22] P. B. Gilkey, *The spectral geometry of a Riemannian manifold*, J. Diff. Geo **10** (1975), 601-618.
[23] P. B. Gilkey, *Recursion relations and the asymptotic behavior of the eigenvalues of the Laplacian*, Compositio Math. 38 (1979), 201-240.
[24] P. B. Gilkey, **Invariance Theory, the Heat Equation, and the Atiyah-Singer Index Theorem**, Publish or Perish, Wilmington, Delaware, 1984.
[25] P. B. Gilkey, *Leading terms in the asymptotics of the heat equation*, Contemporary Math. **73** (1988), 79-85.
[26] P. B. Gilkey, *The geometrical index theorem for Clifford modules*, **Topics in Mathematical Analysis**, edited by T. M. Rassias, (1989) World Scientific Press, 315-327.
[27] P. B. Gilkey, *On the index of geometrical operators for Riemannian manifolds with boundary*, Advances in Mathematics (to appear).

[28] P. Greiner,*An asymptotic expansion for the heat equation*, Arch. Rat. Mech. Anal. 41 (1971), 163-218.

[29] G. Grubb, **Functional calculus of pseudo differential boundary problems**, Progress in Math., **65**, Birkhäuser, Boston (1986).

[30] G. Grubb, *Heat operator trace expansions and index for general Atiyah-Patodi-Singer boundary problems*, (preprint); see also *Le problème d'Atiyah-Patodi-Singer pour les variétés a bord générales (non cylindriques)* Proceedings of Journées EDP, Saint Jean de Monts, France June 1991 (to appear).

[31] P. Gunther, **Huygen's principle and hyperbolic equations**, Perspectives in Mathematics, Academic Press (1988).

[32] V. P. Gusynin, E. V. Gorbar, and V. V. Romankov (1991), *Heat kernel expansion for nonminimal differential operators and manifolds with torsion*, Nuclear Physics B362, 449-471.

[33] J. Hadamard, *Le Problème de Cauchy et les Èquations aux Dérivées Partielles Linéaires Hyperboliques*, Hermann et Cie, Paris, 1932.

[34] G. Kennedy, R. Critchley, and J. S. Dowker, *Finite Temperature Field Theory with Boundaries: Stress Tensor and Surface Action Renormalization,* Annals of Physics **125** (1980), 346-400.

[35] H. Luckock and I. Moss, The Quantum Geometry of Random Surfaces and Spinning Membranes, Class. Quantum Gravity **8** (1001) 1445.

[36] D. M. McAvity, *Heat kernel asymptotics for mixed boundary conditions*, (preprint); see also *Surface energy from heat content asymptotics*, (preprint).

[37] D. M. McAvity and H. Osborn, *Asymptotic expansion of the heat kernel for generalised boundary conditions* (preprint); *A DeWitt expansion of the heat kernel for manifolds with a boundary*, Class. Quantum Gravity **8** (1991), 603-638.

[38] H.P. McKean and I.M. Singer, *Curvature and the eigenvalues of the Laplacian*, J. Diff. Geo. **1** (1967), 43-69.

[39] J. Melmed, *Conformal invariance and the regularized one loop effective action*, J. Phys. A. **21** (1989), L1131-1134.

[40] S. Minakshisundaram, *A generalization of Epstein zeta functions*, Can. J. Math. **1** (1949), 320-329.

[41] S. Minakshisundaram, *Eigenfunctions on Riemannian manifolds*, J. Indian Math. Soc. **17** (1953), 158-165.

[42] S. Minakshisundaram and Å. Pleijel, *Some properties of the eigenfunctions of the Laplace operator on Riemannian manifolds*, Can. J. Math. **1** (1949) 242-256.

[43] I. Moss, *Boundary terms in the heat kernel expansion*, Class quantum gravity **6** (1989), 759-765.

[44] I. Moss and J. S. Dowker, *The correct B_4 coefficient*, Phys. Letts. B **229** (1989), 261-263.

[45] B. Ørsted and A. Pierzchalski (1990), *The Ahlfors Laplacian on a Riemannian manifold with boundary*, (preprint).
[46] B. Osgood, R. Phillips, and P. Sarnak, *Compact isospectral sets of surfaces*, J. Funct. Anal. **80** (1988), 212-234.
[47] V. K. Patodi, *Curvature and the fundamental solution of the heat operator*, Journal of the Indian Math. Soc. **34** (1970), 269-285.
[48] C. G. Phillips and K. M. Jansons, *The short-time transient of diffusion outside a conducting body*, Proc. Royal Soc. London A **428**, 431-449 (1990).
[49] Å. Pleijel, *A Study of Certain Green's functions with Applications in the Theory of Vibrating Membranes*, Ark. Mat. **2** (1954), 553-569; *On Green's Functions for Elastic Plates with Clamped, Supported and Free Edges*, Proceedings of Symp. on Spectral Theory and Differential Problems, Stillwater, Oklahoma (1951), 413-454; *On Green's Functions and the Eigenvalue Distribution of the Three Dimensional Membrane Equation*, Tolfte Skand. Matematikerkongressen i Lund (1953), 222-240.
[50] T. Sakai, *On eigenvalues of Laplacian and curvature of Riemannian manifold*, Tohuku Math. J. **23** (1971), 589-603.
[51] R. Seeley, *Complex powers of an elliptic operator*, Amer. Math. Soc. Proc. Symp. Pure Math **10**, 288-307; see also *Topics in pseudo differential operators*, CIME Conference on pseudo differential operators 1968, Edizioni Cremonese Roma (1969), 169-305.; *Singular integrals and boundary problems*, Amer. J. Math. **88** (1966), 781-809; *Analytic extension of the trace associated with elliptic boundary problems*, Amer. J. Math **91** (1969), 963-983.
[52] L. Smith, *The asymptotics of the heat equation for a boundary value problem*, Invent. Math. **63** (1981), 467-493.
[53] K. Stewartson and R.T. Waechter, *On hearing the shape of a drum*, Proc. Cambridge Philosophical Society **69**, 353-363 (1971).
[54] H. Weyl, *Das asymptotische Verteilungsgesetz der Eigenschwingungen eines beliebig gestalten elastischen Körpers*, Rendiconti del Circolo Mat. di Palermo **39** (1915), 1-49; **The classical groups**, Princeton University Press, Princeton (1946).

P. Gilkey, Department of Mathematics, University of Oregon, Eugene Oregon 97403, USA. e-mail: gilkey@bright.math.uoregon.edu.

Geometry in Partial Differential Equations (pp. 114–154)
eds. A. Pràstaro and Th. M. Rassias

DEFORMATIONS AND RECURSION OPERATORS FOR EVOLUTION EQUATIONS

I. S. KRASIL'SHCHIK
1 Tverskoy-Yamskoy per., 14, Apt. 45, 125047 Moscow, Russia

and

P. H. M. KERSTEN
University of Twente, Department of Applied Mathematics, P.O. Box 217, 7500 AE Enschede, The Netherlands
Email address: kerstenmath.utwente.nl

ABSTRACT. In the differential geometric framework an exposition of Richardson-Nijenhuis and Frölicher-Nijenhuis brackets is given. The theory of deformations in exposed and deformations of a number of classical integrable equations are constructed leading to recursion operators for these equations.

CONTENTS

INTRODUCTION

Recursion operators [7] play an important role in the theory of socalled *completely integrable* equations, i.e. differential equations possessing infinite series of symmetries (e.g. soliton type equations). In what follows we show that they are of cohomological nature. To be more precise, with every formally integrable equation $\mathcal{E}$ cohomological groups $H^i_{\mathcal{C}}(\mathcal{E})$, $i = 0, 1, \ldots$, are associated [3] which are constructed by means of the Frölicher-Nijenhuis bracket techniques [1, 6] applied to the Cartan connection on $\mathcal{E}^{(\infty)}$ [4]. For this theory $H^0_{\mathcal{C}}(\mathcal{E})$ coincides with the Lie algebra sym($\mathcal{E}$) of higher symmetries for $\mathcal{E}$ and $H^1_{\mathcal{C}}(\mathcal{E})$ is identified with the equivalence classes of non trivial deformations of the equation structure. Recursion operators arise due to the natural action of $H^1_{\mathcal{C}}(\mathcal{E})$ onto $H^0_{\mathcal{C}}(\mathcal{E})$ with respect to contraction operators.

We also study extensions of cohomological invariants onto coverings of $\mathcal{E}$ (nonlocal theory [5]) and show that obstructions to such extensions live in the E_1 term of Vinogradov's $\mathcal{C}$-spectral sequence [8].

A series of computational examples is exposed in section 5 where we discuss deformations of Burgers' equation including a nonlocal variable, KdV-equation, Nonlinear

Schrödinger equation and classical Boussinesq equation; all in the setting of coverings. For these examples recursion operators are obtained from the results in a straight forward way.

To make exposition self-contained we give brief reviews of the basic theories used in the text, e.g., geometry of differential equations, Frölicher-Nijenhuis brackets, $\mathcal{C}$-cohomologies, and coverings.

1. Jet manifolds and infinitely prolonged differential equations

In this section we define the basic concepts of the geometrical theory of differential equations (see [4]).

1.1. Jets. Let M be a smooth manifold and $\pi : E \to M$ be a smooth fibre bundle over M. Denote by $\Gamma(\pi)$ the set of sections of π and by $\Gamma_x(\pi)$, $x \in M$, the set of all local sections at x. As a rule, everywhere below we shall use the notation $\Gamma(\pi)$ having in mind that it should be changed by $\Gamma_x(\pi)$ if necessary.

Let $f \in \Gamma_x(\pi)$ and $y = f(x) \in E$. *The k-th jet of f at the point x* is the class of all local sections $f' \in \Gamma_x(\pi)$ such that (a) $f'(x) = y$ and (b) $f'(U')$ is tangent to $f(U)$ at the point y with the order k, where U and U' are neighborhoods of x on which f and f' are defined respectively. Denote this class by $[f]_x^k$.

The set

$$J^k(\pi) = \left\{ [f]_x^k \mid f \in \Gamma_x(\pi), \quad x \in M \right\}$$

carries the natural structure of a smooth manifold and is called the *manifold of k-th jets for the bundle π*. Natural mappings

$$\begin{aligned} &\pi_{k,\ell} : J^k(\pi) \to J^\ell(\pi), \ \pi_{k\ell}[f]_x^k = [f]_x^\ell, \ k \geq \ell, \\ &\pi_k : J^k(\pi) \to M, \qquad \pi_k[f]_x^k = x \end{aligned} \tag{1.1}$$

are smooth fibre bundle projections with respect to this structure. The bundle π_k is called *the bundle of k-th jets for π*. Obviously, one has $\pi_{\ell,s} \cdot \pi_{k,\ell} = \pi_{k,s}$ and $\pi_\ell \cdot \pi_{k,\ell} = \pi_k$ for all $k \geq \ell \geq s$.

The inverse limit with respect to the system of mappings (1.1) is denoted by $J^\infty(\pi)$ and is called *the manifold of infinite jets for π*.

By definition, one has natural projections

$$\pi_{\infty,k} : J^\infty(\pi) \to J^k(\pi), \ \pi_\infty : J^\infty(\pi) \to M,$$

and $\pi_{k,\ell} \circ \pi_{\infty,k} = \pi_{\infty,\ell}$, $\pi_k \circ \pi_{\infty,k} = \pi_\infty$, $k \geq \ell$.

1.2. Calculus on $J^\infty(\pi)$. Since $J^\infty(\pi)$ is an infinite dimensional manifold it needs to be clarified what are the objects of standard calculus on it.

Denote the algebra $C^\infty(J^k(\pi))$ of smooth functions on $J^k(\pi)$ by $\mathcal{F}_k(\pi)$. Then one has the system of embeddings

$$\pi^*_{k,\ell}: \quad \mathcal{F}_\ell(\pi) \to \mathcal{F}_k(\pi),\ k \ge \ell$$
$$\pi^*_k : C^\infty(M) \to \mathcal{F}_k(\pi) \tag{1.2}$$

with the identities $\pi^*_{k,\ell} \circ \pi^*_{\ell,s} = \pi^*_{k,s}$, $\pi^*_{k,\ell} \circ \pi^*_\ell = \pi^*_k$ for $k \ge \ell \ge s$.
The direct limit with respect to the system of mappings (1.2) is denoted by $\mathcal{F}(\pi)$ and its elements are called *smooth functions on* $J^\infty(\pi)$. By definition, one has the embeddings

$$\pi^*_{\infty,k} : \mathcal{F}_k(\pi) \to \mathcal{F}(\pi),\ \pi^*_\infty : C^\infty(M) \to \mathcal{F}(\pi) \tag{1.3}$$

such that $\pi^*_{\infty,k} \circ \pi^*_{\infty,\ell}$ and $\pi^*_{\infty,k} \circ \pi^*_k = \pi^*_\infty$. Identifying $\mathcal{F}_k(\pi)$ and $C^\infty(M)$ with their images under (1.3) one can write down

$$\mathcal{F}(\pi) = \bigcup_{k \ge 0} \mathcal{F}_k(\pi)\ ,\ \mathcal{F}_k(\pi) \subset \mathcal{F}_{k+1}(\pi).$$

In the same manner one can define *the modules of i-forms on* $J^\infty(\pi)$, $i \ge 0$, as

$$\Lambda^i(\pi) = \bigcup_{k \ge 0} \Lambda^i(J^k(\pi)), \qquad \Lambda^i(J^k(\pi)) \subset \Lambda^i(J^{k+1}(\pi)),$$

and consider *the graded algebra*

$$\Lambda^*(\pi) = \sum_{i \ge 0} \Lambda^i(\pi)$$

of differential forms on $J^\infty(\pi)$
We say that an $\mathbb{R}$-linear mapping $X : \mathcal{F}(\pi) \to \mathcal{F}(\pi)$ is *a vector field of degree* $s \in \mathbb{Z}$, if

$$X(\varphi\psi) = \varphi X(\psi) + \psi\, X(\varphi)$$

for any $\varphi, \psi \in \mathcal{F}(\pi)$ and there exists such $r \ge 0$ that

$$X(\mathcal{F}_k(\pi)) \subset \mathcal{F}_{k+s}(\pi)$$

for all $s \ge r$. Denote by $D^{(s)}(\pi)$ the set of such fields and by

$$D(\pi) = \bigcup_{s \in \mathbb{Z}} D^{(s)}(\pi)$$

the set of all vector fields on $J^\infty(\pi)$. Obviously, $D^{(s)}(\pi) \subset D^{(s+1)}(\pi)$, $D(\pi)$ is a filtered $\mathcal{F}(\pi)$-module and Lie algebra over $\mathbb{R}$ with respect to commutator

$$[X, Y] = X \circ Y - Y \circ X.$$

With these definitions at hand, one can use all infinitesimal constructions of classical differential geometry on $J^\infty(\pi)$ as if it was an "ordinary" finite dimensional manifold.

1.3. Differential operators. Let $\pi : E \to M$ be a fibre bundle. Consider a section $f \in \Gamma(\pi)$ and a correspondence $j_k(f) : M \to J^k(\pi)$, $k = 0, 1, \dots, \infty$, defined by

$$j_k(f)(x) = [f]^k_x,\ x \in M.$$

It can be shown that $j_k(f)$ is smooth for any $f \in \Gamma(\pi)$, and from (1.1) it follows that

$$\pi_{k,\ell} \circ j_k(f) = j_\ell(f), \; \pi_k \circ j_k(f) = id_M, \; k \geq \ell.$$

Thus $j_k(f) \in \Gamma(\pi_k)$. In particular, $j_0(f) = f$.

Let $\pi^1 : E^1 \to M$ be another fibre bundle. A correspondent $\Delta : \Gamma(\pi) \to \Gamma(\pi')$ is said to be *a differential operator of the order* $k \geq 0$, if there exists a fibre bundle morphism $\mu_\Delta : J^k(\pi) \to E'$ such that

$$\Delta(f) = \mu_\Delta \circ j_k(f)$$

for all $f \in \Gamma(\pi)$. In particular, $j_k = j_k^\pi : \Gamma(\pi) \to \Gamma(\pi_k)$ is a differential operator of the order k for which $\mu_{j_k} = \mathrm{id}_{J^k(\pi)}$.

Consider a pull-back of the bundle π' onto $J^k(\pi)$ with respect to π_k:

$$\begin{array}{ccc} \pi_k^*(E') & \xrightarrow{\bar{\pi}_k} & E' \\ {\scriptstyle \pi_k^*(\pi')}\downarrow & & \downarrow{\scriptstyle \pi'} \\ J^k(\pi) & \xrightarrow{\pi_k} & M \end{array}$$

(recall that $\pi_k^*(E')$ is a subset of $J^k(\pi) \times E'$ consisting of the pairs (y, y'), $y \in J^k(\pi)$, $y' \in E'$, such that $\pi_k(y) = \pi'(y')$). Denote the set $\Gamma(\pi_k^*(\pi'))$ by $\mathcal{F}_k(\pi, \pi')$. Any differential operator $\Delta : \Gamma(\pi) \to \Gamma(\pi')$ of the order k determines an element $\varphi_\Delta \in \mathcal{F}_k(\pi, \pi')$ by

$$\varphi_\Delta(y) = (y, \mu_\Delta(y)), \; y \in J^k(\pi).$$

Conversely, any section $\varphi \in \mathcal{F}_k(\pi, \pi')$ determines an operator $\Delta = \Delta_\varphi : \Gamma(\pi) \to \Gamma(\pi')$ for which

$$\mu_\Delta = \bar{\pi}_k \circ \varphi.$$

The bundle $\pi_\infty^*(\pi')$ coincides with the inverse limit of bundles $\pi_k^*(\pi')$ existing due to commutative diagrams

$$\begin{array}{ccc} \pi_k^*(E') & \xrightarrow{\bar{\pi}_{k,\ell}} & \pi_\ell^*(E') \\ {\scriptstyle \pi_k^*(\pi')}\downarrow & & \downarrow{\scriptstyle \pi_\ell^*(\pi')} \\ J^k(\pi) & \xrightarrow{\pi_{k,\ell}} & J^\ell(\pi) \end{array} \tag{1.4}$$

$k \geq \ell$. Diagram (1.4) determines embeddings $\pi_{k,\ell}^* : \mathcal{F}_\ell(\pi, \pi') \to \mathcal{F}_k(\pi, \pi')$, and we define the set of sections for $\pi_\infty^*(\pi')$ as

$$\mathcal{F}(\pi, \pi') = \bigcup_{k \geq 0} \mathcal{F}_k(\pi, \pi')$$

Thus $\mathcal{F}(\pi, \pi')$ can be identified with all differential operators acting from π onto π'.

In the case, when π' is a vector bundle, $\mathcal{F}(\pi, \pi')$ becomes a filtered module over $\mathcal{F}(\pi)$.

1.4. Prolongations and compositions. Let $\mu : J^k(\pi) \to E'$ be a fibre bundle morphism of π_k and π'. Define *its s-th prolongation* $\mu^{(s)} : J^{k+s}(\pi) \to J^s(\pi')$ by

$$\mu^{(s)}[f]_x^{k+s} = [\mu \circ j_k(f)]_x^s, \ f \in \Gamma(\pi), \ x \in M.$$

Then *the s-th prolongation* $\Delta^{(s)} : \Gamma(\pi) \to \Gamma(\pi'_s)$ *of an operator* $\Delta : \Gamma(\pi) \to \Gamma(\pi')$ can be defined by

$$\mu_{\Delta^{(s)}} = \mu_\Delta^{(s)}$$

Consider a fibre bundle $\pi'' : E'' \to M$ and an operator $\Delta' : \Gamma(\pi') \to \Gamma(\pi')$ of the order s. *The composition* $\Delta' \circ \Delta : \Gamma(\pi) \to \Gamma(\pi'')$ *of* Δ *and* Δ' is the operator of the order $k + s$ which is defined by

$$\mu_{\Delta' \circ \Delta} = \mu_{\Delta'} \circ \mu_\Delta^{(s)}.$$

In particular, one can see that $\Delta^{(s)} = j_s^{\pi'} \circ \Delta$.

1.5. Differential equations and prolongations. *A differential equation* of the order k in the fibre bundle π is a submanifold $\mathcal{E} \subset J^k(\pi)$. A section $h \in \Gamma(\pi)$ is *a solution* of $\mathcal{E}$ if $j_k(h)(M) \subset \mathcal{E}$.

Consider a point $y \in \mathcal{E}$ and present it in the form $y = [f]_x^k$, $f \in \Gamma(\pi)$, $x \in M$. Then the following subset in $J^{k+s}(\pi)$ is defined:

$$\mathcal{E}_y^{(s)} = \{[f]_x^{k+s} | [f]_x^k = y, \ j_k(f)(M) \text{ is tangent to } \mathcal{E} \text{ with the order } s \text{ at } y\};$$

the set $\mathcal{E}^{(s)} = \bigcup_{y \in \mathcal{E}} \mathcal{E}_y^{(s)} \subset J^{k+s}(\pi)$ is called the *s-th prolongation* of the equation $\mathcal{E}$. We also let $\mathcal{E}^{(-s)} = \pi_{k,k-s}(\mathcal{E}) \subset J^{k-s}(\pi), s = 1, \dots, k$.

With any vector bundle $\pi' : E' \to M$ and with an operator $\Delta : \Gamma(\pi) \to \Gamma(\pi')$ of the order k a differential equation $\mathcal{E}_\Delta \subset J^k(\pi)$ is associated in the following way

$$\mathcal{E}_\Delta = \{y \in J^k(\pi) | \varphi_\Delta(y) = 0\}.$$

Conversely, for any equation $\mathcal{E} \subset J^k(\pi)$ there exist a bundle π' and an operator Δ such that $\mathcal{E} = \mathcal{E}_\Delta$. One can show that prolongations of $\mathcal{E}$ and Δ are related in a natural way:

$$\mathcal{E}_\Delta^{(s)} = \mathcal{E}_{\Delta^{(s)}}, \ s \geq 0.$$

An equation $\mathcal{E} \subset J^k(\pi)$ is called *formally integrable*, if all $\mathcal{E}^{(s)}$ are smooth submanifolds in $J^{k+s}(\pi)$ and all the mappings $n_{k+s+1,k+s} : \mathcal{E}^{(s+1)} \to \mathcal{E}^{(s)}$ are smooth fibre bundle projections.

1.6. Calculus on $\mathcal{E}^{(\infty)}$. Let $\mathcal{E} \subset J^k(\pi)$ and $\mathcal{E}^{(s)} \subset J^{k+s}(\pi)$. Consider the ideal $I_s(\mathcal{E})$ consisting of the functions vanishing on $\mathcal{E}^{(s)}$:

$$I_s(\mathcal{E}) = \{\varphi \in \mathcal{F}_{k+s}(\pi) | \ \varphi \big|_{\mathcal{E}^{(s)}} = 0\}$$

Define the algebra of smooth functions on $\mathcal{E}^{(s)}$ by

$$\mathcal{F}_s(\mathcal{E}) = \mathcal{F}_{k+s}(\pi)/I_s(\mathcal{E}).$$

Obviously, for any s one has the commutative diagram

$$\begin{array}{ccccccccc}
0 & \longrightarrow & I_{s+1}(\mathcal{E}) & \longrightarrow & \mathcal{F}_{k+s+1}(\pi) & \longrightarrow & F_{s+1}(\mathcal{E}) & \longrightarrow & 0 \\
 & & \uparrow & & \uparrow & & \uparrow{\scriptstyle \pi^*_{s+1,s}} & & \\
0 & \longrightarrow & I_s(\mathcal{E}) & \longrightarrow & \mathcal{F}_{k+s}(\pi) & \longrightarrow & \mathcal{F}_s(\mathcal{E}) & \longrightarrow & 0 \\
 & & \uparrow & & \uparrow & & & & \\
 & & 0 & & 0 & & & &
\end{array}$$

The direct limit of the system $\{\mathcal{F}_s(\mathcal{E}), \pi^*_{s+1,s}\}$ is denoted by $\mathcal{F}(\mathcal{E})$ and is called *the algebra of smooth functions on* $\mathcal{E}^{(\infty)}$. For any s one has a homomorphism $\pi^*_{\infty,s}$: $\mathcal{F}_s(\mathcal{E}) \to \mathcal{F}(\mathcal{E})$, and $\mathcal{F}(\mathcal{E})$ is filtered by the images of these homomorphisms

If $\pi' : E' \to M$ is another fibre bundle, then we consider the modules

$$\mathcal{F}_s(\mathcal{E}, \pi') = \mathcal{F}_{k+s}(\mathcal{E}, \pi')/I_s(\mathcal{E}) \circ \mathcal{F}_{k+s}(\mathcal{E}, \pi')$$

together with the $\mathcal{F}_s(\mathcal{E})$ homomorphisms $\pi^*_{s+1,s} : \mathcal{F}_s(\mathcal{E}, \pi') \to \mathcal{F}_{k+s}(\mathcal{E}, \pi')$. The direct limit of $\{\mathcal{F}_s(\mathcal{E}, \pi'), \pi^*_{s+1,s}\}$ is denoted by $\mathcal{F}(\mathcal{E}, \pi')$ and is filtered by its submodules of the form $\pi^*_{\infty,s}(\mathcal{F}_s, (\mathcal{E}, \pi'))$. It is a filtered $\mathcal{F}(\mathcal{E})$ module.

With these definitions at hand, calculus on $\mathcal{E}^{(\infty)}$ is constructed precisely in the same way as it was done in subsection 1.2 for $J^\infty(\pi)$. In what follows $\Lambda^*(\mathcal{E}) = \sum\limits_{i \geq 0} \Lambda^i(\mathcal{E})$ the graded algebra of differential forms on $\mathcal{E}^{(\infty)}$ is denoted and by $D(\mathcal{E})$ we denote the Lie algebra of vector fields on $\mathcal{E}^{(\infty)}$.

Remark 1.1. In the case when $\mathcal{E}$ is a formally integrable equation, the algebras $\mathcal{F}_s(\mathcal{E})$ coincide with $C^\infty(\mathcal{E}^{(s)})$, $\mathcal{F}_s(\mathcal{E}, \pi')$ are the modules of smooth sections of the bundles $\pi^*_{k+s}(\pi')$ restricted onto $\mathcal{E}^{(s)} \subset J^{k+s}(\pi)$, and all $\pi^*_{s+1,s}$ are embeddings.

2. Cartan distribution, linearizations, and symmetries

Here we continue the exposition of geometry of differential equation following, as before, the point of view of [4].

2.1. Liftings. Let $\pi : E \to M$ be a smooth fibre bundle and $\pi' : E' \to M$ and $\pi'' : E'' \to M$ be two vector bundles. Consider a linear differential operator $\Delta : \Gamma(\pi') \to \Gamma(\pi'')$ of the order k and define linear differential operators

$$\Delta_s : \mathcal{F}_s(\pi, \pi') \to \mathcal{F}_{k+s}(\pi, \pi'')$$

for any $s \geq 0$ by

$$\Delta_s(\varphi) \circ j_{k+s}(f) = \Delta(\varphi \circ j_s(f)) \tag{2.1}$$

for any $\varphi \in \mathcal{F}_s(\pi, \pi')$, $f \in \Gamma(\pi)$. One can see that Δ_s is well-defined by (2.1); it is called *the s-th lifting of* Δ. A point-wise analogue of (2.1) looks as follows:

$$\Delta_s(\varphi)|_{[f]_x^{k+s}} = \Delta(\varphi \circ j_s(f))|_x,$$

where $x \in M$.

For any linear $\Delta : \Gamma(\pi') \to \Gamma(\pi'')$ the system of operators $\{\Delta_s\}_{s\geq 0}$ can be looked at as a linear differential operator $\mathcal{F}(\pi, \pi') \to \mathcal{F}(\pi, \pi'')$ which agrees with the filtrations. It is denoted by $\mathcal{C}\Delta$ and is called *the infinite lifting of* Δ, or simply *the lifting of* Δ.

The basic properties of the operator $\mathcal{C}$ are

(i) if $\Delta : \Gamma(\pi') \to \Gamma(\pi'')$ is a linear differential operator of the order k, then $\mathcal{C}\Delta : \mathcal{F}(\pi, \pi') \to \mathcal{F}(\pi, \pi'')$ is a linear operator of the order k as well,
(ii) if $f \in C^\infty(M)$, then $\mathcal{C}(f\Delta) = f\mathcal{C}\Delta$,
(iii) if $\Delta, \Delta' : \Gamma(\pi') \to \Gamma(\pi'')$ are two operators, then $\mathcal{C}(\Delta + \Delta') = \mathcal{C}\Delta + \mathcal{C}\Delta'$,
(iv) if $\pi''' : E''' \to M$ is a vector bundle and $\nabla : \Gamma(\pi'') \to \Gamma(\pi''')$ is a linear operator, then $\mathcal{C}(\nabla \circ \Delta) = \mathcal{C}(\nabla) \circ \mathcal{C}(\Delta)$.

2.2. Total derivatives and the Cartan distribution. Consider a vector field $X \in D(M)$ on M. It is an operator of the first order acting from $C^\infty(M)$ to $C^\infty(M)$. Hence, the operator $\mathcal{C}X : \mathcal{F}(\pi) \to \mathcal{F}(\pi)$ is defined. From (2.1) it follows that $\mathcal{C}X$ is a filtered derivation of the algebra $\mathcal{F}(\pi)$, i.e. it is a vector field on $J^\infty(\pi)$. This vector field is called *the total derivation in the direction of* X. Thus we have a correspondence

$$\mathcal{C} : D(M) \to D(\pi)$$

which possesses the following properties

(i) $\mathcal{C}(fX + gY) = f\mathcal{C}X + g\mathcal{C}Y$,
(ii) $\mathcal{C}[X, Y] = [\mathcal{C}X, \mathcal{C}Y]$,
and, as it can be easily shown,
(iii) $\mathcal{C}X(\pi_\infty^* f) = \pi_\infty^* X(f)$,

where $f, g \in C^\infty(M)$, $X, Y \in D(M)$.

Consider a submodule $\mathcal{C}D(\pi) \subset D(\pi)$ generated by the fields of the form $\mathcal{C}X$, $X \in D(M)$, i.e.

$$\mathcal{C}D(\pi) = \{\sum_i \varphi_i \mathcal{C}X_i / \varphi_i \in \mathcal{F}(\pi),\ x_i \in D(M)\}$$

Thus we have a distribution on $J^\infty(\pi)$ which is called *the Cartan distribution.* From (ii) it follows that

$$[\mathcal{C}D(\pi),\ \mathcal{C}D(\pi)] \subset \mathcal{C}D(\pi).$$

It means that the Cartan distribution satisfies Frobenius integrability conditions. Keeping this in mind we say that it is *completely integrable.*

The structure of integral manifolds of the Cartan distribution is described by

Theorem 2.1. *Let* $f \in \Gamma(\pi)$. *Then*

$$j_\infty(f)(M) \subset J^\infty(\pi) \tag{2.2}$$

is a maximal integral manifold of the Cartan distribution on $J^\infty(\pi)$. *Conversely, every local maximal manifold is of the form (2.2) for some local section* f *of* π.

2.3. Restriction onto $\mathcal{E}^{(\infty)}$. Let $\mathcal{E} \subset J^k(\pi)$ be an equation and $\mathcal{E}^{(\infty)} \subset J^\infty(\pi)$ be its infinite prolongation. Consider ideals $I_s(\mathcal{E}) \subset \mathcal{F}_{k+s}(\pi)$ and the ideal

$$I(\mathcal{E}) = \bigcup_{s \geq 0} I_s(\mathcal{E}) \subset \mathcal{F}(\pi).$$

Then one can show that for any $X \in D(M)$ the field $\mathcal{C}X$ preserves $I(\mathcal{E})$:

$$\mathcal{C}X(I(\mathcal{E})) \subset I(\mathcal{E})$$

It means that $\mathcal{C}X$ is tangent to $\mathcal{E}^{(\infty)}$ and can be restricted onto this manifold. Thus for any $\mathcal{E}$ we get a correspondence

$$\mathcal{C} = \mathcal{C}_\mathcal{E} : D(M) \to D(\mathcal{E}) \tag{2.3}$$

which possesses the properties (i-iii) from subsection 2.2. Just as in the case of $J^\infty(\pi)$ one can consider the module

$$\mathcal{C}D(\mathcal{E}) = \{\sum_i \varphi_i \mathcal{C}X_i / \varphi_i \in \mathcal{F}(\mathcal{E}),\ x_i \in D(M)\}.$$

Corresponding distribution is called *the Cartan distribution on* $\mathcal{E}^{(\infty)}$; it is completely integrable and its local maximal integral manifolds are of the form $j_\infty(f)(M)$, where f is a local solution of $\mathcal{E}$.

2.4. Splittings. A vector field $X \in D(\mathcal{E})$ is called *vertical*, if $X(C^\infty(M)) = 0$, or, which is the same, if X is tangent to the fibres of the bundle $\pi_\infty : \mathcal{E}^{(\infty)} \to M$. The set of all such fields forms a Lie subalgebra in $D(\mathcal{E})$ and is denoted by $D^v(\mathcal{E})$. Moreover, $D(\mathcal{E})$ splits into the direct sum

$$D(\mathcal{E}) = D^v(\mathcal{E}) \oplus \mathcal{C}D(\mathcal{E}) \tag{2.4}$$

$\mathcal{C}D(\mathcal{E})$ being an ideal in $D(\mathcal{E})$.

A form $\omega \in \Lambda^i(\mathcal{E})$ is said to be *horizontal*, if $X \rfloor \omega = 0$ for any $X \in D^v(\mathcal{E})$. Horizontal forms of degree i constitute a submodule in $\Lambda^i(\mathcal{E})$ which is denoted by $\Lambda_0^i(\mathcal{E})$. For any $i \leq 0$ one has

$$\Lambda_0^i(\mathcal{E}) = \Lambda_0^1(\mathcal{E}) \wedge \ldots \wedge \Lambda_0^1(\mathcal{E}) \qquad (i \text{ times}),$$

and the direct sum

$$\Lambda_0^*(\mathcal{E}) = \sum_{i \geq 0} \Lambda_0^i(\mathcal{E})$$

is a graded algebra with respect to the wedge product

$$\wedge : \Lambda_0^i(\mathcal{E}) \times \Lambda_0^j(\mathcal{E}) \to \Lambda_0^{i+j}(\mathcal{E}).$$

We say that $\omega \in \Lambda^i(\mathcal{E})$ is *a Cartan form*, if $X \rfloor \omega = 0$ for any $X \in \mathcal{C}D(\mathcal{E})$. Cartan forms of a degree i constitute a submodule in $\Lambda^i(\mathcal{E})$ which is denoted by $\mathcal{C}^i\Lambda(\mathcal{E})$. For any $i \geq 0$ one has

$$\mathcal{C}^i\Lambda(\mathcal{E}) = \mathcal{C}^i\Lambda(\mathcal{E}) \wedge \ldots \wedge \mathcal{C}^i\Lambda(\mathcal{E}) \qquad (i \text{ times}),$$

and the direct sum

$$\mathcal{C}^*\Lambda(\mathcal{E}) = \sum_{i \geq 0} C^i\Lambda(\mathcal{E})$$

is a graded algebra with respect to the wedge product

$$\wedge : \mathcal{C}^i\Lambda(\mathcal{E}) \times \mathcal{C}^j\Lambda(\mathcal{E}) \to \mathcal{C}^{i+j}\Lambda(\mathcal{E}).$$

The module of 1-forms on $\mathcal{E}^{(\infty)}$ splits into the direct sum

$$\Lambda^i(\mathcal{E}) = \Lambda_0^1(\mathcal{E}) \oplus \mathcal{C}^i\Lambda(\mathcal{E}) \tag{2.5}$$

this splitting being dual to (2.4). From (2.5) it follows that any $\Lambda^i(\mathcal{E})$, $i > 0$, splits into

$$\Lambda^i(\mathcal{E}) = \sum_{\alpha+\beta=i} \Lambda_0^\alpha(\mathcal{E}) \times \mathcal{C}^\beta\Lambda(\mathcal{E}).$$

Let $d : \Lambda^i(\mathcal{E}) \to \Lambda^{i+1}(\mathcal{E})$ be the de Rham differential on $\mathcal{E}^{(\infty)}$. Then it can be shown that if $\omega \in \Lambda_0^\alpha(\mathcal{E}) \times \mathcal{C}^\beta\Lambda(\mathcal{E})$, then

$$d\omega = \omega_0 + \omega_{\mathcal{C}},$$

where $\omega_0 \in \Lambda_0^{\alpha+1}(\mathcal{E}) \wedge \mathcal{C}^\beta\Lambda(\mathcal{E})$, $\omega_{\mathcal{C}} \in \Lambda_0^\alpha(\mathcal{E}) \wedge \mathcal{C}^{\beta+1}\Lambda(\mathcal{E})$. This decomposition allows one to introduce the mappings

$$d_{\mathcal{C}} : \Lambda_0^\alpha(\mathcal{E}) \wedge \mathcal{C}^\beta\Lambda(\mathcal{E}) \to \Lambda_0^\alpha(\mathcal{E}) \wedge C^{\beta+1}\Lambda(\mathcal{E}),$$

$$d_h : \Lambda_0^\alpha(\mathcal{E}) \wedge C^\beta\Lambda(\mathcal{E}) \to \Lambda_0^{\alpha+1}(\mathcal{E}) \wedge C^\beta\Lambda(\mathcal{E})$$

by setting

$$d_{\mathcal{C}}(\omega) = \omega_{\mathcal{C}},\ d_h(\omega) = \omega_0.$$

In fact, one has $d_{\mathcal{C}}^2 = 0, d_h^2 = 0$, and, by definition, $d = d_{\mathcal{C}} + d_h$.
Thus the system

$$\{\Lambda_0^\alpha(\mathcal{E}) \wedge C^\beta\Lambda(\mathcal{E}), d_{\mathcal{C}}, d_h\}$$

forms a bicomplex with the total differential d. Denote by $E_1^{\beta,\alpha}(\mathcal{E})$ the cohomology of d_h at the term $\Lambda_0^\alpha \otimes C^\beta\Lambda$.

In particular, we have two complexes

$$0 \to \mathcal{F}(\mathcal{E}) \xrightarrow{d_{\mathcal{C}}} \mathcal{C}^1\Lambda(\mathcal{E}) \to \ldots \to \mathcal{C}^i\Lambda(\mathcal{E}) \xrightarrow{d_{\mathcal{C}}} \mathcal{C}^{i+1}\Lambda(\mathcal{E}) \to \ldots \tag{2.6}$$

and

$$0 \to \mathcal{F}(\mathcal{E}) \xrightarrow{d_h} \Lambda_0^1(\mathcal{E}) \to \ldots \to \Lambda_0^i(\mathcal{E}) \xrightarrow{d_h} \Lambda_0^{i+1}(\mathcal{E}) \to \ldots \tag{2.7}$$

The first one is called *the Cartan complex* for the equation $\mathcal{E}$, the second one is said to be *the horizontal complex* for $\mathcal{E}$.

Remark 2.2. Consider the cotangent bundle $\tau : T^*(M) \to M$ and its i-th external power $\Lambda^i\tau : \Lambda^i T^*(M) \to M$. Then the following identification takes place

$$\Lambda_0^i(\mathcal{E}) = \mathcal{F}(\mathcal{E}, \Lambda^i\tau).$$

Moreover, consider the de Rham differential on M as the first order linear differential operator

$$d : \Gamma(\Lambda^i\tau) \to (\Lambda^{i+1}\tau).$$

Then its lifting

$$\mathcal{C}d : \mathcal{F}(\mathcal{E}, \Lambda^i\tau) \to \mathcal{F}(\mathcal{E}, \Lambda^{i+1}\tau)$$

coincides with the horizontal differential (2.7).

2.5. Symmetries. Let $\mathcal{B}$ be a distribution on a smooth manifold P. Let $\mathcal{B}D(P) \subset D(P)$ be a submodule determining $\mathcal{B}$. Suppose $\mathcal{B}$ to be completely integrable, i.e. $[\mathcal{B}D(P), \mathcal{B}D(P)] \subset \mathcal{B}D(P)$.
An infinitesimal symmetry of $\mathcal{B}$ is a vector field $X \in D(P)$ such that

$$[X, \mathcal{B}D(P)] \subset \mathcal{B}D(P).$$

The set $D_{\mathcal{B}}(P)$ of such fields forms a Lie subalgebra in $D(P), \mathcal{B}D(P)$ being its ideal. Let $X \in D_{\mathcal{B}}(P)$ and A_t be the one-parameter group of transformations corresponding to X. Then if $S \subset P$ is a maximal integral manifold of $\mathcal{B}$, then $A_t(S)$ is a maximal integral manifold as well. If $X \in \mathcal{B}D(P)$, then $A_t(S) \subset S$, at least for sufficiently small t. By this reason the elements of $\mathcal{B}D(P)$ are called *trivial symmetries* and the quotient Lie algebra

$$\mathrm{sym}(\mathcal{B}) = D_{\mathcal{B}}(P)/\mathcal{B}D(P)$$

is said to be *the algebra of (nontrivial) symmetries of* $\mathcal{B}$.
In particular, if $P = \mathcal{E}^\infty$ and $\mathcal{B} = \mathcal{C}$ is the Cartan distribution on $\mathcal{E}^{(\infty)}$, then the algebra

$$\mathrm{sym}(\mathcal{E}) = \mathrm{sym}(\mathcal{C}) = D_{\mathcal{C}}(\mathcal{E})/\mathcal{C}D(\mathcal{E})$$

is called *the algebra of (higher infinitesimal) symmetries* of $\mathcal{E}$. When $\mathcal{E}^{(\infty)} = J^\infty(\pi)$, this algebra is denoted by

$$\mathrm{sym}(\pi) = D_{\mathcal{C}}(\pi)/\mathcal{C}D(\pi).$$

Due to decomposition (2.4), each equivalence class from $\mathrm{sym}(\mathcal{E})$ contains a unique element from $D^v(\mathcal{E})$. Thus $\mathrm{sym}(\mathcal{E})$ can be identified with

$$D^v_{\mathcal{C}}(\mathcal{E}) = \{X \in D^v(\mathcal{E}) | [X, \mathcal{C}D(\mathcal{E})] \subset \mathcal{C}D(\mathcal{E})\}.$$

Respectively,

$$\mathrm{sym}(\pi) = D^v_{\mathcal{C}} = \{X \in D^v(\pi) | [X, \mathcal{C}D(\pi)] \subset \mathcal{C}D(\pi)\}.$$

By this reason the elements of $D^v_{\mathcal{C}}(\mathcal{E})$(resp, of $D^v_{\mathcal{C}}(\pi)$) are called *(higher infinitesimal) symmetries of the equation* $\mathcal{E}$ (resp. of $J^\infty(\pi)$).

2.6. Linearizations and evolutions. Let $\pi : E \to M, \pi' : E' \to M$ be vector bundles and $\Delta \quad \Gamma(\pi) \to \Gamma(\pi')$ be a differential operator of the order k. Consider a section $f \in \Gamma(\pi)$ and a smooth family of sections $f_t \in \Gamma(\pi)$, $t \in \mathbb{R}$, such that $f_0 = f$. Then it can be shown that a section

$$\frac{d}{dt}\Delta(f_t)\Big|_{t=0} \in \Gamma(\pi')$$

does not depend on f_t and depends on f and $\frac{df_t}{dt}\Big|_{t=0} \in \Gamma(\pi)$ only. Thus we get a correspondence

$$\ell_{\Delta,f} : \Gamma(\pi) \to \Gamma(\pi')$$

which is a linear differential operator of the order k and which is called *the linearization of* Δ *along the section* f.
Moreover, there exists uniquely defined linear differential operator $\ell_\Delta : \mathcal{F}(\pi, \pi) \to \mathcal{F}(\pi, \pi')$ of the order k such that

$$(j_\infty(f))^*\ell_\Delta(\varphi) = \ell_{\Delta,f}((j_\infty(f))^*\varphi)$$

for any $f \in \Gamma(\pi), \varphi \in \mathcal{F}(\pi, \pi)$. This operator is called *the universal linearization of* Δ.
Fix a section $\varphi \in \mathcal{F}(\pi, \pi)$ and consider a mapping

$$\ni^{\pi'}_\varphi : \mathcal{F}(\pi, \pi') \to \mathcal{F}(\pi, \pi')$$

defined by

$$\ni^{\pi'}_\varphi (\psi) = \ell_\Delta(\varphi),$$

where $\psi \in \mathcal{F}(\pi, \pi')$ and $\psi = \varphi_\Delta$. This mapping is $\mathcal{C}^\infty(M)$-linear and possesses the following property

$$\ni^{\pi'}_\varphi (h\psi) = \ni_\varphi (h)\psi + h \ni^{\pi'}_\varphi (\psi),$$

where $h \in \mathcal{F}(\pi)$ and $\ni_\varphi = \ni^{\mathbb{I}}_\varphi$ for the trivial one-dimensional bundle $\mathbb{I} : M \times \mathbb{R} \to M$. In particular, $\ni_\varphi : \mathcal{F}(\pi) \to \mathcal{F}(\pi)$ is a vertical derivation of $\mathcal{F}(\pi)$ which is called *the evolutionary derivation with the generating section* φ.

Theorem 2.3. *Let $\pi : E \to M$ be a vector bundle. Then the set of all evolutionary derivations coincides with $D^v_C(\pi)$, i.e. any element $X \in sym(\pi)$ is of the form $X = \ni_\varphi$ for some $\varphi \in \mathcal{F}(\pi, \pi)$ and vice versa.*
Let $X = \ni_\varphi$ and $Y = \ni_\varphi$ be two symmetries. Then their bracket $[X, Y]$ is a symmetry again and thus is to be of the form $\ni_\rho$ for some $\rho \in \mathcal{F}(\pi, \pi)$. The element ρ is denoted by $\{\varphi, \psi\}$ and is called the higher Jacobi bracket of φ and ψ. By definition one has

$$\ni_{\{\varphi,\psi\}} = [\ni_\varphi, \ni_\psi],$$

and one can show that

$$\{\varphi, \psi\} = \ni^\pi_\varphi (\psi) - \ni^\pi_\psi (\varphi).$$

2.7. Description of sym($\mathcal{E}$). Let $\mathcal{E} \subset J^k(\pi)$ be a differential equation and $\Delta : \Gamma(\pi) \to \Gamma(\pi')$ be an operator determining $\mathcal{E}$. It can be shown that universal linearization ℓ_Δ of Δ admits a restriction onto $\mathcal{E}^{(\infty)}$, i.e. there exists a linear differential operator $\ell_\mathcal{E} : \mathcal{F}(\mathcal{E}, \pi) \to \mathcal{F}(\mathcal{E}, \pi')$ such that the diagram

$$\begin{array}{ccc} \ell_\Delta : \mathcal{F}(\pi, \pi) & \longrightarrow & \mathcal{F}(\pi, \pi') \\ \downarrow & & \downarrow \\ \ell_\mathcal{E} : \mathcal{F}(\mathcal{E}, \pi) & \longrightarrow & \mathcal{F}(\mathcal{E}, \pi') \end{array}$$

is commutative (her vertical arrows denote natural restriction of the sections from $J^\infty(\pi)$ onto $\mathcal{E}^{(\infty)}$).

Theorem 2.4. *Let $\mathcal{E} \subset J^k(\pi)$ be a differential equation. The following statements hold:*

(i) *Any element $X \in sym(\mathcal{E}) = D^v_\mathcal{C}(\mathcal{E})$ is a restriction of some evolutionary derivation $\ni_\varphi \in sym(\pi)$.*
(ii) *If $X = \ni_\varphi \big|_{\mathcal{E}(\infty)}$, then X is uniquely determined by the restriction $\overline{\varphi} = \varphi\big|_{\mathcal{E}(\infty)} \in \mathcal{F}(\mathcal{E}, \pi)$, and we write $X = \ni_{\overline{\varphi}}$.*
(iii) *$X \in sym(\mathcal{E})$ iff $\ell_\mathcal{E}(\overline{\varphi}) = 0$, or*

$$sym(\mathcal{E}) = \ker \ell_\mathcal{E}.$$

Note that the Jacobi bracket admits a restriction onto $\mathcal{E}^{(\infty)}$ and determines a Lie algebra structure in $\ker \ell_\mathcal{E}$.

2.8. Local coordinates. Choose a coordinate neighborhood U in M in such a way that restrictions of all the bundles under consideration are trivial. Let $x_1, \ldots, x_n$ be local coordinates in U and $u^1, \ldots, u^m$ be coordinates along the fibre of π in $\pi^{-1}(U) \subset E$. Then the following local coordinates in $J^k(\pi)$ are naturally associated with $(x_1, \ldots, x_n, u^1, \ldots, u^m)$.
Let $\sigma = (\sigma_1, \ldots, \sigma_n)$ be a multi-index and $|\sigma| = \sigma_1 + \ldots + \sigma_n$. Define coordinate functions $p^j_\sigma, j = 1, \ldots, m, |\sigma| \le k$, by

$$j_k(f)^*(p^j_\sigma) = \frac{\partial^{|\sigma|} f^j}{\partial x_1^{\sigma_1} \dots \partial x_n^{\sigma_n}}$$

for any local section $f = \{u^j = f^j(x_1, \dots, x_n)\}$ of π over U. It can be shown that the system of neighborhoods $\pi_k^{-1}(U)$ and functions $\{p^j_\sigma\}$ form a smooth atlas in $J^k(\pi)$. To get coordinate functions in $J^\infty(\pi)$ it needs to consider p^j_σ for all j and σ.

Let us write down the basic constructions introduced above in local coordinates

(i) If $\partial_i = \partial/\partial x_i$, then

$$\mathcal{C}\partial_i = D_i = \partial/\partial x_i + \sum_{j,\sigma} p^j_{\sigma+1_i}\, \partial/\partial p^j_\sigma$$

is the i-th total derivative, where $i = 1, \dots, n$ and $\sigma + 1_i = (\sigma_1, \dots, \sigma_i + 1, \dots, \sigma_k)$.

(ii) If $(v^1, \dots, v^r)$ are coordinates along the fibre of π' in $(\pi')^{-1}(U)$ and $V^\ell = F^\ell(x_1, \dots, x_n, \dots, p^j_\sigma, \dots)$, $\ell = 1, \dots, r$, is a local representation of the operator $\Delta(\pi) \to \Gamma(\pi')$ and $\{q^\ell_\tau\}$ are coordinates in $J^\infty(\pi')$, then the s-th prolongation of Δ is represented by

$$q^\ell_\tau = D_\tau\, F^\ell, |\tau| \le s,\ \ell = 1, \dots, r,$$

where $\tau = (\tau_1, \dots, \tau_n)$ and $D_\tau = D_\tau^{\tau_1} \circ \dots \circ D_n^{\tau_n}$.

(iii) If the equation $\mathcal{E} \subset J^k(\pi)$ is determined by

$$\begin{cases} F^1(x_1, \dots, x_n, \dots, p^j_\sigma, \dots) = 0, \\ \dots \\ F^r(x_1, \dots, x_n, \dots, p^j_\sigma, \dots) = 0, \end{cases}$$

then its s-th prolongation is

$$\begin{cases} D_\tau F^j = 0, \\ |\tau| \le s,\ j = 1, \dots, r. \end{cases}$$

(iv) The elements of $\mathcal{C}D(\pi)$ are of the form

$$X = \sum_i \varphi_i\, D_i,\ \varphi_i \in \mathcal{F}(\pi),$$

while vertical vector fields are represented as

$$Y = \sum_{|\sigma| \ge 0, j} Y^j_\sigma\, \partial/\partial p^j_\sigma,\ Y^j_\sigma \in \mathcal{F}(\pi).$$

(v) Horizontal 1-forms an represented as

$$\omega = \sum_i \varphi_i\, dx_i,\ \varphi_i \in \mathcal{F}(\pi)$$

Cartan 1-forms are of the form

$$\theta = \sum_{j,\sigma} \theta^j_\sigma\, \omega^j_\sigma,\ \theta^j_\sigma \in \mathcal{F}(\pi),$$

where $\omega^j_\sigma = dp^j_\sigma - \sum_i p^j_{\sigma+1_i}\, dx_i$ and only finite number of θ^j_σ are non-zero.

(vi) Differentials d_h and $d_\mathcal{C}$ are completely determined by the following conditions

$$\begin{cases} d_\mathcal{C}\varphi = \sum_{j,\sigma} \frac{\partial\varphi}{\partial p^j_\sigma}\, \omega^j_\sigma, \\ d_\mathcal{C}\omega^j_\sigma = 0, \\ d_\mathcal{C}(dx_i) = 0, \end{cases}$$

and

$$\begin{cases} d_h\varphi = \sum_i D_i(\varphi)dx_i, \\ d_h\omega^j_b = -\sum_i \omega^j_{b+1_i} \wedge dx_i, \\ d_h(dx_i) = 0, \end{cases}$$

where $\varphi \in \mathcal{F}(\pi)$.

(vii) Evolutionary derivation with the generating section $\varphi = (\varphi^1, \ldots, \varphi^m)$ is of the form

$$\ni_\varphi = \sum_{j,\sigma} D_\sigma(\varphi^j)\partial/\partial p^j_\sigma. \tag{2.8}$$

If

$$v^\ell = F^\ell(x_1, \ldots, x_n, \ldots, p^j_\sigma, \ldots), \ell = 1, \ldots, r,$$

is a differential operator, then

$$\ell_F = \| \sum_\sigma \frac{\partial F^\ell}{\partial p^j_\sigma} D_\sigma \|. \tag{2.9}$$

3. Frölicher-Nijenhuis bracket and $\mathcal{C}$-cohomologies

The Frölicher-Nijenhuis bracket [4, 6] is introduced and $\mathcal{C}$-cohomologics [3] are defined. The basic formulae are given, both in invariant and local coordinates form.

3.1. Vector valued forms. Let R be a commutative ring with the unit and A be a commutative unitary R-algebra. By $D_i = D_i(A)$ the module of *i-derivations of A* is denoted, $\Lambda^i = \Lambda^i(A)$ denotes the module of *i-forms over A*. We also use the notations $D_* = \sum_{i\geq 0} D_i(A)$ and $\Lambda^* = \sum_{i\geq 0} \Lambda^i(A)$. We formally set $\Lambda^i = 0$ and $D_i = 0$ for $i < 0$. All tensor products below are taken over R.

Both D_* and Λ^* are commutative graded algebras with respect to *wedge product*

$$\wedge : D_i \times D_j \to D_{i+j}$$

and

$$\wedge : \Lambda^i \times \Lambda^j \to \Lambda^{i+j}$$

For any i and j *the contraction operation*

$$\lrcorner : D_i \times \Lambda^j \to \Lambda^{j-i}$$

is defined. This operation can be extended onto the module $\Lambda^* \otimes D_*$,

$$\lrcorner : \Lambda^i \otimes D_k \times \Lambda^j \otimes D_\ell \to \Lambda^{i+j-k} \otimes D_\ell,$$

by setting

$$\Omega \lrcorner \Theta = i_\Omega(\Theta) = \omega \wedge (X \lrcorner\, \theta) \otimes Y$$

for any homogeneous decomposable elements $\Omega = \omega \otimes X$ and $\Theta = \theta \otimes Y$, $\omega, \theta \in \Lambda^*$, $X, Y \in D_*$. We also define the wedge product

$$\wedge : \Lambda^i \otimes D_k \times \Lambda^j \otimes D_\ell \to \Lambda^{i+j} \otimes D_{k+\ell}$$

by

$$\Omega \wedge \Theta = \omega \wedge \theta \otimes X \wedge Y.$$

With respect to this operation $\Lambda^* \otimes D_*$ becomes a bigraded commutative algebra, i.e.

$$\Omega \wedge \Theta = (-1)^{ij+k\ell} \Theta \wedge \Omega$$

for homogeneous $\Omega \in \Lambda^i \otimes D_k$ and $\Theta \in \Lambda^j \otimes D_\ell$. Moreover, if $\Xi \in \Lambda^* \otimes D_*$, then

$$\Omega \lrcorner (\Theta \wedge \Xi) = (\Omega \lrcorner \Theta) \wedge \Xi + (-1)^{(i+1)j} \Theta \wedge (\Omega \lrcorner \Xi) \tag{3.1}$$

and

$$\Omega \lrcorner (\Theta \lrcorner \Xi) = (\Omega \lrcorner \Theta) \lrcorner \Xi + (-1)^j (\Omega \wedge \Theta) \lrcorner \Xi \tag{3.2}$$

From now on we restrict ourselves to the graded module $\Lambda^* \otimes D_1 = \sum_{i \geq 0} \Lambda^i \otimes D_1$.

3.2. Richardson-Nijenhuis bracket.

Proposition 3.1. *Let $\Omega, \Theta \in \Lambda^* \otimes D_1$ be homogeneous elements. Then there exists a unique element $\Xi \in \Lambda^* \otimes D_1$ such that*

$$[i_\Omega, i_\Theta] = i_\Omega \circ i_\Theta - (-1)^{\overline{\Omega}\overline{\Theta}} i_\Theta \circ i_\Omega = i_\Xi. \tag{3.3}$$

The element Ξ is denoted by $[\![\Omega, \Theta]\!]^r$ and is called the Richardson-Nijenhuis bracket *of Ω and Θ.*

Remark 3.2. In (3.3) and everywhere below we use the following notation: If $A = \sum_i A_i$ and $a \in A_i$, then $(-1)^a = (-1)^i$ and $(-1)^{\overline{a}} = (-1)^{i+1}$.
Thus we have a correspondence

$$[\![\cdot,\cdot]\!]^r \cdot \Lambda^i \otimes D_1 \times \Lambda^j \otimes D_1 \to \Lambda^{i+j-1} \otimes D_1$$

With the following basic properties.

Proposition 3.3. *Let $\Omega, \Theta, \Xi \in \Lambda^* \otimes D_1$ be homogeneous elements. Then*

(i)
$$[\![\Omega,\Theta]\!]^r = \Omega \lrcorner \Theta - (-1)^{\overline{\Omega\Theta}} \Theta \lrcorner \Omega, \tag{3.4}$$

(ii)
$$[\![\Omega,\Theta]\!]^r + (-1)^{\overline{\Omega\Theta}} [\![\Theta,\Omega]\!]^r = 0, \tag{3.5}$$

(iii)
$$\oint (-1)^{(\Omega+\Xi)\overline{\Theta}} [\![\Omega, [\![\Theta,\Xi]\!]^r]\!]^r = 0, \tag{3.6}$$

where $\oint$ denotes the sum of all cyclic permutations,

(iv)
$$[\![\Omega, \rho \wedge \Theta]\!]^r = (\Omega \lrcorner \rho) \wedge \Theta + (-1)^{\rho\overline{\Omega}} \rho \wedge [\![\Omega,\Theta]\!]^r, \tag{3.7}$$

where ρ is a form.

Remark 3.4.
1. From (3.2) it follows that $\Lambda^1 \otimes D_1$ is an associative algebra with respect to contraction while

$$\Omega \mapsto i_\Omega \in \mathrm{End}_R(D_1)$$

is its representation.
2. Let $\Lambda^{*+1} \otimes D_1 = \sum_{i \geq -1} \Lambda^{i+1} \otimes D_1$ be the module $\Lambda^* \otimes D_1$ with a shifted grading. Then from (3.5) and (3.6) it follows that $\Lambda^{*+1} \otimes D_1$ is a graded Lie algebra with respect to the Richardson-Nijenhuis bracket.

3.3. Frölicher-Nijenhuis bracket. From now on we consider a smooth manifold P and take $C^\infty(P)$ for the algebra A. Hence, $D_i = D_i(P)$ and $\Lambda^j = \Lambda^j(P)$.
Let $\Omega = \omega \otimes X \in \Lambda^* \otimes D_1$. Define an action of Ω on Λ^* by

$$L_\Omega(\rho) = \omega \wedge X(\rho) + (-1)^\omega d\omega \wedge (X \lrcorner \rho),$$

where $\rho \in \Lambda^*$ and $X(\rho)$ is the Lie derivation

Proposition 3.5. *For any homogeneous $\Omega \in \Lambda^* \otimes D_1$ the correspondence $L_\Omega : \Lambda^* \to \Lambda^*$ is a graded derivation, i.e.*

$$L_\Omega(\rho \wedge \rho') = L_\Omega(\rho) \wedge \rho' + (-1)^{\rho\Omega} \rho \wedge L_\Omega(\rho'), \qquad \rho, \rho' \in \Lambda^*.$$

For $\Omega \in \Lambda^\circ \otimes D_1 = D_1$ it coincides with the Lie derivation.
If $\Omega, \Theta \in \Lambda^* \otimes D_1$ then there exists a unique element $\Xi \in \Lambda^* \otimes D_1$, such that

$$[\![L_\Omega, L_\Theta]\!] = L_\Omega \circ L_\Theta - (-1)^{\Omega\Theta} L_\Theta \circ L_\Omega = L_\Xi.$$

This element is denoted by $[\![\Omega,\Theta]\!]$ and is called the Frölicher-Nijenhuis bracket *of Ω and Θ.*

Thus one has a correspondence

$$[\![\cdot,\cdot]\!] : \Lambda^i \otimes D_1 \times \Lambda^j \otimes D_1 \to \Lambda^{i+j} \otimes D_1$$

the main properties of which are described by

Proposition 3.6. *Let $\Omega, \Theta, \Xi \in \Lambda^* \otimes D_1$ be homogeneous elements. Then*

(i) *For decomposable $\Omega = \omega \otimes X$ and $\Theta = \theta \otimes Y$ one has*

$$\begin{aligned}[\![\Omega,\Theta]\!] &= \omega \wedge \Theta \otimes [X,Y] + L_\Omega(\theta) \otimes Y - (-1)^{\Omega\Theta} L_\Theta(\omega) \otimes X = \\ &= \omega \wedge \Theta \otimes [X,Y] + \omega \wedge X(\theta) \otimes Y + (-1)^\Omega d\omega \wedge (X \lrcorner \theta) \otimes Y - \\ &- Y(\omega) \wedge \theta \otimes X + (-1)^\Omega (Y \lrcorner \omega) \wedge d\theta \otimes X, \end{aligned} \tag{3.8}$$

in particular, $[\![X,Y]\!] = [X,Y]$,

(ii)

$$[\![\Omega,\Theta]\!] + (-1)^{\Omega\Theta}[\![\Theta,\Omega]\!] = 0, \tag{3.9}$$

(iii)

$$\oint (-1)^{(\Omega+\Xi)\Theta}[\![\Omega,[\![\Theta,\Xi]\!]]\!] = 0, \tag{3.10}$$

(iv)

$$[\![\Omega,\rho\wedge\Theta]\!] = L_\Omega(\rho)\wedge\Theta - (-1)^{\overline{\Omega}(\rho+\Theta)} d\rho \wedge \Theta \lrcorner \Omega + (-1)^{\Omega\rho}\rho \wedge [\![\Omega,\Theta]\!], \tag{3.11}$$

where ρ is a form,

(v)

$$\Xi \lrcorner [\![\Omega,\Theta]\!] + (-1)^{\overline{\Omega}\Theta}[\![\Xi,\Theta]\!]\lrcorner\Omega = \tag{3.12}$$

$$= (-1)^{\overline{\Xi}\Omega}[\![\Omega,\Xi\lrcorner\Theta]\!] + [\![\Xi\lrcorner\Omega,\Theta]\!] + (-1)^\Omega[\![\Xi,\Omega]\!]\lrcorner\Theta.$$

Note that due to (3.9) and (3.10) $\Lambda^* \otimes D_1$ is a graded Lie algebra with respect to the Frölicher-Nijenhuis bracket.

3.4. Integrable elements and associated complexes. We say that an element $U \in \Lambda^1 \otimes D_1$ is *integrable*, if

$$[\![U,U]\!] = 0. \tag{3.13}$$

For any $U \in \Lambda^1 \otimes D_1$ one can define a mapping

$$\partial = \partial_U^i : \Lambda^i \otimes D_1 \to \Lambda^{i+1} \otimes D_1,\ i = 0,1,\dots,$$

by $\partial_U^i(\Omega) = [\![U,\Omega]\!]$, $\Omega \in \Lambda^* \otimes D_1$. In the case of integrable U, from (3.10) and (3.13) it follows that $\partial^2 = 0$, and thus one gets a complex

$$0 \to D_1 \to \Lambda^1 \otimes D_1 \to \dots \to \Lambda^i \otimes D_1 \xrightarrow{\partial_U^i} \Lambda^{i+1} \otimes D_1 \to \dots \tag{3.14}$$

Denote by

$$H^i_U(A) = \ker \partial^i_U / \mathrm{im}\ \partial^{i-1}_U$$

its cohomology at the i-th term.
An immediate corollary of proposition 3.3 is

Proposition 3.7. *For any homogeneous $\Omega, \Theta \in \Lambda^* \otimes D_1$ and $\rho \in \Lambda^*$ one has*

(i)

$$\partial[\![\Omega, \Theta]\!] = [\![\partial\Omega, \Theta]\!] + (-1)^{\Omega}[\![\Omega, \partial\Theta]\!], \tag{3.15}$$

(ii)

$$\partial(\rho \wedge \Omega) = L_U(\rho) \wedge \Omega - d\rho \wedge (\Omega \lrcorner U) + (-1)^{\rho} \rho \wedge \partial\Omega, \tag{3.16}$$

(iii)

$$\Omega \lrcorner \partial\Theta + (-1)^{\Omega}\partial(\Omega \lrcorner \Theta) + [\![\Omega, \Theta]\!] \lrcorner U = \tag{3.17}$$

$$= [\![\Omega \lrcorner U, \Theta]\!] + \partial(\Omega) \lrcorner \Theta.$$

Identity (3.15) shows that $H^*_U(A) = \sum_{i \geq 0} H^i_U(A)$ inherits the Frölicher-Nijenhuis bracket from $\Lambda^* \otimes D_1$ and thus becomes a graded Lie algebra with respect to this bracket.

3.5. Connections. Let M be a smooth manifold and $\xi : P \to M$ be a smooth fibre bundle. We consider $C^\infty(M)$ to be embedded into $C^\infty(P)$ and the latter to be $C^\infty(M)$-algebra.
A connection in the bundle ξ is a $C^\infty(M)$-linear mapping $\nabla : D(M) \to D(P)$, $\nabla(X) = \nabla_X$, such that

$$\nabla_X(f) = X(f),\ X \in D(M), \tag{3.18}$$

for any $f \in C^\infty(M)$.
Let $X, Y \in D(M)$ be two vector fields. Then one can consider an element

$$R_\nabla(X, Y) = [\nabla_X, \nabla_Y] - \nabla_{[X,Y]} \in D(P).$$

Thus we get a vector valued 2-form $R_\nabla \in \Lambda^2(M) \otimes_{C^\infty(M)} D(P)$ which is called *the curvature form* of the connection ∇.
For any connection ∇ consider in $D(P)$ a $C^\infty(P)$-submodule generated by the fields ∇_x, $x \in D(M)$, and denote it by $\nabla D(P)$, then $\nabla D(P)$ determines on P a distribution. Due to (3.18), the planes of this distribution are of the dimension $n = \dim M$ and are in general position with the bundles of ξ. It is easy to show that $\nabla D(P)$ is integrable iff $R_\nabla = 0$. In this case the connection ∇ is called *flat*.
Let $x \in P$, $v \in T_x(P)$, ξ_x be the fibre of ξ passing through x, and ∇_x be the plane $\nabla D(P)$ at the point x. Then one can consider the projection $U_x(v)$ of the vector v onto $T_x(\xi_x)$ along ∇_x. Respectively, if $X \in D(P)$ is a vector field, one can consider the projections $U_x(X_x)$ at each $x \in P$, which gives us a mapping

$$U = U_\nabla : D(P) \to D(P),$$

or, in other words, a one-form

$$U = U_\nabla \in \Lambda^1(P) \otimes D(P)$$

which is called *the connection form* for ∇.
With a connection at hand, one has a direct sum decomposition

$$D(P) = D^v(P) \otimes \nabla D(P) \tag{3.19}$$

which is given by the identity

$$X = U_\nabla(x) + (X - U_\nabla(x)), \tag{3.20}$$

$D^v(P)$ in (3.19) denotes the algebra of *ξ-vertical fields:*

$$D^v(P) = \{X \in D(P) | X(f) = 0, \ f \in C^\infty(M)\}$$

Proposition 3.8. *Let $\xi : P \to M$ be a fibre bundle and $\nabla : D(M) \to D(P)$ be a connection in ξ. Then:*

(i) *for any ξ-vertical field X^v one has*

$$X^v \lrcorner [\![U_\nabla, U_\nabla]\!] = 0;$$

(ii) *if $X^h, Y^h \in \nabla D(P)$ and $X^h = \sum_i f_i \nabla_{x_i}$, $Y^h = \sum_j g_j \nabla_{Y_j}$, $X_i, Y_j \in D(M)$, then*

$$Y^h \lrcorner X^h \lrcorner [\![U_\nabla, U_\nabla]\!] = 2 \sum_{i,j} f_i g_j R_\nabla(X_i, Y_j),$$

or, to be short,

$$[\![U_\nabla, U_\nabla]\!] = 2R_\nabla. \tag{3.21}$$

Hence, if the connection ∇ is flat, then U_∇ is integrable, and thus with any flat connection (see subsection 3.4) one can associate the differential $\partial_\nabla = [\![U_\nabla, \cdot]\!]$, the complex

$$0 \to D(P) \to \Lambda^1(P) \otimes D(P) \to \ldots \to \Lambda^2(P) \otimes D(P) \xrightarrow{\partial_\nabla} \Lambda^{i+1}(P) \otimes D(P) \to \ldots,$$

and cohomology modules

$$H^i_\nabla(P) = \ker \partial^i_\nabla / \operatorname{im} \partial^{i-1}_\nabla.$$

3.6. Vertical subtheory. Consider a submodule $\Lambda^*(P) \otimes D^v(P)$ in $\Lambda^*(P) \otimes D(P)$. Then the following statements are obvious:

(1) $[\![\nabla^*(P) \otimes D^v(P), \Lambda^*(P) \otimes D^v(P)]\!] \subset \Lambda^*(P) \otimes D^v(P)$,
(2) $U_\nabla \in \Lambda^1(P) \otimes D^v(P)$,
(3) $\Lambda^*(P) \wedge (\Lambda^*(P) \otimes D^v(P)) \subset \Lambda^*(P) \otimes D^v(P)$,
(4) $(\Lambda^*(P) \otimes D^v(P)) \lrcorner (\Lambda^*(P) \otimes D^v(P)) \subset \Lambda^*(P) \otimes D^v(P)$.

Hence, the theory developed in previous subsections, can be restricted onto $\Lambda^*(P) \otimes D^v(P)$. In particular, one has a complex

$$0 \to D^v(P) \to \Lambda^1(P) \otimes D^v(P) \to \ldots \to \Lambda^i(P) \otimes D^v(P) \xrightarrow{\partial_\nabla} \Lambda^{i+1}(P) \otimes D^v(P) \to \ldots \tag{3.22}$$

and corresponding cohomologies which we also denote by $H^i_\nabla(P)$.
Note that for any element $\Omega \in \Lambda^*(P) \otimes D^v(P)$ the identity

$$\Omega \lrcorner U_\nabla = \Omega \tag{3.23}$$

holds. Due to (3.23) equalities (3.16) and (3.17) transforms into

$$\partial_\nabla(\rho \wedge \Omega) = (L_{U_\nabla}(\rho) - d\rho) \wedge \Omega + (-1)^\rho \rho \wedge \partial_\nabla(\Omega) \tag{3.24}$$

and

$$\Omega \lrcorner \partial\Theta + (-1)^\Omega \partial(\Omega \lrcorner \Theta) = \partial(\Omega) \lrcorner \Theta \tag{3.25}$$

respectively.
From the last two equalities it follows that in the vertial subtheory the cohomology module $H^*_\nabla(P)$ possesses two additional structures. Namely, consider the mapping

$$d_h = d - L_{U_\nabla} : \Lambda^i(P) \to \Lambda^{i+1}(P),\ i = 0, 1, \ldots ,$$

It can be easily seen that $d_h^2 = 0$, and from (3.24) one can see that $H^*_\nabla(P)$ is a graded module over $H^*_0(P)$, where

$$H^*_0(P) = \ker d_h / \operatorname{im} d_h.$$

Secondly, (3.25) implies that $H^*_\nabla(P)$ inherits from $\Lambda^*(P) \otimes D^v(P)$ the contraction operation

$$\lrcorner : H^i_\nabla(P) \times H^j_\nabla(P) \to H^{i+j-1}_\nabla(P)$$

and the Richardson-Nijenhuis bracket

$$[\![\cdot,\cdot]\!]^r : H^i_\nabla(P) \times H^j_\nabla(P) \to H^{i+j-1}_\nabla(P).$$

3.7. $\mathcal{C}$-cohomologies of differential equations. From now on we consider a formally integrable equation $\mathcal{E} \subset J^k(\pi)$ and take the projection $\pi_\infty : \mathcal{E}^{(\infty)} \to M$ for the bundle ξ. From subsection 2.2 it follows that the operation $\mathcal{C} : D(M) \to D(\mathcal{E})$ is a flat connection in the bundle π_∞; it is called *the Cartan connection.* Hence, the theory of subsections 3.5-3.6 can be applied to the case of infinitely prolonged formally integrable equations. It gives one the complex

$$0 \to D^v(\mathcal{E}) \to \Lambda^1(\mathcal{E}) \otimes D^v(\mathcal{E}) \to \ldots \to \Lambda^i(\mathcal{E}) \otimes D^v(\mathcal{E}) \to \Lambda^{i+1}(\mathcal{E}) \otimes D^v(\mathcal{E}) \to \ldots$$

and cohomology modules

$$H^i_C(\mathcal{E}) = \ker \partial^i_\mathcal{E} / \operatorname{im} \partial^{i-1}_\mathcal{E}$$

which are called $\mathcal{C}$-cohomologies of the equation $\mathcal{E}$.
Note that in the case under consideration the differential d_h coincides with the horizontal differential of the equation $\mathcal{E}$ defined in subsection 2.4. Thus the following result takes place:

Theorem 3.9. *Let $\mathcal{E} \subset J^k(\pi)$ be a formally integrable equation. Then*

(i) *The cohomology module $H_{\mathcal{C}}^*(\mathcal{E})$ is a graded Lie algebra with respect to the Frölicher-Nijenhuis bracket $[\![\cdot,\cdot]\!]$;*
(ii) *$H_{\mathcal{C}}^*(\mathcal{E})$ is a graded module over cohomologies $E_0^{**}(\mathcal{E}) = \sum_{ij} E_0^{ij}(\mathcal{E})$ of the horizontal differential d_h;*
(iii) *The module $H_{\mathcal{C}}^{*+1}(\mathcal{E})$ with the shifted grading is a graded Lie algebra with respect to the Richardson-Nijenhuis bracket $[\![\cdot,\cdot]\!]^r$,*
(iv) *$H_{\mathcal{C}}^1(\mathcal{E})$ is an associative algebra with respect to contraction,*
(v) *The contraction of $H_{\mathcal{C}}^1(\mathcal{E})$ into $H_{\mathcal{C}}^0(\mathcal{E})$ determines a representation of this algebra as endomorphisms of $H_{\mathcal{C}}^0(\mathcal{E})$.*

Let $\varepsilon \in \mathbb{R}$ be a parameter. A smooth family $U(\varepsilon) \in \Lambda^1(\mathcal{E}) \otimes D^v(\mathcal{E})$ is called *a (vertial) deformation of the equation structure on $\mathcal{E}^{(\infty)}$*, if $U(0) = U_{\mathcal{C}}$ and $[\![U(\varepsilon), U(\varepsilon)]\!] = 0$ for any ε. A deformation is called *trivial*, if it originates from a one-parameter group $A_\varepsilon : \mathcal{E}^{(\infty)} \to \mathcal{E}^{(\infty)}$ of transformations of $\mathcal{E}^{(\infty)}$.
By expanding $U(\varepsilon)$ into a formal series in ε and considering the coefficient at the first power of ε in this expansion, one gets the notion of *infinitesimal deformation*: an element $U_1 \in \Lambda^1(\mathcal{E}) \otimes D^v(\mathcal{E})$ is an infinitesimal deformation of the equation structure, if $[\![U_{\mathcal{C}}, U_1]\!] = 0$.

Theorem 3.10. *For any formally integrable equation $\mathcal{E} \subset J^k(\pi)$*

(i) *$H_{\mathcal{C}}^0(\mathcal{E})$ coincides with the Lie algebra sym$(\mathcal{E})$;*
(ii) *$H_{\mathcal{C}}^1(\mathcal{E})$ is the set of equivalence classes of nontrivial infinitesimal deformations of the equation structure on $\mathcal{E}^{(\infty)}$.*

Thus, if Ω is a nontrivial deformation of $\mathcal{E}^{(\infty)}$ and X is a higher symmetry of $\mathcal{E}$, then $X \lrcorner \Omega$ is a symmetry again.

3.8. Some computational results and local coordinates representation. Let $\pi : E \to M$ be a vector bundle and $\mathcal{E}$ be the "empty" equation, i.e. let $\mathcal{E}^{(\infty)} = J^\infty(\pi)$. Denote $\mathcal{C}$-cohomologies by $H_{\mathcal{C}}^i(\pi)$ in this case.

Theorem 3.11. *For any vector bundle $\pi : E \to M$ and $i \geq 0$ one has*

$$H_{\mathcal{C}}^p(\pi) \approx F(\pi,\pi) \otimes_{F(\pi)} \mathcal{C}^p\Lambda(\pi)$$

(for the definition of $\mathcal{C}^p\Lambda(\pi)$ see subsection 2.4). This isomorphism is given by the correspondence

$$\ni : F(\pi,\pi) \otimes_{F(\pi)} \mathcal{C}^p\Lambda(\pi) \to \Lambda^p(\pi) \otimes D^v(\pi)$$

which in local coordinates is represented as

$$\ni_\Omega = \sum_{j,\sigma} D_\sigma(\Omega^j) \otimes \partial/\partial p_\sigma^j,$$

where $\Omega = (\Omega^1, \dots, \Omega^m)$, $\Omega^1, \dots, \Omega^m \in \mathcal{C}^1\Lambda(\pi)$.

Let Ω be an element of $H_{\mathcal{C}}^1(\pi)$ and

$$\Omega^j = \sum_{s,\sigma} \varphi_{s,\sigma}^j \cdot \omega_\sigma^s, \qquad j = 1, \dots, m,$$

be its representation in local coordinates, $\omega^s_\sigma = dp^s_\sigma - \sum_i p^s_{\sigma+1_i} dx_i$ being the basis of Cartan forms in $\mathcal{C}^1\Lambda(\pi)$. Then the action of Ω on $H^0_{\mathcal{C}}(\pi) \approx F(\pi,\pi)$ given by contraction is described as

$$Q_\Omega(\psi) = \psi \lrcorner \Omega = \left(\sum_{s,\sigma} \varphi^1_{s,\sigma} D_\sigma(\psi^s), \dots, \sum_{s,\sigma} \varphi^m_{s,\sigma} D_\sigma(\psi^s) \right), \tag{3.26}$$

where $\psi = (\psi^1, \dots, \psi^m) \in F(\pi, \pi)$.
Let now $\mathcal{E} \subset J^k(\pi)$ be a system of evolution equations

$$\begin{cases} \frac{\partial u^1}{\partial t} = f^1(t, x, \dots, \frac{\partial^{|\sigma|} u^j}{\partial x_\sigma}, \dots), \\ \dots \\ \frac{\partial u^m}{\partial t} = f^m(t, x, \dots, \frac{\partial^{|\sigma|} u^j}{\partial x_\sigma}, \dots), \end{cases} \tag{3.27}$$

where $x = (x_1, \dots, x_n)$. Consider functions $t, x_1, \dots, x_n$ and p^j_σ, $j = 1, \dots, m$, $|\sigma| \geq 0$, for the *internal coordinates* on $\mathcal{E}^{(\infty)}$.
For any $\Omega \in J(\mathcal{E}, \pi) \otimes \mathcal{C}^p\Lambda(\mathcal{E})$ one can define an element $\ell^{(p)}_{\mathcal{E}}(\Omega) \in F(\mathcal{E}, \pi) \otimes \mathcal{C}^p\Lambda(\mathcal{E})$ by setting

$$\left(\ell^{(p)}_{\mathcal{E}}(\Omega)\right)^j = \sum_{s,\sigma} \frac{\partial f^j}{\partial p^s_\sigma} D_\sigma(\Omega^s) - D_t(\Omega^j), \qquad j = 1, \dots, m. \tag{3.28}$$

The mapping $\ell^{(p)}_{\mathcal{E}} : F(\mathcal{E}, \pi) \otimes \mathcal{C}^p\Lambda(\mathcal{E}) \to F(\mathcal{E}, \pi) \otimes \mathcal{C}^p\Lambda(\mathcal{E})$ is well-defined and can be viewed as the extension of the universal linearization operator for $\mathcal{E}$ from $\mathcal{F}(\mathcal{E}, \pi)$ to $F(\mathcal{E}, \pi) \otimes \mathcal{C}^p\Lambda(\mathcal{E})$.

Theorem 3.12. *For any evolution equation of the form (3.27) one has*

$$H^p_{\mathcal{C}}(\mathcal{E}) = \begin{cases} \ker \ell^{(0)}_{\mathcal{E}}, & p = 0 \\ \ker \ell^{(p)}_{\mathcal{E}} \oplus dt \wedge coker\ \ell^{(p-1)}_{\mathcal{E}}, & p > 0 \end{cases}$$

where the action of $\ell^{(p)}_{\mathcal{E}}$ *is given by (3.28).*

Keeping in mind the action of $H^1_{\mathcal{C}} = \ker \ell^{(1)}_{\mathcal{E}} \oplus dt \wedge \operatorname{coker} \ell^{(0)}_{\mathcal{E}}$ one can see that only the first summand in this decomposition acts nontrivially on $H^0_{\mathcal{C}}(\mathcal{E}) = \operatorname{sym} \mathcal{E}$. Let us write down the equations for computing $\ker \ell^{(1)}_{\mathcal{E}}$ in the case $m = 1$, $n = 1$, i.e. when $\mathcal{E}$ is a scalar equation in one time and or space variables

$$\frac{\partial u}{\partial t} = f(t, x, u, \dots, \frac{\partial^k u}{\partial x^k}).$$

For the internal coordinates on $\mathcal{E}^{(\infty)}$ we take the functions $t, x, p_0, \dots, p_s, \dots$, where p_s corresponds to $\frac{\partial^s u}{\partial x^s}$. Let $\omega_0 = dp_0 - p_1 dx - f dt, \dots, \omega_s = dp_s - p_{s+1} dx - D^s_x(f) dt, \dots$, be the basis of Cartan form in $\mathcal{C}^1\Lambda(\mathcal{E})$ and let

$$\Omega = \sum_{s=0}^{N} \varphi_s \cdot \omega_s \in \mathcal{C}^1\Lambda(\mathcal{E}).$$

Then $\Omega \in \ker \ell_{\mathcal{E}}^{(1)}$ iff

$$\ell_{\mathcal{E}}(\varphi_s) = \sum_{i=0}^{N} \varphi_i D_x^i(f_s) + \sum_{j=1}^{s} \left(\sum_{i=j}^{N} \binom{i}{j} \varphi_i D_x^{i-j}(f_{s-j}) - \sum_{r=j}^{k} f_r D_x^{r-j}(\varphi_{s-j}) \right), \quad (3.29)$$

where $s = 0, 1, \dots, N$, $f_\alpha = \frac{\partial f}{\partial p_\alpha}$.

4. Coverings and recursion operators

Here we briefly expose the nonlocal theory of differential equations (the theory of coverings [5]). A nonlocal version of $\mathcal{C}$-cohomologies is developed and the results concerning the extension of cohomological invariants onto coverings are proved (cf. [2, 5]).
It is shown that there exist obstructions to such extensions and these obstructions are interpreted in terms of Vinogradov's $\mathcal{C}$ spectral sequence [8].

4.1. Coverings. Let $\mathcal{E} \subset J^k(\pi)$ be an equation and $\mathcal{E}^{(\infty)} \subset J^\infty(\pi)$ be its infinite prolongation. Consider a fibre bundle $\tau : P \to \mathcal{E}^{(\infty)}$ *A covering structure* is determined in τ (or τ is *a covering* over $\mathcal{E}^{(\infty)}$), if

(i) a flat connection $\mathcal{C}_\tau : D(M) \to D(P)$ is given in the bundle $\pi_\tau = \pi_\infty \circ \tau : P \to M$,
(ii) this connection is in agreement with the Cartan connection $\mathcal{C}$ in $\pi_\infty : \mathcal{E}^{(\infty)} \to M$, i.e. $\tau_*(\mathcal{C}_\tau(X)) = \mathcal{C}X$ for any $X \in D(M)$.

Two coverings $\tau : P \to \mathcal{E}^{(\infty)}$ and $\tau' : P' \to \mathcal{E}^{(\infty)}$ are said to be *equivalent*, if there exists a diffeomorphism $\varphi : P \to P'$ such that

(i) $\tau = \tau' \circ \varphi$,
(ii) $\varphi_* \mathcal{C}_\tau(X) = \mathcal{C}_{\tau'}(X)$ for any $X \in D(M)$.

For any $\mathcal{E} \subset J^k(\pi)$ and $N \in \mathbb{Z}_+$ *a standard N-dimensional covering* $\tau_N : P_N = \mathcal{E}^{(\infty)} \times \mathbb{R}^N \to \mathcal{E}^{(\infty)}$ is defined. Namely, we take the projection on the first factor for τ_N and set

$$\mathcal{C}_{\tau_N}(X) = \mathcal{C}X, \qquad X \in D(M).$$

Or, to be more formal, we consider the connection $\nabla : D(\mathcal{E}^{(\infty)}) \to D(P_N)$ naturally associated with the direct product structure in P_N and set $\mathcal{C}_{\tau_N} = \nabla \circ \mathcal{C}$. A covering τ is called *trivial*, if it is equivalent to some τ_N.

Let $\tau' : P' \to \mathcal{E}^{(\infty)}$, $\tau'' : P'' \to \mathcal{E}^{(\infty)}$ be two coverings. A covering $\tau : P \to \mathcal{E}^{(\infty)}$ is called *the Whitney product* of τ' and τ'', if

(i) *the bundle* τ is the Whitney product of *the bundles* τ' and τ'' (and is denoted by $\tau = \tau'' \oplus \tau''$), thus the following commutative diagram takes place

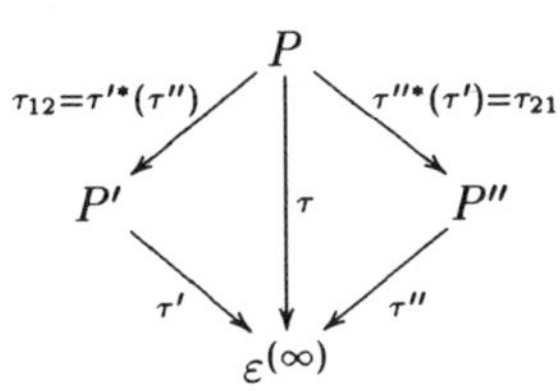

(ii) for any $f' \in C^\infty(P')$ and $f'' \in C^\infty(P'')$ one has

$$\mathcal{C}_\tau(X)(\tau_{12}^* f') = \mathcal{C}_{\tau'}(X)(f')$$

and

$$\mathcal{C}_\tau(X)(\tau_{21}^* f'') = \mathcal{C}_{\tau''}(X)(f''),$$

where $X \in D(M)$.

A covering τ is called *reducible*, if there exists a covering τ' such that τ is equivalent to $\tau' \oplus \tau_N$, τ_N being a trivial covering. Otherwise, τ is called *irreducible*.

Local coordinate representation of coverings is as follows. Consider a local trivialization of τ and let $w^1, \dots, w^s, \dots$ be coordinates along the fibre of τ in this trivialization. Let

$$D_1 = \mathcal{C}(\partial/\partial x_1), \dots, D_n = \mathcal{C}(\partial/\partial x_n)$$

be total derivatives on $\mathcal{E}^{(\infty)}$. Then a covering structure in τ is determined by the system of vector fields

$$\widetilde{D}_i = D_i + X_i, \qquad i = 1, \dots, n, \tag{4.1}$$

where $X_i = \sum_s X_i^s \partial/\partial w^s$, satisfying the integrability conditions

$$0 = [\widetilde{D}_i, \widetilde{D}_j] = D_i(X_j) - D_j(X_i) + [X_i, X_j], \qquad i, j = 1, \dots, n \tag{4.2}$$

(we use the notation $D_i(X_j) = \sum_s D_i(X_j^s)\partial/\partial w^s$).

Recall that a π_τ-vertical vector field $Y \in D(P)$ is called *a nonlocal symmetry of the type* τ, or a *τ-symmetry*, if $[Y, \mathcal{C}_\tau(X)] = 0$ for any $X \in D(M)$. The set of all τ-symmetries is denoted by $\mathrm{sym}_\tau(\mathcal{E})$ and forms a Lie algebra with respect to commutator of vector fields. The following local description of τ-symmetries is useful in applications.

Theorem 4.1. *Let $\mathcal{E} \subset J^k(\pi)$ be an equation and $\tau : P \to \mathcal{E}^{(\infty)}$ be its covering. Consider an extension $\widetilde{\ell}_\mathcal{E}$ of the operator $\ell_\mathcal{E}$ onto P locally determined by*

$$\widetilde{\ell}_\mathcal{E} = \| \sum_\sigma \frac{\partial F^\ell}{\partial p_\sigma^j} \widetilde{D}_\sigma \|_P$$

(see (2.9). Let $\varphi \in \Gamma(\pi_\tau^*(\pi))$ and

$$\widetilde{\ni}_\varphi = \sum_{j,\sigma} \widetilde{D}_\sigma(\varphi^j)\partial/\partial p_\sigma^j.$$

then a vector field $Y \in D(P)$ lies in $\mathrm{sym}_\tau(\mathcal{E})$ if and only if

$$Y = \widetilde{\ni}_\varphi + \sum_s Y^s \partial/\partial w^s,$$

where

$$\widetilde{\ell}_\mathcal{E}(\varphi) = 0 \tag{4.3}$$

and

$$\tilde{\ni}_\varphi(X_i^s) + \sum_j Y^j \frac{\partial X_i^s}{\partial w^j} = \widetilde{D}_i(Y^s), \qquad i = 1, \dots, n \tag{4.4}$$

(see (4.1))

The solutions of (4.3) are called *the shadows* of nonlocal symmetries. In general, if one has a shadow φ in a covering τ, equations (4.4) are not necessary solvable, i.e. not every shadow is extendable to a real nonlocal symmetry in a particular covering. But one can prove the following result.

Theorem 4.2. *Let $\tau : P \to \mathcal{E}^{(\infty)}$ be a covering and φ be a shadow in this covering. Then there exist a covering $\tau' : P' \to \mathcal{E}^{(\infty)}$ and a projection $\nu : P' \to P$ such that*

(i) *$\tau' = \tau \circ \nu$,*
(ii) *equations (4.4) are solvable for the covering τ'.*

In particular, for any local symmetry $\varphi \in sym(\mathcal{E})$ there exists a covering τ such that φ is extendable to a nonlocal symmetry in this covering.

4.2. Extending $\mathcal{C}$-cohomologies onto coverings. Let $\mathcal{E} \subset J^k(\pi)$ be an equation and $\tau : P \to \mathcal{E}^{(\infty)}$ be a covering. Then one has a flat connection $\nabla = \mathcal{C}_\tau : D(M) \to D(P)$ with which corresponding cohomologies $H^p_\nabla(P)$ are associated. We denote these cohomologies by $H^p_\mathcal{C}(\mathcal{E};\tau)$ and call them *$\mathcal{C}$-cohomologies of the equation $\mathcal{E}$ with the coefficients in the covering τ*. Repeating literally considerations of subsection 3.7 one can prove that

$$H^0_\mathcal{C}(\mathcal{E};\tau) = \mathrm{sym}_\tau(\mathcal{E})$$

while $H^1_\mathcal{C}(\mathcal{E};\tau)$ coincides with non-trivial equivalence classes of τ-deformations for the equation structure on $\mathcal{E}^{(\infty)}$.

From now on we confine ourselves to the case of a scalar evolution equation in one space and one time variable. Let

$$u_t = f(t, x, u, \dots, \partial^k u \big/ \partial x^k)$$

be such an equation. Consider a covering $\tau : P \to \mathcal{E}^{(\infty)}$, and let $(t, x, p_0, \dots, p_i, \dots, w^1, \dots, w^s, \dots)$ be local coordinates in P. Then τ can be determined by the fields

$$\widetilde{D}_x = D_x + X, \quad \widetilde{D}_\tau = D_t + T,$$

where

$$X = \sum_s X^s \partial \big/ \partial w^s, \quad T = \sum_s T^s \partial \big/ \partial w^s$$

are τ-vertical fields satisfying the relations

$$\begin{cases} D_x(T^s) - D_t(X^s) + \sum_\beta (X^\beta \frac{\partial T^s}{\partial w^\beta} - T^\beta \frac{\partial X^s}{\partial w^\beta}) = 0 \\ s = 1, 2, \dots \end{cases}$$

Consider the forms

$$\omega_i = dp_i - p_{i+1}dx - D_x^i(f)dt, \qquad i = 0, 1, \dots,$$
$$\theta^s = dw^s - X^s dx - T^s dt, \qquad s = 1, 2, \dots,$$

which constitute a basis in $\mathcal{C}_\tau^1\Lambda^1(P)$, the module of 1-forms vanishing on the Cartan distribution on P. The connection form for $\mathcal{C}_\tau$ is

$$U_\tau = \sum_i \omega_i \otimes \partial/_{\partial p_i} + \sum_s \theta^s \otimes \partial/_{\partial w^s}$$

and for any

$$\Phi = \sum_i \Phi_i \otimes \partial/_{\partial p_i} + \sum_s \varphi^s \otimes \partial/_{\partial w^s} \in \Lambda^p(P) \otimes D^v(P)$$

one has

$$\partial_\tau(\Phi) = [\![U_\tau, \Phi]\!] = S_p(\Phi) + R_p(\Phi), \tag{4.5}$$

where

$$S_p(\Phi) = \sum_i \{dx \wedge (\Phi_{i+1} - \widetilde{D}_x(\Phi_i)) \tag{4.6}$$
$$+ dt \wedge (\sum_\alpha \frac{\partial}{\partial p_\alpha}(D_x^i f)\Phi_\alpha - \widetilde{D}_t(\Phi_i))\} \otimes \partial/_{\partial p_i},$$
$$R_p(\Phi) = \sum_s \{dx \wedge (\sum_\alpha \frac{\partial X^s}{\partial p_\alpha}\Phi_\alpha + \sum_\beta \frac{\partial X^s}{\partial w^p}\varphi^\beta - \widetilde{D}_x(\varphi^s)) \tag{4.7}$$
$$+ dt \wedge (\sum_\lambda \frac{\partial T^s}{\partial p_\alpha}\Phi_\alpha + \sum_\beta \frac{\partial T^s}{\partial w^\beta}\varphi^\beta - \widetilde{D}_t(\varphi^s))\} \otimes \partial/_{\partial w^s},$$

Just like it was done in subsection (2.4), one can split the modules $\Lambda^p(P)$ as

$$\Lambda^p(P) = \sum_{\alpha+\beta=p} \Lambda_0^\alpha(P) \wedge \mathcal{C}_\tau^\beta\Lambda(P),$$

where $\Lambda_0^\alpha(P)$ is the module of $\pi_\infty \circ \tau$-horizontal forms on P and $\mathcal{C}_\tau^\beta\Lambda(P) = \mathcal{C}_\tau^1\Lambda(P) \wedge \ldots \wedge \mathcal{C}_\tau^1\Lambda(P)$ (β times). From (4.5)-(4.7) it is clear that

$$\partial_\tau(\Lambda_0^\alpha(P) \wedge \mathcal{C}_\tau^\beta\Lambda(P) \otimes D^v(P)) \subset \Lambda_0^{\alpha+1}(P) \wedge \mathcal{C}_\tau^\beta\Lambda(P) \otimes D^v(P)$$

and thus

$$H_{\mathcal{C}}^p(\mathcal{E};\tau) = \sum_{\alpha+\beta=p} H_{\mathcal{C}}^{\alpha,\beta}(\mathcal{E};\tau), \tag{4.8}$$

where

$$H_{\mathcal{C}}^{\alpha,\beta}(\mathcal{E};\tau) = \ker\partial_\tau^{\alpha,\beta}/_{\operatorname{im}\,\partial_\tau^{\alpha-1,\beta}}$$

and

$$\partial_\tau^{\alpha,\beta} = \partial_\tau\Big/_{\Lambda_0^\alpha(P)\wedge \mathcal{C}_\tau^\beta\Lambda(P)\otimes D^v(P)}.$$

Consider the contraction operation induced in $H_\mathcal{C}^*(\mathcal{E},\tau)$. Then one has

$$H_\mathcal{C}^{\alpha,\beta}(\mathcal{E};\tau)\lrcorner H_\mathcal{C}^{\gamma,\delta}(\mathcal{E};\tau)\subset H_\mathcal{C}^{\alpha+\beta,\beta+\delta-1}(\mathcal{E};\tau).$$

In particular,

$$H_\mathcal{C}^1(\mathcal{E};\tau) = H_\mathcal{C}^{1,0}(\mathcal{E};\tau)\otimes H_\mathcal{C}^{0,1}(\mathcal{E};\tau)$$

and

$$H_\mathcal{C}^{\alpha,\beta}(\mathcal{E};\tau)\lrcorner H_\mathcal{C}^{1,0}(\mathcal{E};\tau)\subset H_\mathcal{C}^{\alpha+1,\beta-1}(\mathcal{E};\tau),$$
$$H_\mathcal{C}^{\alpha,\beta}(\mathcal{E};\tau)\lrcorner H_\mathcal{C}^{0,1}(\mathcal{E};\tau)\subset H_\mathcal{C}^{\alpha,\beta}(\mathcal{E};\tau).$$

Hence, only the $H_\mathcal{C}^{0,1}(\mathcal{E};\tau)$ summand of (4.8) decomposition for $H_\mathcal{C}^1(\mathcal{E};\tau)$ act non-trivially on $\mathrm{sym}_\tau(\mathcal{E}) = H_\mathcal{C}^{0,0}(\mathcal{E};\tau)$ by contraction. The following statement describes the elements of the $H_\mathcal{C}^{0,p}$-component in (4.8).

Proposition 4.3. *For any evolution equation*

$$\mathcal{E} : u_t = f(x,u,\dots,\partial^k u\big/_{\partial x^k})$$

and its covering $\tau : P\to\mathcal{E}^{(\infty)}$ the module $H_\mathcal{C}^{0,p}(\mathcal{E};\tau)$, $p\geq 0$, consists of the elements $\Phi = \sum_i \Phi_i\otimes\partial\big/_{\partial p_i} + \sum_s \varphi^s\otimes\partial\big/_{\partial w^s}$, $\Phi_i,\varphi^s\in\mathcal{C}_\tau^p\Lambda(P)$, such that

$$\Phi_i = \widetilde{D}_x^i\Phi_0$$

and

$$\begin{cases} \widetilde{\ell}_\mathcal{E}^{[p]}(\Phi_0) = 0, & (4.9)\\ \displaystyle\sum_\alpha \frac{\partial X^s}{\partial p_\alpha}\widetilde{D}_x^\alpha(\Phi_0) + \sum_\beta \frac{\partial X^s}{\partial w^\beta}\varphi^\beta = \widetilde{D}_x(\varphi^s), & (4.10)\\ \displaystyle\sum_\alpha \frac{\partial T^s}{\partial p_\alpha}\widetilde{D}_x^\alpha(\Phi_0) + \sum_\beta \frac{\partial T^s}{\partial w^\beta}\varphi^\beta = \widetilde{D}_t(\varphi^s), & (4.11)\\ s = 1,2,\dots, \end{cases}$$

where $\widetilde{\ell}_\mathcal{E}^{[p]}$ is the natural extension of $\widetilde{\ell}_\mathcal{E} = \widetilde{\ell}_\mathcal{E}^{[0]}$ onto $\mathcal{C}_\tau^p\Lambda(P)$.

We call (4.9) the shadow equation and (4.10), (4.11) the relation equations in the covering τ. We also call the solutions of (4.9) the p-shadows in τ. For the case $p=0$ (symmetries) we saw that for any covering τ ans a shadow Φ_0 there existed a covering τ' such that the relation equations were solvable on τ'. It seems that it is not true for arbitrary p, and we shall consider some obstructions for reconstructing elements of $H_{\mathcal{C}}^{0,p}(\mathcal{E};\tau)$ by their shadows in the next subsection. Before doing this we prove a useful result concerning the action of 1-shadows onto nonlocal symmetries.

Note first that form (4.5) – (4.7) it follows that an element Φ is a shadow iff $\partial_\tau(\Phi)$ is a τ-vertical element (i.e. its coefficients at $\partial/\partial p_i$, $i=0,1,\ldots,$ vanish).

Let now Φ be a p-shadow, i.e. $S_p(\Phi)=0$, and $\Psi \in H_{\mathcal{C}}^{0,q}(\mathcal{E};\tau) = \ker \partial_\tau^{0,q}$. Then, due to (3.25), one has

$$\partial_\tau(\Psi \lrcorner \Phi) = (-1)^{\Psi}\{\partial_\tau(\Psi)\lrcorner\Phi - \Psi\lrcorner\partial_\tau\Phi\} = -(-1)^{\Psi}\Psi\lrcorner\partial_\tau\Phi.$$

But since Φ is a shadow, $\partial_\tau(\Phi)$ is τ-vertical. Hence, $\Psi\lrcorner\partial_\tau\Phi$ is τ-vertical as well. It means that $\Psi\lrcorner\Phi$ is a $(p+q-1)$-shadow. Taking $p=1$ and $q=0$ one has the following

Theorem 4.4. *Let Ψ be a τ-symmetry and $\Phi \in \mathcal{C}_\tau^1\Lambda(P)\otimes D^v(P)$ be a shadow of τ-deformation for $\mathcal{E}$, i.e. $\widetilde{\ell}_{\mathcal{E}}^{[1]}(\Phi)=0$. Then $\Psi\lrcorner\Phi$ is a shadow of a nonlocal symmetry in τ, i.e. $\widetilde{\ell}_{\mathcal{E}}^{[0]}(\Psi\lrcorner\Phi)=0$, and hence there exists a covering τ' over $\mathcal{E}$, such that the diagram*

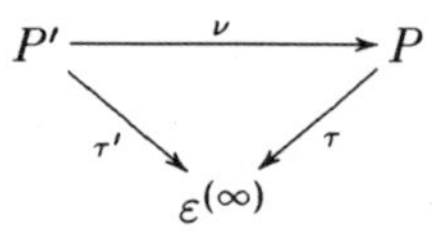

is commutative, onto which $\Psi\lrcorner\Phi$ can be extended.

4.3. Vinogradov's $\mathcal{C}$-spectral sequence and obstructions. Let P be a smooth manifold and $\mathcal{D}\Lambda^1(P)\subset\Lambda^1(P)$ be a submodule consisting of 1-forms vanishing on an integrable distribution $\mathcal{D}$ on P. Then one can considers the ideal $\mathcal{D}\Lambda^*(P)$ generated by $\mathcal{D}\Lambda^1(P)$ and its powers

$$(\mathcal{D}\Lambda^*(P))^p = \mathcal{D}\Lambda^*(P)\wedge\ldots\wedge\mathcal{D}\Lambda^*(P) \qquad (p \text{ times})$$

Obviously,

$$\Lambda^*(P)\supset\mathcal{D}\Lambda^*(P)\supset\ldots\supset(\mathcal{D}\Lambda^*(P))^p\supset(\mathcal{D}\Lambda^*(P))^{p+1}\supset\ldots$$

and

$$d(\mathcal{D}\Lambda^*(P))^p\subset(\mathcal{D}\Lambda^*(P))^p.$$

Thus a spectral sequence is defined which converges to the de Rham cohomologies of P. This is the general formulation for the Vinogradov's $\mathcal{C}$-spectral sequence.

In the case $P=\mathcal{E}^{(\infty)}$ and $\mathcal{D}$ being the Cartan distribution on $\mathcal{E}^{(\infty)}$ this spectral sequence coincides with that associated with the bi-complex $\{\Lambda_0^q(\mathcal{E})\wedge\mathcal{C}^p\Lambda(\mathcal{E}), d_{\mathcal{C}}, d_h\}$

defined in subsection 2.4. The differential $d_0^{p,q}$ for this spectral sequence is the horizontal differential

$$d_h = d_h^{p,q} : \Lambda_0^q(\mathcal{E}) \wedge \mathcal{C}^p\Lambda(\mathcal{E}) \to \Lambda_0^{q+1}(\mathcal{E}) \wedge \mathcal{C}^p\Lambda(\mathcal{E}).$$

The term $E_1^{p,q}$ coincides with the cohomology

$$H^{p,q}(\mathcal{E}) = \ker d_h^{p,q} \Big/ \operatorname{im} d_h^{p,q-1}$$

and the differentials

$$d_1^{p,q} : E_1^{p,q}(\mathcal{E}) \to E_1^{p+1,q}(\mathcal{E})$$

are induced by the Cartan differential $d_{\mathcal{C}}$.

Quite in the same way one can construct the $\mathcal{C}$-spectral sequence for any covering $\tau : P \to \mathcal{E}^{(\infty)}$ taking the Cartan distribution $\mathcal{C}_\tau(P)$ for $\mathcal{D}$. Its terms will be denoted by $E_h^{p,q}(\mathcal{E};\tau)$. In this situation one has the analogues for the Cartan and the horizontal differentials which will be denoted by $\tilde{d}_{\mathcal{C}}$ and $\tilde{d}_h$ respectively. In the sequel we shall suppose the coverings under consideration to satisfy *the 2 lines hypothesis: all the terms $E_1^{p,q}(\mathcal{E};\tau)$ are trivial except for the cases $q = n-1$, n* or $p = 0$ (in the local setting this is the theorem valid for the equations "in general position" and proved by A.M. Vinogradov). Thus the configuration of the E_1-term should be as follows: Let now Φ be a p-shadow in a covering τ of the evolution equation $\mathcal{E}$. Then from

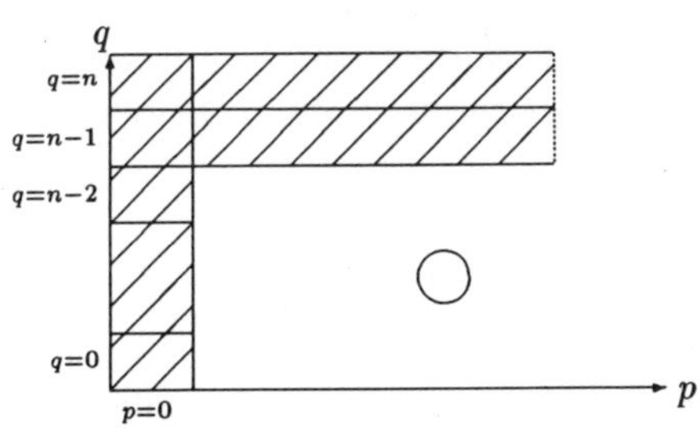

FIGURE 1. Configuration of the E_1 term

(4.5) – (4.7) it follows that

$$\partial_\tau(\Phi) = R_p(\Phi) = \sum_s R_{p,s}(\Phi) \otimes \partial \Big/ \partial w^s,$$

where

$$R_{p,s}(\Phi) = dx \wedge \left(\sum_\alpha \frac{\partial X^s}{\partial p_\alpha}\Phi_\alpha + \sum_\beta \frac{\partial X^s}{\partial w^\beta}\varphi^\beta - \widetilde{D}_x(\varphi^s) \right) + dt \wedge \left(\sum_\alpha \frac{\partial T^s}{\partial p_\alpha}\Phi_\alpha + \sum_\beta \frac{\partial T^s}{\partial w^\beta}\varphi^\beta - \widetilde{D}_t(\varphi^s) \right) \tag{4.12}$$

Then from (3.24) rewritten for the nonlocal case one has

$$0 = \partial_\tau^2(\Phi) = -\sum_s \tilde{d}_h(R_{p,s}\Phi) \otimes \partial/\partial w^s.$$

Hence, $R_{p,s}\Phi$ are $\tilde{d}_h$-closed forms and thus determine some elements $[\Phi_s^p]$ of the term $E_1^{p,1}(\mathcal{E};\tau)$. Obviously, one has the following

Proposition 4.5. *In the case when the coefficients $X^1, T^1, X^2, T^2, \ldots$ do not depend on nonlocal variables $w^1, w^2, \ldots$ a p-shadow Φ can be extended up to an element of $H_{\mathcal{C}}^{0,p}(\mathcal{E};\tau)$ iff $[\Phi_s^p] = 0$, $s = 1, 2, \ldots$.*

In particular, one can see that the obstructions to extending the shadows of τ-deformations within the same covering τ lie in the term $E_1^{1,0}(\mathcal{E};\tau)$.

Consider now the elements $\partial_1^{1,1}[\Phi_1^1] \in E_1^{2,1}(\mathcal{E};\tau)$. In the case they vanish, $[\Phi_s^1]$ determine the elements $[[\Phi_s^1]] \in E_2^{1,1}(\mathcal{E};\tau)$. Under the 2 lines hypothesis, the latter should be trivial: in fact, otherwise they will survive in $E_\infty^{1,1}(\mathcal{E};\tau)$ thus determining nontrivial elements in the de Rham cohomology group $H^2(P)$. But this group is trivial due to homotopical triviality of $P = \mathbb{R}^\infty$. Hence, there exist elements $[\Psi_s^1] \in E_1^{0,1}(\mathcal{E};\tau)$ such that $\partial_1^{0,1}[\Psi_s^1] = [\Phi_s^1]$. But the term $E_1^{0,1}(\mathcal{E};\tau)$ coincides with the first horizontal cohomology group for the covering τ. Therefore, one can consider a covering τ' corresponding to these elements. For this covering a commutative diagram exists.

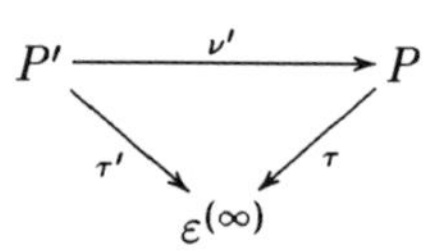

while nonlocal coordinates $\{w_s^1\}$ along the fibre of y^1 are determined by the relations

$$\tilde{d}_h w_s^1 = \Psi_s^1, \qquad s = 1, 2, \ldots$$

By taking

$$\theta_s^{1,1} = \tilde{d}_{\mathcal{C}} w_s^1, \qquad s = 1, 2, \ldots$$

one can see that

$$d_0^{1,0}\theta_s^1 = R_{1,s}\Phi, \qquad s = 1, 2, \ldots$$

Hence, the elements $[\Phi_s^1]$ become trivial on τ'. It means that the shadow Φ in the covering τ has been extended up to a shadow in the covering τ'

Proposition 4.6. *If the covering τ with the coefficients not depending on nonlocal variables satisfies the 2 lines hypothesis and $E_1^{2,1}(\mathcal{E};\tau) = 0$, then for any 1-shadow Φ there exists a covering τ^1 such that Φ can be extended up to a 1-shadow Φ^1 in τ^1.*

If one meets no obstructions in τ^1, Φ^1 can be extended to some Φ^2 in τ^2, and so on. As the inverse limit one will get a covering τ^∞ and the element $\Phi^\infty \in H_C^{0,1}(\mathcal{E};\tau^\infty)$. *By contraction Φ^∞ will act as a recursion operator on* $\mathrm{sym}_{\tau^\infty}(\mathcal{E})$. Thus the process of extension for deformations is analogous to that one for symmetries, but on each step obstructions in $E_1^{2,1}(\mathcal{E},\tau)$ may arise.

5. Applications

In this section we shall discuss and compute deformations for several evolution equations and demonstrate how the classical recursion operators for integrable systems can be directly copied form these results.
This will be demonstrated for Burgers' equation, KdV -equation, nonlinear Schrödinger equations and classical Boussinesq equations. From the resulting structure of the overdetermined systems the ansatz for recursion operators can be formulated [9].
Moreover it is shown that deformations of evolution equations are just examples of symmetries in graded coverings of the equations at hand, i.e. in effect we are constructing symmetries in the Cartan-covering of the associated equation.

5.1. Deformations of Burgers' equation.

The deformation of Burgers' equation

$$u_t = uu_1 + u_2 \tag{5.1}$$

shall be discussed from the theory of deformations in coverings. In [3] the following theorem has been proved

Theorem 5.1. *The only solution of the deformation equation*

$$\ell_{\mathcal{E}}^{(1)}(\omega) = 0$$

for the Burgers' equation (5.1) is $\omega = \alpha\omega_0$ where α is a constant and

$$\omega_0 = du - u_1 dx - (uu_1 + u_2)dt$$

i.e., Cartan form associated to u.
This leads to the trivial deformation of U (3.3).

In order to find and discuss nontrivial deformations for Burgers' equation we start from the covering $\mathcal{E}^1 = \mathcal{E} \otimes \mathbb{R}$ where the Cartan distribution is given by

$$\begin{aligned} \widetilde{D}_x &= D_x + u\partial_y \\ \widetilde{D}_t &= D_t + (\frac{1}{2}u^2 + u_1)\partial_y \end{aligned} \tag{5.2}$$

where y is the (formal) nonlocal variable $y = \int^x u dx$ with the associated Cartan form is defined by

$$\omega_{-1} = dy - udx - (\frac{1}{2}u^2 + u_1)dt. \tag{5.3}$$

Local coordinates in $\mathcal{E}^1$ are given by

$$(x, t, y, u, u_1, \dots)$$

and associated Cartan forms are

$$\omega_{-1}, \omega_0, \omega_1 = D_x\omega_0, \dots$$

We introduce the characteristic of the deformation

$$U_1 = \sum D_x^i(\omega) \otimes \partial_{u_i} \tag{5.4}$$

by setting the characteristic function

$$\omega = \sum_{i=0}^{8} F^i\omega_i + F^9\omega_{-1} \tag{5.5}$$

where $F(i)$ $(i = 0, \dots, 9)$ are functions dependent on $x, t, u, \dots, u_9, y$.
The deformation equation (4.3)

$$\ell^1_{\mathcal{E}^1}(\omega) = 0$$

results in an overdetermined system of partial differential equations for F^i $(i = 0, \dots, 9)$.
Note that due to the structure of (3.4), the resulting condition on F^9 is just the classical symmetry condition for this component. The solution of the overdetermined system is generated by 15 indeependent solutions. The first few of them are

$$\begin{aligned}
W^1 &:= \omega_0 \\
W^2 &:= +u_1\omega_{-1} + u_0\omega_0 + 2\omega_1 \\
W^3 &:= +2(uu_1 + u_2)\omega_{-1} + (u^2 + 6u_1)\omega_0 + 4u\omega_1 + 4\omega_2 \\
W^4 &:= (+3u^2u_1 + 6uu_2 + 6u_1^2 + 4u_3)\omega_{-1} + (u^3 + 18uu_1 + 16u_2)\omega_0 \\
&\quad + 6(u_1^2 + 4u_1)\omega_1 + 12u\omega_2 + 8\omega_3 \\
W^5 &:= +(tu_1 + 1)\omega_{-1} + (tu + x)\omega_0 + 2t\omega_1 \\
W^6 &:= +(2tuu_1 + 2tu_2 + xu_1 + u)\omega_{-1} \\
&\quad + (tu^2 + 6tu_1 + xu)\omega_0 + (4tu + 2x)\omega_1 + 4t\omega_2 \\
W^7 &:= +(2t^2(uu_1 + u_2) + 2txu_1 + 2tu + 2x)\omega_{-1} \\
&\quad + (t^2(uu^2 + 6u_1) + 2txu + 6t + x^2)\omega_0 \\
&\quad + (4t^2u + 4tx)\omega_1 + 4t^2\omega_2
\end{aligned} \tag{5.6}$$

In order to compute the recursion operator for symmetries resulting from the deformations (5.4) induced by the characteristic functions $W_1, W_2, \ldots$ we use theorem 3.10. Suppose we start at a (nonlocal) symmetry $\ni_X$ of Burgers' equations; its presentation is

$$\ni_X = X_{-1}\partial_Y + \sum_i D_x^i(X)\partial_{u_i}. \tag{5.7}$$

The nonlocal component X_{-1} is obtained from the invariance of the equations, cf. (5.2)

$$\begin{aligned} Y_x &= u \\ Y_t &= \frac{1}{2}u^2 + u_1 \\ D_x(X_{-1}) &= X, \end{aligned} \tag{5.8}$$

i.e. from which we have

$$X_{-1} = D_x^{-1}(X). \tag{5.9}$$

Theorem 3.10, stating that $\ni_X \lrcorner U_1$ is a symmetry yields for the component ∂_u,

$$\begin{aligned} \ni_X \lrcorner W^2 &= u_1 X_{-1} + uX + 2D_x X \\ &= (u_1 D_x^{-1} + u + 2D_x)X \end{aligned} \tag{5.10}$$

and similar for W^3

$$\begin{aligned} \ni_X \lrcorner W^3 &= (tu_1 + 1)X_{-1} + (tu + x)X + 2tD_x X \\ &= ((tu_1 + 1)D_x^{-1} + (tu + x) + 2tD_x)X. \end{aligned} \tag{5.11}$$

From formulae (5.10), (5.11) and similar results with respect to $W_4, \ldots, W_7$ we arrive in a straightforward way at the recursion operators

$$\begin{aligned} R_1 &= id \\ R_2 &= u_1 D_x^{-1} + u + 2D_x \\ R_3 &= t(u_1 D_x^{-1} + u + 2D_x) + x + D_x^{-1} \\ R_4 &= 2(uu_1 + u_2)D_x^{-1} + (u^2 + 6u_1) + 4uD_x + 4D_x^2 \\ R_5 &= t[(2uu_1 + 2u_2)D_x^{-1} + (u^2 + 6u_1) + 4uD_x + 4D_x^2] \\ &\quad + x[u_1 D_x^{-1} + u + 2D_x] + uD_x^{-1} \\ R_6 &= t^2[(2uu_1 + 2u_2)D_x^{-1} + (u^2 + 6u_1) + 4uD_x + 4D_x^2] \\ &\quad + 2tx[u_1 D_x^{-1} + u + 2D_x] \\ &\quad + x^2 \\ &\quad + t(2uD_x^{-1} + 6) + 2xD_x^{-1} \\ R_7 &= (3u^2 u_1 + 6uu_2 + 6u_1^2 + 4u_3)D_x^{-1} + (u^3 + 18uu_1 + 16u_2) \\ &\quad + 6(u^2 + 4u_1)D_x + 12uD_x^2 + 8D_x^3. \end{aligned} \tag{5.12}$$

This application shows that from the deformations of Burgers' equation one arrives in a straightforward way at the recursion operators for symmetries.
Moreover the appearance of formal integrals in these operators is clearified by their derivation.

5.2. Deformations of KdV-equation. Motivated by the results obtained for Burgers' equation we search for deformations in coverings. We start at the four dimensional covering $\mathcal{E}^\infty \otimes \mathbb{R}^4$ of the KdV -equation $\mathcal{E}^\infty$ defined by

$$u_t = uu_1 + u_3, \tag{5.13}$$

and its prolongation to $\mathcal{E}^\infty \otimes \mathbb{R}^4$, where the Cartan distribution is given by

$$\begin{aligned} \widetilde{D}_x &= D_x + u\partial_{y_1} + \frac{1}{2}u^2\partial_{y_2} + y_1\partial_{y_3} + (u^3 - 3u_1^2)\partial_{y_4} \\ \widetilde{D}_t &= D_t + (\frac{1}{2}u^2 + u_2)\partial_{y_1} + (\frac{1}{3}u^3 - \frac{1}{2}u_1^2 + uu_2)\partial_{y_2} \\ &\quad + (u_1 + y_2)\partial_{y_3} + (\frac{3}{4}u^2 + 3u^2u_2 - 6uu_1^2 - 6u_1u_3 + 3u_2^2)\partial_{y_3} \end{aligned} \tag{5.14}$$

where D_x, D_t are the total partial derivative operators on $\mathcal{E}^\infty$. In fact y_1, y_2, y_4 are just potentials for KdV , i.e.,

$$\begin{aligned} y_1 &= \int^x u dx \\ y_2 &= \int^x \frac{1}{2}u^2 dx \\ y_4 &= \int^x u^3 - 3u_1^2 dx \end{aligned} \tag{5.15}$$

while the nonlocal potentials y_3

$$y_3 = \int^x y_1 dx \tag{5.15b}$$

is formally equivalent to an (x, t) dependent local one.
The Cartan forms associated to $y_1, \ldots, y_4$ are denoted $\omega_{-1}, \ldots, \omega_{-4}$ while $\omega_0, \omega_1, \ldots$ are the Cartan forms associated to $u, u_1, \ldots$. The characteristic function for the deformation U_1 is defined by

$$W = \sum_{i=0}^{6} F^i\omega_i + F^7\omega_{-1} + F^8\omega_{-2} + F^9\omega_{-3} + F^{10}\omega_{-4} \tag{5.16}$$

where F^i $(i = 0, \ldots, 10)$ are dependent on

$$x, t, u, \ldots, u_7, y_1, \ldots, y_4.$$

The overdetermined system of partial differential equations resulting from the deformation equation can be solved in an obvious way which yields the following characteristic functions

$$
\begin{aligned}
W_0 &= \omega_0 \\
W_1 &= (\frac{2}{3}u)\omega_0 + \omega_2 + \frac{1}{3}u_1\omega_{-1} \\
W_2 &= (\frac{4}{9}u^2 + \frac{4}{3}u_2)\omega_0 + 2u_1\omega_1 + \frac{4}{3}u\omega_2 + \omega_4 \\
&\quad + \frac{1}{3}(uu_1 + u_3)\omega_{-1} + \frac{1}{9}u_1\omega_{-2} \\
W_3 &= (\frac{8}{27}u^3 + \frac{8}{3}uu_2 + 2u_1^2 + 2u_4)\omega_0 + (4uu_1 + 5u_3)\omega_1 \\
&\quad (\frac{4}{3}u^2 + \frac{20}{3}u_2)\omega_2 + 5u_1\omega_3 + 2u\omega_4 + \omega_6 \\
&\quad + \frac{1}{18}(5u^2u_1 + 10uu_3 + 20u_1u_2 + 6u_5)\omega_{-1} + \frac{1}{9}(uu_1 + u_3)\omega_{-2} + \frac{1}{54}u_1\omega_{-4}
\end{aligned}
\tag{5.17}
$$

Note that the coefficients of $\omega_{-1}, \omega_{-2}, \omega_{-4}$ in (5.17) are just classical generalized symmetries in a agreement with the remark for Burgers' equation.
From these results it is straightforward to obtain recursion operators for KdV , i.e.,

$$
\begin{aligned}
\tilde{R}_1 &= \frac{2}{3}u + D_x^2 + \frac{1}{3}u_1D_x^{-1} \\
\tilde{R}_2 &= (\frac{4}{9}u^2 + \frac{4}{3}u_2) + 2u_1D_x + \frac{4}{3}uD_x^2 + D_x^4 \\
&\quad + \frac{1}{3}(uu_1 + u_3)D_x^{-1} + \frac{1}{9}u_1D_x^{-1}u
\end{aligned}
$$

while

$$
\begin{aligned}
\tilde{R}_3 &= (\frac{8}{27}u^3 + \frac{8}{3}uu_2 + 2u_1^2 + 2y_4) + (4uu_1 + 5u_3)D_x + (\frac{4}{3}u^2 + \frac{20}{3}u_2)D_x^2 \\
&\quad + 5u_1D_x^3 + 2uD_x^4 + D_x^6 + \frac{1}{18}(5u^2u_1 + 10uu_3 + 20u_1u_2 + 6u_5)D_x^{-1} \\
&\quad + \frac{1}{9}(uu_1 + u_3)D_x^{-1}u + \frac{1}{54}u_1D_x^{-1}(3u^2 - 6u_1D_x).
\end{aligned}
\tag{5.18}
$$

The last term in $\tilde{R}_2$ and the last two terms in $\tilde{R}_3$ arise due to the invariance of

$$
\begin{aligned}
y_2 &= D_x^{-1}(\frac{1}{2}u^2) \\
y_4 &= D_x^{-1}(u^3 - 3u_1^2).
\end{aligned}
\tag{5.19}
$$

The operators $\widetilde{R}_1$, $\widetilde{R}_2$, $\widetilde{R}_3$ are just the classical recursion operators for the KdV equations (5.15).

Deformations of KdV equations in the linear covering and the prolongation coverings were discussed too.

(1) Linear covering $\mathcal{E}^\infty \otimes \mathbb{R}^2$.
Local coordinates are $(x, t, u, u_1, \ldots, s_1, s_2)$ while the Cartan distribution is given by [5]

$$\begin{aligned}\widetilde{D}_x &= D_x + \frac{1}{6}s_2\partial_{s_1} + (-\frac{1}{6}s_2u_1 + \frac{1}{18}s_2u - \frac{1}{9}\lambda s_2)\partial_{s_2} \\ \widetilde{D}_t &= D_t - (\lambda + u)\partial_{s_1} + (-s_1u_2 + \frac{1}{6}s_2u_1 - \frac{1}{3}s_1u^2 + \frac{1}{3}\lambda s_1u + \frac{2}{3}\lambda^2 s_1)\partial_{s_2}\end{aligned} \tag{5.20}$$

The only deformation admitted here is the trivial one. There is however a yet unknown symmetry in this case, i.e.,

$$V = s_1s_2\partial_u \tag{5.21}$$

(2) Prolongation covering $\mathcal{E}^\infty \otimes \mathbb{R}^1$.
In this case the Cartan distribution is given by

$$\begin{aligned}\widetilde{D}_x &= D_x + (u + \frac{1}{6}q^2 + \alpha)\partial_q \\ \widetilde{D}_t &= D_t + (u_2 + \frac{1}{3}qu_1 + \frac{1}{3}u^2 + u(\frac{1}{8}q^2 - \frac{1}{3}\alpha) - \frac{2}{3}\alpha(\frac{1}{6}q^2 + \alpha))\partial_q\end{aligned} \tag{5.22}$$

But here no nontrivial results were obtained.

5.3. Deformations of nonlinear Schrödinger equation. In this section deformations and recusion operators of the NLS equation

$$\begin{aligned}u_t &= -v_2 + kv(u^2 + v^2) \\ v_t &= +u_2 - ku(u^2 + v^2)\end{aligned} \tag{5.23}$$

will be discussed in coverings.
We introduce the nonlocal variables y_1, y_2, y_3 associated to the conservation laws of

NLS and given by

$$y_{1x} = u^2 + v^2$$
$$y_{1t} = 2(-uv_1 + vu_1)$$

$$y_{2x} = uv_1$$
$$y_{2t} = -\frac{3}{4}ku^4 - \frac{1}{2}ku^2v^2 + \frac{1}{4}kv^4 + 4uu_2 - 2u_1^2 - 2v_1^2 \tag{5.24}$$

and

$$y_{3x} = k(u^2 + v^2)^2 + 2u_1^2 + 2v_1^2$$
$$y_{3t} = 4((-kuv_1 + kvu_1)(u^2 + v^2) - u_1v_2 + v_1u_2)$$

In the three dimensional covering $\mathcal{E}^\infty \otimes \mathbb{R}^3$ of the NLS-equation the Cartan distribution is given by

$$\begin{aligned} \widetilde{D}_x &= D_x + y_{ix}\partial_{y_i} \\ \widetilde{D}_t &= D_t + y_{it}\partial_{y_i} \end{aligned} \tag{5.25}$$

Now in order to construct a deformation of NLS we construct a tuple of characteristic functions

$$\begin{aligned} W^u &= \sum_{i=0}^{3}(f^i\omega_i^u + \overline{f}^i\omega_i^0) + \sum_{i=1}^{3}\overline{\overline{f}}^i\omega_{y_i} \\ W^v &= \sum_{i=0}^{3}(g^i\omega_i^u + \overline{g}^i\omega_i^0) + \sum_{i=1}^{3}\overline{\overline{g}}^i\omega_{y_i} \end{aligned} \tag{5.26}$$

where (5.3.4) ω_i^u, ω_i^v, ω_{y_i} are the Cartan forms associated to u_i, v_i, y_i respectively; de coefficients $f^i, \overline{f}^i, \overline{\overline{f}}^i, g^i, \overline{g}^i, \overline{\overline{g}}^i$ are dependent on

$$x, t, u, v, \ldots, u_4, v_4, y_1, y_2, y_3.$$

The solution constructed from the deformation equation (4.3) leads to the following nontrivial results.

$$
\begin{aligned}
W_1^u &= \frac{1}{k}\omega_1^v - v\omega_{y_1} \\
W_1^v &= -\frac{1}{k}\omega_1^u + u\omega_{y_1} \\
\\
W_2^u &= (u^2+v^2)\omega_0^u + uv\omega_0^v - \frac{1}{2k}\omega_2^u + \frac{1}{2}u_1\omega_{y_1} - v\omega_{y_2} \\
W_2^v &= v^2\omega_0^v - \frac{1}{2k}\omega_2^v + \frac{1}{2}v_1\omega_{y_1} + u\omega_{y_2} \\
\\
W_3^u &= 8uv_1\omega_0^u + 12vv_1\omega_0^v + 4uv\omega_1^u + (4u^2+8v^2)\omega_1^v - \frac{2}{k}\omega_3^v \\
&\quad + 2(-k(u^2+v^2)v + v_2)\omega_{y_1} + 4u_1\omega_{y_2} - v\omega_{y_3} \\
W_3^v &= (-12uu_1 - 4vv_1)\omega_0^u + (-4uv_1 - 8vu_1)\omega_0^v \\
&\quad + (-8u^2 - 4v^2)\omega_1^u - 4uv\omega_1^v + \frac{2}{k}\omega_3^u \\
&\quad + 2(k(u^2+v^2)u - 2u_2)\omega_{y_1} + 4v_1\omega_{y_2} + u\omega_{y_3}
\end{aligned}
\tag{5.27}
$$

Suppose we have a shadow of a nonlocal symmetry $X = X^u\partial_u + X^v\partial_v$ then the nonlocal component X_{-1} associated to y_1 is obtained from the invariance of the equations

$$
\begin{aligned}
y_{1x} &= u^2 + v^2 \\
y_{1t} &= 2(-uv_1 + vu_1)
\end{aligned}
\tag{5.28}
$$

So from (5.28) we arrive at the following condition

$$D_x(X_{-1}) = 2uX^u + 2vX^v$$

or formally

$$X_{-1} = D_x^{-1}(2uX^u + 2vX^v) \tag{5.29}$$

Using this results we arrive from W_1^u, W_1^v in a straightforward way at the wellknown recusrion operator

$$R_1 = \begin{bmatrix} -vD_x^{-1}(2u) & -vD_x^{-1}(2v) + \frac{1}{k}D_x \\ +uD_x^{-1}(2u) - \frac{1}{k}D_x & +uD_x^{-1}(2v) \end{bmatrix} \tag{5.30}$$

Results for recusion operators resulting from W_i^u, W_i^v $(i = 2, 3, \dots)$ can be obtained similarly

5.4. Deformations of classical Boussinesq equation. As a last application we discuss deformations of classical Boussinesq equation

$$\begin{aligned} v_t &= u_1 + vv_1 \\ u_t &= u_1 v + uv_1 + \sigma v_3 \end{aligned} \tag{5.31}$$

To this end we start at a fourdimensional covering $\mathcal{E}^\infty \otimes \mathbb{R}^4$ of the Boussinesq equation, where local coordinates are given by

$$(x, t, v, u, \ldots, y_1, \ldots, y_4)$$

with the Cartan distribution

$$\begin{aligned} \widetilde{D}_x &= D_x + v\partial_{y_1} + u\partial_{y_2} + uv\partial_{y_3} + (u^2 + uv^2 + vv_3\sigma)\partial_{y_4} \\ \widetilde{D}_t &= D_t + (u + \frac{1}{2}v^2)\partial_{y_1} + (uv + v_2\sigma)\partial_{y_2} + (\frac{1}{2}u^2 + uv^2 + vv_2\sigma - \frac{1}{2}v_1^2\sigma)\partial_{y_3} \\ &\quad + (2u^2 v + uv^3 + 2\sigma uv_2 + 2\sigma v^2 v_2 + \sigma vu_2 - \sigma v_1 u_1)\partial_{y_4}. \end{aligned} \tag{5.32}$$

We assume the characteristic functions W^v, W^u to be dependent on $\omega_0^v, \omega_0^u, \ldots, \omega_5^v, \omega_5^u$, $\omega_{y_1}, \ldots, \omega_{y_4}$, whereas the coefficients are required to be dependent of $x, t, v, u, \ldots$, $v_5, u_5, y_1, \ldots, y_4$.
Solving the overdetermined system of partial differential equations resulting from the deformation condition (4.3) we arrive at the following non-trivial characteristic functions

$$\begin{aligned} W_1^v &= v\omega_0^v + 2\omega_0^u + v_1\omega_{y_1} \\ W_1^u &= 2u\omega_0^v + v\omega_0^u + 2\sigma\omega_2^v + u_1\omega_{y_1} \\ W_2^v &= (4u + v^2)\omega_0^v + 4v\omega_0^u + 4\sigma\omega_2^v + (2vv_1 + 2u_1)\omega_{y_1} + 2v_1\omega_{y_2} \\ W_2^u &= (4uv + 6\sigma v_2)\omega_0^v + (4u + v^2)\omega_0^u + 6v_1\sigma\omega_1^v + 4\sigma v\omega_2^v + 4\sigma\omega_2^u \\ &\quad + (2uv_1 + 2vu_1 + 2\sigma v_3)\omega_{y_1} + 2u_1\omega_{y_2} \end{aligned} \tag{5.33}$$

and two more deformations.
As in the preceding subsection we use the invariance of the equations

$$\begin{aligned} (y_1)_x &= v \\ (y_2)_x &= u \end{aligned} \tag{5.34}$$

to arrive at the associated recursion operators

$$R_1 = \begin{bmatrix} v + v_1 D_x^{-1} & 2 \\ 2u + 2\sigma D_x^2 + u_1 D_x^{-1} & v \end{bmatrix} \tag{5.35}$$

and

$$R_2 = \left[\begin{array}{l|l} (4u+v^2)+4\sigma D_x^2+(2vv_1+2u_1)D_x^{-1} & 4v+2v_1D_x^{-1} \\ \hline (4uv+6\sigma_2 v_2)+6\sigma v_1 D_x & (4u+v^2)+4\sigma D_x^2 \\ +4\sigma v D_x & \\ +(2uv_1+2vu_1+2\sigma v_3)D_x^{-1} & +2u_1 D_x^{-1} \end{array}\right] \qquad (5.36)$$

REFERENCES

[1] Frolicher, A., Nijenhuis, A., *Derivations in the graded ring of differential forms*, Indagnationes Math. **18** (1956) 338–359.

[2] Khor'kova, N.G., *On the C-spectral sequence for evolution equations*, Ph.D. Thesis, Moscow State University, Moscow (1991) 77 pp. (in Russian).

[3] Krasil'shchik, I.S., *Some new cohomological invariants for nonlinear differential equations*, Diff. Geom. and Its Appl. (to appear).

[4] Krasil'shchik, I.S., Lychagan, V.V., Vinogradov, A.M., *Geometry of Jet Spaces and Nonlinear Partial Differential Equations*, Gordon and Breach, New York (1986).

[5] Krasil'shchik, I.S., Vinogradov, A.M., *Nonlocal trends in the geometry of differential equations: symmetries, conservation laws and Bäcklund transformations*, Acta Appl. Math. **15** (1989) 161–209.

[6] Nijenhuis, A., Richardson, R.W. Jr., *Deformations of Lie algebra structures*, J. Math. and Mech. **17** (1967) 89–105.

[7] Olver, P., *Evolution equations possessing infinitely many symmetries*, J. Math. Phys. **18** (1977) 1212–1215.

[8] Vinogradov, A.M., *The C-spectral sequence, Lagrangian formalism and conservation laws*, J. Math. Anal. Appl. **100** (1984) 2–129.

[9] Bemmelen, T. van, Gragert, P.K.H., Kersten, P.H.M., Sym, A.M., *Symmetries and conservations laws of the system:* $u_x = vw_x$, $v_y = uw_y$, $uv + w_{xx} + w_{yy} = 0$, University of Twente, Department of Applied Mathematics, Memorandum No. 184, 1992.

Geometry in Partial Differential Equations (pp. 155–172)
eds. A. Pràstaro and Th. M. Rassias

GEOMETRIC HAMILTONIAN FORMS FOR THE KADOMTSEV-PETVIASHVILI AND ZABOLOTSKAYA-KHOKHLOV EQUATIONS

B. A. KUPERSHMIDT
School of Mathematics, Institute for Advanced Study
Princeton, NJ 08540, USA
and
The University of Tennessee Space Institute
Tullahoma, TN 37388, USA

ABSTRACT

We first construct, from the point of view of the modified Kadomtsev–Petviashvili (mKP) hierarchy, the mKP equation together with its map into the KP one, and a deformed KP equation together with its contraction onto the KP equation itself. Similar formulae are deduced for the Zabolotskaya–Khokhlov (ZK) equation. The rest of the paper addresses the problem of Hamiltonian forms for the constructed equations. The algebraic approach fails here, and we have to take up the geometric point of view by treating our equations as submanifolds in 4-jet bundles. In this new language, purely local Hamiltonian forms are found for the KP and ZK equations; first of these reduces to a novel Hamiltonian structure for the Boussinesq equation.

§ 1. Introduction

The Kadomtsev–Petviashvili (KP) equation plays in 2 + 1-dimensions the same rôle that the Korteweg–deVries (KdV) equation plays in 1 + 1-dimensions: it is one of the universal models of wave propagation in fluids and plasmas [**1**]. By itself, this equation has been extensively studied and it has a great number of applications to numerous mathematical problems, see e.g. [**2**] and references therein. It is instructive, nevertheless, to examine for the KP equation 2 + 1-dimensional analogs of the most basic properties of its 1 + 1-dimensional degeneration, the KdV equation, namely the following: a modified form, a Miura-type map, a Gardner-type deformation, and Hamiltonian structures. To fix the normalization constants, let us write the KP equation in one of the two equivalent forms:

$$u_{tx} = \left(\frac{u^2}{2} + \frac{u_{xx}}{12}\right)_{xx} + u_{yy} \tag{1}$$

or

$$u_t = \left(\frac{u^2}{2} + \frac{u_{xx}}{12}\right)_x + \partial_x^{-1}(u_{yy}); \tag{2}$$

when the y-dependence is ignored, the KP equation (2) reduces to the KdV equation

$$u_t = \left(\frac{u^2}{2} + \frac{u_{xx}}{12}\right)_x . \tag{3}$$

Now, one knows from the classical $1+1$-dimensional theory that if v satisfies the modified KdV (mKdV) equation

$$v_t = \left(-\frac{v^3}{6} + \frac{v_{xx}}{12}\right)_x , \tag{4}$$

then the Miura map [**3**]

$$u = -\frac{1}{2}(v_x + v^2) \tag{5}$$

sends the solutions of the mKdV equation (4) into solutions of the KdV equation (3); moreover, combining Miura map (5) with the Galilean invariance of the KdV equation (3), one obtains the Gardner deformation [**4**] of the KdV equation:

$$w_t = \left(\frac{w^2}{2} + \frac{w_{xx}}{12} - \frac{\epsilon^2 w^3}{6}\right)_x , \tag{6}$$

together with the contraction map

$$u = w - \frac{\epsilon}{2} w_x - \frac{\epsilon^2}{2} w^2 \tag{7}$$

from the deformed KdV (6) into the undeformed one (3). These maps and equations have the following basic properties:

(8i) The mKdV equation (4) is Hamiltonian, the KdV equation (3) is bi-Hamiltonian, and the Miura map (5) is a Hamiltonian map between the Hamiltonian structure of the mKdV equation and the second Hamiltonian structure of the KdV equation [**5**];

(8ii) The deformed KdV equation (6) is Hamiltonian, and the contraction (8) is a Hamiltonian map into a linear combination of the two Hamiltonian structures of the KdV equation [**6**];

(8iii) Since the deformed KdV equation (6) depends only upon ϵ^2, one gets a Bäcklund transformation of the KdV equation by sending w via the contraction map (8) into a pair of u's, with the parameters ϵ and $-\epsilon$ respectively, and then eliminating the intermediate object w. (This procedure works for all the higher KdV equations [**7**].)

There are a number of results known to date about $2+1$-dimensional analogs of the basic structures, described above, of the KdV equation. Namely,using gauge transformations of the Manakov Lax pair [**8**], Konopelchenko derived [**9**], [**10**] a mKP equation together with its map into the KP one, and also a Gardner-like deformed KP equation together with its contraction onto the KP equation. Now, it is well known [**11**] that the KP equation is related to the fundamental KP hierarchy [**12**] as the integrability conditions of the first two nontrivial flows in the hierarchy. It is also known that there exists a separate mKP hierarchy together with its Miura-like map onto the KP hierarchy [**13**], [**14**]. It is interesting then to determine what kind of a mKP equation together with its Miura-like map into the KP equation (2), a deformed KP equation together with its contraction, and the resulting Bäcklund transformation (similar to the one described in (8iii) above but with a new twist) for the KP equation one obtains as the integrability conditions of the first two nontrivial flows in the mKP hierarchy. This is done in Section 2 using the non-standard integrable systems from [**13**]. The results found turn out to be essentially those of Konopelchenko. The problem of the (local) Hamiltonian formalism is more interesting since the KP equation (1) is not of evolution type. To circumvent this problem, we take the *geometric* point of view by looking at the KP equation (1) as a submanifold in a 4-jet bundle. Now that the independent variables t, x, y are all on the same footing, we choose x as a new evolution parameter. This allows one to find a purely local Hamiltonian form of the KP equation (1); this described in Section 3. Modified and deformed equations, together with their Miura-like and contraction maps, for the Zabolotskaya–Khokhlov (ZK) equation of nonlinear optics [**15**] which results from the KP (1) equation upon dropping off the higher-order dispersion terms:

$$u_{tx} = \left(\frac{u^2}{2}\right)_{xx} + u_{yy}, \tag{9}$$

are studied in Section 4, where a local Hamiltonian form of the ZK equation (9) can also be found.

§ 2. The mKP Equation

The classical theory of the KdV equation (3) is based on the Lax reresentation

$$\overline{L}_t = [\overline{P}, \overline{L}], \tag{10}$$

where

$$\overline{L} = \partial^2 + 2u, \quad \overline{P} = \frac{1}{3}\partial^3 + u\partial + \frac{1}{2}u_x, \tag{11}$$

and $\partial = \partial_x$ stands for $\partial/\partial x$. Then the mKdV equation (4) and the Miura map (5) result at once from the factorization [**16**]

$$\overline{L} = \partial^2 + 2u = (\partial + v)(\partial - v) = \overline{l}_1 \overline{l}_2. \tag{12}$$

For the KP equation (2), Manakov [**8**] found a $2+1$-dimensional analog of formula (11):

$$\tilde{L} = \partial^2 + 2u - \sqrt{-4}\partial_y, \quad \tilde{P} = \frac{1}{3}\partial^3 + u\partial + \frac{1}{2}u_x + \sqrt{-1}\partial^{-1}(u_y). \tag{13}$$

Factorizing $\tilde{L}$ we get

$$\tilde{L} = \partial^2 + 2u - \sqrt{-4}\partial_y = (\partial + v + \sqrt[4]{-4}\mathcal{D}_y)(\partial - v - \sqrt[4]{-4}\mathcal{D}_y) = \tilde{l}_1 \tilde{l}_2, \tag{14}$$

where

$$\mathcal{D}_y = \sqrt{\partial_y} = \frac{\partial}{\partial \theta} + \theta \partial_y,$$

θ being an odd Grassmann variable; however, no modified equation results from the factorization (14).

Another interpretation [**11**] of the KP equation is the integrability condition for the first two nontrivial flows in the KP hierarchy:

$$\mathcal{L}_y = \left[\frac{1}{2}(\mathcal{L}^2)_+, \mathcal{L}\right], \tag{15}$$

$$\mathcal{L}_t = \left[\frac{1}{3}(\mathcal{L}^3)_+, \mathcal{L}\right], \tag{16}$$

where

$$\mathcal{L} = \partial + \sum_{i=0}^{\infty} A_i \partial^{-i-1}. \tag{17}$$

For A_0 and A_1 we get from (15) – (17) (see detailed calculation in e.g. [**17**], [**18**])

$$A_{0,y} = A_{1,x} + \frac{1}{2}A_{0,xx}, \tag{18}$$

$$A_{1,y} = A_{2,x} + \frac{1}{2}A_{1,xx} + A_0 A_{0,x}, \tag{19}$$

$$A_{0,t} = A_{2,x} + A_{1,xx} + \frac{1}{3}A_{0,xxx} + 2A_0 A_{0,x}. \tag{20}$$

Substituting (19) into (20) and using (18) in the form

$$A_1 = -\frac{1}{2}A_{0,x} + \partial^{-1}(A_{0,y}) \tag{21}$$

we obtain the KP equation (2) for

$$u := A_0.$$

Let us use the mKP hierarchy in a similar fashion to get the desired mKP equation. Recall [**13**] that the mKP hierarchy has the form

$$\frac{\partial L}{\partial t_n} = \left[\frac{1}{n}(L^n)_{\geq 1}, L\right] = \left[L, \frac{1}{n}(L^n)_{\leq 0}\right], \quad n \in \mathbf{N}, \tag{22}$$

where

$$L = \partial + \sum_{i=0}^{\infty} a_i \partial^{-1}, \tag{23}$$

$$\left(\sum p_j \partial^j\right)_{\geq k} := \sum_{j \geq k} p_j \partial^j, \quad k \in \mathbf{Z}, \tag{24}$$

and similar convention applies to the notation $(\quad)_{\leq k}$. There is a map which sends the mKP hierarchy (22) into the KP hierarchy

$$\frac{\partial \mathcal{L}}{\partial t_n} = \left[\frac{1}{n}(\mathcal{L}^n)_{\geq 0}, \mathcal{L}\right] = \left[\mathcal{L}, \frac{1}{n}(\mathcal{L}^n)_{\leq -1}\right], \quad n \in \mathbf{N}, \tag{25}$$

namely [**13**]

$$L \mapsto \mathcal{L} = exp(-\textstyle\int a_0)\, L\, exp(\textstyle\int a_0) = \partial + \displaystyle\sum_{i=1}^{\infty} a_j(\partial - a_0)^{-i}. \tag{26}$$

In particular,

$$A_0 = a_1. \tag{27}$$

The latter innocent-looking formula is going to be our $2+1$-dimensional analog of the $1+1$-dimensional Miura map (5). To see how this comes about, let us consider the first two nontrivial flows in the mKP hierarchy (22):

$$L_y = \left[\frac{1}{2}(L^2)_{\geq 1}, L\right], \tag{28}$$

$$L_t = \left[\frac{1}{3}(L^3)_{\geq 1}, L\right]. \tag{29}$$

Since

$$(L^2)_{\geq 1} = \partial^2 + 2a_0\partial, \quad (L^3)_{\geq 1} = \partial^3 + 3a_0\partial^2 + 3(a_0^2 + a_{0,x} + a_1)\partial,$$

from (28) and (29) we obtain for a_0 and a_1:

$$a_{0,y} = \left[a_1 + \frac{1}{2}(a_{0,x} + a_0^2)\right]_x, \tag{30}$$

$$a_{1,y} = \left(a_2 + a_0 a_1 + \frac{1}{2}a_{1,x}\right)_x, \tag{31}$$

$$a_{0,t} = \left(a_2 + a_{1,x} + 2a_0 a_1 + a_0 a_0, x + \frac{1}{3}a_0^3 + \frac{1}{3}a_{0,xx}\right)_x. \tag{32}$$

Substituting (31) into (32) we get

$$a_{0,t} = a_{1,y} + \left(\frac{1}{2}a_{1,x} + a_0 a_1 + a_0 a_{0,x} + \frac{1}{3}a_0^3 + \frac{1}{3}a_{0,xx}\right)_x, \tag{33}$$

and substituting (30), in the form

$$a_1 = -\frac{1}{2}\left(a_{0,x} + a_0^2\right) + \partial^{-1}(a_{0,y}), \tag{34}$$

into (33) we finally obtain for

$$v := a_0 \quad : \tag{35}$$

$$v_t = \left(-\frac{v^3}{6} + \frac{v_{xx}}{12}\right)_x + \partial^{-1}(v_{yy}) + v_x\partial^{-1}(v_y). \tag{36}$$

This is our mKP equation; it reduces to the mKdV equation 94) when the y-dependence is neglected. The map (27) becomes, in view of formula (34),

$$u = -\frac{1}{2}(v_x + v^2) + \partial^{-1}(v_y). \tag{37}$$

This is our $2+1$-dimensional analog of the $1+1$-dimensional Miura map (5). The map (37) sends the solutions of the mKP equation (36) into solutions of the KP equation (2).

This follows from the general theory in [**13**]; alternatively, the reader can verify this fact directly.

To get a deformed KP equation we argue along the classical lines. The same map (37) sends the solutions of the equation

$$v_t = \left(-\frac{v^3}{6} + \frac{v_{xx}}{12}\right)_x + \partial^{-1}(v_{yy}) + v_x\partial^{-1}(v_y) + \lambda v_x \tag{38}$$

into solutions of the equation

$$u_t = \left(\frac{u^2}{2} + \frac{u_{xx}}{12}\right)_x + \partial^{-1}(u_{yy}) + \lambda u_x, \tag{39}$$

for arbitrary constant λ. Taking

$$v = \epsilon w - \frac{1}{\epsilon}, \quad u = \overline{u} - \frac{1}{2\epsilon^2}, \quad \lambda = \frac{1}{2\epsilon^2}, \tag{40}$$

and dropping the bar off u we find that the contraction map

$$u = w - \frac{\epsilon}{2}w_x - \frac{\epsilon^2}{2}w^2 + \epsilon\partial^{-1}(w_y) \tag{41}$$

sends the solutions of the deformed KP equation

$$w_t = \left(\frac{w^2}{2} + \frac{w_{xx}}{12} - \frac{\epsilon^2 w^3}{6}\right)_x + \partial^{-1}(w_{yy}) + \epsilon w_x\partial^{-1}(w_y) \tag{42}$$

into solutions of the KP equation (2). When the y-dependence is neglected, formulae (41) and (42) reduce to formulae (7) and (6) respectively. Notice that, in contrast to the $1+1$-dimensional case of the equation (6), the deformed KP equation (42) is not invariant under the change $\epsilon \mapsto -\epsilon$. However, the invariance is restored if we simultaneously change y into $-y$. Since the latter change leaves the KP equation invariant, we get the desired (infinitesimal) Bäcklund transformation of the KP equation by eliminating w from the diagram

$$\begin{array}{ccccc} w(y;\epsilon) & \overset{(41)(\epsilon)}{\longmapsto} & u(y) & & \\ \downarrow & & & \nwarrow & \text{Bäcklund} \\ \downarrow & & & & \searrow \\ w(-y;-\epsilon) & \overset{(41)(-\epsilon)}{\longmapsto} & \tilde{u}(-y) & \overset{y\longmapsto -y}{\longmapsto} & \tilde{u}(y) \end{array} \tag{43}$$

§ 3. Geometric Hamiltonian Forms

Let us consider the problem of a local Hamiltonian representation for the KP equation. Since the form (2) is nonlocal, we have to work with the form (1). However, the form (1) is not of an evolutionary form: we do not have u_t as a local expression but only u_{tx}. This is similar to the sine-Gordon case

$$u_{xt} = \sin u$$

for which a quasi-Hamiltonian formalism is available [**19**], [**20**]. But since $(u^2/2)_{xx}$ is not a variational derivative of anything, we can't write the KP equation (1) in the quasi-Hamiltonian form

$$\partial_t(u_x) = \delta H/\delta u.$$

This suggests that we should abandon the algebraic point of view held till now. Noticing that the highest-order derivative in x, in the term $u_{xxxx}/12$, is not mixed with ∂_y- and ∂_t-derivatives, we take up the *geometric point of view* and consider x *as a new time parameter*. In other words, we no longer treat ∂_t as a derivation of a differential ring $\mathbf{C}[u, u_x, u_y, \ldots]$. Instead, we look at our equation as a submanifold in a 4-jet bundle over $\mathbf{R}^3 = \{x, t, y\}$, and then treat x as a new evolution parameter. (The same argument applies to the highest ∂_y-derivative, in the term u_{yy}, but this route leads nowhere.) Since our equation is of the 4th-order in x, to convert it to a dynamical (in x) system, it must be first rewritten as a 4-component 1st-order system. Denoting

$$\psi := A_1 + u_x/2, \tag{44}$$

we can use the equation (18) to rewrite the scalar equation (1) as a 2-component system

$$u_y = \psi_x, \quad \psi_y = u_t - \left(u^2/2 + u_{xx}/12\right)_x, \tag{45}$$

and then as a 4-component one:

$$\psi_x = u_y, \quad u_x = u_1, \quad u_{1,x} = u_2, \quad u_{2,x} = 12(u_t - uu_1 - \psi_y). \tag{46}$$

From (46) we find:

$$(u_2 + 6u^2)_x = (-12\psi)_y + (12u)_t, \tag{47}$$

$$\left(\frac{\psi^2}{2}-\frac{u^3}{3}-\frac{1}{12}uu_2+\frac{1}{24}u_1^2\right)_x=(\psi u)_y-\left(\frac{1}{2}u^2\right)_t. \tag{48}$$

Hence, we get two local conserved densities

$$H_1=u_2+6u^2, \tag{49}$$
$$H_2=\frac{1}{2}\psi^2-\frac{1}{3}u^3-\frac{1}{12}uu_2+\frac{1}{24}u_1^2. \tag{50}$$

Let B_1 be the following matrix

$$B_1:=\begin{pmatrix} 0 & 0 & 0 & -12\partial_y \\ 0 & 0 & 12 & 0 \\ 0 & -12 & 0 & 144u \\ -12\partial_y & 0 & -144u & -144\partial_t \end{pmatrix}. \tag{51}$$

Since the vector of variational derivatives of the Hamiltonian H_2, with respect to the variables $\{\psi,u,u_1,u_2\}$, is:

$$\delta(H_2)=\begin{pmatrix} \psi \\ -u^2-\frac{1}{12}u \\ \frac{1}{12}u_1 \\ -\frac{1}{12}u \end{pmatrix}, \tag{52}$$

we see at once that the KP equation (46) can be written in the form

$$\begin{pmatrix} \psi \\ u \\ u_1 \\ u_2 \end{pmatrix}_x=B_1\delta(H_2). \tag{53}$$

Let us show that the matrix B_1 (51) is Hamiltonian. Since it is an affine skewsymmetric matrix, it must correspond to a (generalized) two-coycle on a Lie algrbra [**21**]. The Lie algebra in question, $\mathbf{g}$, comes out of the linear part B_1^{lin} of B_1, via the defining relation

$$Y^tB_1^{\text{lin}}(X)\sim\sum_{k=1}^{4}q_k[X,Y]_k, \tag{54}$$

where $\mathbf{q} := (\psi, u, u_1, u_2)^t$, and "$\sim$" means "equality modulo $\{\mathrm{Im}(\partial_y) + \mathrm{Im}(\partial_t)\}$". Thus,

$$\left[\begin{pmatrix} X_1 \\ X_2 \\ X_3 \\ X_4 \end{pmatrix}, \begin{pmatrix} Y_1 \\ Y_2 \\ Y_3 \\ Y_4 \end{pmatrix}\right] = 144 \begin{pmatrix} 0 \\ Y_3X_4 - Y_4X_3 \\ 0 \\ 0 \end{pmatrix}, \tag{55}$$

and the Jacobi identity for the commutator (55) in $\mathbf{g}$ is obviously satisfied. Further, the following skewsymmetric bilinear forms on the Lie algebra $\mathbf{g}$:

$$\omega_1(X, Y) = -12(X_1 Y_{4,y} + X_4 Y_{1,y}), \tag{56}$$

$$\omega_2(X, Y) = 12(X_2 Y_3 - X_3 Y_2), \tag{57}$$

$$\omega_3(X, Y) = -144 X_4 Y_{4,t}, \tag{58}$$

which come out of the constant (i.e. $\mathbf{q}$-independent) part of the affine matrix B_1 (51), are obviously two-cocycles on $\mathbf{g}$, i.e., each of these bilinear forms satisfies the two-cocycle condition

$$\omega([X, Y], Z) + \omega([Y, Z], X) + \omega([Z, X], Y) \sim 0, \tag{59}$$

for any $X, Y, Z, \in \mathbf{g}$. Thus the matrix B_1 is Hamiltonian. An alternative proof is even shorter. For any function $f = f(u)$ consider the following slight generalization of the KP equation in the forms (1), (45), (46) respectively:

$$u_{tx} = \left(f + \frac{u_{xx}}{12}\right)_{xx} + u_{yy}, \tag{60}$$

$$\begin{cases} u_y = & \psi_x \\ \psi_y = & u_t - (f + \frac{1}{12} u_{xx})_x, \end{cases} \tag{61}$$

$$\begin{cases} \psi_x = & u_y \\ u_x = & u_1 \\ u_{1,x} = & u_2 \\ u_{2,x} = & 12(u_t - f_u u_1 - \psi_y). \end{cases} \tag{62}$$

Then

$$(u_2 + 12f)_x = (-12\psi)_y + (12u)_t, \tag{63}$$

$$\left(\frac{\psi^2}{2} - F(u) - \frac{1}{12}uu_2 + \frac{1}{24}u_1^2\right)_x = (\psi u)_y - (\frac{1}{2}u^2)_t, \tag{64}$$

where $F = F(u)$ is such that

$$F_u = uf_u. \tag{65}$$

Taking

$$\overline{H}_1 = u_2 + 12f, \tag{66}$$

$$\overline{H}_2 = \frac{1}{2}\psi^2 - F - \frac{1}{12}uu_2 + \frac{1}{24}u_1^2, \tag{67}$$

and defining the skewsymmetric matrix

$$\overline{B}_1 := \begin{pmatrix} 0 & 0 & 0 & -12\partial_y \\ 0 & 0 & 12 & 0 \\ 0 & -12 & 0 & 144f_u \\ -12\partial_y & 0 & -144f_u & -144\partial_t \end{pmatrix}, \tag{68}$$

we immediately see that the system (62) results upon applying the matrix $\overline{B}_1$ (68) to the vector

$$\delta(\overline{H}_2) = \begin{pmatrix} \psi \\ -uf_u - \frac{1}{12}u_2 \\ \frac{1}{12}u_1 \\ -\frac{1}{12}u \end{pmatrix} \tag{69}$$

of variational derivatives of the Hamiltonian $\overline{H}_2$. To check the Hamiltonian character of the matrix $\overline{B}_1$ (68) we have to verify the identity **[22]**

$$\overline{B}_1\delta[Y^t\overline{B}_1(X)] = D[\overline{B}_1(Y)]\overline{B}_1(X) - D[\overline{B}_1(X)]\overline{B}_1(Y) \tag{70}$$

for arbitrary vectors Y, X, where D is the operator of the Fréchet derivative. The identity (70) is quite obvious, since both sides of it are quickly seen to be

$$-12^3(Y_3X_4 - Y_4X_3)\begin{pmatrix}0\\0\\1\\0\end{pmatrix}. \tag{71}$$

Thus the matrix $\overline{B}_1$ (68) and the system (62) are Hamiltonian. In particular, when the y-dependence is neglected, formulae (60), (62) – (64), and (67) become respectively:

$$u_t = \left(f + \frac{1}{12}u_{xx}\right)_x, \tag{72}$$

$$\begin{cases} u_x = & u_1 \\ u_{1,x} = & u_2 \\ u_{2,x} = & 12(u_t - f_u u_1), \end{cases} \tag{73}$$

$$(u_2 + 12f)_x = (12u)_t, \tag{74}$$

$$\left(-F - \frac{1}{12}uu_2 + \frac{1}{24}u_1^2\right)_x = \left(-\frac{1}{2}u^2\right)_t, \tag{75}$$

$$\widetilde{H}_2 = -F - \frac{1}{12}uu_2 + \frac{1}{24}u_1^2. \tag{76}$$

Since the lower-right 3×3 submatrix $\tilde{B}_1$ of the matrix $\overline{B}_1$ (68):

$$\tilde{B}_1 := \begin{pmatrix} 0 & 12 & 0 \\ -12 & 0 & 144f_u \\ 0 & -144f_u & -144\partial_t \end{pmatrix}, \tag{77}$$

is Hamiltonian by the same argument as applies to the matrix $\overline{B}_1$, we see that the system (73) is Hamiltonian with the Hamiltonian matrix $\tilde{B}_1$ (77) and the Hamiltonian $\widetilde{H}_2$ (76). For the KdV equation (3), with $f = u^2/2$, this fact had been found by Tsarev **[23]** (who also found a second Hamiltonian form of the KdV equation in the x-representation); our argument, being general, applies also to the mKdV equation (4), in which case one simply takes $f = -u^3/6$. (The second Hamiltonian form of the KdV equation, like in the classical case, is then induced at once by the Miura map written in the x-representation form as $u = -\frac{1}{2}(v_1 + v^2)$, from the Hamiltonian structure of the mKdV equation, $(77)|_{f=-u^3/6}$.)

What about a second Hamiltonian form of the KP equation (46)? I couldn't find it and it probably doesn't exist (as a local object), even though we have the candidate for the Hamiltonian: H_1 (49). (Notice that

$$B_1\delta(H_1) = \overline{B}_1\delta(\overline{H}_1) = \widetilde{B}_1\delta\widetilde{H}_1) = \mathbf{0}, \tag{78}$$

where $\widetilde{H}_1 := \overline{H}_1$.) The mKP equation (36) and the deformed KP equation (42) are of no help either. Written in the x-representation as the systems, respectively

$$\begin{cases} v_y = & \varphi_x \\ v_t = & (\frac{1}{12}v_{xx} - \frac{1}{6}v^3)_x + \varphi_y + v_x\varphi \ . \end{cases} \tag{79}$$

$$\begin{cases} \varphi_x = & v_y \\ v_x = & v_1 \\ v_{1,x} = & v_2 \\ v_{2,x} = & 12(v_t + \frac{1}{2}v^2 v_1 - \varphi_y - \varphi v_1), \end{cases} \tag{80}$$

and

$$\begin{cases} w_y = & \chi_x \\ w_t = & \left(\frac{1}{12}w_{xx} + \frac{1}{2}w^2 - \frac{\epsilon^2}{6}w^3\right)_x + \chi_y + \epsilon w_x \chi, \end{cases} \tag{81}$$

$$\begin{cases} \chi_x = & w_y \\ w_x = & w_1 \\ w_{1,x} = & w_2 \\ w_{2,x} = & 12\left(w_t - ww_1 - \chi_y - \epsilon w_1 \chi + \frac{\epsilon^2}{2}w^2 w_1\right), \end{cases} \tag{82}$$

they do have each a pair of (local) conservation laws:

$$(v_2 + 12v\varphi - 2v^3)_x = (6v^2 - 12\varphi)_y + (12v)_t, \tag{83}$$

$$\left(\frac{1}{2}\varphi^2 - \frac{1}{2}v^2\varphi + \frac{1}{8}v^4 - \frac{1}{12}vv_2 + \frac{1}{24}v_1^2\right)_x = \left(v\varphi - \frac{1}{6}v^3\right)_y - \left(\frac{1}{2}v^2\right)_t, \tag{84}$$

$$(w_2 + 6w^2 + 12\epsilon w\chi - 2\epsilon^2 w^3)_x = (-12\chi + 6\epsilon w^2)_y + (12w)_t \tag{85}$$

$$\left(\frac{1}{2}\chi^2 - \frac{1}{3}w^3 - \frac{1}{12}ww_2 + \frac{1}{24}w_1^2 - \frac{\epsilon}{2}w^2\chi + \frac{\epsilon^2}{8}w^4\right)_x = \left(w\chi - \frac{\epsilon}{6}w^3\right)_y - \left(\frac{1}{2}w^2\right)_t. \tag{86}$$

But neither of the systems (80) or (82) seems to have a local Hamiltonian form. Besides, even if such a form were to be found it will be of little use in determing a second Hamiltonian form of the KP system (46) since neither of the maps (37) or (41) can be made local in the 4-component x-representation.

Remark. It is well known that if one neglects the t-dependence in the KP equation (1) one gets the Boussinesq equation

$$u_{yy} = -\left(\frac{1}{2}u^2 + \frac{1}{12}u_{xx}\right)_{xx},$$

where y is now understood as the new time variable. In this case, when the t-dependence is neglected everywhere, the mKP equation (36) reduces to the modified Boussinesq system (the remnant of the system (79))

$$\begin{cases} v_y = & \varphi_x \\ \varphi_y = & \left(\dfrac{v^3}{6} - \dfrac{v_{xx}}{12}\right)_x - v_x\varphi \end{cases}$$

which comes from the minimal factorization [**24**] of the 3rd order Lax operator for the Boussinesq equation:

$$\partial^3 + u\partial + \overline{u} = (\partial + \tilde{u})(\partial^2 - \tilde{u}\partial + \tilde{\tilde{u}}).$$

More interestingly, removing the $-144\partial_t$ term from the Hamiltonian matrix B_1 (51) of the KP equation we obtain the (first) Hamiltonian form of the Boussinesq equation in the x-representation:

$$\begin{pmatrix} 0 & 0 & 0 & \partial_y \\ 0 & 0 & -1 & 0 \\ 0 & 1 & 0 & -12u \\ \partial_y & 0 & 12u & 0 \end{pmatrix}.$$

§ 4. The ZK Equation

If we rescale all the independent variables t, x, y, by a common factor ℓ^{-1} and then keep in the equations and maps of Section 2 only the lowest order terms in ℓ, we obtain: the ZK equation

$$u_{tx} = \left(\frac{1}{2}u^2\right)_{xx} + u_{yy} \tag{87}$$

$$u_t = \left(\frac{1}{2}u^2\right)_x + \partial^{-1}(u_{yy}); \tag{88}$$

the mZK equation

$$v_t = \left(-\frac{1}{6}v^3\right)_x + \partial^{-1}(v_{yy}) + v_x\partial^{-1}(v_y); \tag{89}$$

the map from the mZK equation (89) to the ZK equation (88)

$$u = -\frac{1}{2}v^2 + \partial^{-1}(v_y); \tag{90}$$

the deformed ZK equation

$$w_t = \left(\frac{1}{2}w^2 - \frac{\epsilon^2}{6}w^3\right)_x + \partial^{-1}(w_{yy}) + \epsilon w_x\partial^{-1}(w_y) \tag{91}$$

and its contraction into the ZK equation (88)

$$u = w - \frac{\epsilon^2}{2}w^2 + \epsilon\partial^{-1}(w_y). \tag{92}$$

We can repeat the Hamiltonian arguments at the beginning of Section 3 and move on to the x-representation for the equations (86) – (92). Since we have discarded higher-order terms, all our systems will be 2-component ones rather than the 4-component systems we have dealt with previously. In particular, we can't use for the ZK equation the Hamiltonian form of the KP equation found in the previous Section; we shall have to start from scratch.

We have then, for the ZK equation:

$$\begin{cases} \psi_x &= u_y \\ u_x &= u^{-1}(u_t - \psi_y), \end{cases} \tag{93}$$

$$(6u^2)_x = (-12\psi)_y + (12u)_t, \tag{94}$$

$$\left(\frac{1}{2}\psi^2 - \frac{1}{3}u^3\right)_x = (\psi u)_y - \left(\frac{1}{2}u^2\right)_t, \tag{95}$$

$$H_1 = 6u^2, \tag{96}$$

$$H_2 = \frac{1}{2}\psi^2 - \frac{1}{3}u^3. \tag{97}$$

For the mZK equation we have, similarly:

$$\begin{cases} \varphi_x = v_y \\ v_x = \left(\varphi - \frac{v^2}{2}\right)^{-1} (v_t - \varphi_y) \end{cases}, \tag{98}$$

$$\left(v\varphi - \frac{1}{6}v^3\right)_x = \left(\frac{1}{2}v^2 - \varphi\right)_y + v_t, \tag{99}$$

$$\left(\frac{1}{2}\varphi^2 - \frac{1}{2}\varphi v^2 + \frac{1}{8}v^4\right)_x = \left(v\varphi - \frac{1}{6}v^3\right)_y - \left(\frac{1}{2}v^2\right)_t. \tag{100}$$

Finally, for the deformed ZK equation we have:

$$\begin{cases} \chi_x = w_y \\ w_x = \left(w - \frac{\epsilon^2}{2}w^2 + \epsilon\chi\right)^{-1} (w_t - \chi_y) \end{cases}, \tag{101}$$

$$\left(\frac{1}{2}w^2 + \epsilon w\chi - \frac{\epsilon^2}{6}w^3\right)_x = \left(-\chi + \frac{\epsilon}{2}w^2\right)_y + w_t, \tag{102}$$

$$\left(\frac{1}{2}\chi^2 - \frac{1}{3}w^3 - \frac{\epsilon}{2}w^2\chi + \frac{\epsilon^2}{8}w^4\right)_x = \left(w\chi - \frac{\epsilon}{6}w^3\right)_y - \left(\frac{1}{2}w^2\right)_t. \tag{103}$$

(Notice that, as in the previous Section, ψ, φ, χ are also conserved densitites.)

While the Hamiltonian status of the mZK system (98) and the deformed ZK system (101) is uncertain at the moment, the ZK system (93) has the following Hamiltonian form:

$$\begin{pmatrix} \psi \\ u \end{pmatrix}_x = \begin{pmatrix} 0 & \partial_y u^{-1} \\ u^{-1}\partial_y & u^{-1}\partial_t u^{-1} \end{pmatrix} \begin{pmatrix} \delta H/\delta\psi \\ \delta H/\delta u \end{pmatrix} \tag{104}$$

where

$$H = -\frac{1}{2}\psi^2 + \frac{1}{3}u^3 = -H_2. \tag{105}$$

To see that the matrix in the right-hand-side of formula (104) is indeed Hamiltonian it is enough to notice that in the variables $\{\psi, \tilde{u} := u^2/2\}$ this matrix becomes

$$\begin{pmatrix} 0 & \partial_y \\ \partial_y & \partial_t \end{pmatrix}, \tag{106}$$

i.e., skewsymmetric constant-coefficient.

Acknowledgement

Discussions with Allan Fordy and Boris Konopelchenko were helpful to me. The hospitality of the Institute for Advanced Study is much appreciated. This work was partially supported by the NSF.

References

[1] B. B. Kadomtsev and V. I. Petviashvili, Sov. Phys. Dokl. **15** (1970) 539.

[2] E. Previato and G. Wilson, *Vector bundles over curves and solutions of the KP equation*, in Proc. Symp. Pure Math. **49** Part I (1989) 553.

[3] R. M. Miura, J. Math. Phys. **9** (1968) 1201.

[4] R. M. Miura, C. S. Gardner, and M. D. Kruskal, J. Math. Phys. **9** (1968) 1204.

[5] B. A. Kupershmidt and G. Wilson, Invent. Math. **62** (1981) 403.

[6] B. A. Kupershmidt, J. Math. Phys. **22** (1981) 449.

[7] B. A. Kupershmidt, J. Math. Phys. **23** (1982) 1427.

[8] S. V. Manakov, Phyica D **3** (1981) 420.

[9] B. G. Konopelchenko, Phys. Lett. A **92** (1982) 323.

[10] B. G. Konopelchenko and V. G. Dubrovsky, Phys. Lett. A **102** (1984) 15.

[11] G. Wilson, Lecture at the Soliton meeting at the Clarkson College, Potsdam, NY (1979).

[**12**] G. Wilson, Quart. J. Math. Oxford (2) **32** (1981) 491.

[**13**] B. A. Kupershmidt, Comm. Math. Phys. **99** (1985) 51.

[**14**] B. A. Kupershmidt, J. Phys. A **23** (1990) 871.

[**15**] E. A. Zabolotskaya and R. V. Khokhlov, Sov. Phys. Acoust. **15** (1969) 35.

[**16**] M. Adler and J. Moser, Comm. Math. Phys. **61** (1978) 219.

[**17**] F. A. Grünbaum, Phys. Lett. A **139** (1989) 146.

[**18**] B. A. Kupershmidt, Lett. Math. Phys. **20** (1990) 19.

[**19**] B. A. Kupershmidt and G. Wilson, Comm. Math. Phys. **81** (1981) 189.

[**20**] G. Wilson, Phys. Lett. A **132** (1988) 445.

[**21**] B. A. Kupershmidt, *Discrete Lax Equations and Differential-Difference Calculus*, Asterisque (Paris, 1985).

[**22**] B. A. Kupershmidt, Lett. Math. Phys. **9** (1985) 323–330.

[**23**] S. P. Tsarev, Math. Zametky **46** N1 (1989) 105.

[**24**] B. A. Kupershmidt, Proc. Roy. Irish Acad. A **83** (1983) 45.

Geometry in Partial Differential Equations (pp. 173–188)
eds. A. Pràstaro and Th. M. Rassias

CLASSIFICATION OF MIXED TYPE MONGE–AMPERE EQUATIONS

ALEXEY KUSHNER
International Sophus Lie Center,
ul.Mechnikova, 2–2, Astrahan, 414000, Russia

ABSTRACT

We present the classification of mixed type Monge-Ampere equations on a plane.

Introduction

0.1. The problem of classification of the Monge-Ampere equations up to the group of contact diffeomorphisms was put by Sophus Lie in 1874 [1]. In 1979 T. Morimoto [2] announced some classification results for equations of special kind using the theory of G-structures. The Monge-Ampere equations reducible to equations with constant coefficients were classified in [3] by V. Lychagin, V. Rubtsov, I. Tchekalov. Their results were based on the relations between the Monge-Ampere equations and an effective differential forms defined by V. Lychagin [4].

The problem of local equivalence of Monge-Ampere equations on a plane which are not reduced to equations with constant coefficients are solved in this work. The equations considered here, are mixed type equations, which are arise in gas dynamics, non-linear acoustics, physics of black holes.

0.2. Let M be a 2-dimensional smooth manifold, J^1M be a manifold of 1-jet of smooth functions on M, U_1 be Cartan form on J^1M (in a special coordinate system q_1, q_2, u, p_1, p_2 on J^1M, $U_1 = du - p_1dq_1 - p_2dq_2$) [5]. Following [4], for any differential 2-form ω on J^1M construct a non-linear differential operator $\Delta_\omega : C^\infty(M) \to \Lambda^2(M)$ acting via the formula $\Delta_\omega(h) = \sigma^*_{j_1(h)}(\omega)$, where

$$\sigma_{j_1(h)} : M \ni x \to j_1(h) \mid_x \in J^1M$$

is the section determined by a function $h \in C^\infty(M)$. A restriction $dU_{1,x} \mid_{\operatorname{Ker} U_{1,x}}$ get a symplectic structure Ω_x on $\operatorname{Ker} U_{1,x} \subset T_x(J^1M)$. Let X_1 be a contact vector field on J^1M with a generating function 1. A differential form $\omega \in \Lambda^2(J^1M)$ is called an effective 2-form if hold the following conditions:

1) $i_{X_1}\omega = 0$,
2) $\omega \wedge \Omega = 0$.

One can suppose ω is an effective form on J^1M [4]. It is easy to prove that in a local coordinate system an equation $E_\omega = \{\Delta_\omega = 0\} \subset J^1M$ has the form

$$a\frac{\partial^2 u}{\partial q_1^2} + 2b\frac{\partial^2 u}{\partial q_1 \partial q_2} + c\frac{\partial^2 u}{\partial q_2^2} + e\,\mathrm{Hess}\,u + f = 0,$$

$$\text{where } a, b, c, e, f \in C^\infty(J^1M),\ \mathrm{Hess}\,u = \frac{\partial^2 u}{\partial q_1^2}\frac{\partial^2 u}{\partial q_2^2} - \left(\frac{\partial^2 u}{\partial q_1 \partial q_2}\right)^2.$$

The last two terms of this form a classic Monge-Ampere equation on the plane.

Suppose in a neighbourhood of a point $m \in J^1M$ the operator Δ_ω has a contact infinitesimal symmetry X_f $(f(m) \neq 0)$. Then there is exist a local contact diffeomorphism ϕ, such that $\phi_*(X_f) = X_1$ [3]. Therefore we assume $L_{X_f}(\omega) = 0$ (L is Lie derivative) and ω is the 2-form on the cotangent bundle T^*M. The problem of equivalence of Monge-Ampere equations is a problem of the equivalence of differential effective forms on T^*M.

The 2-form Ω is a symplectic structure on T^*M. Let $\omega \in \Lambda^2(J^1M)$ is effective differential form, $m \in J^1M$. Remember [3], that the number $Pf(\omega_m)$ on T^*M is called *pfaffian* of the form $\omega_m \in \Lambda^2(T^*_mM)$ if

$$\omega_m \wedge \omega_m = Pf(\omega_m)\Omega_m \wedge \Omega_m$$

A differential 2-form $\omega \in \Lambda^2(J^1M)$ in a point $m \in J^1M$ is called
1) elliptic, if $Pf(\omega_m) > 0$;
2) hyperbolic, if $Pf(\omega_m) < 0$;
3) parabolic, if $Pf(\omega_m) = 0$.

The equation E_ω is called elliptic, hyperbolic, parabolic in the point m, correspondingly.

From [3] it follows that the equation E_ω can be reduced to an equation with constant coefficients in a neighbourhood of m, where $Pf(\omega_m) \neq 0$, if and only if

$$2d\omega = d\ln|Pf(\omega)| \wedge \omega.$$

The equation reduces in this case to Laplace equation or to the wave equation. If the equation E_ω changes a type in the point m, then E_ω is non-reducible to an equation with constant coefficients. For example Tricomi equation

$$q_1\frac{\partial^2 u}{\partial q_2^2} + \frac{\partial^2 u}{\partial q_1^2} = 0$$

in the point $m \in \{q_1 = 0\}$.

0.3. The equations E_ω and $E_{\omega_0}(\omega, \omega_0 \in \Lambda^2(T^*M))$ are symplectically equivalent if and only if exist a symplectic diffeomorphism $\phi : T^*M \to T^*M$, such that

$\phi^*(\omega) = \lambda\omega_0$ $(\lambda \in C^\infty(T^*M))$. If $\phi^*(\omega) = \omega_0$, then $\phi^*(Pf(\omega)) = Pf(\omega_0)$. Let m, m_0 be such that $m \in \{Pf(\omega) = 0\}$, $m_0 \in \{Pf(\omega_0) = 0\}$. There exists a symplectic diffeomorphism ϕ, such that $\phi(m) = m$. Therefore suppose that $m = m_0$. We can suppose also $Pf(\omega) = Pf(\omega_0)$. Really, if in the point m $Pf(\omega_m) = 0$ and $dPf(\omega)_m \neq 0$, then we can suppose that in a neighbourhood of the point m $Pf(\omega) = q_1$. In a more general case such that a level surface $\{Pf(\omega) = 0\}$ is a smooth manifold, we can suppose $Pf(\omega) = q_1^n$, $n \in \mathbb{N}$. It is clear that if $\phi^*(\omega) = \omega_0$ and $Pf(\omega) = q_1^n$, $Pf(\omega_0) = q_1^k$, then $n = k$.

Definition. Two differential 2-forms ω and ω_0 are called *local symplectically equivalent* at the point m, if exists diffeomorphism ϕ in a neighbourhood of the point m, such that $\phi^*(\omega) = \omega_0$.

Let us formulate the problem of local equivalence of Monge-Ampere equations in the terms of differential forms: when two effective forms ω and ω_0 are local symplectically equivalent?

1. Equivalence of Monge-Ampere Equations

1.1. Let $a \in T^*M$, $A_{\omega,a}$ be a linear operator $A_{\omega,a} : T_a(T^*M) \to T_a(T^*M)$, such that

$$A_{\omega,a}X_a \lrcorner \Omega_a = X_a \lrcorner \omega_a$$

for any $X_a \in T_a(T^*M)$. We have a field of linear operators A_ω on T^*M. It is not hard to prove that A_ω^2 is a scalar matrix, $A_\omega^2 + Pf(\omega) = 0$ and $\Omega(A_\omega X, Y) = \Omega(X, A_\omega Y)$ for any vector fields X, Y.

Let S_m be a level surface of the function $Pf(\omega)$, $m \in S_m$. Suppose in the neighbourhood $o(m)$ of the point m the surface S_m is a smooth manifold. Then in the neighbourhood $o(m)$ we have $Pf(\omega) = F^n$, where $n \in \mathbb{N}$, $dF_m \neq 0$.

Let X_F be hamiltonian vector field with hamiltonian $F : X_F \lrcorner \Omega = -dF$. Define in the neighbourhood $o(m)$ the vector fields $Z_\omega, W_\omega, V_\omega$:

$$W_\omega \lrcorner \Omega^2 = 2d\omega,\ Z_\omega = A_\omega X_F,\ V_\omega = A_\omega W_\omega,$$

where $\Omega^2 = \Omega \wedge \Omega$. Define the differential 1-forms

$$\mu_\omega = W_\omega \lrcorner \Omega,\ \tau_\omega = Z_\omega \lrcorner \Omega,\ \eta_\omega = V_\omega \lrcorner \Omega.$$

It is not hard to check that
1) $X_F \lrcorner \omega = \tau_\omega$, 2) $Z_\omega \lrcorner \omega = F^n dF$, 3) $V_\omega \lrcorner \omega = -F^n \mu_\omega$, 4) $W_\omega \lrcorner \omega = \eta_\omega$.
By definition, put $v = V(F)$, $w = W(F)$. The values of the forms dF, τ_ω, μ_ω, η_ω on the vector fields are enumerated in the Table 1.

Table 1.

	X_F	Z_ω	W_ω	V_ω
dF	0	0	w	v
τ_ω	0	0	$-v$	$F^n w$
μ_ω	w	v	0	0
η_ω	v	$-F^n w$	0	0

Lemma 1.

$$dw = \mu_\omega \wedge \Omega, \tag{1}$$
$$\tau_\omega \wedge \Omega = dF \wedge \omega, \tag{2}$$
$$\eta_\omega \wedge \Omega = -\mu_\omega \wedge \omega, \tag{3}$$
$$\eta_\omega \wedge \omega = F^n \mu_\omega \wedge \Omega, \tag{4}$$
$$\tau_\omega \wedge \mu_\omega \wedge \Omega = \eta_\omega \wedge dF \wedge \Omega, \tag{5}$$
$$\tau_\omega \wedge \eta_\omega \wedge \Omega = -F^n \mu_\omega \wedge dF \wedge \Omega, \tag{6}$$
$$\tau_\omega \wedge \mu_\omega \wedge dF = -w\tau_\omega \wedge \Omega - v dF \wedge \Omega. \tag{7}$$

Proof.
1) $2d\omega = W_\omega \lrcorner \Omega^2 = 2(W_\omega \lrcorner \Omega) \wedge \Omega = 2\mu_\omega \wedge \Omega$;
2) Since the form ω is effective, we see that $\omega \wedge \Omega = 0$. Then $0 = X_F \lrcorner (\omega \wedge \Omega) = \tau_\omega \wedge \Omega - dF \wedge \omega$;
3) $0 = W_\omega \lrcorner (\omega \wedge \Omega) = \eta_\omega \wedge \Omega + \omega \wedge \mu_\omega$;
4) $0 = V_\omega \lrcorner (\omega \wedge \Omega) = -F^n \mu_\omega \wedge \Omega + \omega \wedge \eta_\omega$;
5) $\mu_\omega \wedge \tau_\omega \wedge \Omega = \mu_\omega \wedge dF \wedge \omega = -dF \wedge \mu_\omega \wedge \omega = dF \wedge \eta_\omega \wedge \Omega$;
6) $-\eta_\omega \wedge \tau_\omega \wedge \Omega = -\eta_\omega \wedge dF \wedge \omega = dF \wedge \eta_\omega \wedge \omega = F^n dF \wedge \mu_\omega \wedge \Omega$;
7) From Eq. 5 it follows that

$$X_F \lrcorner (\tau_\omega \wedge \mu_\omega \wedge \Omega) = X_F \lrcorner (\eta_\omega \wedge dF \wedge \Omega).$$

Then

$$v dF \wedge \Omega = -w\tau_\omega \wedge \Omega - \tau_\omega \wedge \mu_\omega \wedge dF.$$

Lemma 2.

$$2dF \wedge \tau_\omega \wedge \eta_\omega \wedge \mu_\omega = (F^n w^2 + v^2)\Omega^2.$$

Proof. Multiply Eq. 7 by V_ω, we get

$$(F^n w^2 + v^2)\Omega = v dF \wedge \eta_\omega + w\tau_\omega \wedge \eta_\omega - v\tau_\omega \wedge \mu_\omega - F^n w \mu_\omega \wedge dF. \tag{8}$$

Multiply Eq. 8 by Ω. From Eqs. 5, 6 we obtain:

$$(F^n w^2 + v^2)\Omega^2 = 2(w\tau_\omega \wedge \eta_\omega \wedge \Omega - v\tau_\omega \wedge \mu_\omega \wedge \Omega).$$

It remains to prove that

$$dF \wedge \tau_\omega \wedge \eta_\omega \wedge \mu_\omega = w\tau_\omega \wedge \eta_\omega \wedge \Omega - v\tau_\omega \wedge \mu_\omega \wedge \Omega.$$

But this follows from Eqs. 7, 8.

By definition, put $\delta = F^n w^2 + v^2$.

Remark. Since Lemma 1 we see that the vectors $X_{F,m}$, $Z_{\omega,m}, W_{\omega,m}, V_{\omega,m}$ (where $F(m) = 0$) are linearly independent if and only if $v(m) = 0$.

1.3. Suppose $\delta(m) \neq 0$. Then

$$\operatorname{rank}\{X_{F,m}, Z_{\omega,m}, W_{\omega,m}, V_{\omega,m}\} = 4.$$

The vector fields

$$\begin{aligned} X_1 &= \delta^{-1}(F^n w W_\omega + v V_\omega), \\ X_2 &= X_F, \\ X_3 &= \delta^{-1}(v W_\omega - w V_\omega), \\ X_4 &= Z_\omega \end{aligned}$$

are called a local free *canonical basis of the vector fields module* $\mathcal{D}(T^*M)$. The differential 1-forms

$$\begin{aligned} \theta^1 &= dF, \\ \theta^2 &= \delta^{-1}(F^n w \mu_\omega + v \eta_\omega), \\ \theta^3 &= -\tau_\omega, \\ \theta^4 &= \delta^{-1}(v \mu_\omega - w \eta_\omega) \end{aligned}$$

are called *the canonical basis of the module* $\Lambda^1(T^*M)$.

The bases $\{X_i\}$ and $\{\theta^i\}$ are dual and

$$A_\omega X_3 = X_1, \ A_\omega X_2 = X_4,$$

$$X_1 \lrcorner \Omega = \theta^2, \ X_2 \lrcorner \Omega = -\theta^1, \ X_3 \lrcorner \Omega = \theta^4, \ X_4 \lrcorner \Omega = -\theta^3.$$

Easily check that

$$\Omega = \theta^1 \wedge \theta^2 + \theta^3 \wedge \theta^4, \tag{9}$$

$$\omega = \theta^3 \wedge \theta^2 + F^n \theta^4 \wedge \theta^1. \tag{10}$$

1.4. Suppose $\mathfrak{g}_\omega = \bigoplus_{i=1}^{4} \mathbb{R}X_i$ is a Lie algebra:

$$[X_i, X_j] = \sum_{k=1}^{4} c_{ij}^k X_k \quad (c_{ij}^k \in \mathbb{R}; \quad i,j = 1,\ldots,4).$$

We have

$$d\theta^k(X_i, X_j) = -\theta([X_i, X_j]) = -c_{ij}^k. \tag{11}$$

The Maurer-Cartan equation

$$d\theta^k = -\sum_{1\leqslant i\leqslant j\leqslant 4} c_{ij}^k \theta^i \wedge \theta^j$$

follows from Eq. 11.

Let ω_0 be a differential effective 2-form on the manifold T^*M and $S_m^{\omega_0}$ a level surface of the function $Pf(\omega_0)$, $m \in S_m^{\omega_0}$. Suppose in the neighbourhood $\mathcal{O}_m$ of the point m the surface $S_m^{\omega_0}$ is a smooth manifold. Then $Pf(\omega_0) = F_0^n$, where $dF_{0,m} \neq 0$. Let $\delta_{\omega_0}(m) \neq 0$, $\mathfrak{g}_{\omega_0}$ be the Lie algebra an $F_0 = F$.

Theorem 1. *The effective differential 2-forms ω and ω_0 are locally symplectically equivalent at the point m if and only if the corresponding structure constants of the Lie algebras $\mathfrak{g}_\omega$ and $\mathfrak{g}_{\omega_0}$ coincide in the canonical bases.*

Proof. Necessity is evident. Sufficiency. There exist diffeomorphism $\phi : T^*M \to T^*M$, such that $\phi(m) = m$, $\phi^*(\theta_\omega^i) = \theta_{\omega_0}^i$ $(i = 1, \ldots, 4)$. From Eq. 9 it follows that ϕ is a symplectic diffeomorphism. Since $\theta_\omega^1 = \theta_{\omega_0}^1 = dF$, $\phi(m) = m$, we see that $\phi^*(F) = F$. From Eq. 10 we obtain $\phi^*(\omega) = \omega_0$.

1.5. Suppose that a system of the vector fields $X_1, \ldots, X_4$ do not form a Lie algebra, but will generate a finite-dimensional Lie algebra. We denote the obtained Lie algebra by $\mathfrak{g}_\omega$ too. Let $\mathfrak{h}_\omega$ be a stabilizer of the point m. The pair of the algebras $(\mathfrak{g}_\omega, \mathfrak{h}_\omega)$ gives a homogeneous space.

Theorem 2. *The effective differential 2-form ω and ω_0 are locally symplectically equivalent at the point m if and only if structure constants of the pairs of the Lie algebras $(\mathfrak{g}_\omega, \mathfrak{h}_\omega)$ and $(\mathfrak{g}_{\omega_0}, \mathfrak{h}_{\omega_0})$ are coincide.*

2. Classification of an Effective Differential Form

2.1. Let us apply the Theorem 1 to the symplectic classification of effective 2-forms. Suppose $\mathfrak{g}_\omega = \bigoplus_{i=1}^{4} \mathbb{R}X_i$ is a Lie algebra. From $d\Omega = 0$, Eq. 1, 9 it follows that

$$\begin{aligned}
&c_{23}^2 + c_{12}^4 = 0, \qquad c_{24}^2 - c_{12}^3 = 0, \qquad c_{23}^3 + c_{24}^4 = 0,\\
&c_{13}^3 + c_{12}^2 = 0, \qquad c_{34}^3 + c_{24}^2 = -w,\ c_{34}^2 - c_{13}^3 - c_{14}^4 = 0,\\
&c_{14}^3 - F^n c_{24}^4 = v,\ c_{23}^4 = c_{14}^2 = c_{34}^4 = 0, \qquad c_{ij}^1 = 0\ (i,j = 1, \ldots, 4).
\end{aligned}$$

From $c_{34}^3 + c_{24}^2 = -w$ it follows that $w = const$. By definition, put

$$a_1 = c_{12}^2,\ a_2 = c_{13}^2,\ a_3 = c_{23}^2, a_4 = c_{12}^3,\ a_5 = c_{34}^2,\ a_6 = c_{23}^3,$$

$$a_7 = c_{24}^3,\ a_8 = c_{13}^4,\ a_9 = c_{14}^3.$$

Then $v = a_9 + F^n a_6$. We obtain anticommutative algebra. The multiplication is given in the Table 2.

Table 2.

	X_1	X_2	X_3	X_4
X_1	0	$a_1X_2+a_4X_3-a_3X_4$	$a_2X_2-a_1X_3+a_8X_4$	$a_9X_3+(a_1+a_5)X_4$
X_2		0	$a_3X_2+a_6X_3$	$a_4X_2+a_7X_3-a_6X_4$
X_3			0	$a_5X_2-(w+a_4)X_3$
X_4				0

A necessary condition for the anticommutative algebra to be a Lie algebra is that its elements satisfy the Jacobi identity. Using the Jacobi identity, we see that Table 2 gives 8 Lie algebras $\mathfrak{g}_1, \ldots, \mathfrak{g}_8$. Their commutation relations are indicated in the Tables 4.1–4.8 respectively. In order to find out which of these Lie algebras corresponds to a given 2-form ω, one can use the cases shown on Table 3. For example, if ω satisfies the condition $a_6 = a_8 = w = 0$ and $a_4 \neq 0$ then the corresponding Lie algebra is $\mathfrak{g}_4$.

Remark. For equation of mixed type $(F(m) = 0)$ $a_9 \neq 0$.

Table 3.

	$\mathfrak{g}_1$	$\mathfrak{g}_2$		$\mathfrak{g}_3$	$\mathfrak{g}_4$	$\mathfrak{g}_5$		$\mathfrak{g}_6$		$\mathfrak{g}_7$	$\mathfrak{g}_8$
a_1	$\neq 0$	$= 0$									
a_4	$= 0$	$= 0$	$= 0$	$\neq 0$	$\neq 0$	$= 0$		$= 0$			
a_5						$\neq 0$	$\neq 0$	$= 0$	$= 0$		
a_6	$= 0$	$= 0$	$= 0$	$= 0$	$= 0$	$= 0$	$= 0$	$= 0$	$= 0$	$\neq 0$	$\neq 0$
a_7			$\neq 0$			$= 0$		$= 0$			
a_8	$= 0$	$= 0$	$= 0$	$= 0$	$= 0$	$= 0$	$\neq 0$	$= 0$	$\neq 0$		
a_9										$\neq 0$	$= 0$
w	$\neq 0$	$\neq 0$	$= 0$	$\neq 0$	$= 0$	$= 0$		$= 0$			

Table 4.1. ($a_1 \neq 0,\ w \neq 0$)

	X_1	X_2	X_3	X_4
X_1	0	a_1X_2	$-2a_1^2w^{-1}X_2 - a_1X_3$	a_9X_3
X_2	$-a_1X_2$	0	0	0
X_3	$2a_1^2w^{-1}X_2 + a_1X_3$	0	0	$-a_1X_2 - wX_3$
X_4	$-a_9X_3$	0	$a_1X_2 + wX_3$	0

Table 4.2.

	X_1	X_2	X_3	X_4
X_1	0	0	0	a_9X_3
X_2	0	0	0	a_7X_3
X_3	0	0	0	$-wX_3$
X_4	$-a_9X_3$	$-a_7X_3$	wX_3	0

Table 4.3. ($a_4 \neq 0,\ \beta = 2\lambda^2a_2 \neq 2a_4$)

	X_1	X_2	X_3	X_4
X_1	0	$-\lambda a_2X_2 + a_4X_3$	$a_2X_2 + \lambda a_2X_3$	a_9X_3
X_2	$\lambda a_2X_2 - a_4X_3$	0	0	$a_4X_2 + \lambda a_4X_3$
X_3	$-a_2X_2 - \lambda a_2X_3$	0	0	$\lambda a_2X_2 + (\beta+a_4)X_3$
X_4	$-a_9X_3$	$-a_4X_2 - \lambda a_4X_3$	$-\lambda a_2X_2 - (\beta+a_4)X_3$	0

Table 4.4. ($a_7 \neq 0,\ \sigma \neq 0$)

	X_1	X_2	X_3	X_4
X_1	0	$\sigma^2a_7X_2+\sigma a_7X_3$	$-\sigma^3a_7X_2-\sigma^2a_7X_3$	a_9X_3
X_2	$-\sigma^2a_7X_2-\sigma a_7X_3$	0	0	$\sigma a_7X_2+a_7X_3$
X_3	$\sigma^3a_7X_2+\sigma^2a_7X_3$	0	0	$-\sigma^2a_7X_2-\sigma a_7X_3$
X_4	$-a_9X_3$	$-\sigma a_7X_2-a_7X_3$	$\sigma^2a_7X_2+\sigma a_7X_3$	0

Table 4.5. $(a_1 \neq 0)$

	X_1	X_2	X_3	X_4
X_1	0	a_1X_2	$a_2X_2 - a_1X_3 + a_8X_4$	$a_9X_3 + 2a_1X_4$
X_2	$-a_1X_2$	0	0	0
X_3	$-a_2X_2 + a_1X_3 - a_8X_4$	0	0	a_1X_2
X_4	$-a_9X_3 - 2a_1X_4$	0	$-a_1X_2$	0

Table 4.6.

	X_1	X_2	X_3	X_4
X_1	0	a_1X_2	$a_2X_2 - a_1X_3 + a_8X_4$	$a_9X_3 + a_1X_4$
X_2	$-a_1X_2$	0	0	0
X_3	$-a_2X_2 + a_1X_3 - a_8X_4$	0	0	0
X_4	$-a_9X_3 - a_1X_4$	0	0	0

Table 4.7. $(a_4 \neq 0,\ a_9 \neq 0,\ \gamma = a_4^2 a_9^{-1})$

	X_1	X_2	X_3	X_4
X_1	0	a_4X_3	0	a_9X_3
X_2	$-a_4X_3$	0	γX_3	$a_4X_2 + a_7X_3 - \gamma X_4$
X_3	0	$-\gamma X_3$	0	$-a_4X_3$
X_4	$-a_9X_3$	$-a_4X_2 - a_7X_3 + \gamma X_4$	a_4X_3	0

Table 4.8. $(a_6 \neq 0)$

	X_1	X_2	X_3	X_4
X_1	0	0	0	0
X_2	0	0	a_6X_3	$a_7X_3 - a_6X_4$
X_3	0	$-a_6X_3$	0	0
X_4	0	$-a_7X_3 + a_6X_4$	0	0

Let us write out the representations of the Lie algebras $\mathfrak{g}_i$, by vector fields (the representation of $\mathfrak{g}_1$ has been given to me by B.Doubrov).

$\mathfrak{g}_1$

$$X_1 = \frac{\partial}{\partial q_1} - (a_1p_1 + 2a_1^2w^{-1}p_2)\frac{\partial}{\partial p_1} + (a_1p_2 + a_9q_2)\frac{\partial}{\partial p_2}$$
$$X_2 = \frac{\partial}{\partial p_1}$$

$$X_3 = -\frac{\partial}{\partial p_2}$$
$$X_4 = \frac{\partial}{\partial q_2} + (a_1 p_2 + a_9 q_2)\frac{\partial}{\partial p_1} - (w p_2 + a_9 a_1^{-1} w q_2)\frac{\partial}{\partial p_2}$$

$\mathfrak{g}_2$

$$X_1 = \frac{\partial}{\partial q_1}$$
$$X_2 = \frac{\partial}{\partial p_1}$$
$$X_3 = -\frac{\partial}{\partial p_2}$$
$$X_4 = \frac{\partial}{\partial q_2} - (a_9 q_1 + a_7 p_1 + w p_2)\frac{\partial}{\partial p_2}$$

$\mathfrak{g}_3$

$$X_1 = \frac{\partial}{\partial q_1} + (\lambda a_2 p_1 + a_2 p_2 - a_2 a_9 a_4^{-1} q_1)\frac{\partial}{\partial p_1} + (a_4 p_1 - \lambda a_2 p_2)\frac{\partial}{\partial p_2}$$
$$X_2 = \frac{\partial}{\partial p_1}$$
$$X_3 = -\frac{\partial}{\partial p_2}$$
$$X_4 = \frac{\partial}{\partial q_2} + (a_4 p_1 - \lambda a_2 p_2)\frac{\partial}{\partial p_1} + (-\lambda a_4 p_1 + (2\lambda^2 a_2 + a_4) p_2 - a_9 q_1)\frac{\partial}{\partial p_2}$$

$\mathfrak{g}_4$

$$X_1 = \frac{\partial}{\partial q_1} + \sigma^2(a_9 q_1 - a_7 p_1 - \sigma a_7 p_2)\frac{\partial}{\partial p_1} + \sigma a_7 (p_1 + \sigma p_2)\frac{\partial}{\partial p_2}$$
$$X_2 = \frac{\partial}{\partial p_1}$$
$$X_3 = -\frac{\partial}{\partial p_2}$$
$$X_4 = \frac{\partial}{\partial q_2} + \sigma a_7 (p_1 + \sigma p_2)\frac{\partial}{\partial p_1} - (a_7 p_1 + \sigma a_7 p_2 + a_9 q_1)\frac{\partial}{\partial p_2}$$

$\mathfrak{g}_5$

$$X_1 = \frac{\partial}{\partial q_1} + (a_8 p_2 - 2a_1 q_2)\frac{\partial}{\partial q_2} + (a_2 p_2 - a_1 p_1)\frac{\partial}{\partial p_1} + (a_9 q_2 + a_1 p_2)\frac{\partial}{\partial p_2}$$
$$X_2 = \frac{\partial}{\partial p_1}$$
$$X_3 = (a_8 p_2 - 2a_1 q_2)\frac{\partial}{\partial p_1} - \frac{\partial}{\partial p_2}$$
$$X_4 = \frac{\partial}{\partial q_2} + (a_9 q_2 + a_1 p_2)\frac{\partial}{\partial p_1}$$

$\mathfrak{g}_6$

$$X_1 = \frac{\partial}{\partial q_1} + (a_8 p_2 - a_1 q_2)\frac{\partial}{\partial q_2} +$$
$$+ (-a_2 p_2 - a_1 p_1 + \tfrac{1}{2} a_1 a_9 q_2^2 - \tfrac{1}{2} a_1 a_8 p_2^2 + a_1^2 p_2 q_2)\frac{\partial}{\partial p_1} + (a_9 q_2 + a_1 p_2)\frac{\partial}{\partial p_2}$$

$$X_2 = \frac{\partial}{\partial p_1}$$
$$X_3 = (a_1 q_2 - a_8 p_2)\frac{\partial}{\partial p_1} + \frac{\partial}{\partial p_2}$$
$$X_4 = -\frac{\partial}{\partial q_2} - (a_9 q_2 + a_1 p_2)\frac{\partial}{\partial p_1}$$

$\mathfrak{g}_7$

$$X_1 = -a_9 \frac{\partial}{\partial q_2} - a_4 a_9 p_2 \frac{\partial}{\partial p_1}$$
$$X_2 = a_4 a_9^{-1} \frac{\partial}{\partial q_1} - a_4^2 (a_9^{-1} q_2 + a_7 p_2)\frac{\partial}{\partial q_2} + a_9^{-1}(a_4^2 p_2 - 1)\frac{\partial}{\partial p_1}$$
$$X_3 = a_4 \frac{\partial}{\partial q_2} + (a_4^2 p_2 - 1)\frac{\partial}{\partial p_1}$$
$$X_4 = \frac{\partial}{\partial q_1} - (a_4 q_2 + a_4 a_7 a_9 p_2)\frac{\partial}{\partial q_2} + (a_7 a_9 p_2 + q_2)\frac{\partial}{\partial p_1} + a_4 p_2 \frac{\partial}{\partial p_2}$$

$\mathfrak{g}_8$

$$X_1 = \frac{\partial}{\partial q_1}$$
$$X_2 = \frac{\partial}{\partial p_1} + (a_7 q_2 - a_6 p_2)\frac{\partial}{\partial p_2} + a_6 q_2 \frac{\partial}{\partial q_2}$$
$$X_3 = -\frac{\partial}{\partial p_2} - a_6 q_2 \frac{\partial}{\partial q_1}$$
$$X_4 = (a_6 p_2 - a_7 q_2)\frac{\partial}{\partial q_1} + \frac{\partial}{\partial q_2}$$

Let us construct normal forms of effective differential 2-forms. Consider, for example, the Lie algebra $\mathfrak{g}_2$. In specific coordinates $\Omega = dq_1 \wedge dp_1 + dq_2 \wedge dp_2$. Using the mentioned representation of $\mathfrak{g}_2$ by vector fields, we construct the differential 1-forms $\theta^1, \ldots, \theta^2$: $\theta^1 = -X_2 \lrcorner \Omega = dq_1$, $\theta^2 = X_1 \lrcorner \Omega = dp_1$, $\theta^3 = -X_4 \lrcorner \Omega = -dp_2 - (a_9 q_1 + a_7 p_1 + w p_2) dq_2$, $\theta^4 = X_3 \lrcorner \Omega = dq_2$. Note that $\theta^1 \wedge \theta^2 + \theta^3 \wedge \theta^4 = dq_1 \wedge dp_1 + dq_2 \wedge dp_2 = \Omega$. Since $\theta^1 = dF = dq_1$, let $F = q_1$. By formula Eq. 10 we obtain the following normal form:

$$\omega_2 = dp_1 \wedge dp_2 - (a_9 q_1 + a_7 p_1 + w p_2) dq_2 \wedge dp_1 - q_1^n dq_1 \wedge dq_2.$$

Below we write out all normal forms. Here ω_i corresponds to the Lie algebra $\mathfrak{g}_i$.

The normal forms of the effective differential 2-forms.

$$\omega_1 = (-q_1^n + a_1^2 p_2^2 - a_9^2 q_2^2 + w a_1 p_1 p_2 + a_9 w q_2 p_1) dq_1 \wedge dq_2 + \\ + (a_1 p_2 + a_9 q_2) dq_1 \wedge dp_1 - (w p_2 + a_9 a_1^{-1} w q_2) dq_2 \wedge dp_1 + \\ + (a_1 p_1 + 2 a_1^2 w p_2) dq_1 \wedge dp_2 - (a_1 p_2 + a_9 q_2) dq_2 \wedge dp_2 + dp_1 \wedge dp_2;$$
$$\omega_2 = dp_1 \wedge dp_2 - (a_9 q_1 + a_7 p_1 + w p_2) dq_2 \wedge dp_1 - q_1^n dq_1 \wedge dq_2;$$
$$\omega_3 = (-q_1^n + a_2 a_9^2 a_4^{-1} q_1^2 - \alpha a_4 p_1^2 + \alpha a_2 p_2^2 + 2\alpha\lambda a_2 p_1 p_2 - 2\alpha a_2 a_9 a_4^{-1} q_1 p_2) dq_1 \wedge dq_2 + \\ + (a_4 p_1 - \lambda a_2 p_2) dq_1 \wedge dp_1 + (a_2 a_9 a_4^{-1} q_1 - \lambda a_2 p_1 - a_2 p_2) dq_1 \wedge dp_2 + \\ + ((2\lambda^2 a_2 + a_4) p_2 - \lambda a_4 p_1 - a_9 q_1) dq_2 \wedge dp_1 - (a_4 p_1 - \lambda a_2 p_2) dq_2 \wedge dp_2 + dp_1 \wedge dp_2 \\ (\alpha = a_4 + \lambda^2 a_2);$$

$$\omega_4 = -(q_1^n + \sigma^2 a_9 q_1)dq_1 \wedge dq_2 + \sigma a_7(p_1 + \sigma p_2)dq_1 \wedge dp_1 -$$
$$- \sigma a_7(p_1 + \sigma p_2)dq_2 \wedge dp_2 + \sigma^2(a_7 p_1 + \sigma a_7 p_2 - a_9 q_1)dq_1 \wedge dp_2 -$$
$$- (a_7 p_1 + \sigma a_7 p_2 + a_9 q_1)dq_2 \wedge dp_1 + dp_1 \wedge dp_2;$$
$$\omega_5 = -(q_1^n + (a_9 q_2 + a_1 p_2)^2)dq_1 \wedge dq_2 + (a_9 q_2 + a_1 p_2)dq_1 \wedge dp_1 + dp_1 \wedge dp_2 +$$
$$+ (a_1 p_1 - a_2 p_2 + (a_8 a_9 - 2a_1^2)q_2 p_2 - 2a_1 a_9 q_2^2 + a_1 a_8 p_2^2)dq_1 \wedge dp_2 -$$
$$- (a_9 q_2 + a_1 p_2)dq_2 \wedge dp_2;$$
$$\omega_6 = (q_1^n + (a_9 q_2 + a_1 p_2)^2)dq_1 \wedge dq_2 - (a_9 q_2 + a_1 p_2)dq_1 \wedge dp_1 +$$
$$+ (a_9 q_2 + a_1 p_2)dq_2 \wedge dp_2 + ((2a_1^2 - a_8 a_9)q_2 p_2 + \tfrac{3}{2}a_1 a_9 q_2^2 -$$
$$- \tfrac{3}{2}a_1 a_8 p_2^2 - a_1 p_1 - a_2 p_2)dq_1 \wedge dp_2 - dp_1 \wedge dp_2;$$
$$\omega_7 = (-a_4^2 a_9 p_2^2 - a_9^{-1}(1 - a_4^2 p_2)^2 F^n)dq_1 \wedge dq_2 +$$
$$+ (a_4 a_9 p_2 - a_4 a_9^{-1}(1 - a_4^2 p_2)F^n)dq_1 \wedge dp_1 +$$
$$+ (q_2 + a_7 a_9 p_2)(1 - a_4^2 p_2)(a_4^2 a_9^{-1} F^n - a_9)dq_1 \wedge dp_2 + (a_9 + a_4^2 a_9^{-1} F^n)dp_1 \wedge dp_2 +$$
$$+ (-a_4 a_9 p_2 + a_4 a_9^{-1}(1 - a_4^2 p_2)F^n)dq_2 \wedge dp_2,$$
where $F = -a_9^{-1} q_2 + a_4^2 a_9^{-1} p_2 q_2 - a_4 a_9^{-1} p_1 + \frac{1}{2} a_4^2 a_7 p_2^2$;
$$\omega_8 = F^n(-dq_1 \wedge dq_2 + a_6 q_2 dq_1 \wedge dp_1 + a_6 q_2(a_7 q_2 - a_6 p_2)dq_2 \wedge dp_1 -$$
$$- a_6 q_2 dq_2 \wedge dp_2) + (F^n a_6^2 q_2^2 + 1)dp_1 \wedge dp_2,$$
where $F = q_1 + \frac{1}{2} a_7 q_2^2 - a_6 p_2 q_2$.

Remark. We have $\omega(X_1, X_3) = \Omega(X_1, X_3) = 0$ and $\omega(X_2, X_4) = \Omega(X_2, X_4) = 0$. If $\{X_1, X_3\}$ or $\{X_2, X_4\}$ is the Lie algebra, then the distribution $\mathcal{F}(X_1, X_3)$ or $\mathcal{F}(X_2, X_4)$ is a complete integrable distribution. The 2-dimensional integral manifold $\mathcal{F}$ are solution of Monge-Ampere equation. Therefore in the case $\mathfrak{g}_1$, $\mathfrak{g}_2$, $\mathfrak{g}_5$, $\mathfrak{g}_6$, $\mathfrak{g}_7$, $\mathfrak{g}_8$ equations E_{ω_i} are equivalent to quasilinear ones.

3. Minimal Rank

3.1. Suppose $F(m) = 0$, $v = w = 0$.

Lemma 3. *If* $F(m) = 0$ *and* $X_{F,m} \notin \operatorname{Ker} A_{\omega,m}$, *then* $\operatorname{rank}\{X_{F,m}, Z_{\omega,m}\} = 2$.

Proof. Suppose $Z_{\omega,m} = \lambda X_{F,m}$ ($\lambda \in \mathbb{R}$). Then λ is a proper value of the operator $A_{\omega,m}$. But $\det(A_{\omega,m} - \lambda E) = (Pf(\omega_m) + \lambda^2)^2 = \lambda^4$. Then $\lambda = 0$.

Lemma 4. *If* $v(a) = w(a) = 0$, *then*

1) the vectors $X_{F,a}, Z_{\omega,a}, W_{\omega,a}$ *are linearly dependent;*

2) the vectors $X_{F,a}, Z_{\omega,a}, V_{\omega,a}$ *are linearly dependent.*

Proof. 1) Multiply the both parts of Eq. 5 considering at the point a, by the vector $X_{F,a}$, we obtain:

$$v(a)dF_a \wedge \Omega_a = -w(a)\tau_{\omega,a} \wedge \Omega_a - dF_a \wedge \tau_{\omega,a} \wedge \mu_{\omega,a}.$$

But $v(a) = w(a) = 0$. Then $dF_a \wedge \tau_{\omega,a} \wedge \mu_{\omega,a} = 0$.

2) Multiply Eq. 5 by the vector $Z_{\omega,a}$:

$$-v(a)\tau_{\omega,a} \wedge \Omega_a = -F^n(a)w(a)\Omega_a \wedge dF_a + \tau_{\omega,a} \wedge \eta_{\omega,a} \wedge dF_a.$$

Then $\tau_{\omega,a} \wedge \eta_{\omega,a} \wedge dF_a = 0$.

Suppose $X_{F,m} \notin \operatorname{Ker} A_{\omega,m}$. Since the Lemmas 3,4, we see that

$$\operatorname{rank}\{X_F, Z_\omega, V_\omega, W_\omega\} = 2.$$

Then

$$W_\omega = \alpha Z_\omega + \beta X_F,\ V_\omega = \alpha_1 Z_\omega + \beta_1 X_F,$$

where $\alpha, \alpha_1, \beta, \beta_1$ are smooth functions. The functions $\alpha_1 = \beta,\ \beta_1 = -\alpha F^n$. In fact,

$$V_\omega \lrcorner \Omega = W_\omega \lrcorner \omega = (\beta X_F + \alpha Z_\omega) \lrcorner \omega = \beta \tau_\omega + \alpha F^n dF = (\beta Z_\omega - \alpha F^n X_F) \lrcorner \Omega.$$

We have $V_\omega = -\alpha F^n X_F + \beta Z_\omega$.

3.2. Suppose α and β are constants.

Theorem 3. *The effective differential 2-form ω at the point m is locally symplectically equivalent up to a sign to the form*

$$\omega_0 = q_1^n dq_1 \wedge dp_2 - dq_2 \wedge dp_1 + (ap_1 + bp_2) dq_1 \wedge dq_2$$

if and only if
1) the vector fields Z_ω is a hamiltonian and the point m is a regular point of Z_ω;
2) $W_\omega + \alpha Z_\omega + \beta X_F = 0$.

Proof. Suppose $Pf(\omega) = q_1^n$. First of all we prove that the symplectic diffeomorphism ϕ exists, such that $\phi(m) = m$, $\phi_*(Z_\omega) = Z_{\omega_0}$, $\phi^*(q_1) = q_1$. Using the method of the path lifting [6]. By definition, put

$$Z_t = Z_{\omega_0} + t(Z_\omega - Z_{\omega_0}).$$

We prove that exists a 1-parametric family of diffeomorphisms $\{\phi_t\}$, such that

$$\phi_{t_*}(Z_t) = Z_{\omega_0} \tag{12}$$

for any $t \in [0,1]$. Suppose ϕ is a shift along the trajectory of the vector fields X_{H_t} from $t = 0$ to t. From Eq. 12 it follows that

$$[X_{H_t}, Z_t] = \frac{\partial}{\partial t} Z_t \tag{13}$$

Let h_t be a hamilton function of the vector field Z_t. Then Eq. 13 is equivalent to

$$[H_t, h_t] = \frac{\partial}{\partial t} h_t, \tag{14}$$

where [,] is Lagrange bracket. The Eq. 14 has the form

$$X_{h_t}(H_t) + \frac{\partial}{\partial t} h_t = 0. \tag{15}$$

Since m is a regular point of Z_ω, we see that the Eq. 15 has smooth solution H_t ($t \in [0,1]$) in the neighbourhood of the point m. It is not hard to prove that if $Z_\omega = Z_{\omega_0} = -\frac{\partial}{\partial p_2}$, $Pf(\omega) = q_1^n$, $W_\omega = -\alpha Z_\omega - \beta X_F$, the 2-form ω has the form

$$\omega = \omega_0 + k(q_1, q_2) dq_1 \wedge dq_2,$$

where k is a smooth function. Let the functions $\xi = \xi(q_1, q_2)$, $\chi = \chi(q_1, q_2)$ are the solutions of a system

$$\frac{\partial \xi}{\partial q_1} = k - q_1^n \frac{\partial \chi}{\partial q_2} - \alpha \xi - \beta \chi, \quad \frac{\partial \chi}{\partial q_1} = \frac{\partial \xi}{\partial q_2}$$

such that $\xi(m) = \chi(m) = 0$. By definition, put

$$\nu : (q_1, q_2, p_1, p_2) \mapsto (q_1, q_2, p_1 + \xi(q_1, q_2), p_2 + \chi(q_1, q_2))$$

Then $\nu^*(\omega_0) = \omega$.

Remark. The form ω_0 is corresponding to the Tricomi equation

$$q_1^n \frac{\partial^2 u}{\partial q_2^2} + \frac{\partial^2 u}{\partial q_1^2} + \alpha \frac{\partial u}{\partial q_1} + \beta \frac{\partial u}{\partial q_2} = 0.$$

3.3. Let α and β are not constants.

Lemma 5. *If vector fields X, Y are hamiltonian and $\Omega(X,Y) = 0$, then $[X,Y] = 0$.*

Proof. $0 = X \lrcorner L_Y(\Omega) = X \lrcorner d(Y \lrcorner \Omega) = L_X(Y \lrcorner \Omega) = L_X(Y) \lrcorner \Omega + Y \lrcorner L_x(\Omega) = [X,Y] \lrcorner \Omega$.

From definition of the vector field Z_ω follows that $\Omega(X_F, Z_\omega) = 0$. Suppose $L_{Z_\omega}(\Omega) = 0$. Since the Lemma 5, we see that $[X_F, Z_\omega] = 0$. Then a distribution $\mathcal{F}(X_F, Z_\omega)$ is a complete integrable distribution. A 2-dimensional integral manifold of $\mathcal{F}$ gives a Lagrangian bundle π_ω.

Let ω_0 be effective differential 2-form, $Pf(\omega_0) = F^n$, $v^{\omega_0} = w^{\omega_0} = 0$, $Z_{\omega_0, m} \neq 0$, $L_{Z_{\omega_0}}(\Omega) = 0$. There exists a symplectic diffeomorphism, transposing π_{ω_0} to π_ω. Therefore suppose $\pi_{\omega_0} = \pi_\omega$. If exists symplectic diffeomorphism ϕ, such that $\phi^*(\omega) = \omega_0$, then $\phi_*(Z_{\omega_0}) = Z_\omega$, $\phi_*(W_{\omega_0}) = W_\omega$. From the proof of Theorem 3 it follows that there exists a symplectic diffeomorphism χ, such that up to a sign $\chi(m) = m$, $\chi^*(F) = F$, $\chi_*(Z_{\omega_0}) = Z_\omega$. Therefore suppose $Z_\omega = Z_{\omega_0}$. Let a be any point in the neighbourhood of the point m, $\mathcal{E}_a$ be a fiber of the bundle π_ω ($a \in \mathcal{E}_a$). By definition, put $X^a = X \mid_{\mathcal{E}_a}$, $f^a = f \mid_{\mathcal{E}_a}$. The vector fields $X_F, Z_{\omega(\omega_0)}$ give the local system of coordinates on $\mathcal{E}_a$. Write the functions α^a_ω, β^a_ω, $\alpha^a_{\omega_0}$, $\beta^a_{\omega_0}$ in this coordinates:

$$W^a_{\omega_0} = \beta^a_{\omega_0} X_F + \alpha^a_{\omega_0} Z_{\omega_0}, \; W^a_\omega = \beta^a_\omega X_F + \alpha^a_\omega Z_\omega.$$

If $\phi^*(\omega) = \omega_0$, then $\alpha^a_{\omega_0} = \alpha^a_\omega$, $\beta^a_{\omega_0} = \beta^a_\omega$. Therefore in this case we have functional modules.

Remark. From $\omega(X_F, Z_\omega) = 0$ it follows that the fibers of π_ω are solutions of the equation E_ω.

4. Application

Consider one equation of gas dynamics, which describes two-dimensional stationary flow of n-component gas mixture [7]:

$$(b\frac{\partial u}{\partial q_1} + a)\frac{\partial^2 u}{\partial q_1^2} - c^2\frac{\partial^2 u}{\partial q_2^2} + k\frac{\partial u}{\partial q_1} = 0.$$

Here u is the velocity potential; in other words, $\frac{\partial u}{\partial q_i}$ is the axial velocity component corresponding to the q_i-axis; $a, b, c, k \in \mathbb{R}$, $c \neq 0$, $b \neq 0$. The effective form is

$$\omega = -(bp_1 + a)dq_2 \wedge dp_1 - c^2 dq_1 \wedge dp_2 + kp_1 dq_1 \wedge dq_2$$

The pfaffian $Pf(\omega) = F = -c^2(bp_1 + a)$. We have:

$$X_F = c^2 b\frac{\partial}{\partial q_1}, \; Z_\omega = -c^4 b\frac{\partial}{\partial q_2} - kc^2 bp_1\frac{\partial}{\partial p_2},$$

$$W_\omega = k\frac{\partial}{\partial p_2}, \; V_\omega = -kc^2\frac{\partial}{\partial p_1}.$$

Let $m \in \{F = 0\}$. If $k = 0$, then $v = w = 0$, the vector field Z_ω is hamiltonian. From Theorem 3 we see that the normal form is the Tricomi form

$$\omega = q_1 dq_1 \wedge dp_2 - dq_2 \wedge dp_1.$$

If $k \neq 0$ then the canonical basis has the form:

$$X_1 = -b^{-1}c^{-2}\frac{\partial}{\partial p_1}, \; X_2 = bc^2\frac{\partial}{\partial q_1},$$

$$X_3 = b^{-1}c^{-4}\frac{\partial}{\partial p_2}, \; X_4 = -bc^4\frac{\partial}{\partial q_2} - kc^2 bp_1\frac{\partial}{\partial p_2}.$$

The algebra $\mathfrak{g}_\omega$ is $\mathfrak{g}_6$. The normal form is

$$\frac{\partial^2 u}{\partial q_2^2} + \frac{\partial u}{\partial q_1}\frac{\partial^2 u}{\partial q_1^2} + \lambda\frac{\partial u}{\partial q_1} = 0.$$

The author would like to express his sincere appreciation to B. P. Komrakov and V. V. Lychagin. Thanks are also due to D. Shabes for the help in preparing the manuscript.

REFERENCES

1. S. Lie, *Gesamette Ahandlungen*, vol. 1–7, Leipzig–Oslo, 1922–1935.
2. T. Morimoto, *La problem d'equivalence des equationes de Monge-Ampere*, Comptes Rendus Acad. Sci. **289, 1–2** (1979).
3. V. Lychagin, V. Rubtsov, I. Tchekalov, *A local classification of Monge-Ampere equations*, Annales Scient. Ecole Normale Superieur **1** (1993).
4. V. Lychagin, *Contact Geometry and Non-linear Second order Differential Equations*, Russian Math. Surveys **34** (1979), 149–180.
5. V. Lychagin, *The Local Classification of Non-linear Differential Equations in First Order Partial Derivatives*, Russian Math. Surveys **30** (1975), 105–175.
6. A. Weinstein, *Symplectic manifolds and their Lagrangian submanifolds*, Adw. Math. **6** (1971), 329–346.
7. N. Larkin, *The smooth solutions of the equations of the transonic gas dynamics*, Novosibirsk, Nauka, 1991. (Russian)

Geometry in Partial Differential Equations (pp. 189–214)
eds. A. Pràstaro and Th. M. Rassias

NON-HOLONOMIC FILTRATION: ALGEBRAIC AND GEOMETRIC ASPECTS OF NON-INTEGRABILITY

VALENTIN LYCHAGIN

and

VLADIMIR RUBTSOV

"Sophus Lie" Center, Moscow branch krasnokazarmennaya, 6, Moscow 111250, Russia

ABSTRACT

We present various algebraic and geometric applications of a spectral sequence approach to non-integrability.

0 - Introduction.

The main constructions in the theory of differential equations and differential geometry are intimately connected and are both based on an analysis of singularities. We mean symbols, characteristics and transport operators in the theory of differential equations and their "counterparts" in differential geometry-metrics, geodesic curves, connections.

The approach to the singularities of the solutions of differential equation (proposed in [L1]) unifies the point of view of all this concepts, expressing them in the language of spectral sequences. A filtration providing a given singularity type plays the main role here. We call this filtration "holonomic" (or "integrable") because the corresponding singularities arise on submanifolds (see [L1]).

In this paper we transfer the ideas of [L1] to a "non-holonomic" (non-integrable) case. This generalization is motivated on the one hand by an attempt to describe a suitable super-analogue of classical mechanics (and geometry) as the propagation of non-holonomic singularities and on the other hand - clarify the algebraic nature of non-integrability.

The paper is organized as follows. In section 1 we give the definition and basic properties of the non-holonomic filtration for an arbitrary differential graded algebra, our main example being, of course, the De Rham algebra. We formulate the theorem on the structure of the initial terms of the spectral sequence associated with a non-holonomic filtration. Section 2 describes a non-holonomic filtration which arose in a Lie algebra - context. It is worth noting that the non-holonomic filtration may be used to define a generalization of the classical Serre-Hochschild spectral sequence, when $\mathfrak{H}$ is only a subspace in a Lie algebra $\mathcal{G}$. We indicate a connection with the Konstant-Sternberg super-Poisson structure and reformulate in the language of spectral sequences the definitions of the classical objects related to Drinfeld's (quasi) Hopf algebras [D] given by Y. Kosmann-Schwarzbach [5K-S].

In Section 3 we describe a situation with an almost complex structure. Besides a theorem on the structure of the spectral sequence we obtain a full description existing only as a folk theorem up to now of the derivations of the exterior algebra of the exterior algebra of an almost complex manifold.

The most interesting geometric example - a contact manifold with the natural distribution is discussed in Sect. 4. We calculate a spectral sequence, based on the non-holonomic filtration and show that the terms E_1 and E_2 relate to the Hodge-Lepage complexes of effective forms.

We are grateful to prof. Y. Kosmann-Schwarzbach for helpful discussions and for pointing out some incorrections, Vladimir Rubtsov would like to thank IHES for hospitality while the last part of this work was done.

1. Non-holonomic filtration : algebraic description.

1.0. All algebras under consideration are over a field k of characteristic 0.

If M is a (bi)-graded module over a (bi)-graded algebra $A = \bigoplus_{p \in \mathbb{Z}} A_p$ then any k-linear map $\overline{\partial} : M \to M$ satisfying the generalized Leibnitz rule

$$\overline{\partial}(a_p m) = a_p \overline{\partial} m + (-1)^{\varphi(r)\varphi(p)} \partial a_p \cdot m, a_p \in A, m \in M,$$

is called a *derivation or order r or M over the derivation* $\partial : A \to A$.

Here $\varphi : \mathbb{Z} \to \mathbb{Z}_2$ (in the bi-graded case $\varphi : \mathbb{Z} \oplus \mathbb{Z} \to \mathbb{Z}_2$) is a "parity" homomorphism.

1.1. Let $A = \bigoplus_{p\geq 0, p\in \mathbb{Z}}$ be a differential graded algebra where $d : A_p \to A_{p+1}, p \geq 0$ is a differential of degree one, $\varphi : \mathbb{N} \to \mathbb{Z}_2$ is a "parity" : for each $a \in A$ if $a \in A_p$ the parity of a is $\varphi(p) = \widetilde{p} = \begin{cases} 0 \\ 1 \end{cases}$

We will assume that A is a "super-commutative algebra, i.e. $a_p a_\ell = (-1)^{\widetilde{p}\widetilde{\ell}} a_\ell a_p, a_\ell \in A_\ell, a_p \in A_p$.

Let $J = \bigoplus_p J_p = J_0 = 0$ be a graded differential ideal, i.e. $dJ \subset J, J$ generated by the space J_1 of the first order elements.

1.2. Consider a decreasing filtration (F) in the algebra $A = \bigoplus_{p\geq 0} A_p$ defined by the ideal J :

$$F_{p,q} = J_1^{-q} \otimes A_{p+2q}, p \geq 0, \quad q < 0.$$

We will refer to (F) as the *non-holonomic filtration.* This fiftration is compatible with the differential d in A, i.e. $dF_{p,q} \subset F_{p,q+1}$.

The filtration (F) is a particular case of a more general filtration which is defined in the case where the ideal J is generated by elements of order greater than one : (r for instance)

$$F_{p,q} = \bigoplus_{i_1<\cdots<i_r} J_1^{i_1} \otimes \cdots J_r^{i_r} \otimes A_{p+q-i},$$

where $i_1 + \cdots + i_r \geq p, i = i_1 + 2i_2 + \cdots + ri_r$.

Then $F_{p,q} \subset A_{p+q}, F_{0,q} = A_q$ and $F_{p+1,q-1} \subset F_{p,q}$. Moreover, $d(F_{p,q}) \subset F_{p,q+1}$. Hence $F_{p,q}$ is a filtration in the algebra A compatible with d.

Consider the spectral sequence of the forth quadrant $\{E_r^{p,q}, d_r^{p,q}\}$ base on the non-holonomic filtration and convergent to the cohomology of the differential graded algebra A. The terms of $\{E_r^{p,q}, d_r^{p,q}\}$ are situated in the interior of the parallelogramm $0 \leq p + 2q \leq n; p \geq 0, q \leq 0$. The graded algebra structure of A induces a natural bi-graded multiplicative structure on the spectral sequence :

$$E_r^{p,q} \otimes E_r^{p',q'} \to E_r^{p+p',q+q'}, r \geq 0.$$

The E_0-terms $E_0^{p,q} = F_{p,q}/F_{p+1,q-1}$ are A_0-modules and the differentials d_0 are A_0-homomorphisms. We will call the operators $d_0^{p,q}$ the *Nijenhuis operators of the ideal J*.

The E_1-terms have an A_0-module structure too, but the differential $d_1^{p,q}$ are first order differential operators, because

$$d_1^{p,q}(f\omega) = f d_1^{p,q}(\omega) + \Pi(df) \otimes \omega,$$

where $f \in A_0, \omega \in E_1^{p,q}$ and $\Pi : A_1 \to E_1^{1,0}$ is a natural projection.

1.3. Assume that the algebra A_0 is a smooth commutative algebra over the field k of characteristic 0. Let $A_i = \Lambda^i(A_0), i \geq 0$ and let $\{\Lambda^\bullet(A_0), d\}$ be the algebraic de Rham complex of A_0 [KLV]. We will also assume that the graded ideal J is projective and generated by the module $J_1 \subset \Lambda^1(A_0)$.

In this case we are able to describe the structure of the first terms of the sepctral sequence $E_r^{p,q}$, based on the non-holonomic filtration.

Denote by D_J the annihilator of J_1 in the A_0-module $D(A_0)$ of derivations of A_0. Then $D_J^* = \Lambda^1(A_0)/J_1$ and we obtain the following short exact sequences of A_0-modules :

$$0 \to J_1 \to \Lambda^1(A_0) \to D_J^* \to 0,$$

$$0 \to D_J \to D(A_0) \to (J_1)^* \to 0.$$

Therefore

$$E_0^{p,q} = \Lambda^{-q}(J_1) \otimes \Lambda^{p+2q}(D_J^*), p \geq 0, q \leq 0,$$

and the operators

$$d_0^{p,q} : \Lambda^{-q}(J_1) \otimes \Lambda^{p+2q}(D_J^*) \to \Lambda^{-q-1}(J_1) \otimes \Lambda^{p+2q+2}(D_J^*)$$

are of bi-degree $(-1,2)$.

The homomorphism $d_0 := d_0^{2,-1} : J_1 \to \Lambda^2(D_J^*)$ will be called the *Nijenhuis tensor of the ideal* J_1. This homomorphism defines a two-form $d_0 \in \Lambda^2(D_J^*) \otimes J_1^*$ on D_J with values in the module J_1^*. By definition the ideal J is *"integrable"* if its Nijenhuis tensor $d_0 \equiv 0$.

Proposition. For $\omega_j \in J_1, \Theta_k \in D_J^*, 1 \leq j \leq -q, 1 \leq k \leq p+2q, q \leq 0$

$$d_0(\omega_1 \wedge \cdots \wedge \omega_{-q} \otimes \Theta_1 \wedge \cdots \wedge \Theta_{p+2q})$$

$$= \sum_{i \geq 0} (-1)^i \omega_1 \wedge \cdots \wedge \widehat{\omega}_i \wedge \cdots \wedge \omega_{-q} \otimes (d_0 \omega_i) \wedge \Theta_1 \wedge \cdots \wedge \Theta_{p+2q}$$

The proof is evident.

1.4. Let us consider the non-holonomic filtration in a de Rham algebra from a dual point of view. We keep all assumptions and notations of subsect 1.3. Fix a splitting $D(A_0) = D_J \oplus \widetilde{D}_J$ which implies the splitting $\Lambda^1(A_0) = J_1 \oplus \widetilde{J}_1 \left(\widetilde{D}_J^* = \Lambda^1(A_0)/\widetilde{J}_1\right)$. It means the choice of a direct summand $\widetilde{D}_J$ as a complement to the annihilator. Then the splitting of the exterior powers

$$\Lambda^k(A_0) = \bigoplus_{p+q=k} \left(\Lambda^p(D_j^*) \otimes \Lambda^q\left(\widetilde{D}_j^*\right)\right), \qquad k \geq 1$$

induces a bi-grading

$$\Lambda^{p,q}(A_0) = \Lambda^p(D_j^*) \otimes \Lambda^q\left(\widetilde{D}_j^*\right),$$

$$\Lambda^*(A_0) = \bigoplus_{k \geq 0} \bigoplus_{p+q=k} \Lambda^{p,q}(A_0)$$

A differential d is represented in this situation as a sum of four components $d = \partial + \overline{\partial} + \partial' + \partial''$ of bi-degrees $(1,0), (0,1), (2,-1), (-1,2)$ respectively.

Proposition 1

1° ∂' and ∂'' are A_0-homomorphisms ;

2° $\partial^2 + \overline{\partial}\partial' + \partial'\overline{\partial} = \overline{\partial}^2 + \partial\partial'' + \partial''\partial = 0$;

3° $\partial'^2 = \partial''^2 = \partial\partial' + \partial'\partial = \overline{\partial}\partial'' + \partial''\overline{\partial} = 0$;

4° $\partial'\partial'' + \partial''\partial' + \partial\overline{\partial} + \overline{\partial}\partial = 0$.

Each of the operators in this decomposition defines a derivation of the corresponding bi-degree in the bi-graded algebra $\Lambda^*(A_0)$.

The spectral sequence constructed on the non-holonomic filtration (F) coincides in this situation with a spectral sequence based on the following filtration (F') of the algebra $\Lambda^*(A_0)$.

$$F'_p\left(\Lambda^{p+q}(A_0)\right) = \bigoplus_{\substack{r+s=p+q \\ r\geq -q}} \Lambda^{r,s}(A_0) = \bigoplus_{\substack{r+s=p+q \\ r\geq -q}} \Lambda^r\left(D_j^*\right) \otimes \Lambda^s\left(\widetilde{D}_j^*\right), p \geq 0, q \leq 0.$$

The first terms in this spectral sequence are described in the

Theorem 1. The spectral sequence $\{E_r^{p,q}, dr\}$ constructed with the filtration (F) or (F') has the following initial terms.

1° The term $E_0^{p,q}$ is isomorphic (as an A_0-module) with the module $\Lambda^{-q,p+2q}(A_0) = \Lambda^{-q}\left(D_j^*\right) \otimes \Lambda^{p+2q}\left(\widetilde{D}_j^*\right)$;

2° The term $E_1^{p,q} = H_{\partial''}^{-q,p+2q}(A_0)$, where $H_{\partial''}^{r,s}(A_0)$ is the cohomology of the complex

$$\cdots \to \Lambda^{r,s}(A_0) \xrightarrow{\partial''} \Lambda^{r-1,s+2}(A_0) \xrightarrow{\partial''} \Lambda^{r-2,s+4}(A_0) \to \cdots \tag{1}$$

3° The term $E_2^{p,q} = H_{\overline{\partial}}^{-q,p+2q}(A_0)$, where $H_{\overline{\partial}}^{r,s}(A_0)$ is the cohomology of the complex

$$\cdots \to H_{\partial''}^{s+2}(A_0) \xrightarrow{\overline{\partial}} H_{\partial''}^{r,s+1}(A_0) \xrightarrow{\overline{\partial}} H_{\partial''}^{r,s+2}(A_0) \to \cdots \tag{2}$$

and we denote by $\overline{\partial}$ which by Proposition 1, is induced by $\overline{\partial}$ on the cohomology of the complex (1).

4° The term $E_3^{p,q} = H_{\partial''}^{r,s+1}(A_0)$, where $H_{\partial}^{r,s}(A_0)$ is the cohomologie of the complex

$$\cdots \to H_{\overline{\partial}}^{r,s}(A_0) \xrightarrow{\partial} H_{\overline{\partial}}^{r+1,s}(A_0) \xrightarrow{\partial} H_{\overline{\partial}}^{r+2,s}(A_0) \to \cdots \tag{3}$$

and the operator ∂ is induced by ∂ on the cohomology of the complex (2) by Proposition 1.

5° The term $E_4^{p,q} = H_{\partial'}^{r,s+1}(A_0)$ is the cohomology of the complex

$$\cdots \to H_{\partial}^{r,s}(A_0) \xrightarrow{\partial'} H_{\overline{\partial}}^{r+2,s-1}(A_0) \xrightarrow{\partial} H_{\overline{\partial}}^{r+4,s-2}(A_0) \to \cdots \tag{4}$$

and the operator ∂' is induced by ∂' on the cohomology of the complex (3) by the Proposition 1.

Proof. Straightforward computations based on the definitions of the terms E_1, E_2, E_3, E_4 and on Proposition.

Now we compare the results with the "integrable" or "'holonomic" case. This is the case when the Nijenhuis operator d_0 is identically 0. The submodule $D_J \subset D(A_0)$ is closed under the Lie bracket of derivations of the algebra A_0 (by the Frobenius theorem). Hence the following modification of Proposition 1 holds :

Proposition 1'. In the "integrable" case, the differentail d is decomposed into a sum of three summands :

$$d = \partial + \overline{\partial} + \partial'$$

of bi-degres $(1,0), (0,1)$ and $(2,-1)$ respectively, corresponding to a bi-grading in $\Lambda^*(A_0)$, generated by the splitting of $\Lambda^1(A_0) : \Lambda^1(A_0) = J_1 \oplus \widetilde{J}_1$ and

$$\overline{\partial}^2 = \partial'^2 = 0, \partial^2 + \overline{\partial}\partial' + \partial'\overline{\partial} = \partial\partial + \partial\overline{\partial} = \partial\partial' + \partial'\partial = 0.$$

Theorem 1'. In the "integrabl" case the corresponding spectral sequence has the following initial terms :

1° $E_0^{p,q} = E_1^{p,q} = \Lambda^{-q,p+2q}(A_0)$;

2° The term $E_2^{p,q} = \mathcal{H}_{\overline{\partial}}^{-q,p+2q}(A_0)$ where $\mathcal{H}_{\overline{\partial}}^{-r,s}$ is the cohomology of the complex :

$$\cdots \to \Lambda_{\partial}^{r,s}(A_0) \xrightarrow{\overline{\partial}} \Lambda^{r,s+1}(A_0) \xrightarrow{\overline{\partial}} \Lambda^{r,s+2}(A_0) \to \cdots \tag{5}$$

3° The term $E_3^{p,q} = \mathcal{H}_{\partial}^{-q,p+2q}(A_0)$ where $\mathcal{H}_{\partial}^{q,s}$ is the cohomology of the complex :

$$\cdots \to H_{\overline{\partial}}^{r,s}(A_0) \xrightarrow{\partial} H_{\overline{\partial}}^{r+1,s}(A_0) \xrightarrow{\partial} H_{\overline{\partial}}^{r+2,s}(A_0) \to \cdots \tag{6}$$

and the operator ∂ is induced on the cohomology of the complex (5) due to Proposition 1'.

4° The term $E_4^{p,q} = \mathcal{H}_{\partial'}^{-q,p+2q}(A_0)$, where $\mathcal{H}_{\partial'}^{r,s}$ - is the cohomology of the complex :

$$\cdots \to H_{\partial}^{r,s}(A_0) \xrightarrow{\partial'} H_{\partial}^{r+2,s-1}(A_0) \xrightarrow{\partial''} H_{\partial}^{r+4,s-2}(A_0) \to \cdots \qquad (7)$$

and the operator ∂' is induced by ∂' on the cohomology of the complex (6) due to Proposition 1'.

But the situation becomes much more simple if we assume the "integrability" of the submodule $\widetilde{J}_1(not J_1)$. In this case $d = \partial + \overline{\partial} + \partial''$ and we obtain :

Theorem 1". If $\widetilde{J}_1$ is "integrable", then the initial theorem of the spectral sequence based on the non-holonomic filtration have the following description :

1° The terms $E_0^{p,q}$ is described by 1° in Theorem 1.

2° The terms $E_1^{p,q}$ and $E_2^{p,q}$ are described by 2° and 3° of Theorem 1.

3° $E_3^{p,q} = E_4^{p,q} = H_{\partial}^{-q,p+2}(A_0)$, where $H_{\partial}^{r,s}(A_0)$ is the cohomology of the complex (3).

The cases we considered in theorems 1' and 1" should be called actually "semi-integrable" in distinction with the "'completely integrable" case, when both D_J and $\widetilde{D}_J$ are Lie subalgebras in $D(A_0)$. In this case $d = \partial + \overline{\partial}$ and holds.

Theorem 1"'.

1° The terms $E_0^{p,q} = E_1^{p,q} = \Lambda^{-q,p+2q}(A_0)$;

2° The terms $E_2^{p,q} = \mathcal{H}_{\overline{\partial}}^{-q,p+2q}(A_0)$, where $\mathcal{H}_{\overline{\partial}}^{r,s}(A_0)$ is the cohomology of the complex (5).

3° The terms $E_3^{p,q} = E_4^{p,q} = \mathcal{H}_{\partial}^{-q,p+2q}(A_0)$, where $\mathcal{H}_{\partial}^{-q,p+2q}(A_0)$ is the cohomology of the complex (6).

2. Non-holonomic filtration and Lie algebras.

In this section we will transfer the ideas of the previous section to the Lie algebra case. The language of the spectras sequences gives an alternative description of the connections between Lie algebra cohomology and the classical objects corresponding to Drinfeld's theory of (quasi) Hopf algebras [D] - the so-called "proto-Lie bialbebras" * Let

* In the short note [GLR] we used another terminology introduced by Y. Kosmann-Schwarzbach [K-S] : quasi-Lie quasi-bialgebras - instead of proto-Lie bialgebras, quasi-Lie bialgebras - instead of Co-Jacobian quasi-bialgebras and Lie quasi-bialgebras - instead of Jacobian quasi-bialgebras. Here we adopt the new terminology which she used in her subsequent publication [K-S1].

$J^r_{\mathfrak{H}} = J^1_{\mathfrak{H}} \wedge \cdots \wedge J^1_{\mathfrak{H}}$ times and define a filtration $F^{p,q}$ in $C^\bullet(\mathcal{G}, M)$ by the formula

$$F^{p,q} = J^{-q}_{\mathcal{G}} \otimes C^{p+2q}(\mathcal{G}, M), \quad p \geq 0, q \leq 0. \tag{F''}$$

The filtration (F") will be called the "*non-holomic filtration* " in the module $C^\bullet(\mathcal{G}, M)$ of cochains. This filtration is compatible with the Chevalley-Eilenberg differential and leads to a spectral sequence of the fourth quadrant $\{{}^M E^{p,q}_r, {}^M d_r\}$ which converges to the Lie algebra cohomology of $\mathcal{G}$ with a coefficients in the $\mathcal{G}$-module M. This spectral sequence lies completely in the parallelogramm $0 \leq p + 2q \leq n, p \geq 0, q \leq 0$, where $n = \dim \mathcal{G}$. Moreover, it is a bi-graded module over the bigraded algebra $\{E^{p,q}_r, dr\}$ corresponding to the trivial $\mathcal{G}$-module k. The differential ${}^M d_r$ corresponding to the trivial $\mathcal{G}$-module k. The differentials ${}^M d_r$ are derivations of modules over the derivations d_r of the algebra E_r (cf 0.1).

2.2. An easy calculation shows that the following proposition holds :

Proposition. The term ${}^M E^{p,q}_0 = \Lambda^{-q}\left(J^1_{\mathfrak{H}}\right) \otimes \mathrm{Hom}\left(\Lambda^{p+2q}(\mathfrak{H}), M\right)$. The differential $d^{2,-1}_0 : \Lambda^1\left(J^1_{\mathfrak{H}}\right) \to \mathrm{Hom}\left(\Lambda^2(\mathfrak{H}), k\right)$ will be called the *Nijenhuis operator of the subspace* $\mathfrak{H}$ and the subspace $\mathfrak{H}$ will be referred to "integrable" if its Nijenhuis operator $d^{2,-1}_0 \equiv 0$ or equivalently, if $\mathfrak{H}$ is a Lie subalgebra in $\mathcal{G}$.

2.3 The "integrable case". If $\mathfrak{H} \subset \mathcal{G}$ is a Lie subalgebra, the operator ${}^M d^{2,-1}_0$: $\Lambda^1\left(\mathcal{J}^1_{\mathfrak{H}}\right) \otimes M \to \mathrm{Hom}\left(\Lambda^2(\mathfrak{H}), M\right)$ acting as

$${}^M d^{2,-1}_0(\xi \otimes m) = \left(d^{2,-1}_0 \xi\right) \otimes m; \left(d^{2,-1}_0 \xi\right) \otimes m(X \wedge Y) = d^{2,-1}_0 \xi(X \wedge Y) \cdot m$$

for every $\xi \in \Lambda^1\left(J^1_{\mathfrak{H}}\right), X, Y \in \mathfrak{H}, m \in M$, is equal zero identically. Therefore

$${}^M E^{p,q}_0 = {}^M E^{p,q}_1 = \Lambda^1\left(J^\ell_{\mathfrak{H}}\right) \otimes C^{p+2q}(\mathfrak{H}, M)$$

and the differential ${}^M d^{p,q}_1 = 1 \otimes \delta$, where δ is the Chevalley-Eilenberg differential. Hence, the term ${}^M E^{p,q}_2 = \Lambda^{-q}\left(J^\ell_{\mathfrak{H}}\right) \otimes H^{p+2q}(\mathfrak{H}, M)$, where $H^{p+2q}(\mathfrak{H}, M)$ is the cohomology of the Lie algebra $\mathfrak{H}$ with the coefficients in M, $p \geq 0, q \leq 0$.

If the subalgebra $\mathfrak{H} \subset \mathcal{G}$ is an ideal, $\mathcal{G}/\mathfrak{H}$ is a Lie algebra and ${}^M E^{p,q}_3 = H^{-q}$ $(\mathcal{G}/\mathfrak{H}, H^{p+2q}(\mathfrak{H}, M))$ because $H^{p+2q}(\mathfrak{H}, M)$ is a trivial $\mathfrak{H}$-module and hence a $\mathcal{G}/\mathfrak{H}$-module.

2.4. Super Poisson structures. Let $\mathcal{G} = \mathfrak{H} \oplus \mathfrak{H}^*$ be a direct sum of vector spaces.

The pairing $\mathfrak{H}^* \times \mathfrak{H} \to k$ defines on $\mathfrak{H}$ a scalar product denoted by $\langle\ ,\ \rangle$.

The exterior algebra $\Lambda^\bullet(\mathcal{G})$ in this case is a super Poisson algebra. This fact is widely known (cf. [KS], [LR]) and the structure is given by so-called "grand crochet" ([K-S])

$$[\ ,\ ] : \Lambda^\bullet(\mathfrak{H}^* \oplus \mathfrak{H}) \times \Lambda^\bullet(\mathfrak{H}^* \oplus \mathfrak{H}) \to \Lambda^\bullet(\mathfrak{H}^* \oplus \mathfrak{H}),$$

$$\Lambda^\bullet(\mathfrak{H}^* \oplus \mathfrak{H}) = \bigoplus_{\substack{p+q=r \\ p\geq -1, q\geq -1}} \left(\Lambda^{q+1}(\mathfrak{H}^*) \otimes \Lambda^{p+1}(\mathfrak{H})\right),$$

and for a pair of decomposible elements $\xi \otimes X$ and $\eta \otimes Y$,

$$\xi \otimes X \in \Lambda^{q+1}(\mathfrak{H}^*) \otimes \Lambda^{p+1}(\mathfrak{H}), \eta \otimes Y \in \Lambda^{s+1}(\mathfrak{H}^*) \otimes \Lambda^{r+1}(\mathfrak{H})$$

$$[\xi \otimes X, \eta \otimes Y] = (-1)^{p+1} \xi \otimes K(X,\eta)Y + (-1)^{(p+q)(r+s)+1}(-1)^{s+1} \eta \otimes K(Y,\xi)X$$

where K is given by

$$K(X,\xi) = \sum_k \left(i_{g_k^*} X\right) \otimes (i_{g_k} \xi)$$

in any basis $\{g_k\}$ of $\mathfrak{H}$ and dual basis $\{g_k^*\}$ of $\mathfrak{H}^*$ ([LR]).

Consider an element M of the degree 1 in $\Lambda^\bullet(\mathfrak{H}^* \oplus \mathfrak{H})$, which satisfies the "Master equation" $[M,M] = 0$. If the scalar product $\langle\ ,\ \rangle$ is invariant, such element M can be defined and conversely, each solution of "Master equation" defines a metrizable Lie algebra structure on $\mathcal{G} = \mathfrak{H}^* \oplus \mathfrak{H}$.

Connect with the element M above a cochain comples

$$\cdots \to \Lambda^\bullet(\mathcal{G}) \xrightarrow{d_\mu} \Lambda^{\bullet+1}(\mathcal{G}) \xrightarrow{d_\mu} \Lambda^{\bullet+2}(\mathcal{G}) \to \cdots,$$

$d_\mu = [M,\ ]$ and $d_M^2 = 0$ due to the "Master equation". Equip $\Lambda^\bullet(\mathcal{G})$ with "the non-holonomic" filtration :

$$F^{p,q} = J_{\mathfrak{H}}^{-q} \otimes \Lambda^{p+2q}(\mathfrak{H}^* \oplus \mathfrak{H}), q \leq 0, p \geq 0.$$

The differential d_M is compatible with this filtration. Having in mind a description of the corresponding spectral sequence, consider a decomposition of the differential $d_\mu = \partial_\mu + \overline{\partial}_\mu + \partial'_\mu + \partial''_\mu$, where the operators $\partial_\mu, \overline{\partial}_\mu, \partial'_\mu, \partial''_\mu$ correspond to the decomposition of M into the element of the bidegres $(1,0),(0,1),(2,-1),(1,2)$:

$$M = m + \overline{m} + m' + m''$$

$$\partial_\mu = [m,\ ], \overline{\partial}_\mu = [\overline{m},\ ], \partial'_\mu = [m',\ ], \partial''_\mu = [m'',\ ]$$

The following proposition (analogue to the proposition1 of 1.4) is true : (see also [K-S])

Proposition. Operator $\partial_\mu, \overline{\partial}_\mu, \partial'_\mu, \partial''_\mu$ satisfy the following relations :

1° $\partial_\mu^2 + \overline{\partial}_\mu \partial'_\mu + \partial'_\mu \overline{\partial}_\mu = \overline{\partial}_\mu^2 + \partial_\mu \partial''_\mu + \partial''_\mu \partial_\mu = 0;$

2° $\partial'^2_\mu = \partial''^2_\mu = \partial_\mu \partial'_\mu + \partial'_\mu = \partial_\mu \partial''_\mu + \partial''_\mu \overline{\partial}_\mu = 0;$

3° $\partial'_\mu \partial''_\mu + \partial''_\mu \partial'_\mu + \partial_\mu \overline{\partial}_\mu + \overline{\partial}_\mu \partial_\mu = 0.$

Remark. The spectral sequence based on the "non-holonomic" filtration coincides with the spectral sequence constructed from the filtration.

$$F'_p\left(\Lambda^{p+q}(\mathfrak{H}^* \oplus \mathfrak{H})\right) = \sum_{\substack{r+s=p+q \\ r\geq -q}} \Lambda^{r,s}(\mathfrak{H}^* \oplus \mathfrak{H}) = \sum_{\substack{r+s=p+q \\ r\geq -q}} \Lambda^r(\mathfrak{H}^*) \otimes \Lambda^s(\mathfrak{H}) \quad q \leq 0.$$

and analogue of theorem 1 is still valid.

Theorem. The spectral sequence based on the above filtration converges to the cohomology of the complex $\{\Lambda^\bullet(\mathcal{G}), d_\mu\}$ and has the following initial terms :

1° $E_0^{p,q} \simeq \Lambda^{-q}(\mathfrak{H}) \otimes \mathrm{Hom}\left(\Lambda^{p+2q}(\mathfrak{H}), k\right)$

2° $E_1^{p,q} \simeq H^{-q,p+2q}_{\partial''_\mu}(\mathfrak{H} \oplus \mathfrak{H}^*)$ where $H^{r,s}_{\partial''}(\mathfrak{H} \oplus \mathfrak{H}^*)$ is the cohomology of the complex

$$\cdots \to \Lambda^{*r}(\mathfrak{H}) \otimes \Lambda^s(\mathfrak{H}^*) \xrightarrow{\partial''_\mu} \Lambda^{*r-1}(\mathfrak{H}) \otimes \Lambda^{s+2}(\mathfrak{H}^*) \xrightarrow{\partial''_\mu} \Lambda^{*r-2}(\mathcal{G}) \otimes \Lambda^{s+4}(\mathfrak{H}^*) \to \cdots$$

3° $E_2^{p,q} \simeq H_{\overline{\partial}_\mu}^{-q,p+2q}(\mathfrak{H} \oplus \mathfrak{H}^*)$ where $H_{\overline{\partial}}^{r,s}(\mathfrak{H} \oplus \mathfrak{H}^*)$ is the cohomology of the complex

$$\cdots \to H_{\partial''_\mu}^{r,s}(\mathfrak{H} \oplus \mathfrak{H}^*) \xrightarrow{\overline{\partial}_\mu} H_{\partial''_\mu}^{r,s+1}(\mathfrak{H} \oplus \mathfrak{H}^*) \xrightarrow{\overline{\partial}_\mu} H_{\partial''_\mu}^{r,s+2}(\mathfrak{H} \oplus \mathfrak{H}^*) \to \cdots$$

4° $E_3^{p,q} \simeq H_{\partial_\mu}^{-q,p+2q}(\mathfrak{H} \oplus \mathfrak{H}^*)$ where $H_{\partial_\mu}^{r,s}(\mathfrak{H} \oplus \mathfrak{H}^*)$ is the cohomology of the complex

$$\cdots \to H_{\overline{\partial}_\mu}^{r,s}(\mathfrak{H} \oplus \mathfrak{H}^*) \xrightarrow{\partial_\mu} H_{\overline{\partial}_\mu}^{r+1,s}(\mathfrak{H} \oplus \mathfrak{H}^*) \xrightarrow{\partial_\mu} H_{\overline{\partial}_\mu}^{r+2,s}(\mathfrak{H} \oplus \mathfrak{H}^*) \to \cdots$$

5° $E_4^{p,q} \simeq H_{\partial'_\mu}^{-q,p+2q}(\mathfrak{H} \oplus \mathfrak{H}^*)$ where $H_{\partial'_\mu}^{r,s}(\mathfrak{H} \oplus \mathfrak{H}^*)$ is the cohomology of the complex

$$\cdots \to H_{\partial'_\mu}^{r,s}(\mathfrak{H} \oplus \mathfrak{H}^*) \xrightarrow{\partial'_\mu} H_{\partial'_\mu}^{r+2,s-2}(\mathfrak{H} \oplus \mathfrak{H}^*) \xrightarrow{\partial'_\mu} H_{\partial'_\mu}^{r+4,s-2}(\mathfrak{H} \oplus \mathfrak{H}^*) \to \cdots$$

6° The "grand crochet" induces on each term E_r a super-Poisson algebra structure.

Corollary

a) If $\mathfrak{H}$ be a Lie subalgebra of $\mathcal{G}$, then $d_0 = d_3 \equiv 0$ and as in 2.3,

$$E_0^{p,q} = E_1^{p,q} = \Lambda^{-q}(\mathfrak{H}) \otimes \mathbb{C}^{p+2q}(\mathfrak{H}),$$

where $\mathbb{C}^{p+2q}(\mathfrak{H})$ is the space of Lie algebra cochains with trivial coefficients.

$$E_2^{p,q} = \Lambda^{-q}(\mathfrak{H}) \otimes H^{p+2q}(\mathfrak{H}),$$

where $H^{p+2q}(\mathcal{G})$ is the Lie-algebra cohomology ;

b) If $\mathfrak{H}$ is an ideal in $\mathcal{G}$, then $\mathfrak{H}^*$ is a Lie algebra of $\mathcal{G}$ and we have (due to the triviality of $\mathfrak{H}$-module $H^{p+q}(\mathfrak{H})$

$$E_3^{p,q} = H^{-q}\left(\mathfrak{H}^*, H^{p+2q}(\mathfrak{H})\right),$$

is cohomology (of the Lie algebra $\mathfrak{H}^*$) with coefficients in the $\mathfrak{H}^*$-module $H^{p+2q}(\mathfrak{H})$.

$E_4^{p,q} \simeq E_3^{p,q}$, because $d_3 = \partial'_\mu$ is trivial.

2.5. Non-holomic filtration and proto-Lie bialgebras

Recall the definition of proto-Lie bialgebra $\mathcal{G}$ ([KS]). This a vector space equipped with a bracket $\overline{m} \in \Lambda^2(\mathcal{G}^*) \otimes \mathcal{G}$, co-bracket $m \in \mathcal{G}^* \otimes \Lambda^2(\mathcal{G})$ and two elements m' and m'', such that $m' \in \Lambda^3(\mathcal{G}^*)$, $m'' \in \Lambda^3(\mathcal{G})$ and the relations of proposition 2.4 are valid.

Consider the spectral sequence converging to the cohomology of the complex $\{\Lambda^\bullet(\mathcal{G} \oplus \mathcal{G}^*), d_M\}$, where $M = m + \overline{m} + m' + m''$ is a solution of the "Master equaiton". Then the previous subsection theorem and corollaries imply :

a) Let $\mathcal{G}$ be a Co-Jacobian quasi-bialgebra, then by definition $m'' = 0$, therefore $d_0 \equiv 0$ and hence

$$E_0^{p,q} \simeq E_1^{p,q} = \Lambda^{-q}(\mathcal{G}) \otimes \Lambda^{p+2q}(\mathcal{G}^*) = \Lambda^{-q}(\mathcal{G}) \otimes C^{p+2q}(\mathcal{G}),$$

$$E_2^{p,q} = \Lambda^{-q}(\mathcal{G}) \otimes \Lambda^{p+2q}(\mathcal{G}).$$

$\mathcal{G}^*$ is a Jacobian quasi-bialgebra. Conversely, if $\mathcal{G}$ is a Jacobian quasi-bialgebra then $m' = 0$ and $d_3 \equiv 0$, i.e. $E_3^{p,q} = E_4^{p,q}$ and $\mathcal{G}^*$ is a Co-Jacobian quasi-bialgebra.

6° Let $\mathcal{G}$ be a Lie bialgebra. Then $m' = m'' = 0, d_0 = d_3 = 0$. $\mathcal{G}^*$ is an subalgebra Lie, hence $E_0^{p,q} = E_1^{p,q} = \Lambda^{-q}(\mathcal{G}) \otimes C^{p+2q}(\mathcal{G}), E_0^{p,q} = \Lambda^{-q}(\mathcal{G}) \otimes H^{p+2q}(\mathcal{G}), E_3^{p,q} = E_4^{p,q} = H^{-q}(\mathcal{G}^*), H^{p+q}(\mathcal{G})$.

Remark. It is possible to use the standard filtration Serre-Hohchild in the case of Jacobian quasi bialgebra, but in the general case of a proto-Lie bialgebras the standard filtration is not compatible with the differential and the non-holonomic filtration is the only possible one.

3. Almost-complex structures.

3.1. We briefly recall some generalities about almost-complex structures.

An almost-complex structure J on a smooth manifold M is a smooth field of complex structures on the tangent spaces $T_m = T_m M, m \in M$, i.e. a smooth field of automorphisms $J_m : T_m \to T_m$ such that $J_m^2 = -1$ for every $m \in M$. The filed of adjoint automorphism $J_m^* : T_m^* \to T_m^*$ defines the complex structures in the cotangent spaces and $J_m^*(h)(v) = h(J_m(v))$, for $h \in T_m^*, v \in T_m$.

A space T_m (considered as a complex vector space with the structure J_m) is denoted by T_m^J and the space T_m^* with the structure J_m^* is denoted by T_m^{*J} correspondingly.

We will identify a complexification $T_m^* \otimes \mathbb{C}$ of T_m^* with the space of all $\mathbb{R}$-linear $\mathbb{C}$-valued functions on T_m.

This space on an almost-complex manifold M (= a manifold M with an almost-complex structure J) is decomposed into the vector space sum of

$$T_m^* \otimes \mathbb{C} = T_m^{0,1} \oplus T_m^{1,0},$$

where

$$T_m^{1,0} = \left\{ f : T_m \to \mathbb{C} \mid f(Jv) = \sqrt{-1} f(v) \right\},$$

$$T_m^{1,0} = \left\{ f : T_m \to \mathbb{C} \mid f(Jv) = -\sqrt{-1} f(v) \right\}, v \in T_m$$

The conjugation $Q : T_m^* \otimes \mathbb{C} \to T_m^* \otimes \mathbb{C}, Q(f)(v) = \overline{f(v)}$ defines an anti-isomorphism of the complex vector spaces $T_m^{1,0}$ and $T_m^{0,1}$. Note, that the space T_m^{*J} is isomorphic (over $\mathbb{C}$) to the space $T_m^{1,0}$ and is anti-isomorphic to the space $T_m^{0,1}$.

3.2. An exterior power $\Lambda^k (T_m^* \otimes \mathbb{C})$ (over $\mathbb{C}$) is decomposed into the sum of the tensor products of the exterior powers $\Lambda^p \left(T_m^{1,0}\right)$ and $\Lambda^q \left(T_m^{1,0}\right)$,

$$\Lambda^k (T_m^* \otimes \mathbb{C}) = \sum_{p+q=k} \Lambda^{p,q} (T_m^*)$$

where $\Lambda^{p,q} (T_m^*) = \Lambda^p \left(T_m^{1,0}\right) \otimes \Lambda^q \left(T_m^{0,1}\right)$.

The conjugation Q is obvioulsy extended to the conjugation on $\Lambda^k (T_m^* \oplus \mathbb{C})$ and defines an anti-isomorphism $Q : \Lambda^{p,q} (T_m^*) \to \Lambda^{q,p} (T_m^*)$

3.3. Denote by $\Omega^k(M, \mathbb{C})$ the smooth section of the module of the bundle $\Lambda^k(\tau^* \otimes \mathbb{C})$, where $\tau^* : T^*M \to M$ is te cotangent bundle of M. We will denote this module as Ω^k when it dos not lead to confusion. Correspondingly, denote by $\Omega^{p,q}(M)$ (or $\Omega^{p,q}$ the modules of smooth section of the bundles $\Lambda^{p,q}(\tau^*)$ over an almost-complex manifold M.

The modules Ω^k and $\Omega^{p,q}$ are $\mathbb{C}^\infty(M, \mathbb{C})$-modules. Here $\mathbb{C}^\infty(M, \mathbb{C})$ is the algebra of $\mathbb{C}$-valued smooth functions on M.

The decomposition

$$\Omega^k = \sum_{p,q} \Omega^{p,q} \tag{1}$$

holds (due to the decompositon of 3.2).

The exterior differential $d : \Omega^k \to \Omega^{k+1}, 0 \leq k \leq \dim M = 2n$ acts on each homogeneous component $\Omega^{p,q}$ as

$$d : \Omega^{p,q} \to \Omega^{p+2,q-1} \oplus \Omega^{p,q+1} \oplus \Omega^{p+1,q} \oplus \Omega^{p-1,q+2} \tag{2}$$

and hence is decomposed in the sum of operators

$$d = \partial + \overline{\partial} + \partial' + \partial'', \tag{3}$$

where

$$\partial : \Omega^{p,q} \to \Omega^{p+1,q}, \overline{\partial} : \Omega^{p,q} \to \Omega^{p,q+1}$$

$$\partial' : \Omega^{p,q} \to \Omega^{p+2,q-1}, \partial'' : \Omega^{p,q} \to \Omega^{p-1,q+2}.$$

The usual properties of d (with the decomposition (3) lead to the following properties of $\partial, \overline{\partial}, \partial', \partial''$:

1) The operators ∂ and $\overline{\partial}$ are $\mathbb{C}$-linear and the operators ∂' and ∂'' are $\mathbb{C}^\infty(M, \mathbb{C})$-linear;

2) The operators ∂ and $\overline{\partial}$ are first order differential operators :

$$\partial(f\omega) = f\partial\omega + \omega\partial f \wedge \omega,$$

$$\overline{\partial}(f\omega) = f\overline{\partial}\omega + \overline{f} \wedge \omega, f \in \mathbb{C}^\infty(M, \mathbb{C}), \omega \in \Omega^{p,q}$$

3) The operators $\partial, \overline{\partial}, \partial', \partial''$ are exterior algebra derivations :

$$\Delta(\Theta \wedge \omega) = \Delta(\Theta) \wedge \omega + (-1)^{|\Theta|}\Theta \wedge \Delta(\omega),$$

where Δ is any of $\partial, \overline{\partial}, \partial', \partial''$ and $\Theta \in \Omega^{a,b}, \omega \in \Omega^{p,q}, |\Theta| = a + b$;

4) The relation $dQ = Qd$ implies relations

$$\partial' = Q\partial''Q, \overline{\partial} = Q\partial Q, \partial'' = Q\partial' Q, \partial = Q\overline{\partial}Q.$$

for the conjugation Q.

3.4. Let $\Omega = \sum_{p,q} \Omega^{p,q}$ be the bi-graded module of forms on an almost-complex manifold M. Call a $\mathbb{C}$-linear bi-graded map $\mathcal{D} : \Omega \to \Omega$ a *derivation of i-degree* (p, q) if $\mathcal{D} : \Omega^{r,s} \to \Omega^{p+r,q+s}, \forall (r, s) \in \mathbb{Z} \times \mathbb{Z}$ and the Leibnitz rule

$$\mathcal{D}(\Theta \wedge \omega) = \mathcal{D}\Theta \wedge \omega + (-1)^{|\theta||\mathcal{D}|}\theta \wedge \mathcal{D}\omega$$

holds for all $\Theta \in \Omega^{r,s}, \omega \in \Omega$.

The set of all derivations of bi-degree (p, q) is denoted by $\mathrm{Der}_{p,q}\,\Omega$ and $\mathrm{Der}\,\Omega = \sum_{p,q} \mathrm{Der}_{p,q}\,\Omega$.

The commutator

$$[\mathcal{D}_1, \mathcal{D}_2] = \mathcal{D}_1 \circ \mathcal{D}_2 - (-1)^{(p+q)(r+s)}\mathcal{D}_1 \circ \mathcal{D}_2$$

of derivations $\mathcal{D}_1 \in \mathrm{Der}_{p,q}\,\Omega, \mathcal{D}_2 \in \mathrm{Der}_{r,s}\,\Omega$ equips $\mathrm{Der}_{p,q}$ with the structure of a bi-graded Lie algebra. Note, that the components $\partial, \overline{\partial}, \partial', \partial''$ of the operator d define the derivations

$$\partial \in \mathrm{Der}_{1,0}\,\Omega, \overline{\partial} \in \mathrm{Der}_{0,1}\,\Omega, \partial' \in \mathrm{Der}_{2,-1}\,\Omega, \partial'' \in \mathrm{Der}_{1,-2}\,\Omega$$

correspondingly.

Having in mind a description of all derivations of Ω, decompose $T_m^* \otimes \mathbb{C}$ as in 3.2 and note that there exists a decomposition of the complexified tangent space $T_m \otimes \mathbb{C} = T_{m(1,0)} \oplus T_{m(1,0)}$, where $T_{m(1,0)} = \mathrm{Hom}_{\mathbb{C}}\left(T_m^{1,0}, \mathbb{C}\right), T_{m(1,0)} = \mathrm{Hom}_{\mathbb{C}}\left(T_m^{0,1}, \mathbb{C}\right)$ and the corresponding decomposition of the tangent bundle into the Whitney sum $\tau \otimes \mathbb{C} = \tau_{1,0} \oplus \tau_{0,1}$ of the bundles $\tau_{1,0}$ and $\tau_{0,1}$ with fibers $T_{m(1,0)}$ and $T_{m(0,1)}$ in each $m \in M$. The $C^\infty(M, \mathbb{C})$-module of smooth sections of $\tau_{1,0}(\tau_{0,1})$ is denoted by $\mathcal{D}_{1,0}(\mathcal{D}_{0,1})$.

Each element $A \in \Omega^{p,q} \otimes D_{1,0}$ defines an inner derivation $i_A \in \mathrm{Der}_{p-1,q}\,\Omega$: for $A = \omega \otimes X, \omega \in \Omega^{p,q}, X \in D_{1,0}$ and $\Theta \in \Omega^{r,s}$ we have

$$i_A\Theta = \omega \wedge i_X\Theta$$

and we extends this definition by linearity over all tensors A. Each element $B \in \Omega^{p,q} \otimes D_{0,1}$ defines a derivation $i_B \in \mathrm{Der}_{p,q-1}\,\Omega^1$ in the same manner.

Theorem 1. Each derivation $\mathcal{D} \in \mathrm{Der}_{p,q}\,\Omega$ on an almost complex manifold M may be represented as a sum $\mathcal{D} = [i_A, \partial] + [i_B, \overline{\partial}] + i_c + i_{\widetilde{c}}$ in a unique way with $A \in \Omega^{p,q} \otimes \mathcal{D}_{1,0}, B \in \Omega^{p,q} \otimes \mathcal{D}_{0,1}, C \in \Omega^{p+1,q} \otimes \mathcal{D}_{1,0}, \widetilde{C} \in \Omega^{p,q+1} \otimes \mathcal{D}_{0,1}$.

Proof. We begin with the *algebraic* derivations $\mathcal{D}$ (i.e. $\mathcal{D}(f\omega) = f\mathcal{D}\omega$ for all $f \in C^\infty(M, \mathbb{C}), \omega \in \Omega$). Each algebraic derivation is defined completely by homomorphism $F : \Omega^{1,0} \to \Omega^{p+1,q}$ and $G : \Omega^{0,1} \to \Omega^{p,q+1}$. Choose tensors $C \in \Omega^{p+1,q} \otimes \mathcal{D}_{1,0}$ and $\widetilde{C} \in \Omega^{p,q+1} \otimes \mathcal{D}_{1,0}$ such that $F = i_C$ and $G = i_{\widetilde{C}}. Then \mathcal{D} = i_C + i_{\widetilde{C}}$. Now consider an arbitrary derivation $\mathcal{D}$. The symbol $\sigma(\mathcal{D})$ of this derivation is a homomorphism

$$\sigma(\mathcal{D}) : \Omega^1 = \Omega^{1,0} \oplus \Omega^{0,1} \to \Omega^{p,q}.$$

Choose tensors A and B, $A \in \Omega^{p,q} \otimes D_{1,0}, B \in \Omega^{p,q} \otimes D_{0,1}$ such that $\sigma(\mathcal{D}) = i_A + i_B$. Then the symbol of the derivation $\widetilde{\mathcal{D}} = \mathcal{D} - [i_A, \partial] - [i_B, \overline{\partial}]$ equals 0 and hence, $\widetilde{\mathcal{D}}$ is an algebraic derivation. This completes the proof.

For $A \in \Omega^{p,q} \otimes D_{1,0}$ denote by d_A the derivation $[i_A, \partial]$ for $B \in \Omega^{p,q} \otimes D_{0,1}$ denote by $\overline{d}_B$ the derivation $[i_B, \overline{\partial}]$.

The decomposition of the Theorem can be rewritten as

$$\mathcal{D} = d_A + \overline{d}_B + i_{c+\widetilde{c}}$$

The derivations ∂' and ∂'', a propos, can be represented as $\partial' = i_c, \partial'' = i_{\widetilde{c}}$ for $C \in \Omega^{2,0} \otimes D_{0,1}$ and $\widetilde{C} \in \Omega^{0,2} \otimes D_{1,0}$.

The tensor C is nothing else than the Nijenhuis tensor, or torsion of the almost-complex structure.

An almost-complex structure J may be regarded as a tensor $J \in \Omega^{1,0} \otimes D_{1,0}$. The corresponding derivation d_J can be expresed through the operators $\partial, \overline{\partial}, \partial', \partial''$ as

$$d_J = -\sqrt{-1}(\partial - \overline{\partial} + 3\partial' - 3\partial'').$$

3.5. Proposition. The relations

a) $\partial^2 + [\overline{\partial}, \partial'] = \overline{\partial}^2 + [\partial, \partial''] = 0$

b) $[\partial, \partial''] + [\partial, \overline{\partial}] = 0$;

c) $\partial'^2 = \partial''^2 = [\partial, \partial'] = [\overline{\partial}, \partial''] = 0$

are true for any almost-complex manifold.

To prove this gather all terms of equal bi-degree in $d^2 = [d, d] = 0$.

3.6. If M is equipped with a non-integrable almost complex structure, then $\overline{\partial}^2 \neq 0$ and the $\overline{\partial}$-complex doesn't exists. We propose below some natural complexes generalizing the Dolbeault complex in the non-integrable situation.

The operators $\partial' : \Omega^{p,q} \to \Omega^{p+2,q-1}$ define a complex with the cohomology $H'_{p,q}(M)$ in the term $\Omega^{p,q}$

$$H'_{p,q}(M) = \frac{\operatorname{Ker}\left(\partial' : \Omega^{p,q} \to \Omega^{p+2,q-1}\right)}{\operatorname{Im}\left(\partial' : \Omega^{p-2,q+1} \to \Omega^{p,q}\right)}$$

We will suppose that the $C^\infty(M, \mathbb{C})$-homomorphism ∂' has constant rank and, hence, we may consider $H'_{p,q}(M)$ as the module of smooth sections for a corresponding bundles.

A module $H''_{p,q}(M)$ is defined in analogous way as the cohomology module for the differential ∂'' in the term $\Omega^{p,q}$.

Proposition 3.5 implies that the operator $\overline{\partial}$ reduces to an operator acting on the modules $H'_{p,q}(M)$

$$\cdots \to H'_{p,q}(M) \xrightarrow{\overline{\partial}} H'_{p,q+1}(M) \xrightarrow{\overline{\partial}} H'_{p,q+2}(M) \to \cdots \tag{1}$$

(We denote this operator by $\overline{\partial}$ too).

The operator $\overline{\partial}$ in (1) is a first order differential operator and $\left(\partial_{|H'(M)}\right)^2 = 0$. The cohomology of the complex (1) in the term $H'_{p,q}(M)$ will be denoted by $h'_{p,q}(M)$.

The conjugate complex

$$\cdots \to H''_{p,q}(M) \xrightarrow{\overline{\partial}} H''_{p,q+1}(M) \xrightarrow{\overline{\partial}} H''_{p,q+2}(M) \to \cdots \tag{2}$$

is given by the pair of operators ∂'' and ∂ and we let $h''_{p,q}(M)$ be the cohomology of (2) in the term $H''_{p,q}(M)$. The conjugation Q gives a cohomology isomorphism

$$Q : h'_{p,q}(M) \simeq h''_{p,q}(M).$$

Remark. If the almost-complex structure is integrable, the operator $\partial' \equiv 0$ and the complex (1) coincides with the usual Dolbeault complex.

It is worth to noting also that the complexes (1) and (2) are not elliptic.

3.7. The constructions of 3.6 suggest to find a spectral sequence with initial terms $h'_{p,q}(M)$ and $h''_{p,q}(M)$. Such sequence exists. Consider a filtration in the algebra $\Omega = \sum_{p,q} \Omega^{p,q}$:

$$F_p(\Omega^k) = \sum_{\substack{r+s=k \\ r \geq p-k}} \Omega^{r,s}$$

where $\Omega^k = \sum\limits_{a+b=k} \Omega^{a,b}, F_0(\Omega^k) = \Omega^k$.

Proposition. (F) is a decreasing filtration in Ω, compatible with the de Rham differential

$$d\left(F_p\left(\Omega^k\right)\right) \subset F_p\left(\Omega^{k+1}\right)$$

3.8. Let $\{E_r^{p,q}, d_r\}$ be a spectral sequence based on the filtration (1) of 3.7. Then

$$E_0^{p,q} = \frac{F_p\left(\Omega^{p+q}\right)}{F_{p+1}\left(\Omega^{p+q}\right)} = \frac{\Omega^{-q,p+2q} + \cdots, + \Omega^{n,p+q-n}}{\Omega^{-q+1,p+2q-1} + \cdots, + \Omega^{n,p+q-n}} = \Omega^{-q,p+2q}. \tag{1}$$

The differential $d_0^{p,q} : E_0^{p,q} \to E_0^{p,q+1}$ coincides with the operator $\partial'' : \Omega^{-q,p+2q} \to \Omega^{-q-1,p+2q-2}$ due to proposition of 3.5 and the definition of $E_0^{p,q}$. Hence, a term

$$E_1^{p,q} \simeq H''_{-q,p+2q}(M). \tag{2}$$

Furthermore, the differential $d_1^{p,q} : E_1^{p,q} \to E_1^{p+1,q}$ coincides with the differential $\overline{\partial}$ in the complex (1) of 3.6.

Hence, the term

$$E_2^{p,q} \simeq h''_{-q,p+2q}(M) \tag{3}$$

A differential $d_2^{p,q} : E_2^{p,q} \to E_2^{p+2,q-1}$ is induced by the differential ∂ which reduce to $h''(M)$ via proposition of 3.5. Therefore

$$E_3^{p,q} = \frac{\mathrm{Ker}\,(\partial : h''_{-q,p+2q} \to h''_{-q+1,p+2q})}{\mathrm{Im}\,(\partial : h''_{-q-1,p+2} \to h''_{-q+1,p+2q})} \tag{4}$$

We denote this cohomology by $h'_{-q,p+2q}(h''(M))$. The differential $d_3^{p,q} : E_3^{p,q} \to E_3^{p+3,q-2}$ coincides with a reduction of ∂' to this cohomology :

$$\cdots \xrightarrow{\partial'} h'_{-q,p+2q}(h''(M)) \xrightarrow{\partial'} h'_{-q+2,p+2q-1}(h''(M)) \to \cdots \tag{5}$$

and a term

$$E_q^{p,q} = H'_{-q,p+2q}(h'(h''(M))), \tag{6}$$

where $H'_{-q,p+2q}(h'(h''(M)))$ is a cohomology of (5).

3.9. Theorem. The spectral sequence $\{E_r^{p,q}, d_r^{p,q}\}$ of the previous subsection converges to the de Rham cohomologies of M and has initial terms described by

$$E_0^{p,q} \simeq \Omega^{-q,p+2q}(M), E_1^{p,q} = H''_{-q,p+2q}(M),$$

$$E_2^{p,q} \simeq h''_{-q,p+2q}(M), E_s^{p,q} \simeq h'_{-q,p+2q}(M)(h''(M)),$$

$$E_4^{p,q} \simeq H'_{-q,p+2q}(h'(h''(M))).$$

Remarks.

1° The conjugation isomorphism $Q : \Omega^{p,q} \to \Omega^{q,p}$ defines a conjugate filtration

$$\overline{F}_p(\Omega^k) = \sum_{\substack{a+b=k \\ b\geq p-k}} \Omega^{a,b}.$$

Consider a set $V^{p,q} = F_p(\Omega) \cap \overline{F}_q(\Omega), \Omega = \sum_k \Omega^k$. Then $\sum_{p+q=n} V^{p,q} = \Omega$ and our filtration induces a Hodge structure of weight n ([Dl]).

2° Let M be a smooth manifold and $T_{\mathbb{C}}M$ be a complexification of the tangent bundle TM. Suppose that $T_{\mathbb{C}}M = S + T$ splits into the sum of subbundles S and T and that S is a *complex foliation*, i.e. the module of smooth sections S yields a Lie subalgebra in the Lie algebra $\Gamma(\tau_{\mathbb{C}}M)$ of vector fields. We can take $T = S^{\perp}$ relatively to a fixed Hermitean metric on $T_{\mathbb{C}}M$.

In this case there is a straightforward generalization of the results in section 3 to the case of "S-holomorphic" or "S-Dolbeault complex" ([FW]).

3° Another straightforward generalization is obtained if we consider an almost-product structure instead of an almost-complex one. This situation can be realized on a supermanifold M with splitting $TM = (TM)_0 \oplus (TM)_1$ and the corresponding splitting of super-derivation algebra. We think that our cohomologies may be useful for an algebraic description of the Weyl triviality condition and the anomaly problem in supersymmetric gauge theory [BL].

4. Non-holonomic filtration and contact manifolds

4.1. Let M be a contact manifold. It means an existence of a line nondegenerate $\xi \subset T^*(M)$ of the cotangent bundle on M. In other words there exist a submodule C in $\Omega^1(M)$. We will assume that $C = C^\infty(M)\cdot\omega$, where ω is a 1-form such that $\omega\wedge(d\omega)^n \neq 0, n = (\dim M - 1)/2$.

The contact distribution E on M is the sub-bundle of TM whose fibers consist of those tangent vectors annihilated by ω. In other words $E = \operatorname{Ann}\omega$.

4.2. Consider a non-holonomic filtration in this case:

$$F_{p,q} = C^{-q}\wedge\Omega^{p+2q}(M), q \leq 0, p \geq 0.$$

Since the submodule C is one-dimensional, thus there exist only two lines in $E_0^{p,q}$: $q = 0$ and $q = -1$.

$$E_0^{p,q} = \frac{F_{p,q}}{F_{p+1,q-1}} = \frac{C^{-q}\wedge\Omega^{p+2q}(M)}{C^{-q+1}\wedge\Omega^{p+2q-1}(M)} \simeq \Lambda^{p+2q}(E^*)\otimes C^{-q},$$

where $E^* = \Omega^1(M)/C$

q \ p	p = 0	p = 1	p = 2	p = 3		p = 2n	p = 2n + 1	p = 2n + 2		
q = 0	$\Lambda^0(E^*)$	$\Lambda^1(E^*)$	$\Lambda^2(E^*)$	$\Lambda^3(E^*)$	...	$\Lambda^{2n}(E^*)$	0	0	0	...
q = 1	0	0	C	$C \otimes E^*$	...	$C \otimes \Lambda^{2n-2}(E^*)$	$C \otimes \Lambda^{2n-1}(E^*)$	$C \otimes \Lambda^{2n}(E^*)$	0	...

term$E_0^{p,q}$

4.3. To describe the E_0-terms more precisely recall some facts from symplectic geometry (see [L2]).

The spaces E_x are symplectic vector spaces for each $x \in M$ since the restriction of $(d\omega)_x$ gives a symplectic form on E_x.

In the algebra of exterior forms on the symplectic vector space E_x $\Lambda^*(E_x^*) = \bigoplus_{k \geq 0} \Lambda^*(E_x^*)$ we introduce two operators

$$T : \Lambda^k(E_x^*) \to \Lambda^{k+2}(E_x^*) \quad \text{and} \quad :\perp: \Lambda^k(E_x^*) \to \Lambda^{k-2}(E_x^*)$$

The first is the operator of exterior multiplicatoin by the two-form $d\omega_x, T(\Theta) = d\omega_w \wedge \Theta$ and the second - is the operator of interior multiplication by the canonical bivector $X_{d\omega_x}$ dual to $d\omega_x$, $\perp(\theta) = X_{d\omega_x} \perp \Theta$

Then the differential $d_0^{p,-1} : E_0^{p,1} \to E_0^{p,0}$ or $d_0^{p,-1} : C \otimes \Lambda^{p-2}(E^*) \to \Lambda^p(E^*)$ may be described as follows : $\omega \otimes [\alpha_1 + f_1\omega] \wedge \cdots \wedge [\alpha_{p-2} + f_{p-2}\omega] \stackrel{d_0}{\to} d\omega \wedge \alpha_1 \wedge \cdots \wedge \alpha_{p-2}$, where $\alpha_j \in \Omega^1(M), [\alpha_j + f_j\omega] \in \Omega^1(M)/C, f \in C^\infty(M)$ $1 \leq j \leq p-2$. Hence, $d_0^{p,-1}$ is nothing else than the multiplication by the differential of ω and thus [L2]

$$\operatorname{Ker} d_0^{p,-1} = \begin{cases} 0, & p-2 \le n-1 \\ T^k\left(\Lambda_\varepsilon^{n-k}\right), p-2 = n+k, & k = 0, \cdots, n. \end{cases}$$

Here $T^k = T \circ \cdots \circ T$ k times and Λ_ε^s is a set of *effective* s-forms (a form $\Theta \in \Lambda^k(E_x^*)$, $K \le n$, is said to be the effective if $\perp \Theta = 0$). We obtain

	p = 0	p = 1	p = 2		p = n - 1	p = n	p = n + 1	p = n + 2			p = 2n + 2
q = 0	$\Lambda^0(E^*)$	$\Lambda^1(E^*)$	$\Lambda^2(E^*)$	...	$\Lambda_\varepsilon^{n-1}(E^*)$	$\Lambda_2^n(E^*)$	0	0	0	...	0
q = 1	0	0	0	...	0	0	0	$\Lambda_\varepsilon^n(E^*)$	$T\left(\Lambda_\varepsilon^{n-1}(E^*)\right)$	...	$C \otimes \Lambda^{2n}(E^*)$

The differential $d_1^{p,q} : E_1^{p,q} \to d_1^{p+1,q}$ splits into two components :

$$d_1^{p,q} = d_1' + d_1''.$$

d_1' is the differential of the complex of effective forms

$$0 \to \Lambda^0(E^*) \xrightarrow{d_1'} \Lambda^1(E^*) \xrightarrow{d_1'} \Lambda_\varepsilon^2(E^*) \xrightarrow{d_1'} \cdots \xrightarrow{d_1'} \Lambda_\varepsilon^n(E^*) \to 0$$

(see [L2]).

On the other hand, d_2'' is the differentail of the complex

$$0 \to \Lambda_\varepsilon^k(E^*) \xrightarrow{d_1''} T\left(\Lambda_\varepsilon^{n-1}(E^*) \xrightarrow{d_1''} T^2 \Lambda_\varepsilon^{n-2}(E^*)\right) \to \cdots \to T^n\left(\Lambda^0(E^*)\right)$$

But $T^k : \Lambda^{n-k}(E^*) \to \Lambda^{n+k}(E^*)$ are isomorphism and hence d_1'' is the "effective" differential also. We obtain for tfeh $E_2^{p,q}$ terms

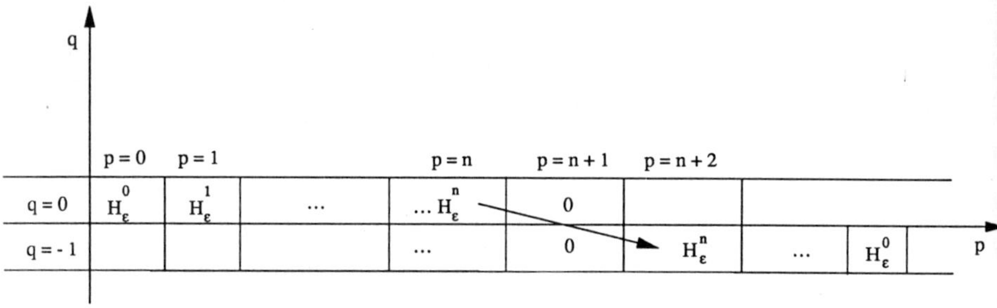

where H_ε^k are the cohomology spaces of the effective complex.

The only non-trivial differential is

$$d_2^{n,0} : H_\varepsilon^n \to H_\varepsilon^n$$

This operator is very often rediscovered *Euler operator* $\varepsilon = L_1 + d_p \circ \perp \circ d_p : \Lambda_\varepsilon^n \to \Lambda_\varepsilon^n$, where L_1 is the Lie derivation along the contact vector field X_1 and $d_p = p \circ d : \Omega^s(M) \to \Lambda^s(E^*)$, $p(\omega)$ is the projection of ω on the E. [L2].

Hence we have a complete description of our spectral sequence :

$$E_3^{p,q} = \begin{cases} E_2^{p,q}, p \neq n, p \neq 0 \\ [\text{divergent } n - \text{forms}], p = n, q = 0 \end{cases}$$

and $E_r^{p,q} = E_{r+1}^{p,q}$ if $r \geq 3$.

Summing up we receive

Theorem. The spectral sequence constructed with the non-holonomic filtration in the de Rham algebra on a contact manifold M collapses in the term E_3 and has the following initial terms :

1°

$$\text{The term} E_0^{p,q} = \begin{cases} \Lambda^p\left(\Omega^1(M)/C\right) & , 0 \leq p \leq 2n, q = 0 \\ C \otimes \Lambda^{p-2}\left(\Omega^1(M)/C\right) & , 2 \leq p \leq 2n+2, q = -1 \\ 0 & , \text{otherwise}. \end{cases}$$

$$\text{The term} E_1^{p,q} = \begin{cases} \Lambda_\varepsilon^p\left(\Omega^1(M)/C\right) & , 0 \le p \le n, q = 0 \\ \Lambda_\varepsilon^{p-2}\left(\Omega^1(M)/C\right) & , n+2 \le p \le 2n+2, q = -1 \\ 0 & , \text{otherwise}. \end{cases}$$

$$\text{The term} E_2^{p,q} = \begin{cases} H_\varepsilon^p & , 0 \le p \le n, q = 0 \\ H_\varepsilon^{p-2} & , n+2 \le p \le 2n+2, q = -1 \\ 0 & , \text{otherwise}. \end{cases}$$

$$\text{The term} E_3^{p,q} = \begin{cases} E_2^{p,q} & , \text{if } \; p \ne n, q \ne 0 \\ \text{set of divergent} n - \text{forms} & , p = n, q = 0. \end{cases}$$

Here C is the canonical contact distribution, and spaces Λ_ε^k, H_ε^k are effective forms and cohomology of effectives forms ([L2]) respectively.

2° The differential $d_2^{n,0} : E_2^{n,0} \to E_2^{n+2,-1}$ is the Euler operator.

REFERENCES

[BL] Bruzzo, V, Landi, G. Simple proof of Weil triviality in supersymmetry gauge theories, Phys. Rev. D 39, NY 1989, 1174-1178

[V] Drinfeld, V., Quazi-Hopf algebras, Algebra i Analiz, 1, n° 6, 1989,

[Dl] Deligne, P., Théorie de Hodge, II, Publ. Math. IHES, N 40, 1971, 5-58

[G] Gourevitch, D., Lychagin, V., Rubtsov, V, Filtration non-holonome et cohomologie d'algèbres de Lie. C.R. Ac. Sci. Paris, t 312, ser. I (5 décembre)

[K-S] Kosmann-Schwarzbach, Y. Grand crochets, crochets de Schouten et cohomologies d'algèbres de Lie, C.R. Ac. Sci, Paris, 312, série I, p. 123-126, 1991

[KS] Kostant, B., Sternberg, S. Symplectic reduction, BRS cohomology and infinite-dimensional Clifford algebras. Annals of Physics, 176, 1987, p 49-113

[KLV] Krasilschik, I., Lychagin, V. Vinogradov, A., Geometry of jet spaces and differential equations, Gordon and Breach, 1986.

[L1] Lychagin, V. Singularities of solutions, spectral sequences and normal forms vector fields Lie algebras, Izvestiya AN SSSR, V. 51, N° 3, 1987, 584-612.

[L2] Lychagin, V. Contact geometry and non-linear second order differential equations, Russ. Math. Surv. 34, N° 1, 1979, 137-165.

[LR] Lecomte, P. Roger, C. Modules et cohomologie des bialgèbres de Lie, C.R. Ac. Sci. Paris, 310, ser. I. 1990 p. 405-410.

Geometry in Partial Differential Equations (pp. 215–252)
eds. A. Pràstaro and Th. M. Rassias

SPENCER COHOMOLOGIES

V. LYCHAGIN
"Sophus Lie" Center (Moscow Branch)
P.B. 546, 119618 Moscow, Russia

and

L. ZILBERGLEIT
"Sophus Lie" Center (Moscow Branch)
P.B. 546, 119618 Moscow, Russia

ABSTRACT

We describe properties of the Spencer complexes and resolve the problem of duality for Spencer cohomologies. We propose construction of spectral sequence converging to Spencer cohomologies. By using symmetry groups of differential equations systems we manage to unify computations by reduction to the invariant systems over a homogeneous space. The conditions of coincidence of Spencer cohomologies with the cohomologies of invariant Spencer complex we obtain from arithmetic of $\mathbb{C}$-characteristic manifold with respect to fundamental weights of the homogeneous space.

Introduction

Stable Spencer cohomologies associated with a formally integrable system of differential equations (SDE) constitute a universal model for differential-geometric constructions of cohomology theories. Spencer cohomologies of a SDE, from the sheaves theory viewpoint, are the cohomologies of the base manifold with coefficients in corresponding solutions sheaf.

We use methods of calculus and homological algebra, which allow us, in particular, to by-pass the question of local exactness of Spencer complex.

Using these methods we resolve the problem of duality for Spencer cohomologies. A posteriory, it appears that dual cohomologies are Spencer cohomologies of a certain (generally speaking different one) SDE. Duality theorems, applied to the differential equations corresponding to classical differential operators, lead to the Poincare and Serre dualities.

We propose construction of spectral sequence determined by natural filtrations in Spencer complex. This sequence for the de Rham operator leads to classical

Leray-Serre theorem. Our result links the stable Spencer cohomologies of a SDE on the space of a noncharacteristic bundle with the de Rham cohomologies of the base with the values in stable Spencer cohomologies of the fiberwise image equations.

By using symmetry groups of SDE we, in the majority of cases, manage to unify computations by reduction to the invariant SDE over a homogeneous space. The coincidence of stable cohomologies with the cohomologies of invariant Spencer complex is the crucial point in computation of Spencer cohomologies. The conditions of coincidence we obtain from arithmetic of (complex) characteristic manifold of a SDE with respect to fundamental weights of a homogeneous space. Hence, the computation problem for stable Spencer cohomologies of invariant SDE over a homogeneous manifold acquires a purely algebraic nature.

Contents

1. Spencer complexes.

This section contains basic definitions and properties relating to Spencer complexes.

1.1. Jets and differential equations

For a smooth manifold M, $\dim M = n$, and a smooth vector bundle $\alpha : E(\alpha) \to M$ by $C^\infty(\alpha)$ we denote the module of smooth sections of bundle α.

Let $\alpha_k : J^k(\alpha) \to M$ be corresponding bundle of k-jets of sections. By $\mathcal{J}^k(\alpha)$ we denote the module of smooth sections of α_k.

For any section $h \in C^\infty(\alpha)$ a section $j_k(h) \in C^\infty(\alpha_k)$ is defined:

$$j_k(h) : x \mapsto [h]_x^k,$$

where $[h]_x^k$ is the k-jet of h at point $x \in M$.

Natural projections

$$\alpha_{k,l} : \ J^k(\alpha) \to J^l(\alpha)$$

define module epimorphisms

$$\alpha_{k,l} : \ \mathcal{J}^k(\alpha) \to \mathcal{J}^l(\alpha),$$

such that $\alpha_{k,l} \circ j_k(h) = j_l(h)$ for all $h \in C^\infty(\alpha)$.

We denote by $\mathrm{Diff}_k(\alpha, \beta)$ the module of linear differential operators of the order k acting from the sections of α into the sections of β.

Any operator $\Delta \in \mathrm{Diff}_k(\alpha, \beta)$ defines a homomorphism $\varphi_\Delta \in \mathrm{Hom}(\alpha_k, \beta)$ by the following rule:

$$\varphi_\Delta([h]_x^k) \ = \ \Delta(h)(x).$$

Any operator $\Delta \in \mathrm{Diff}_k(\alpha, \beta)$ satisfying natural regularity conditions defines a system of differential equetions $R_\Delta = \varphi_\Delta^{-1}(0)$. We shall consider this system as a subbundle r of the bundle α_k:

$$r : \ E(r) = R \to M$$

A section $h \in C^\infty(\alpha)$, such that $j_k(h) \subset C^\infty(r)$, is called a *smooth solution* of R.

1.2. Prolongation and formal integrability

The first step in the investigation SDE solvability consists of the analysis of solutions in formal power series. Such formal solutions naturally lead to the concept of the prolongation.

If the operator $\Delta \in \mathrm{Diff}_k(\alpha, \beta)$ then the operator $\Delta^{(l)} \in \mathrm{Diff}_{k+l}(\alpha, J^l\beta)$

$$\Delta^{(l)}: \; C^\infty(\alpha) \to \mathcal{J}^l\beta$$

$$h \mapsto j_l \circ \Delta(h)$$

is called *l-th prolongation of the operator* Δ.

The l-th prolongation of SDE is a subbundle $r^{(l)} \subset \alpha_{k+l}, \; r^{(l)} : R^{(l)} \to M$ such that

$$R^{(l)} = \{x_{k+l} \in J^{k+l}\alpha \mid x_{k+l} = [h]^{k+l}_x, \; h \in C^\infty(\alpha), j_k(h)(M) \text{ is tangent to } R \textit{ at } x_k = [h]^k_x \text{ with the order} \geq l\}.$$

By definition, $\alpha_{k+l,k+l-1}(R^{(l)}) \subset R^{(l-1)}$ for any $l \geq 1, \; R^{(0)} = R$. The system $R^{(l)}$ can be considered as the system R completed with the compatibility conditions up to the order l.

Prolongations $R^{(l)}$ are associated with prolongations $\Delta^{(l)}$ as follows:

$$R^{(l)} \;=\; \varphi^{-1}_{\Delta^{(l)}}(0).$$

The equation R is called *regular* if both R and all its prolongations are smooth manifolds.

Note that if $r \subset \alpha_k$ is a regular SDE then

$$R^{(l+1)} = [R^{(l)}]^{(1)} \text{ for all } l \geq 0.$$

The system $r \subset \alpha_k$ is called *formally integrable* if r is regular and for all $l \geq 0$

$$\alpha_{k+l+1,k+l}: \; R^{(l+1)} \to R^{(l)}$$

are smooth bundles.

The system $r \subset \alpha_k$ is called *normal*, if the restriction of epimorphism $\alpha_{k,k-1}$: $R \to J^{k-1}\alpha$ is a smooth vector bundle.

1.3. Symbols

Formal integrability conditions for a system of differential equations are determined by symbols of this system and their prolongations. The symbol of an operator $\Delta \in \mathrm{Diff}_k(\alpha, \beta)$ is a morphism of vector bundles

$$\sigma(\Delta): \; S^k\tau^* \otimes \alpha \to \beta,$$

where $\tau^* : E(\tau^*) \to M$ is the cotangent bundle. It is determined by its values as follows:

$$\sigma(\Delta) \;:\; S^k\tau^*_x \otimes \alpha_x \to \beta_x,$$

$$d_x f_1 \circ d_x f_2 \circ \ldots d_x f_k \otimes h_x \mapsto \Delta(f_1 \ldots f_k h)(x).$$

Here $f_i \in C^\infty(M)$, $f_i(x) = 0$, $i = 1, \ldots, k$, $h \in C^\infty(\alpha)$, $\circ$ is the sign of symmetrical product. Similarly the symbol of l-prolongation of Δ is the mapping

$$\sigma(\Delta^{(l)})(x) : S^{k+l}\tau_x^* \otimes \alpha_x \to S^l\tau_x^* \otimes \beta_x$$

$$d_x f_1 \circ d_x f_2 \circ \ldots d_x f_{k+l} \otimes h_x \mapsto [\Delta(f_1 \ldots f_{k+l} h)]_x^l,$$

where $f_i \in C^\infty(M)$, $f_i(x) = 0$, $i = 1, 2, \ldots, k+1$, $h \in C^\infty(\alpha)$.

Example. One consider the exterior differential

$$d : \Lambda^k(M) \to \Lambda^{k+1}(M).$$

Symbols of its l-prolongation

$$\sigma(d^{(l)})(x) : S^{l+1}\tau_x^* \otimes \Lambda^k\tau_x^* \to S^l\tau_x^* \otimes \Lambda^{k+1}\tau_x^*,$$

$$d_x f_1 \circ \ldots \circ d_x f_{l+1} \otimes \omega_k \mapsto \sum_{i=1}^{l+1} d_x f_1 \circ \ldots \circ \widehat{d_x f_i} \circ \ldots \circ d_x f_{l+1} \otimes d_x f_i \wedge \omega_k,\ \omega_k \in \Lambda^k\tau_x^*,$$

are called Spencer δ-operator and are denoted by δ.

A vector bundle $g = \mathrm{Ker}\sigma(\Delta) \subset S^k\tau^* \otimes \alpha$ is called *a symbol bundle of* SDE $r_\Delta \subset \alpha_k$ associated with the operator $\Delta \in \mathrm{Diff}_k(\alpha, \beta)$. We use notations

$$g^l = \mathrm{Ker}\sigma(\Delta^{(l)}) \subset S^{k+l}\tau^* \otimes \alpha$$

1.4. Spencer δ-complexes.

There exists a relation between the symbols of prolongations of SDE which allows one to use the homology algebra methods for the investigation of formal integrability.

Using the fact that $\delta(C^\infty(g^{l-1})) \subset C^\infty(g^{l-1}) \otimes \Lambda^1(M)$ we define a complex

$$0 \to C^\infty(g^l) \xrightarrow{\delta} C^\infty(g^{l-1}) \otimes \Lambda^1(M) \xrightarrow{\delta} \cdots \xrightarrow{\delta} C^\infty(g^{l-n}) \otimes \Lambda^n(M) \to 0$$

which is called the *Spencer δ-complex.*

We denote by $H^{l,k}(g)$ the cohomologies of this complex (i.e. *the Spencer δ-cohomologies* of an SDE r) at the term $C^\infty(g^l) \otimes \Lambda^k(M)$. The system r is called *involutive* if $H^{l,k} = 0$ for any $l \geq 0, k \geq 0$.

Theorem 1.[2,4] *Let* SDE $r \subset \alpha_k$ *satisfy the conditions*

(i) $\alpha_{k+1,k} : R^{(1)} \to R$ *is a smooth bundle;*

(ii) $\alpha_{k,k-1} : R \to J^{k-1}(\alpha)$ *is a smooth bundle;*

(iii) $H^{l,2}(g) = 0$ *for any* $l \geq 0$

Then r *is a formally integrable system.*

Poincare's δ-lemma.[1] *There exist a number* l_0 *such that* $H^{l,i}(g) = 0$ *for any* $l \geq l_0$, $i \geq 0$.

1.5. Characteristics

For any co-vector $\lambda \in \tau_x^*$ and any linear operator $\Delta \in \operatorname{Diff}_k(\alpha, \beta)$ we denote by $\sigma_\lambda(\Delta) : \alpha_x \to \beta_x$ a mapping

$$\sigma_\lambda(\Delta)(h_x) = \sigma(\Delta)\Big[\frac{1}{k!} h_x \otimes \lambda^k\Big], \qquad h_x \in \alpha_x.$$

A nonzero co-vector $\lambda \in \tau_x^*$ is called *characteristic* with respect to the operator Δ, if $\operatorname{Ker}\sigma_\lambda \neq 0$. More generally, a subspace $U \subset S^k\tau_x^*$ will be called *characteristic* with respect to a linear subspace $g_x \subset \alpha_x \otimes S^k\tau_x^*$, if

$$g_x \cap \alpha_x \otimes U \neq 0.$$

Theorem 2 (Quillen-Guillemin[5]). *For involutive symbols over an algebraically closed field a subspace* $U \subset S^k\tau_x^*$ *is characteristic if and only if* U *contains characteristic co-vectors.*

Consider a bundle $\kappa : M \to B$ with a fiber F. The bundle κ is called *noncharacteristic* with respect to a system of differential equations $r \subset \alpha_k$, if all subspaces $\operatorname{Ann}F_{b,y}$ are noncharacteristic, $b \in B$, $y \in F_b$.

1.6. The Spencer operator

The Spencer operator $D : \mathcal{J}^l(\alpha) \otimes \Lambda^k(M) \to \mathcal{J}^{l-1}(\alpha) \otimes \Lambda^{k+1}(M)$ is iniquely determined by the following conditions:

(i) $D(\theta \wedge \omega) = D\theta \wedge \omega + (-1)^k(\alpha_{k,k-1} \otimes 1)(\theta \wedge D\omega)$,
where $\theta \in \mathcal{J}^l(\alpha) \otimes \Lambda^k(M)$, $\omega \in \Lambda^r(M)$

(ii) the sequence

$$0 \to C^\infty(\alpha) \xrightarrow{j_l} \mathcal{J}^l(\alpha) \xrightarrow{D} \mathcal{J}^{l-1}(\alpha) \otimes \Lambda^1(M)$$

is exact.

The Spencer operator D is a natural operator. More exactly, let Ψ be a fiberwise map from the vector bundle α^1 over M into the vector bundle α^2 over M_2:

$$\begin{array}{ccc} E(\alpha^1) & \xrightarrow{\Psi} & E(\alpha^2) \\ \downarrow & & \downarrow \\ M_1 & \xrightarrow{\bar{\Psi}} & M_2, \end{array}$$

i.e. $\Psi : \alpha^1_x \to \alpha^2_{\bar{\Psi}(x)}$ are isomorphisms for all $x \in M$.
Then diagrams

$$\begin{array}{ccc} \mathcal{J}^l(\alpha^2) \otimes \Lambda^k(M_2) & \xrightarrow{\Psi_l^* \otimes \bar{\Psi}^*} & \mathcal{J}^l(\alpha^1) \otimes \Lambda^k(M_1) \\ D \downarrow & & \downarrow D \\ \mathcal{J}^{l-1}(\alpha^2) \otimes \Lambda^{k+1}(M_2) & \xrightarrow{\Psi_{l-1}^* \otimes \bar{\Psi}^*} & \mathcal{J}^{l-1}(\alpha^1) \otimes \Lambda^{k+1}(M_1) \end{array}$$

are commutative. Here $\bar{\Psi}^* : C^\infty(M_2) \to C^\infty(M_1)$, $\Psi_k^* : \mathcal{J}^k(\alpha^2) \to \mathcal{J}^k(\alpha^1)$ are induced mappings.

In particular, if Ψ is a restriction on a submanifold $M_0 \subset M$, $\alpha|_{M_0} = \alpha^0$, the diagrams have the following form

$$\begin{array}{ccc} \mathcal{J}^l(\alpha) \otimes \Lambda^k(M) & \longrightarrow & \mathcal{J}^l(\alpha^0) \otimes \Lambda^k(M_0) \\ D \downarrow & & \downarrow D \\ \mathcal{J}^{l-1}(\alpha) \otimes \Lambda^{k+1}(M) & \longrightarrow & \mathcal{J}^{l-1}(\alpha^0) \otimes \Lambda^{k+1}(M_0). \end{array}$$

Let us clarify the geometrical sense of the Spencer operator. Let $\theta \in \mathcal{J}^{k+1}(\alpha)$. At any point $x \in M$ the $k+1$-jet $x_{k+1} = \theta(x)$ determines a decomposition

$$T_{x_k} J^k \alpha \;=\; L(x_{k+1}) \oplus J^k_x(\alpha),$$

where $x_k = \alpha_{k+1,k}(x_{k+1})$, and, due to linearity of α, the fibre $J^k_x \alpha$ is identified with the tangent space to the fibre $\alpha_k^{-1}(x)$,

$$L(x_{k+1}) = T_{x_k}[j_k(h)], \qquad x_{k+1} = [h]^{k+1}_x, \qquad h \in C^\infty(\alpha).$$

Let $\widehat{\theta} = \alpha_{k+1,k}(\theta) \in \mathcal{J}^k(\alpha)$. Then the submanifold $\widehat{\theta}(M)$ passes through the point x_k and projects diffeomorphicaly on M. Hence,

$$(\alpha_k)_* : \; T_{x_k}(\widehat{\theta}(M)) \to T_x(M)$$

is an isomorphism.

Using this isomorphism one can lift any vector $v \in T_x(M)$ into $T_{x_k}(J^k(\alpha))$ and then project along $L(x_{k+1})$ using the decomposition above. The element constructed in such way we consider as a value of some 1-form $D\theta \in \mathcal{J}^k(\alpha) \otimes \Lambda^1(M)$ on the vector $v \in T_x M$.

1.7. Spencer complexes

The properties (i) and (ii) of Spencer operators yield $D^2 = 0$. Hence the sequence

$$0 \to C^\infty(\alpha) \xrightarrow{j_k} \mathcal{J}^k(\alpha) \xrightarrow{D} \mathcal{J}^{k-1}(\alpha) \otimes \Lambda^1(M) \to \cdots \to \mathcal{J}^{k-n}(\alpha) \otimes \Lambda^n(M) \to 0$$

is a complex which is called the (first, or naive) *Spencer complex.*

Let r be a formal integrable SDE. Restricting the Spencer operator on $\mathcal{R}^{(l)}$ we obtain a differential operator of order 1:

$$D: \mathcal{R}^{(l)} \otimes \Lambda^k(M) \to \mathcal{R}^{(l-1)} \otimes \Lambda^{k+1}(M).$$

Therefore the first Spencer complex admits a restriction on a formally integrable system r. The complex obtained is called the Spencer complex, associated with the system r:

$$0 \to \mathcal{R}^{(l)} \xrightarrow{D} \mathcal{R}^{l-1} \otimes \Lambda^1(M) \to \cdots \to \mathcal{R}^{k-n} \otimes \Lambda^n(M) \to 0 \qquad (1)$$

Consider a commutative diagram

$$\begin{array}{ccccccccc}
 & & 0 & & 0 & & 0 & & \\
 & & \downarrow & & \downarrow & & \downarrow & & \\
0 \to & & C^\infty(g^l) & \to & C^\infty(g^{l-1}) \otimes \Lambda^1(M) & \to \cdots \to & C^\infty(g^{l-n}) \otimes \Lambda^n(M) & \to 0 \\
 & & \downarrow & & \downarrow & & \downarrow & & \\
0 \to & & \mathcal{R}^{(l)} & \to & \mathcal{R}^{(l-1)} \otimes \Lambda^1(M) & \to \cdots \to & \mathcal{R}^{(l-n)} \otimes \Lambda^n(M) & \to 0 \\
 & & \downarrow & & \downarrow & & \downarrow & & \\
0 \to & & \mathcal{R}^{(l-1)} & \to & \mathcal{R}^{(l-2)} \otimes \Lambda^1(M) & \to \cdots \to & \mathcal{R}^{(l-n-1)} \otimes \Lambda^n(M) & \to 0 \\
 & & \downarrow & & \downarrow & & \downarrow & & \\
 & & 0 & & 0 & & 0 & &
\end{array}$$

The columns are exact due to the formal integrability.

From Poincare's δ-lemma it follows that for $l \geq l_0 + n$ the cohomologies of naive Spencer complex become stable. This stable cohomologies are called *Spencer cohomologies* of the formal integrable system r; we denote them by $\mathcal{H}^i(r)$, $i = 0, 1, \ldots$.

The second (sophisticated) Spencer complex is more simple. It arises as follows. For the operator $\Delta_1 : C^\infty(\alpha^1) \to C^\infty(\alpha^2)$ we consider the compatibility operator $\Delta_2 : C^\infty(\alpha^2) \to C^\infty(\alpha^3)$ which determines a complex $C^\infty(\alpha^1) \xrightarrow{\Delta_1} C^\infty(\alpha^2) \xrightarrow{\Delta_2} C^\infty(\alpha^3)$. For the operator Δ_2 we consider the compatibility operator Δ_3, etc.

In general, the complex $C^\infty(\alpha^1) \xrightarrow{\Delta_1} C^\infty(\alpha^2) \xrightarrow{\Delta_2} C^\infty(\alpha^3) \xrightarrow{\Delta_3} \cdots$ does not exist. But for the formally integrable system such complex always exists and is unique in the class of formally exact ones (the Kuranishi theorem[3]).

To define this complex assume that a morphism

$$\Theta : \mathcal{R}^{(l)} \to \mathcal{R}^{(l+1)}, \quad \alpha_{k+l+1,l} \circ \theta = \mathrm{id},$$

exists. We extend the operator D to the operator $D_\Theta = D \circ \Theta$,

$$D_\Theta : \mathcal{R}^{(l)} \otimes \Lambda^s(M) \to \mathcal{R}^{(l)} \otimes \Lambda^{s+1}(M).$$

If a morphism

$$\Theta' : \mathcal{R}^{(l)} \to \mathcal{R}^{(l+1)}$$

is other one, then

$$(D_\Theta - D_{\Theta'})(\mathcal{R}^{(l)} \otimes \Lambda^s(M)) \subset \delta(C^\infty(g^{(l+1)}) \otimes \Lambda^s(M))$$

and consequently the second Spencer operator

$$D \colon C_l^s \to C_l^{s+1},$$

where $C_l^s = \mathcal{R}^{(l)} \otimes \Lambda^s(M) \;/\; \delta(C^\infty(g^{(l+1)}) \otimes \Lambda^{s-1}(M))$, is well-defined.

The second Spencer complex has the form

$$0 \to C_l^0 \xrightarrow{D} C_l^1 \xrightarrow{D} \cdots \xrightarrow{D} C_l^n \to 0. \tag{2}$$

Cohomologies of this complex stabilize for large l's and coincide with the Spencer cohomologies $\mathcal{H}^\bullet(r)$.

The exterior product on differential forms defines the structure of $\Lambda^\bullet(M) = \bigoplus_{i\geq 0} \Lambda^i(M)$-module on $C_l^\bullet = \bigoplus_{i\geq 0} C_l^i$. Passing to the cohomologies, this product generates a module structure in Spencer cohomologies $\mathcal{H}^\bullet(r)$ over the de Rham cohomologies $H^\bullet(M)$.

Below we shall regard large (stable) values of l and shall omit the subscript l in $C_l^k : C^k = C_l^k$.

In terms of sophisticated Spencer complex one can present a convenient criterion for characteristicity of co-vector $\lambda \in \tau^* M$. Denote by σ_λ the value of the sophisticated Spencer operator symbol on the co-vector λ. Then the following theorem is valid.

Theorem (D. Quillen[2]). *Let $r \subset \alpha_k$ be a formally integrable system. Then for any co-vector $\lambda \in T^*M \setminus 0$ the following conditions are equivalent.*

(i) *The co-vector λ is noncharacteristic.*

(ii) *the sequence*

$$0 \to C^0 \xrightarrow{\sigma_\lambda} C^1$$

is exact.

(iii) *The sequence*

$$0 \to C^0 \xrightarrow{\sigma_\lambda} C^1 \xrightarrow{\sigma_\lambda} \cdots \to C^n \to 0$$

is exact.

Examples.

1. The zero group of Spencer cohomologies of formally integrable system r_Δ is isomorphic to the space of solutions:

$$\mathcal{H}^0(r_\Delta) = \mathrm{Ker}\Delta.$$

2. For any determined system $r_\Delta \subset \alpha_k$ one has

$$\mathcal{H}^0(r_\Delta) = \mathrm{Ker}\Delta, \qquad \mathcal{H}^1(r_\Delta) = \mathrm{Coker}\Delta, \qquad \mathcal{H}^i(r_\Delta) = 0 \text{ for all } i > 1.$$

3. Spencer cohomologies of a system determined by the operator

$$d : C^\infty(M) \to \Lambda^1(M)$$

coincide with de Rham's cohomologies.

4. A natural generalization of the previous example is a SDE determined by a flat connection

$$\nabla : C^\infty(\alpha) \to C^\infty(\alpha) \otimes \Lambda^1(M).$$

Spencer cohomologies of this system coincide with de Rham's cohomologies of the manifold M with the coefficients in sections of bundle α.

5. Let us consider a determined system $r_\Delta \subset \alpha_k$ and define a mapping

$$H^1(M) \otimes \mathcal{H}^0(r_\Delta) \to \mathcal{H}^1(r_\Delta).$$

Any element of $H^1(M) \otimes \mathcal{H}^0(r_\Delta)$ can be locally represented in the form $\sum_i df_i \otimes s_i$, where $f_i \in C^\infty(M)$, $s_i \in \mathrm{Ker}\Delta$. Consider elements $\Delta(\sum_i f_i \otimes s_i)$ glue them together over all the local charts and factorise with respect to $\mathrm{Im}\Delta$. As the result we obtain an element of $\mathrm{Coker}\Delta \simeq \mathcal{H}^1(r_\Delta)$.

2. Duality in Spencer cohomologies.

Here we give a construction on the dual equation which leads to duality theorems in Spencer cohomologies. These theorems generalize classical ones of Poincare and Serre.

2.1. Green's formula

Green's formula is the basis for obtaining duality theorems. For our purpose, sufficient to use Green's formula for first-order differential operators.

Firstly, we introduce necessary notations. We shall call a bundle

$$\alpha^t = \mathrm{Hom}(\alpha, \Lambda^n \tau^*)$$

dual to the bundle α.

Denote by

$$\langle , \rangle : C^\infty(\alpha) \otimes C^\infty(\alpha^t) \to \Lambda^n(M)$$

a natural pairing between sections of bundles α and α^t.

For every morphism $s \in \mathrm{Hom}(\alpha, \beta)$ denote by $s^t \in \mathrm{Hom}(\beta^t, \alpha^t)$ a dual morphism defined by relation:

$$\langle s_a, b^t \rangle = \langle a, s^t b^t \rangle ,$$

where $a \in C^\infty(\alpha)$, $b^t \in C^\infty(\alpha^t)$.

Proposition 1. *Let $\nabla \in \mathrm{Diff}_1(\alpha, \beta)$, $\Box \in \mathrm{Diff}_1(\beta^t, \alpha^t)$. We consider a mapping $\gamma(\nabla, \Box)$ associating the pair (a, b^t), $a \in C^\infty(\alpha)$, $b^t \in C^\infty(\beta^t)$ with n-form $\langle \nabla(a), b^t \rangle - \langle a, \Box(b^t) \rangle$. This mapping is a first-order differential operator acting from the bundle $\alpha \otimes \beta^t$ into the bundle $\Lambda^n \tau^*$ iff the equality*

$$\sigma^t_\lambda(\nabla) + \sigma_\lambda(\Box) = 0$$

is satisfied for all 1-forms $\lambda \in C^\infty(\tau^*)$.

If the conditions of the proposition are fulfilled, the symbol of the operator $\gamma(\nabla, \Box)$ defines a homomorphism $w_\nabla : \alpha \otimes \beta^t \to \Lambda^{n-1}\tau^*$, acting in the following way:

$$\sigma_\lambda(\gamma(\nabla, \Box))(a \otimes b^t) \;=\; \langle \sigma_\lambda(\nabla) a, b^t \rangle \;=\; \lambda \wedge w_\nabla(a, b^t).$$

This formula shows that the symbols of the operator $\gamma(\nabla, \Box)$ and $d \circ w_\nabla$ coincide. Thus, the operators are distinguished by a homomorphism. Consequently, replacing operator $\Box$ by the zero-order operator we may say that the operators γ and $d \circ w_\nabla$ coincide. The operator $\Box$, obtained in this way, is denoted by ∇^t and is called a dual operator to the operator ∇. Thus, one has the following

Proposition 2. *For every operator $\nabla \in \mathrm{Diff}_1(\alpha, \beta)$ there exists a unique dual operator $\nabla^t \in \mathrm{Diff}_1(\beta^t, \alpha^t)$ defined by Green's formula:*

$$\langle \nabla(a), b^t \rangle - \langle a, \nabla^t(b^t) \rangle \;=\; dw_\nabla(a, b^t),$$

where $a \in C^\infty(\alpha)$, $b^t \in C^\infty(\beta^t)$.

2.2. Hodge-Spencer Theory

In this section we present the well-known Hodge-Spencer theory in the form convinient for us. The basic result of this section is Theorem 2.2.1 which establishes the isomorphism between cohomologies of an eliptic complex and the complex dual to it.

Further in this chapter M is a compact oriented Riemannian manifold with metric g. Each metric g_α in a bundle α defines an isomorphism

$$g_\alpha\colon \alpha \to \alpha^*,$$

where $\alpha^* = \mathrm{Hom}(\alpha, \mathbb{I})$ is a conjugate bundle. Here $\mathbb{I}\colon M \times R \to M$ is a trivial linear bundle.

Define the operator

$$*_\alpha\colon \Lambda^j\tau^* \otimes \alpha \to \Lambda^{n-j}\tau^* \otimes \alpha^*$$

for all $j = 0, 1, \ldots, n$ by the relation

$$*_\alpha(w \otimes a) \;=\; *w \otimes g_\alpha(a),$$

where $w \in \Lambda^j(M)$, $a \in C^\infty(\alpha)$ and $*\colon \Lambda^j\tau^* \to \Lambda^{n-j}\tau^*$ is the Hodge operator, corresponding to metric g.

Note that $*_{\alpha^t} = *_\alpha^{-1}$.

If bundles α and β are provided with metrics g_α and g_β, then each differential operator $\nabla \in \mathrm{Diff}_1(\alpha, \beta)$ can be put into correspondence, as usually, to the operator $\nabla^* \in \mathrm{Diff}_1(\beta, \alpha)$, conjugate with respect to the metric

$$(\nabla(a), b)_\beta = (a, \nabla^*(b))_\alpha,$$

where $a \in \Gamma(\alpha)$, $b \in \Gamma(\beta)$ and, for example, $(a_1, a_2)_\alpha = \int_M \langle a_1, *_\alpha a_2\rangle$, $a_1, a_2 \in C^\infty(\alpha)$.

The following commutative diagram shows the relation between two types of conjunctions of differential operators:

$$\begin{array}{ccc} \Gamma(\beta) & \xrightarrow{*_\beta} & \Gamma(\beta^t) \\ {\scriptstyle\nabla^*}\downarrow & & \downarrow{\scriptstyle\nabla^t} \\ \Gamma(\alpha) & \xrightarrow{*_\alpha} & \Gamma(\alpha^t) \end{array}$$

The diagram's commutativity is a corollary of Green's formula:

$$(\nabla(a), b)_\beta \;=\; \int_M \langle \nabla(a), *_\beta b\rangle \;=\; \int_M \langle a, \nabla^t(*_\beta b)\rangle \;=$$

$$\int_M \langle a, *_\alpha(*_\alpha^{-1}\nabla^t *_\beta b)\rangle \;=\; (a, *_\alpha^{-1}\nabla^t *_\beta b)_\alpha.$$

From the above it follows that the construction of a dual operator has a functorial nature and the construction of a conjugate one is artificial, since the latter depends on the choice of metric. The duality theorems, which are natural, establish an isomorphism between the cohomologies of some complex and the dual one. This fact explains the importance of the operator ∇^t. Passing to conjugate operators has a purely technical character.

Consider a complex of first-order differential operators

$$0 \to C^\infty(\alpha^0) \xrightarrow{\nabla_1} C^\infty(\alpha^1) \xrightarrow{\nabla_2} C^\infty(\alpha^2) \to \cdots \to C^\infty(\alpha^N) \to 0. \tag{1}$$

Then, passing on to dual bundles, we obtain a complex:

$$0 \leftarrow C^\infty(\alpha^{0t}) \xleftarrow{\nabla_1^t} C^\infty(\alpha^{1t}) \xleftarrow{\nabla_2^t} C^\infty(\alpha^{2t}) \leftarrow \cdots \leftarrow C^\infty(\alpha^{Nt}) \leftarrow 0. \tag{2}$$

If, besides, the bundles under consideration are provided with metrics, then the transition to conjugate operators yields a complex

$$0 \leftarrow C^\infty(\alpha^0) \xleftarrow{\nabla_1^*} C^\infty(\alpha^1) \xleftarrow{\nabla_2^*} C^\infty(\alpha^2) \leftarrow \cdots \leftarrow C^\infty(\alpha^N) \leftarrow 0. \tag{3}$$

The following commutative diagram establishes the relation between complexes 2.2.2 and 2.2.3

$$\begin{array}{ccccccccccc}
0 & \leftarrow & C^\infty(\alpha^0) & \xleftarrow{\nabla_1^*} & C^\infty(\alpha^1) & \xleftarrow{\nabla_2^*} & C^\infty(\alpha^2) & \leftarrow \cdots \leftarrow & C^\infty(\alpha^N) & \leftarrow & 0 \\
 & & \downarrow{\scriptstyle *_{\alpha^0}} & & \downarrow{\scriptstyle *_{\alpha^1}} & & \downarrow{\scriptstyle *_{\alpha^2}} & & \downarrow{\scriptstyle *_{\alpha^N}} & & \\
0 & \leftarrow & C^\infty(\alpha^{0t}) & \xleftarrow{\nabla_1^t} & C^\infty(\alpha^{1t}) & \xleftarrow{\nabla_2^t} & C^\infty(\alpha^{2t}) & \leftarrow \cdots \leftarrow & C^\infty(\alpha^{Nt}) & \leftarrow & 0.
\end{array}$$

Recall that complex 2.2.1 is called elliptic, if the corresponding symbolic complex

$$0 \to C^\infty(\alpha^0) \xrightarrow{\sigma_\lambda(\nabla_1)} C^\infty(\alpha^1) \xrightarrow{\sigma_\lambda(\nabla_2)} C^\infty(\alpha^2) \to \cdots \to C^\infty(\alpha^N) \to 0 \tag{4}$$

is acyclic for any non-zero covector $\lambda \in C^\infty(\tau^*)$.

The ellipticity of complex 2.2.1 is equivalent to the ellipticity of dual complex 2.2.2 or conjugate complex 2.2.3. This follows from the fact that symbolic complex of complex 2.2.2 is obtained from symbolic complex 2.2.4 by applying the functor $\mathrm{Hom}(\bullet, \Lambda^n\tau^*)$.

A proof of duality theorem 1 given below is based on the following two facts: firstly, the basic result of Hodge-Spencer theory, which states that in each class of cohomologies of an elliptic complex on a compact oriented Riemannian manifold there exists an unique harmonic element. The second fact is Proposition 2.2.3 which links Laplace operators of complex 2.2.1 and ones of the dual complex 2.2.2.

For complex 2.2.1 the Laplace operators

$$\Delta_j : C^\infty(\alpha^j) \to C^\infty(\alpha^j)$$

have the form:

$$\Delta_j = \nabla_j^* \circ \nabla_j + \nabla_{j-1} \circ \nabla_{j-1}^*,$$

while for dual complex 2.2.2 one has

$$\Delta_j^t: C^\infty(\alpha^{jt}) \to C^\infty(\alpha^{jt}),$$

$$\nabla_j^t = \nabla_j^t \circ (\nabla_j^t)^* + (\nabla_{j-1}^t)^* \circ \nabla_{j-1}^t.$$

Proposition 3. *The following diagram*

$$\begin{array}{ccc} C^\infty(\alpha^j) & \xrightarrow{*_{\alpha^j}} & C^\infty(\alpha^{jt}) \\ {\scriptstyle \Delta_j}\downarrow & & \downarrow{\scriptstyle \Delta_j^t} \\ C^\infty(\alpha^j) & \xrightarrow{*_{\alpha^j}} & C^\infty(\alpha^{jt}) \end{array}$$

is comutative.

Proof. Rewrite the expressions for operators Δ_j and Δ_j^t:

$$\Delta_j = [*_{\alpha^j}]^{-1}\nabla_j^t *_{\alpha^{j+1}} \nabla_j + \quad \nabla_{j-1}[*_{\alpha^{j-1}}]^{-1}\nabla_{j-1}^t *_{\alpha^j},$$

$$\Delta_j^t = \nabla_j^t[*_{\alpha^{j+1,t}}]^{-1}\nabla_j *_{\alpha^{jt}} + [*_{\alpha^{j,t}}]^{-1}\nabla_{j-1} *_{\alpha^{j-1,t}} \nabla_{j-1}^t$$

$$= \nabla_j^t *_{\alpha^{j+1}} \nabla_j[*\alpha^j]^{-1} + \quad *_{\alpha^j} \nabla_{j-1}[*_{\alpha^{j-1}}]^{-1}\nabla_{j-1}^t.$$

Therefore the equalities

$$*_{\alpha^j}\Delta_j = \nabla_j^t * \alpha^{j+1}\nabla_j + *\alpha^j\nabla_{j-1}[*_{\alpha^{j-1}}]^{-1}\nabla_{j-1}^t *_{\alpha^j} = \Delta_j^t *_{\alpha^j}$$

hold. ∎

Denote by H^k the cohomologies of complex 2.2.1 in the term $C^\infty(\alpha^k)$ and by H_k the cohomologies of dual complex 2.2.2 in the term $C^\infty(\alpha^{kt})$. Then one has

Theorem 1. *On a compact oriented manifold the operators $*_{\alpha^j}: H^j \to H_j$, defined by a metric, give isomorphism for cohomologies of elliptic complexes.*

Complex 2.2.1 will be called a *Poincare complex,* iff there exist isomorphism $A_j: \alpha^j \to \alpha^{N-j\ t}$ establishing the isomorphism of complexes:

$$\begin{array}{ccccccccc} 0 \to & C^\infty(\alpha^0) & \xrightarrow{\nabla_1} & C^\infty(\alpha^1) & \xrightarrow{\nabla_2} & C^\infty(\alpha^2) & \to \cdots \to & C^\infty(\alpha^N) \to 0 \\ & {\scriptstyle A_0}\downarrow hfil & & {\scriptstyle A_1}\downarrow & & {\scriptstyle A_2}\downarrow & & {\scriptstyle A_N}\downarrow \\ 0 \to & C^\infty(\alpha^{N\ t}) & \xrightarrow{\nabla_N^t} & C^\infty(\alpha^{N-1\ t}) & \xrightarrow{\nabla_{N-1}^t} & C^\infty(\alpha^{N-2\ t}) & \to \cdots \to & C^\infty(\alpha^{0\ t}) \to 0 \end{array}$$

One deduces from Theorem 2.2.1 the following corollary.

Corollary (Poincare duality). *Under the conditions of the theorem above, for Poincare complexes mappings*

$$[*_{\alpha^{N-j}}]^{-1} \circ A_j \colon H^j \to H^{N-j}$$

are isomorphisms.

Example. One considers the de Rham complex

$$0 \to \Lambda^0(M) \to \Lambda^1(M) \to \Lambda^2(M) \to \cdots \to \Lambda^n(M) \to 0.$$

The dual complex has the form

$$0 \to \Lambda^0(M) \xrightarrow{(-1)^n d} \Lambda^1(M) \xrightarrow{(-1)^{n-1} d} \Lambda^2(M) \to \cdots \to \Lambda^n(M) \to 0.$$

For $A_j = (-1)^{\nu(n,j)}$, where $\nu(n,j) = [\frac{n}{2}] + [\frac{n+j}{2}]$, we obtain the classical Poincare duality theorem.

2.3. Dual Spencer Complexes

In this section we apply the results above to the stable Spencer complexes.

Let $r \subset \alpha_k$ be a formally integrable SDE. We shall start the construction of dual complexes with the naive Spencer complex:

$$0 \to C^\infty(\alpha) \xrightarrow{j_k} \mathcal{J}^k(\alpha) \xrightarrow{D} \mathcal{J}^{k-1}(\alpha) \otimes_{C^\infty(M)} \Lambda^1(M) \to \cdots$$
$$\to \mathcal{J}^{k-n}(\alpha) \otimes_{C^\infty(M)} \Lambda^n(M) \to 0 \qquad (5)$$

$$(\mathcal{J}^k(\alpha))^t = \mathrm{Hom}(\alpha_k, \Lambda^n\tau^*) = \mathrm{Diff}_k(\alpha, \Lambda^n\tau^*) = \mathrm{Diff}_k(\alpha, \mathbb{I}) \otimes_{C^\infty(M)} \Lambda^n(M)$$

is a module dual to the module of sections $\mathcal{J}^k(\alpha)$.

Therefore,

$$(\mathcal{J}^k(\alpha) \otimes_{C^\infty(M)} \Lambda^i(M))^t = \mathrm{Hom}(\alpha_k \otimes \Lambda^i\tau^*, \Lambda^n\tau^*) =$$

$$\mathrm{Hom}(\alpha_k, \mathrm{Hom}(\Lambda^i\tau^*, \Lambda^n\tau^*)) = \mathrm{Diff}(\alpha, \Lambda^{n-i}\tau^*) = \mathrm{Diff}_k(\alpha, \mathbb{I}) \otimes_{C^\infty(M)} \Lambda^{n-i}(M).$$

Thus the complex, dual to Spencer complex 2.3.5 has the form

$$\cdots \to \mathrm{Diff}_s(\alpha, \mathbb{I}) \otimes_{C^\infty(M)} \Lambda^i(M) \xrightarrow{D^t} \mathrm{Diff}_{s+1}(\alpha, \mathbb{I}) \otimes_{C^\infty(M)} \Lambda^{i+1}(M) \to \cdots \qquad (6)$$

and operator D^t dual to Spencer operator D acts as follows:

$$D^t(\nabla \otimes \omega) = (-1)^{n-i}(d \circ \nabla) \wedge \omega + (-1)^{n-i}\alpha^t_{s,s+1}\nabla \otimes d\omega, \qquad (7)$$

where $\omega \in \Lambda^i(M)$, $\nabla \in \mathrm{Diff}_s(\alpha, \mathbb{I})$, $\alpha^t_{s,s+1}: \mathrm{Diff}_s(\alpha, \mathbb{I}) \to \mathrm{Diff}_{s+1}(\alpha, \mathbb{I})$ is the natural embedding. Note that dual complex 2.3.6, as well as complex 2.3.5, is acyclic. To construct the complex dual to complex 1.7.1 we shall describe the module $\mathcal{R}_{(l)} = (\mathcal{R}^{(l)})^t$. To this end denote by $\mathrm{Ann}\mathcal{R}^{(l)}$ the annihilator of the $R^{(l)}$:

$$\mathrm{Ann}\mathcal{R}^{(l)} = \left\{\nabla \in \mathrm{Diff}_{k+l}(\alpha, \mathbb{I}) \mid \varphi_\nabla(R^{(l)}) = 0\right\}.$$

Then, $\mathcal{R}_{(l)} = \mathrm{Diff}_{k+l}(\alpha, \mathbb{I})/\mathrm{Ann}\mathcal{R}^{(l)}$. Therefore

$$(\mathcal{R}^{(l)} \otimes_{C^\infty(M)} \Lambda^i(M))^t = \mathcal{R}_{(l)} \otimes_{C^\infty(M)} \Lambda^{n-i}(M)$$

and, consequently , the complex dual to complex 1.7.1 has the form

$$\cdots \to \mathcal{R}_{(l)}) \otimes_{C^\infty(M)} \Lambda^i(M) \xrightarrow{D^t} \mathcal{R}_{(l+1)} \otimes_{C^\infty(M)} \Lambda^{i+1}(M) \to \cdots \tag{8}$$

The cohomologies of this complex stabillize for $l \geq l_0$. In fact, passing on to dual complexes and using formula 2.3.7 we obtain from diagram 1.7 the following commutative one:

$$\begin{array}{ccccccc}
 & & 0 & & 0 & & \\
 & & \uparrow & & \uparrow & & \\
\cdots & \leftarrow & C^\infty(g_{l+1}) \otimes \Lambda^{n-i}(M) & \xleftarrow{\delta^t} & C^\infty(g_l) \otimes \Lambda^{n-i-1}(M) & \leftarrow & \cdots \\
 & & \uparrow & & \uparrow & & \\
\cdots & \leftarrow & \mathcal{R}_{(l+1)} \otimes \Lambda^{n-i}(M) & \xleftarrow{D^t} & \mathcal{R}_{(l)} \otimes \Lambda^{n-i-1}(M) & \leftarrow & \cdots \\
 & & \uparrow{\scriptstyle \alpha^t_{l,l+1}\otimes 1} & & \uparrow{\scriptstyle \alpha^t_{l-1,l}\otimes 1} & & \\
\cdots & \leftarrow & \mathcal{R}_{(l)} \otimes \Lambda^{n-i}(M) & \xleftarrow{D^t} & \mathcal{R}_{(l-1)} \otimes \Lambda^{n-i-1}(M) & \leftarrow & \cdots \\
 & & \uparrow & & \uparrow & & \\
 & & 0 & & 0 & &
\end{array}$$

where bundles dual to bundles g^l is denoted by g_l: $g_l = (g^l)^t$.

The first row of the diagram above is a complex dual to Spencer δ-complex. Therefore, it is acyclic for $l \geq l_0$ (Poincare δ-lemma). Hence, one gets the stabilization of Spencer complex 2.3.8 cohomologies.

Denote by $\mathcal{H}_j(r)$ the corresponding stable cohomologies of the complex in the term $\mathcal{R}_{(l-j)} \otimes_{C^\infty(M)} \Lambda^j(M)$.

Note that the ellipticity of a Spencer complex is equivalent to the ellipticity of the corresponding system of differential equations r (Quillen theorem[1,2]).

One deduces from Theorem 2.2.1 the following theorem.

Theorem 2. *Let $r \subset \alpha_k$ be a formally integrable elliptic* SDE *over a compact oriented manifold M. Then*

$$\mathcal{H}^j(r) \rightleftarrows \mathcal{H}_{n-j}(r), \qquad j = 1, 2, \ldots, n.$$

Examples.

1. One considers a differential equation $r \subset J^1(\mathbb{I})$, defined by the operator of the exterior differentiation

$$d: C^\infty(M) \to \Lambda^1(M),$$

$$R_x = \{[f]_x^1 | \, d_x f = 0\}.$$

The equation itself and all its prolongations define a one-dimensional trivial linear bundle over manifold M while the stable Spencer complex for this equation is the de Rham complex. The dual complex is also de Rham complex. Thus theorem 2.3.2 is the classical Poincare duality theorem (cf. example in 2.2).

The following example is a natural continuation of the previous one.

2. Let α be a vector bundle, provide with a flat connection and let $\nabla: C^\infty(\alpha) \to C^\infty(\alpha) \otimes \Lambda^1(M)$ be the corresponding covariant differential. Consider a differential equation $r \subset \alpha_1$ (see ch.1)

$$R_x = \{[h]_x^1 | \nabla_x h = 0\}.$$

The equation itself and all its prolongations are isomorphic to the bundle α, while the stable Spencer complex of this equation is the complex

$$0 \to C^\infty(\alpha) \xrightarrow{\nabla} C^\infty(\alpha) \otimes \Lambda^1(M) \to \cdots \to C^\infty(\alpha) \otimes \Lambda^n(M) \to 0$$

constructed by means of connection ∇. Its cohomologies are denoted by $H^i(M, \alpha)$. Dual to this complex is one constructed by means of flat connection ∇^* on the conjugate bundle α^* (see example in 2.2). Therefore, by Theorem 2.3.2 there are isomorphisms $H^i(M, \alpha) \xrightarrow{\sim} H^{n-i}(M, \alpha^*)$.

3. Let $r \subset \alpha_k$ be a formally integrable SDE of a finite type and let l be such number that $r^{(l+1)} \simeq r^{(l)}$. Then the Spencer operator

$$\mathcal{R}^{(l)} \simeq \mathcal{R}^{(l+1)} \xrightarrow{D} \mathcal{R}^{(l)} \otimes \Lambda^1(M)$$

defines a flat connection in the bundle r^l. Thus the stable Spencer cohomologies are isomorphic to the cohomologies considered in the previous example and the corresponding duality has the form above.

4. Let M be a complex manifold and let α be a holomorphic bundle over M. Denote by $\Lambda^{p,q}(\alpha)$ differential (p,q)-forms on the manifold M with the values in the bundle α and by $\Omega^p(\alpha)$ holomorphic p-forms with the value in the bundle α.

The stable Spencer cohomologies of Cauchy-Riemann differential equation corresponding to the operator

$$\bar{\partial}: \Lambda^{p,0}(\alpha) \to \Lambda^{p,1}(\alpha)$$

are Dolbeault cohomologies $H^j(M, \Omega^p(\alpha))$. It is easy to check that the cohomologies of the dual complex are Dolbeault cohomologies $H^j(M, \Omega^{n-p}(\alpha^*))$ and Theorem 2.3.2 gives us Serre duality theorem.

Note that the more general duality theorems on the manifold provided with polarization can also be obtained from Theorem 2.3.2 by applying it to the corresponding differntial equation[8].

2.4. Dual equations

In this section we determine the conditions the fulfilment of which makes it possible to interpret the cohomologies of the dual complex as the cohomologies of some Spencer complex.

Theorem 3. (On three isomorphisms). *Let r and $\tilde{r}$ be formally integrable elliptic* SDE*'s, and*

$$F_j\colon \mathcal{R}_{(s_0+j)} \to \tilde{\mathcal{R}}^{(N-s_0-j)}, \qquad j = -1, 0, 1,$$

be isomorphisms such that the diagram

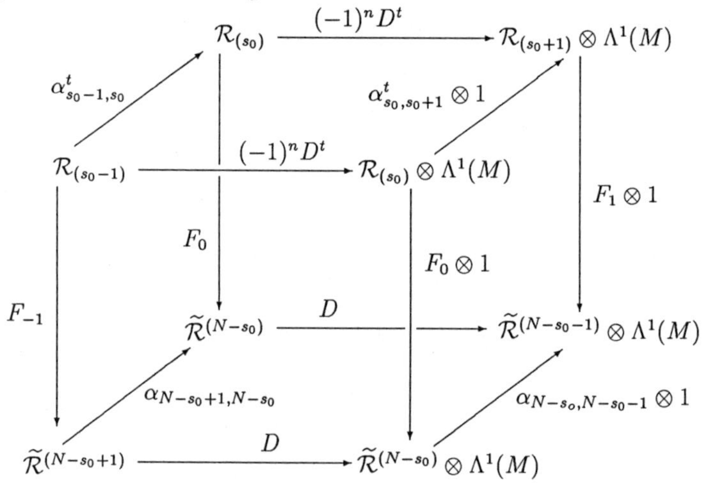

(9)

is commutative for some (stable) values of N and s_0.

Then

$$\mathcal{H}_l(r) \rightrightarrows \mathcal{H}^l(\tilde{r})$$

and, by Theorem 2.3.2

$$\mathcal{H}^l(r) \rightrightarrows \mathcal{H}^{n-l}(\tilde{r}).$$

Proof. To prove the Theorem it is sufficient to check that isomorphisms F_{-1}, F_0 and F_1 for every value of i define isomorphisms

$$F_{ji}: \mathcal{R}_{(s_0+j)} \otimes \Lambda^i(M) \to \widetilde{\mathcal{R}}^{(N-s_0-j)} \otimes \Lambda^i(M),$$

$$F_{ji} = F_j \otimes (-1)^{\nu(n,i)} id; \qquad j = -1, 0, 1, \qquad i = 0, 1, \ldots, n;$$

such that the diagrams

$$\begin{array}{ccccc} \mathcal{R}_{(s_0-1)} \otimes \Lambda^{i-1}(M) & \xrightarrow{D^t} & \mathcal{R}_{(s_0)} \otimes \Lambda^i(M) & \xrightarrow{D^t} & \mathcal{R}_{(s_0+1)} \otimes \Lambda^{i+1}(M) \\ \downarrow F_{-1\ i-1} & & \downarrow F_{0i} & & \downarrow F_{1\ i+1} \\ \widetilde{\mathcal{R}}^{(N-s_0+1)} \otimes \Lambda^{i-1}(M) & \xrightarrow{D^t} & \widetilde{\mathcal{R}}^{(N-s_0)} \otimes \Lambda^i(M) & \xrightarrow{D^t} & \widetilde{\mathcal{R}}^{(N-s_0-1)} \otimes \Lambda^{i+1}(M) \end{array} \tag{10}$$

are commutative.

Let us verify the commutativity of these diagrams.

Let $\theta \otimes \omega \in \mathcal{R}_{(s_0+j)} \otimes \Lambda^l(M)$. Then, using the example from section 2.2, we obtain the equalities

$$\begin{aligned} DF_{jl}(\theta \otimes \omega) &= (-1)^{\nu(n,l)} DF_j(\theta) \otimes \omega + \\ &\quad (-1)^{\nu(n,l)} \alpha_{N-s_0-j, N-s_0-j-1} F_j(\theta) \otimes d\omega, \end{aligned}$$

$$\begin{aligned} F_{j+1\ l+1} D^t(\theta \otimes \omega) &= (-1)^{\nu(n,l)} [(F_{j+1} \otimes 1)((-1)^n D^t\theta)] \wedge \omega + \\ &\quad (-1)^{\nu(n,l)} F_{j+1} \alpha^t_{s_0+j, s_0+j+1}(\theta) \otimes d\omega, \end{aligned}$$

for $j = -1, 0$.

Therefore, the commutativity of diagram 2.4.10 follows from relations

$$DF_j = (F_{j+1} \otimes 1)(-1)^n D^t,$$

$$\alpha_{N-s_0-j, N-s_0-1} \circ F_j = F_{j+1} \circ \alpha^t_{s_0+j, s_0+j+1},$$

$j = -1, 0$.

It remains to note that these relations follow from commutative diagram 2.4.9.

The conditions of Theorem 2.4.3 can be reformulated in another way. To this end we note that every isomorphism $F_j: \mathcal{R}_{(s_0+j)} \to \widetilde{\mathcal{R}}^{(N-s_0-j)}$ can be identified with a non-degenerate bilinear form

$$F_j: \mathcal{R}_{(s_0+j)} \otimes \widetilde{\mathcal{R}}_{(N-s_0-j)} \longrightarrow F,$$

$$(a, b) \longmapsto \langle F_j(a), b \rangle,$$

where $a \in \mathcal{R}_{(s_0+j)}$, $b \in \widetilde{\mathcal{R}}_{(N-s_0-j)}$.

The forms F_j are naturally extended to the forms

$$F_j\colon (\mathcal{R}_{(s_0+j)} \otimes \Lambda^l(M)) \otimes (\widetilde{\mathcal{R}}_{(N-s_0-j)} \otimes \Lambda^q(M)) \longrightarrow \Lambda^{l+q}(M),$$

$$(a \otimes v, b \otimes \omega) \longmapsto \langle F_j(a), b\rangle \otimes (v \wedge \omega),$$

where $v \in \Lambda^l(M)$, $\omega \in \Lambda^q(M)$.

Proposition 4. *The diagram*

$$\begin{array}{ccc} \mathcal{R}_{(s_0+j)} & \xrightarrow{(-1)^n D^t} & \mathcal{R}_{(s_0+j+1)} \otimes \Lambda^1(M) \\ \Big\downarrow F_j & & \Big\downarrow F_{j+1}\otimes 1 \\ \widetilde{\mathcal{R}}^{(N-s_0-j)} & \xrightarrow{D} & \widetilde{\mathcal{R}}^{(N-s_0-j-1)} \otimes \Lambda^1(M), \end{array} \qquad j=-1,0, \tag{11}$$

is commutative iff the equality

$$F_{j+1}((-1)^n D^t a, b) + (-1)^{n+1} F_j(a, (-1)^n D^t b) = dF_j(a, \alpha^t_{N-s_0-j-1,N-s_0-j} b)$$

is fulfilled for all $a \in \mathcal{R}_{(s_0+j)}$, $b \in \widetilde{\mathcal{R}}_{(N-s_0-j-1)}$.

Proof. From the commutativity of diagram 2.4.11 we have the identity

$$\langle (F_{j+1} \otimes 1)((-1)^n D^t a), b\rangle = \langle D(F_j a), b\rangle$$

On the other hand, Green's formula for operator D has the form

$$\langle D(\bar{a}), b \otimes v\rangle - \langle \bar{a}, D^t(b \otimes v)\rangle = d[\langle \bar{a}, \alpha^t_{N-s_0-j-1,N-s_0-j} b\rangle v],$$

where $v \in \Lambda^{n-1}(M)$, $\bar{a} \in \mathcal{R}^{(s_0+j)}$.

Using formula 2.3.7 rewrite Green's formula in the form

$$\langle D(\bar{a}), b\rangle - \langle \bar{a}, D^t(b)\rangle = d\langle \bar{a}, \alpha^t_{N-s_0-j-1,N-s_0-j} b\rangle.$$

Therefore, the equalities

$$\begin{aligned} \langle D(F_j a), b\rangle &= \langle F_j a, D^t b\rangle + d\langle F_j a, \alpha^t_{N-s_0-j-1,N-s_0-j} b\rangle \\ &= (-1)^n F_j(a, (-1)^n D^t b) + dF_j(a, \alpha^t_{N-s_0-j-1,N-s_0-j} b) \end{aligned}$$

hold.

We shall call the system of differential equations *autodual*, if in Theorem 2.4.3 the system $\widetilde{r}$ coincides with system r.

The autoduality conditions can be reformulated as the condition of the existence of three non-degenerate bilinear forms on equation r satisfying the conditions of Proposition 2.4.4.

Thus, for Lie equation r of a finite type, whose fibers $r_x^{(l)}, x \in M$ are semi-simple algebras, such bilinear forms can be constructed by means of Killing's form. In this case Theorem 2.4.3 leads to the duality theorems obtained Goldshmidt in[9].

Note also that for the Cauchy-Riemann equation r in the bundle $\Lambda^{p,0}(\alpha)$ it is possible to choose the Cauchy-Riemann equation in the bundle $\Lambda^{n-p,0}(\alpha^*)$ as the equation $\widetilde{r}$. The corresponding bilinear form is defined by a natural pairing. This gives an alternative proof Serre's duality theorem.

3. Spectral sequences

In this section we expose the construction of Spencer cohomology spectral sequences analogous to the classical Leray-Serre sequence for a fibre bundle.

3.1. D- and δ-Spencer cohomologies.

Let $r \subset \alpha_k$ be a formally integrable system over M.

We use the notations $R^\infty = \lim_{l\to\infty} R^{(l)}$ and $\mathcal{K}^{(l)} = \operatorname{Ker} \alpha_{\infty,l}|_{R^\infty}$.

Introduce a filtration in the first Spencer complex:

$$F_p(\mathcal{R}^{(\infty)} \otimes \Lambda^{p-q}(M)) \quad = \quad \mathcal{K}^{(k+q)} \otimes \Lambda^{p-q}(M), \qquad q \geq -1. \tag{1}$$

Spencer operator D preserves this filtration.

Theorem.[6] *Let $r \subset \alpha_k$ be a formally integrable normal system over a smooth manifold M. Then filtration 3.1.1 determines a spectral sequences converged to Spencer cohomologies of system r. The first term of this sequences has the form:*

$$E_1^{p,q} \quad = \quad H^{q,p-q}(g), \qquad\qquad q \geq 0$$

and $\{E_1^{p,-1}, d_1^{p,-1}\}$ is the following complex:

$$0 \longrightarrow \mathcal{J}^{k-1}(\alpha) \xrightarrow{d_1} \frac{\mathcal{J}^{k-1}(\alpha) \otimes \Lambda^1(M)}{\delta(C^\infty(g))} \longrightarrow \cdots \xrightarrow{d_1^{p-1}} \frac{\mathcal{J}^{k-1}(\alpha) \otimes \Lambda^n(M)}{\delta(C^\infty(g) \otimes \Lambda^{n-1}(M))} \longrightarrow 0 \tag{2}$$

If r is involutive then the term $(E_1^{p,q}, d_1^{p,q})$ coincides with complex 3.1.2 and the spectral sequence degenerates on the second step.

3.2. Systems over fibre bundles.

Let $r \subset \alpha_k$ be a formally integrable normall system over a smooth manifold M. We suppose that M is a total space of a fibre bundle

$$\kappa: M \to B,$$

over a smooth manifold B. We call fibre bundle κ *noncharacteristic* with respect to a system r, if all fibres $F_b = \kappa^{-1}(b), b \in B$ are nonchracteristic.

Keeping in mind the Spencer complex filtration we call a vector field on M vertical, if

$$\kappa_*(X) = 0.$$

Module $\mathcal{D}_\kappa(M)$ of vertical fields is a Lie algebra. A differential i-form $\omega \in \Lambda^i(M)$ is called *q-horizontal* if $X_1 \wedge \ldots \wedge X_{q+1} \lrcorner \omega = 0$ for any $X_1, \ldots, X_{q+1} \in \mathcal{D}_\kappa(M)$. A submodule of q-horizontal forms is denoted by $\Lambda_q^\bullet(M)$.

Let

$$F_{p,q} = F_p(\mathcal{R}^{(l)} \otimes \Lambda^{p+q}(M)) = \mathcal{R}^{(l)} \otimes \Lambda_q^{p+q}(M). \tag{1}$$

Proposition 1. *Modules 3.2.1 define a filtration in Spencer complexes 1.7.1.*

Describe initial terms of the Spencer cohomology spectral sequence associated with the filtration.

Proposition 2. *There exists an isomorphism of $C^\infty(B)$-modules*

$$\Lambda_q^i(M)/\Lambda_{q-1}^i(M) = \Lambda^q(\kappa) \underset{C^\infty(B)}{\otimes} \Lambda^{i-q}(B),$$

$$E_0^{p\,q} = F_{p,q}/F_{p+1,q+1} \simeq \mathcal{R}^l \underset{C^\infty(M)}{\otimes} \Lambda^q(\kappa) \underset{C^\infty(B)}{\otimes} \Lambda^p(B),$$

where $\Lambda^p(\kappa) = \Lambda^p(M)/\kappa^*\Lambda^p(B)$ be differential forms on fibers of bundle κ.

Proposition 3. *The differentials*

$$d_0^{p\,q} \colon E_0^{p\,q} \to E_0^{p\,q+1}$$

are first order fiber-wise differential operators preserving $C^\infty(B)$-module structure:

$$d_0^{p\,q}(fv) = f d_0^{p\,q}(v),$$

$f \in C^\infty(B), \quad v \in E_0^{p\,q}$.

Proof. By the definition of the Spencer operator D we have

$$D(fv) = fDv + df \wedge \alpha_{l+1,l}v.$$

For $df \in \Lambda_0^1(M)$ it follows that

$$df \wedge \alpha_{l+1,l}v \in \mathcal{R}^{l-1} \otimes \Lambda_q^{p+q+1} \subset F_{p+1,q}$$

Let bundle κ be noncharacteristic and

$$\alpha_b\colon E(\alpha_b) \to F_b$$

be the restriction of α on to the fiber F_b. Consider a restriction

$$h_l(b)\colon R_b^l \to (\alpha_b)_{l+k}.$$

The image of $h_l(b)$ is denoted by r_b^l.

Proposition 4. *If κ is a noncharacteristic bundle with respect to r, then $h_l(b)$ is a monomorphism.*

Proof. By formal integrability and normality r, one obtains the result from the following commutative diagram

$$\begin{array}{ccccccccc} 0 & \longrightarrow & g^l & \longrightarrow & r^l & \longrightarrow & r^{l-1} & \longrightarrow & 0 \\ & & \downarrow{\scriptstyle h_l(b)} & & \downarrow{\scriptstyle h_l(b)} & & \downarrow{\scriptstyle h_{l-1}(b)} & & \\ 0 & \longrightarrow & \alpha_b \otimes S^{l+k}\tau^*(B) & \longrightarrow & (\alpha_b)_{l+k} & \longrightarrow & (\alpha_b)_{l+k-1} & \longrightarrow & 0. \end{array}$$

Proposition 5. $(r_b^l)^{(1)} \supset r_b^{l+1}$.

Proof. Let $a \in C^\infty(\alpha)$, $a_b = a|_{F_b}$ and

$$[a_b]_y^{l+k} \in R_b^l, \qquad [a]_y^{l+k} \in R^l.$$

Then the section $j_{l+k}(a)$ is tangent to R^l at the point $[a]_y^{l+k}$ and, consequently, the section $j_{l+k}(a_b)$ is tangent to R_b^l at the point $[a_b]_y^{l+k}$. Therefore $[a_b]_y^{l+k+1} \in (R_b^l)^{(1)}$.

Theorem 1. *Let κ be a noncharacteristic bundle with respect to a formally integrable normal system $r \subset \alpha_k$. Then there exists an integer $l_0 \geq 0$, such that*

$$(r_b^l)^{(1)} = r_b^{l+1}$$

and the image system

$$r_b^{l_0}\colon R_b^{l_0} \to F_b$$

is formally integrable.

Proof. The result follows from the Cartan-Kuranishi theorem[3] and from proposition 5.

Consider a fiber bundle

$$r_\kappa^{l_0}\colon R_\kappa^{l_0} \to M$$

with the fiber $R_{b,y}^{l_0}$ over $y \in F_b$.

There exists a "fiberwise" Spencer complex

$$0 \longrightarrow \mathcal{R}^l_\kappa \xrightarrow{D_\kappa} \mathcal{R}^{l-1}_\kappa \otimes \Lambda^1(\kappa) \xrightarrow{D_\kappa} \cdots \longrightarrow 0. \tag{2}$$

Stable cohomologies of this complex we denote by $\mathcal{H}^\bullet_\kappa(r)$.

From the naturallity of D we obtain the following commutative diagram

$$\begin{array}{ccc} \mathcal{R}^l \otimes \Lambda^q(\kappa) \otimes \Lambda^p(B) & \xrightarrow{d_0^{p\,q}} & \mathcal{R}^{l-1} \otimes \Lambda^{q+1}(\kappa) \otimes \Lambda^p(B) \\ \downarrow{\scriptstyle h_l(\kappa)\otimes 1} & & \downarrow{\scriptstyle h_l(\kappa)\otimes 1} \\ \mathcal{R}^l_\kappa \otimes \Lambda^q(\kappa) \otimes \Lambda^p(B) & \xrightarrow{D_\kappa \otimes 1} & \mathcal{R}^{l+1}_\kappa \otimes \Lambda^{q+1}(\kappa) \otimes \Lambda^p(B) \end{array}$$

Therefore, $E_1^{p\,q}$ can be represented as

$$E_1^{p\,q} = \mathcal{H}^q_\kappa(r) \otimes \Lambda^p(B).$$

Proposition 6. *The differential*

$$d_1^{p\,q} : E_1^{p\,q} \to E_1^{p+1\,q}$$

is defined by a flat connection in the module $\mathcal{H}^\bullet_\kappa(r)$ *:*

$$d_1^{0\,q} : \mathcal{H}^q_\kappa(r) \to \mathcal{H}^q_\kappa(r) \otimes \Lambda^1(B).$$

Putting all this together, we get the following result.

Theorem 2. *Let*

$$\kappa : M \to B$$

be a smooth vector bundle, nonchracteristic with respect to a normal formally integrable system

$$r \subset \alpha_k.$$

Then filtration 3.2.1 determines spectral sequence $(E_r^{p\,q}, d_r^{p\,q})$ *converging to the Spencer cohomologies of* r *and*

1^0. *The terms* $E_1^{p\,q}$ *are isomorphic to*

$$\mathcal{H}^q_\kappa(r) \otimes \Lambda^p(B).$$

2^0. *Differentials*

$$d_1^{0\,q} : \mathcal{H}^q_\kappa(r) \to \mathcal{H}^q_\kappa(r) \otimes \Lambda^1(B)$$

determine a flat connection in the module $\mathcal{H}^\bullet_\kappa(r)$.

3^0. *The terms $E_2^{p\,q}$ are isomorphic to the de Rham cohomologies of manifold B with coeffiticients in the module $\mathcal{H}_\kappa^\bullet(r)$ equipped with the natural flat connection $d_1^{p\,q}$:*

$$E_2^{p\,q} \;=\; H^p(B, \mathcal{H}_\kappa^q(r)).$$

Remarks.

1. Denote by $(\widehat{E}_r^{p\,q}, \widehat{d}_r^{p\,q})$ the Leray-Serre spectral sequence for the de Rham cohomologies of the bundle κ. Then the sequence $(E_r^{p\,q}, d_r^{p\,q})$ is a graded module over $(\widehat{E}_r^{p\,q}, \widehat{d}_r^{p\,q})$.

2. If κ is an elliptic noncharacteristic bundle[10] then Spencer cohomologies are finite-dimensional.

4. Symmetries

We use spectral sequence 3.2 for computation Spencer cogomologies of a SDE with compact connected symmetry group. The crucial point of our computations lies in the term $E_2^{p\,q}$ which, in the case under consideration, reduces to Spencer cohomologies of an invariant SDE over a homogeneous space. Therefore below we accentuate on the systems of such a type.

4.1. Definitions.

Consider a Lie group G and its representation ρ by automorphisms of a fiber bundle $\alpha: E(\alpha) \to M$,

$$\rho: G \to \mathrm{Aut}(\alpha).$$

This representation covers the action of G on M in such a way that the following diagram

$$\begin{array}{ccc} E(\alpha) & \xrightarrow{\rho(g)} & E(\alpha) \\ {\scriptstyle\alpha}\downarrow & & {\scriptstyle\alpha}\downarrow \\ M & \xrightarrow{g} & M \end{array}$$

is commutative for all $g \in G$.

Automorphisms $\rho(g)$ induce automorphisms $\rho^*(G)$ of the module $C^\infty(\alpha)$ over algebra automorphisms

$$(g^{-1})^*: C^\infty(M) \to C^\infty(M).$$

Namely,

$$\rho^*(g)(h) \;=\; \rho(h) \circ h \circ g^{-1},$$
$$\rho^*(g)(fh) \;=\; (g^{-1})^*(f)\rho^*(g)h$$

for any function $f \in C^\infty(M)$ and any section $h \in C^\infty(\alpha)$.

The representation

$$\rho: G \to \mathrm{Aut}(\alpha)$$

is lifted up to a representation

$$\rho^{(k)}: G \to \mathrm{Aut}(\alpha_k)$$

in such a way that

$$\rho^{(k)}(g)([h]_x^k) = [\rho^*(g)(h)]_{g(x)}^k, \qquad h \in C^\infty(\alpha),\ g \in G,\ x \in M.$$

Representations $\rho^{(k)},\ k = 0, 1, \ldots$ commute with projections $\alpha_{k,l}$:

$$\alpha_{k,l} \circ \rho^{(k)}(g) = \rho^{(l)}(g) \circ \alpha_{k,l}$$

for all $g \in G,\ k \geq l$.

We call a representation

$$\rho: G \to \mathrm{Aut}(\alpha)$$

a *symmetry of a linear system* $r \subset \alpha_k$, if

$$\rho^{(k)}(g)(r) = r$$

for all elements $g \in G$.

4.2. Symmetries and Spencer cohomologies.

Let G be a compact connected symmetry group of a system $r \subset \alpha_k$.
Then the representations

$$\rho^{(k)}: G \to \mathrm{Aut}(\alpha_k)$$

induce representations

$$\rho^{(k)}: G \to \mathrm{Aut}(C^i)$$

for all $i = 0, 1, \ldots, n$.

In what follows we consider all the modules and fibre bundles to be complexified.

By this reason any $C^\infty(M)$-module C^k is decomposed in isotypic components of representation $\rho^{(k)}$:

$$C^k = \bigoplus_{\mu \in \mathrm{Ch}(G)} C_\mu^k,$$

where $\mathrm{Ch}(G)$ is the group of characters for group G.

Proposition 1. *The Spencer operator*

$$D: C^k \to C^{k+1}$$

is decomposed into the direct sum of its restrictions

$$D_\mu = D|_{C^k_\mu}: C^k_\mu \to C^{k+1}_\mu,$$

$\mu \in \mathrm{Ch}(G)$.

Proof. Let $v \in C^k_\mu$ be the dominant vector associated with the irreducible representation of character μ, T be maximal torus of group G, B be Borel's subgroup of group G.

To prove of the proposition it is sufficient to check that Dv is the dominant vector in module C^{k+1}_μ.

One can determine v by the conditions

$$\rho^{(k)}(t)v = \mu(t)v,$$

$$\rho^{(k)}(b)v = v,$$

for all $t \in T,\ b \in B$.
Since D is a natural operator (see 1.6), we have

$$\rho^{(k+1)}(t)Dv = \mu(t)Dv$$

$$\rho^{(k+1)}(b)Dv = Dv$$

for all $t \in T,\ b \in B$. ∎

Hence, for every μ one has a complex

$$0 \longrightarrow C^0_\mu \xrightarrow{D_\mu} C^1_\mu \xrightarrow{D_\mu} \cdots \xrightarrow{D_\mu} C^n_\mu \longrightarrow 0 \qquad (1_\mu)$$

Denote by $\mathcal{H}^q_\mu$, $q = 0, 1, \ldots, n$ corresponding cohomology groups.
From proposition 4.2.1 one has

Proposition 2. *The second Spencer complex is the direct sum of complexes 4.2.1$_\mu$ and Spencer cohomology $\mathcal{H}^q(R)$ is the direct sum of cohomologies $\mathcal{H}^q_\mu$*

$$\mathcal{H}^\bullet(r) = \bigoplus_{\mu \in \mathrm{Ch}(G)} \mathcal{H}^\bullet_\mu.$$

We call complex 4.2.1$_e$ associated with the trivial character e invariant Spencer complex; its cohomologies are called invariant cohomologies and are denoted by $\mathcal{H}^\bullet_\mu$.

Corollary. *Let* $\mathcal{H}^q_\mu = 0$ *for all nontrivial characters* $\mu \in \mathrm{Ch}(G)$.
Then $\mathcal{H}^q(r) \simeq \mathcal{H}^q_e, \qquad q = 0, 1, \ldots, n.$

4.3. Differential equations over homogeneous space.

Let α be a vector bundle over a homogeneous space G/G_0, where G is a compact connected Lie group, G_0 is a closed connected subgroup of G, and $R \subset E(\alpha_k)$ be a G-invariant formally integrable system.
One has a bundle

$$\widehat{\alpha} = \kappa^*(\alpha): E(\widehat{\alpha}) \to G$$

induced by the natural projection

$$\kappa: G \to G/G_0.$$

Then the module $C^\infty(\alpha)$ is identified with a submodule of module $C^\infty(\widehat{\alpha})$ with the trivial G_0-action. Conversely, G-fibre bundle $\beta: E(\beta) \to G$ is induced from some fibre bundle $\alpha: E(\alpha) \to G/G_0$ if the action of the stability subgroup G_0 on fibre bundle β is trivial.

The following result is a consequence of the isotypic decomposition.

Proposition 1. *For any G-fibre bundle* $\alpha: E(\alpha) \to G/G_0$ *G-module* $C^\infty(\alpha)$ *is decomposed into the direct sum of isotypic components* $C^\infty(\alpha)_\mu$, *where* μ *are elements of character group* $\mathrm{Ch}(G, G_0)$ *consisting of G-characters trivial on Cartan's subgroup of* G_0.

Using this proposition for differential module $C^\bullet$ we get the following decomposition

$$C^k = \bigoplus_{\mu \in \mathrm{Ch}(G,G_0)} C^k_\mu.$$

Remark. The following result was pointed out by D.P.Jelobenko. Let $\mathcal{G}$ be a reductive Lie algebra over $\mathbb{C}$, $\dim \mathcal{G} < \infty$, $\mathcal{G}_0$ be a regular imbeded subalgebra $\mathcal{G}$, $\mathcal{G} = \mathcal{G}_0 \oplus \mathcal{G}_1$, where $\mathcal{G}_1$ is the $\mathcal{G}_0$-module, and K be a lattice generated by the weights (roots) of module $\mathcal{G}_1$.

Theorem. *A representation* π_λ *of* $\mathcal{G}$ *with the dominant weight* λ $(\dim \pi_\lambda < \infty)$ *contains representation* π_μ *of a subalgebra* $\mathcal{G}_0$ *if and only if*

$$\lambda|_{\mathcal{H}_0} - \mu \in K|_{\mathcal{H}_0}, \tag{1}$$

where $\mathcal{H}_0$ *is Cartan's subalgebra of the algebra* $\mathcal{G}_0$.

Corollary. *Condition 4.3.1 for the representation π_0 of subalgebra $\mathcal{G}_0$ ($\mu = 0$) is*

$$\lambda|_{\mathcal{H}_0} \in K|_{\mathcal{H}_0}.$$

Returning to complexes $4.2.1_\mu$, $\mu \in \mathrm{Ch}(G, G_0)$ we consider operators

$$\partial_\mu : C_e^k \to C_e^{k+1}$$

which are defined by the following commutative diagram:

$$\begin{array}{ccc} C_\mu^k & \xrightarrow{D_\mu} & C_\mu^{k-1} \\ {\scriptstyle \mu^{-1}}\downarrow & & \downarrow{\scriptstyle \mu^{-1}} \\ C_e^k & \xrightarrow{\partial_\mu} & C_e^{k-1} \end{array}$$

Proposition 2. *Operator ∂_μ has the form*

$$\partial_\mu = D_e + \sigma_\mu,$$

where σ_μ is the symbol of D on infinitesimal character $\mu^* = d(\ln \mu) = \mu^{-1} d\mu$.

Proof. Let $\mu \in \mathrm{Ch}(G, G_0)$, $f \in C^\infty(G)_\mu$ and $\omega \in C_e^k$. Then, obviously,

$$D_\mu(f\omega) = d_\mu f \cdot \omega + f \cdot D_e \omega,$$

where $d_\mu : \Lambda_\mu^k(\tau_G^*) \to \Lambda_\mu^{k+1}(\tau_G^*)$ is the isotypic μ-component of de Rham's operator.
Choosing $f = \mu \in \mathrm{Ch}(G, G_0)$ we get

$$[D_\mu(\mu\omega)] = d_\mu \mu \cdot \omega + \mu D_e \omega = \mu[D_e + \sigma_\mu]\omega$$

for all $\omega \in C_e^\bullet$. ■

Proposition 3. *The cochain map μ^{-1}*

$$\begin{array}{ccccccccccc} 0 & \longrightarrow & C_\mu^0 & \xrightarrow{D_\mu} & C_\mu^1 & \xrightarrow{D_\mu} & \cdots & \xrightarrow{D_\mu} & C_\mu^n & \longrightarrow & 0 \\ & & {\scriptstyle \mu^{-1}}\downarrow & & \downarrow{\scriptstyle \mu^{-1}} & & & & \downarrow{\scriptstyle \mu^{-1}} & & \\ 0 & \longrightarrow & C_e^0 & \xrightarrow{\partial_\mu} & C_e^1 & \xrightarrow{\partial_\mu} & \cdots & \xrightarrow{\partial_\mu} & C_e^n & \longrightarrow & 0 \end{array}$$

defines isomorphism of complexes.

The proof is similar to the previous one.
It should be noted that

$$D_e^2 = 0, \qquad \sigma_\mu^2 = 0, \qquad \partial_\mu^2 = 0,$$

and therefore operators σ_μ and D_e anticommute:

$$D_e \circ \sigma_\mu + \sigma_\mu \circ D_e \;=\; 0.$$

4.4. Theorem of coincidence.

As it was proved above, Spencer cohomologies coincide with invariant ones if differential modules $(C_e^\bullet, \partial_\mu)$, where $\partial_\mu = \sigma_0 + \sigma_{\mu^*}$ are acyclic for any nontrivial infinitesimal character $\mu^* \in \mathrm{Ch}(\mathcal{G}, \mathcal{G}_0)$. Here

$$\mathcal{G} = \mathrm{Lie}(G), \qquad \mathcal{G}_0 = \mathrm{Lie}(G_0), \qquad \sigma_0 = D_e, \qquad \mu^* = \frac{d\mu}{\mu}.$$

Fundamental infinitesimal characters $\nu_1, \ldots, \nu_r$ generate space of all infinitesimal characters over $\mathbb{C}$.

Use remark 4.3 and represent any symbol in the form

$$\sigma_{\mu^*} = \sum_{i=1}^{r} k_i \sigma_i, \qquad k_i \in \mathbb{Z},\ \sigma_i = \sigma_{\nu_i}.$$

Therefore, by the results from the Appendix combined with Quillen's theorem[2] we obtain

Theorem. *Let r be a G-invariant formally integrable SDE over a homogeneous space G/G_0, where G is a compact connected Lie group, G_0 is a regular imbedded, connected subgroup of group G. Then Spencer cohomologies of R coincide with invariant ones up to the order m*

$$\mathcal{H}^q(r) \;=\; \mathcal{H}_e^q, \qquad q \le m,$$

if nontrivial characters $\mu^* \in \mathrm{Ch}(\mathcal{G}, \mathcal{G}_0)$ are noncharacteristic over $\mathbb{C}$ and the following conditions hold:

A. *Operators*

$$\Phi_i \circ \sigma_k \colon \mathcal{H}_k^q(C_e/\sigma_i C_e) \to \mathcal{H}_k^q(C_e/\sigma_i C_e),$$

where $k = (1, k_1, \ldots, k_{i-1}, 0, k_{i+1}, \ldots, k_r)$, $k_j \in \mathbb{Z}$, $i = 1, 2, \ldots, r$, $q \le m$ have no nonzero integer eigenvalues.

B. *Differential modules*

$$(\mathcal{H}_k^\bullet, \sigma_i),\ k = (1, k_1, \ldots, k_{i-1}, 0, k_{i+1}, \ldots, k_r),\ k_j \in \mathbb{Z},\ i = 1, 2, \ldots, r$$

are $(m+1)$-acyclic.

C. *Differential modules*

$$(\mathcal{H}_e^\bullet, \sigma_t),\ t = (0, t_1, \ldots, t_r),\ t_j \in \mathbb{Z},\ t \ne 0$$

are $(m+1)$*-acyclic.*

D. *Differential module (see Th. 4 below)*

$$(E \otimes \mathcal{H}_e^\bullet, \sigma)$$

is $(m+1)$*-acyclic.*

Appendix. Degenerated bicomplexes

We discuss here methods of cohomology computations for complexes equipped with a set of anticommuting differentials.

Let

$$P^\bullet = \bigoplus_{s=0}^{n} P^s$$

be a graded module. We suppose that module $P^\bullet$ is equipped with the set of anticommuting operators $\sigma_0, \sigma_1, \ldots, \sigma_r$ of the degree $+1$. Therefore operators

$$\sigma_k = \sum_{i=0}^{r} k_i \sigma_i$$

$$k = (k_0, k_1, \ldots, k_r), \quad k_i \in \mathbb{R}, \quad i = 0, 1, \ldots, r$$

define cohomological complexes. We denote corresponding cohomologies by $H_k^\bullet(P)$.

We find out triviallity conditions for the cohomologies $H_k^\bullet(P)$, $k = (1, k_1, \ldots, k_r)$, $k_i \in \mathbb{R}$, $k \neq 1_0$, where $1_0 = (1, 0, \ldots, 0)$.

For this purpose we consider a submodule $\sigma_i P^\bullet$ of $P^\bullet$:

$$\sigma_i P^\bullet = \bigoplus_{s=1}^{n} \sigma_i P^{s-1},$$

where $i = 1, \ldots, r$.

We suppose cohomologies $H_k^q(P)$ to be trivial for $q \leq m$ and $k = (0, k_1, \ldots, k_r)$, $k \neq 0$.

Then natural mappings

$$\begin{aligned} &\Phi_i \colon \sigma_i P^q \to P^q / \sigma_i P^{q-1}, \\ &\Phi_i \colon \sigma_i(x) \mapsto x \bmod \operatorname{im} \sigma_i, \qquad x \in P^q \end{aligned} \tag{1}$$

are isomorphisms for $i = 1, \ldots, r$ and $q \leq m$.

From the exact differential modules sequence

$$0 \to (\sigma_i P^\bullet, \sigma_k) \to (P^\bullet, \sigma_k) \to (P^\bullet / \sigma_i P^{\bullet - 1}, \sigma_k) \to 0 \tag{2}$$

where $k = (k_0, k_1, \dots, k_{i-1}, 1, k_{i+1}, \dots, k_r)$, $k_j \in \mathbb{R}$, $j = 0, 1, \dots, r$, $k_0 \neq 0$, and from isomorphisms (1) we obtain an exact cohomology sequence

$$\begin{aligned}
&0 \longrightarrow H^0_k(P) \longrightarrow H^0_{k-1_i}(P/\sigma_i P) \xrightarrow{\delta_i} H^0_{k-1_i}(P/\sigma_i P) \longrightarrow \cdots \\
&\longrightarrow H^1_k(P) \longrightarrow \cdots \xrightarrow{\delta_i} H^{m-1}_{k-1_i}(P/\sigma_i P) \longrightarrow H^m_k(P) \longrightarrow \qquad (3)\\
&\longrightarrow H^m_{k-1_i}(P/\sigma_i P) \xrightarrow{\delta_i} H^m_{k-1_i}(P/\sigma_i P) \longrightarrow \cdots,
\end{aligned}$$

where $1_i = (0, 0, \dots, 0, \underset{i}{1}, 0, \dots, 0)$.

The boundary operator δ_i in this sequence is

$$1 + \Phi_i \circ \sigma_{k-1_i}.$$

It should be noted that

$$H^0_{k-1_i}(P/\sigma_i P) = \operatorname{Ker}(\sigma_i \circ \sigma_{k-1_i}),$$

when $n > 1$.
If $m = n$, then

$$\begin{aligned}
H^{n-1}_{k-1_i}(P/\sigma_i P) &= P^{n-1}/\sigma_i P^{n-2} \\
H^n_{k-1_i}(P/\sigma_i P) &= 0.
\end{aligned}$$

The corollary of the exactness for (3) is

Proposition 1. *Cohomologies $H^q_k(P)$, where $k = (k_0, k_1, \dots, k_{i-1}, 1, k_{i+1}, \dots, k_r)$, $k_j \in \mathbb{R}$, $j = 0, 1, \dots, r$, $k_0 \neq 0$, and $q \leq m$, are trivial, if operators*

$$\delta_i\colon H^q_{k-1_i}(P/\sigma_i P) \to H^q_{k-1_i}(P/\sigma_i P)$$

are isomorphisms.

Remark. Condition of proposition 1 means that operators

$$\Phi_i \circ \sigma_{k-1_i}\colon H^q_{k-1_i}(P/\sigma_i P) \to H^q_{k-1_i}(P/\sigma_i P), \qquad q \leq m,$$

have no eigenvalues equal to -1.

Theorem 1. *Let*

$$P^\bullet = \bigoplus_{s=0}^{n} P^s$$

be a graded module equipped with a set of $r+1$ anticommuting operators $\sigma_0, \sigma_1, \dots, \sigma_r$ of the degree $+1$. Cohomologies $H^q_k(P)$ of P with the differential $\sigma_k = \sum_{j=1}^r k_j \sigma_j + \sigma_0$, $k_j \in \mathbb{Z}$, $k = (1, k_1, \dots, k_r)$ are trivial for $k \neq 1_0$ and $q \leq m$, if:

1^0. *Cohomologies $H_t^q(P)$ are trivial for $q \leq m$ and $t = (0, t_1, \ldots, t_r)$, $t_j \in \mathbb{Z}$, $t \neq 0$.*

2^0. *Operators*

$$\Phi_i \circ \delta_k : H_k^q(P/\sigma_i P) \to H_k^q(P/\sigma_i P)$$

for $q \leq m$ and $k = (1, k_1, \ldots, k_{i-1}, 0, k_{i+1}, \ldots, k_r)$, $k_j \in \mathbb{Z}$ for $i = 1, \ldots, r$ have no nonzero integer eigenvalues.

Proof. Consider module P with the differential σ_k, $k = (1, k_1, \ldots, k_r)$, $k \neq 1_0$, and choose a number i, $1 \leq i \leq r$ such that $k_i \neq 0$. Denote by k/k_i a vector $(1/k_i, k_1/k_i, \ldots, k_r/k_i)$.

The next equality is obvious:

$$H_k^\bullet(P) = H_{k/k_i}^\bullet(P)$$

From the proposition 1 applied to cohomologies $H_{k/k_i}^q(P)$ we obtain that they are trivial for $q \leq m$, if operators

$$\delta_i : H_{k/k_i - 1_i}^q(P/\sigma_i P) \to H_{k/k_i - 1_i}^q(P/\sigma_i P)$$

are isomorphisms.
In this case the boundary operators δ_i are equal to $1 + \Phi_i \circ \sigma_{k/k_i - 1_i}$. Using the equality $\sigma_{k/k_i - 1_i} = \frac{1}{k_i} \cdot \sigma_{k-1_i k_i}$ this operators may be expressed in the form

$$1 + \frac{1}{k_i} \Phi \circ \sigma_{k-1_i k_i}.$$

Therefore, the boundary operators δ_i, $i = 1, \ldots, r$, are isomorphisms iff the mappings

$$\Phi_i \circ \sigma_{k-1_i k_i} : H_{k-1_i k_i}^q(P/\sigma_i P) \to H_{k-1_i k_i}^q(P/\sigma_i P), \qquad i = 1, 2, \ldots, r$$

have no nonzero integer eigenvalues. ∎

Keeping in view pure homological conditions for triviallity of $H_k^q(P)$, $q \leq m$, $k = (1, k_1, \ldots, k_r)$, $k_i \in \mathbb{Z}$, $k \neq 1_0$, we consider a differential module $(H_k^\bullet(P), \sigma_i)$, $k = (1, k_1, \ldots, k_{i-1}, 0, k_{i+1}, \ldots, k_r)$, $k_j \in \mathbb{Z}$, $i \neq 0$.

Proposition 2. *If the modules $H_t^q(P)$ are trivial for $q \leq m$ and $t = (0, t_1, \ldots, t_r)$, $t_j \in \mathbb{Z}$, $t \neq 0$, and the modules $(H_k(P), \sigma_i)$, $k = (1, k_1, \ldots, k_{i-1}, 0, k_{i+1}, \ldots, k_r)$, $k_j \in \mathbb{Z}$ are $(m+1)$-acyclic, then*

$$H_k^q(P/\sigma_i P) \simeq H_k^q(P)/\sigma_i H_k^{q-1}(P)$$

for $q \leq m$, $i \neq 0$ and $k = (1, k_1, \ldots, k_{i-1}, 0, k_{i+1}, \ldots, k_r)$, $k_j \in \mathbb{Z}$.

Since differentials σ_i and σ_k anticommute, the operator σ_i, $i \neq 0$, defines a mapping from $H_k^q(P/\sigma_i P)$ into $H_k^{q+1}(P)$. Obviously, the image of this mapping lies in the kernel of σ_i and, moreover these images coincide with the kernels of σ_i. From the conditions of $(m+1)$-acyclicity for the complexes $(H_k^\bullet(P), \sigma_i)$ we obtain that these images are contained in $\sigma_i H_k^q(P)$.

We shall prove that the mappings

$$\bar{\sigma}_i \colon H_k^q(P/\sigma_i P) \to \sigma_i H_k^q(P)$$

are isomorphisms. For this purpose consider an element

$$[x] \in H_k^q(P/\sigma_i P)$$

which belongs to the kernel of $\bar{\sigma}_i$. It means that there exists an element $y \in P^{q-1}$ such that

$$\sigma_i x = \sigma_i \sigma_k y.$$

Hence, there exist $z \in P^q$, such that $x = \sigma_k y + \sigma_i z$.

The last equality means that the class $[x] \in H_k^q(P/\sigma_i P)$ is trivial and therefore $\bar{\sigma}_i$ is an isomorphism.

Conditions of $(m+1)$-acyclicity for the complexes $(H_k^\bullet(P), \sigma_i)$ permit to defines isomorphisms

$$\bar{\Phi}_i \colon \sigma_i H_k^q(P) \to H_k^q(P)/\sigma_i H_k^{q-1}(P).$$

The needed isomorphisms are given by the composition $\bar{\Phi}_i \circ \bar{\sigma}_i$

$$\bar{\Phi}_i \circ \bar{\sigma}_i \colon H_k^q(P/\sigma_i P) \to H_k^q(P)/\sigma_i H_k^{q-1}(P). \qquad \blacksquare$$

From the proposition above it follows

Theorem 2. *Let*

$$P^\bullet = \bigoplus_{s=0}^{n} P^s$$

be a graded module equipped with $r+1$ anticommuting operators $\sigma_0, \sigma_1, \ldots, \sigma_r$ of the degree $+1$. Then cohomologies $H_k^q(P)$ of P with the differential

$$\sigma_k = \sum_{j=1}^{r} k_j \sigma_j + \sigma_0, \; k_j \in \mathbb{Z}, \; k = (1, k_1, \ldots, k_r)$$

are trivial for $k \neq 1_0$ and $q \leq m$, if

1^0. *Cohomologies $H_t^q(P)$ are trivial for $q \leq m$ and $t = (0, t_1, \ldots, t_r)$, $t_j \in \mathbb{Z}$, $t \neq 0$.*

2^0. *Differential modules $(H_k^\bullet(P), \sigma_i)$, $i = 1, 2, \ldots, r$, are $(m+1)$-acyclic for $k = (1, k_1, \ldots, k_{i-1}, 0, k_{i+1}, \ldots, k_r)$, $k_j \in \mathbb{Z}$.*

Proof. Choose a number i, $1 \le i \le r$ such that $k_i \ne 0$. Just like in the proof of theorem triviality of $H_k^q(P)$ for $q \le m$ follows from the fact that the operators

$$\delta_i\colon H^q_{k-k_i 1_i}(P/\sigma_i P) \to H^q_{k-k_i 1_i}(P/\sigma_i P)$$

are isomorphisms. In our case operators δ_i are of the form

$$1 + \frac{1}{k_i} \cdot \Phi_i \circ \sigma_{k-k_i 1_i}.$$

From proposition 2 one obtains that the differentials $\sigma_{k-k_i 1_i}$ act trivially on $H^q_{k-k_i 1_i}(P/\sigma_i P)$ and, therefore, the boundary operators δ_i in exact sequence Eq. 3 are identical

$$\delta_i = \text{id}$$

for $q \le m$. ■

The following result is also valid:

Theorem 3. *Let*

$$P^\bullet = \bigoplus_{s=0}^{n} P^s$$

be a graded module equipped with the set of $r+1$ anticommuting operators $\sigma_0, \sigma_1, \ldots, \sigma_r$ of the degree $+1$. Then cohomologies $H_k^q(P)$ of P with the differential

$$\sigma_k = \sum_{j=1}^{r} k_j \sigma_j + \sigma_0, \; k_j \in \mathbb{Z}, \; k = (1, k_1, \ldots, k_r)$$

are trivial for $k \ne 1_0$ and $q \le m$, if

1^0. *Cohomologies $H_l^q(P)$ are trivial for $q \le m$ and $l = (0, l_1, \ldots, l_r)$, $l_j \in \mathbb{Z}$, $l \ne 0$.*

2^0. *Differential modules $(H^\bullet_{1_0}(P), \sigma_k)$ are $(m+1)$-acyclic for all $k = (0, k_1, \ldots, k_r)$, $k_j \in \mathbb{Z}$, $k \ne 0$.*

Proof. Change operator σ_i to operator σ_k in definition 1, exact sequence 3 and proposition 2. Similar to proof of theorem 2 we obtain that boundary operator

$$\delta_k\colon H^q_{1_0}(P/\sigma_k P) \to H^q_{1_0}(P/\sigma_k P)$$

is identical. ■

Remark. The triviality conditions for $H_k^q(P)$, $q \le m$, and $k = (1, k_1, \ldots, k_r)$, $k \ne 1_0$, are similar to the conditions of "complete intersection"[18].

Let

$$N = \bigoplus_{s=0}^{n} N^s$$

be a graded module equipped with the set of r anticommuting operators $\sigma_1, \ldots, \sigma_r$ of the degree $+1$. Operators

$$\sigma_k = \sum_{j=1}^{r} k_j \sigma_j, \ k_j \in \mathbb{Z}, \ k = (0, k_1, \ldots, k_r)$$

determine cohomological complexes in N. We denote its cohomologies by $H_k^\bullet(N)$.

Intoduce the notations

$$\begin{aligned} E &= \mathbb{Z}[y_1 \ldots y_r] \\ E_k &= \mathbb{Z}[y_1^{k_1} \ldots y_r^{k_r}] \\ k &= (0, k_1, \ldots, k_r) \\ \sigma &= \sum_{j=1}^{r} y_j \frac{\partial}{\partial y_j} \otimes \sigma_j. \end{aligned}$$

Consider differential modules $(E \otimes N^\bullet, \sigma)$ and $(E_k \otimes N^\bullet, \sigma)$ denote by $H^\bullet(E, N)$ and $H^\bullet(E_k, N)$ corresponding cohomologies.

Proposition 3. *Differential modules $(E \otimes N^\bullet, \sigma)$ are decomposed into the direct sum of modules $(E_k \otimes N^\bullet, \sigma)$. Cohomologies $H^\bullet(E, N)$ are decomposed into the direct sum of cohomologies $H^\bullet(E_k, N)$:*

$$H^\bullet(E, N) = \bigoplus_{k \neq 0} H^\bullet(E_k, N)$$

and

$$H^\bullet(E_k, N) = E_k \otimes H_k^\bullet(N), \qquad k \neq 0.$$

Apply this proposition to the module $N^\bullet = H_{1_0}^\bullet(P)$. Using theorem 2 we obtain

Theorem 4. *Let*

$$P^\bullet = \bigoplus_{s=0}^{n} P^s$$

be a graded module equipped with the set of $r+1$ anticommuting operators $\sigma_0, \sigma_1, \ldots, \sigma_r$ of the degree $+1$. Then cohomologies $H_k^q(P)$ of P with the differential

$$\sigma_k = \sum_{j=1}^{r} k_j \sigma_j + \sigma_0, \ k_j \in \mathbb{Z}, \ k = (1, k_1, \ldots, k_r)$$

are trivial for $k \neq 1_0$ and $q \leq m$, if:

1^0. *Cohomologies $H_l^q(P)$ are trivial for $q \leq m$ and $l = (0, l_1, \ldots, l_r)$, $l_j \in \mathbb{Z}$, $l \neq 0$.*

2^0. *Differential module $(E \otimes H_{1_0}(P), \sigma)$ is acyclic.*

Proof. From the condition 2^0 and proposition 3 one obtains

$$H^q(E_k, H_{1_0}(P)) \;=\; 0$$

for all $k = (0, k_1, \ldots, k_r)$, $k \neq 0$, $q \leq m + 1$.
Therefore, differential module $(H^\bullet_{1_0}(P), \sigma_k)$ is $m + 1$ acyclic and the statement of theorem coincides with statement of theorem 3. ■

References

1. D.C. Spencer, *Overdetermined systems of linear partial differential equations, Bull. of the Amer. Math. Soc.* **Vol.75** No. 2 (1969), p.179-239.
2. D.G. Quillen, *Formal properties of overdetermined systems of linear partial differential equations, Thesis*, (Harvard University, Cambridge, Massachusetts 1964), p.96.
3. M. Kuranishi, *On E. Cartan's prolongations theorem of exterior differential systems, Amer. J. Math.* **v.79**, N1 (1957), p.1-47.
4. H. Goldshmidt, *Existence theorems for analytic linear partial differential equations, Ann. of Math.* **Vol.86**, No.2 (1967), p.246-270.
5. V. Guillemin, *Some algebraic results concerning the characteristics of overdetermined partial differential equations, Amer. J. Math.*, **Vol 90** (1968), p.270-284.
6. V.V. Lychagin and L.V. Zil'bergleit, *Spencer Cohomology of differential equations, Lecture Notes in Math.* **Vol. 1453** (1990), p.121-136.
7. V.V. Lychagin and L.V. Zil'bergleit, *Duality in Stable Spencer cohomologies, Lecture Notice in Math.* **Vol.1520** (1992), p.43-55.
8. H.R. Fischer and F.L. Williams, *Complex-foliated structures. I. Cohomology of the Dolbeaut-Kostant Complexes, Trans. of the Amer. Math. Soc.* **Vol.252** (1979), p.163-195.
9. H. Goldshmidt, *Duality theorems in deformation theory, Trans. of the Amer. Math. Soc.* **Vol.292**, N1 (1985), p.1-50.
10. V.V. Lychagin, *Differential operators on fiber bundles, Uspekhi Matematicheskikh Nauk* **Vol.40**, No.2 (1985), p.187-188, (in Russian); *Russian Math. Surveys* **V.40** (1985) (in English).
11. V.V. Lychagin and L.V. Zil'bergleit, *Spencer cohomologies of invariant systems of differential equations above homogeneous spaces, Russian Math. Dokl.* (1992) (in press).

12. V.V. Lychagin and L.V. Zil'bergeit, *Spencer cohomologies and symmetries, Mat. Zametki* (1992) (in press).

13. N. Bourbaki,*Algebre. Chap. X, Algebre homologique* (Masson Paris New York Barcelone Milan, 1980), p.192.

Geometry in Partial Differential Equations (pp. 253–258)
eds. A. Pràstaro and Th. M. Rassias

HAWKING'S RADIATION via FOURIER INTEGRAL OPERATORS*

PHILLIP E. PARKER
Department of Mathematics, The Wichita State University
Wichita KS 67260-0033, USA
pparker@twsuvm.uc.twsu.edu

ABSTRACT

This is a new geometric study in the theory of partial differential equations: an application of Fourier integral operators to equations with distributional coefficients. As an example, a derivation of the existence and spectrum of Hawking's radiation from a collapsing star is given.

1 Introduction

Using techniques of Fourier integral operator theory, one may derive the existence and spectrum of a generalized Hawking's radiation for any pseudoriemannian space in which the metric tensor has distributional coefficients. By regarding the metric tensor as a distributional tensor rather than a smooth tensor, spacetime singularities are included in the smooth manifold, the metric tensor, the curvatures, the d'Alembertian operator for the wave equation, *etc.*[1] Note that this approach will not require the consideration of any topological spaces more general than smooth manifolds. As it turns out, the properties of this generalized Hawking's radiation are intimately connected with certain properties of the solutions of a wave equation having distributional coefficients in its d'Alembertian operator. When applied to the particular example of the spacetime of a collapsing star, the general procedure specializes to yield the original formulas of Hawking.[2]

All manifolds are assumed to be smooth, meaning C^∞, and paracompact. Functions, tensors, *etc.,* are smooth unless explicitly identified as distributional. The signature of a Lorentzian metric tensor is $(+ - \cdots -)$.

2 The General Proceedure

More precisely, let (M, g) be a pseudoriemannian open (noncompact, without boundary) manifold with distributional metric tensor g.[1] We recall that the singular support

*MSC(1991) primary 58G15; secondary 83C75. This paper is in final form and no version of it will be submitted for publication elsewhere.

of a distributional tensor is the set on which the tensor fails to be smooth. The set $\Sigma = \operatorname{singsupp} g$ plays the rôle of the spacetime singularities. Note that the associated d'Alembertian $\Box$ has distributional coefficients which are smooth off Σ. This allows the use of the parametrix construction of Duistermaat and Hörmander[3] when the geometry is disprisoning and pseudoconvex.[4] For wave equations, we recall that a geometry is *disprisoning* when every null geodesic escapes to infinity (i.e., there are no null geodesics imprisoned in compact sets) and is *pseudoconvex* when for every compact set K there exists a compact set K' such that every null geodesic segment with both endpoints in K lies entirely within K'. It is only necessary to apply these definitions off Σ; *i.e.,* on $M - \Sigma$. Thus we shall assume henceforth that M is disprisoning and pseudoconvex with respect to null geodesics. The spacetime of a collapsing star is an example.[2]

Remarks. Trèves[5] is a general introductory reference for Fourier integral operators. In PDE theory, disprisonment is called (real) principal type. The pseudoconvexity condition is a type of interior completeness condition: roughly, there are no holes in the "middle" of the manifold, but it might be incomplete "at infinity." Both conditions are implied by global hyperbolicity, but they are not equivalent to it; see reference 4 and the references therein.

2.1 *Existence*

We shall need to make use of tubular neighborhoods. For the convenience of the reader, we recall that a tubular neighborhood of a submanifold is an embedding of the normal (or conormal) bundle of the submanifold into the ambient manifold. More details may be found, for example, in reference 6.

There are null generalized geodesic rays with endpoint in Σ which are smooth off Σ and that escape to infinity.[1] Let γ_r be one such, oriented outward from Σ, and let γ_d be a timelike geodesic in $M - \Sigma$ which meets γ_r at $p \in M$. Here, r is for *radiation,* since it will be a calculation along γ_r which establishes the existence of our generalized Hawking's radiation, and d is for *detector,* since it will be a calculation along γ_d which yields the spectrum as measured by an observer whose worldline is γ_d. Orient γ_d compatibly with γ_r, so that the chosen causal cones coincide at p. If Σ lies within a horizon (*e.g.,* the boundary of a tubular neighborhood of Σ when Σ is an embedded submanifold and the boundary is a null hypersurface), assume that p is outside it. Compatibly time-orient small tubular neighborhoods of γ_r and γ_d. When we refer to being "near" γ_r or γ_d, we shall mean within these tubular neighborhoods. Now, one may construct global parametrices for $\Box$, namely $E_r^\pm$ and $E_d^\pm$, which are advanced/retarded near γ_r and γ_d, respectively, and which coincide outside the tubular neighborhoods and away from Σ.[3,5] We note that these parametrices play a rôle analogous to that of Green's functions or propagators.[3]

Consider a smooth φ such that $\Box\varphi$ is smooth on *all* of M, including Σ. (This is the distributional equivalent in our parametrix context of a smooth solution of the usual wave equation $\Box\varphi = 0$.) Global smoothness of $\Box\varphi$ in the presence of the distributional coefficients of $\Box$ implies conditions on φ such as $\varphi|\Sigma = 0$ or $d\varphi(\Sigma) \cap WF(g) = \emptyset$. Here, $WF(g)$ is the wave-front set of g.[3,5] (For example, when singsupp $g = \Sigma$ is an embedded submanifold, $WF(g)$ is the conormal bundle of Σ with the 0-section removed.) Then there exists[3,5] a smooth ψ with

$$\varphi = (E_r^+ - E_r^-)\psi = (E_d^+ - E_d^-)\psi, \tag{2.1}$$

whence there exist smooth functions a, b with

$$E_r^+\psi = aE_d^+\psi - bE_d^-\psi. \tag{2.2}$$

Since γ_r is a bicharacteristic of $\Box$, it follows by using the choice of local orientations that b is not rapidly decreasing along γ_r. This is true for every such γ_d, so one says that energy is being radiated along γ_r. We note that a similar analysis along bicharacteristics was used by Dimock in a different context.[7]

2.2 *Spectrum*

Assume now that Σ is an embedded submanifold having a tubular neighborhood with boundary B which is a null hypersurface. Letting $\mathbb{R}_+$ denote the positive real numbers, we may identify $WF(g)/\mathbb{R}_+$ with B by means of the distributional exponential map $\exp_\Sigma$.[1] Fix an affine parameter s for γ_r via this identification of $WF(g)/\mathbb{R}_+$ with B and note that the Fourier transform of $\varphi|\gamma_r$ with respect to s gives an observer-independent frequency density of φ along γ_r.

In order to calculate the spectrum as measured by a detector on γ_d at p, we shall use the *radial part* of a differential operator.[8] (Intuitively, in our context the radial part used will operate *parallel* to γ_r; the remaining part of the operator similarly acts *transverse* to γ_r.) Recall that we have oriented γ_r and γ_d compatibly. If we parallel translate $\dot\gamma_d(p)$ along γ_r, then we can take the radial part $\Box_R$ of the d'Alembertian near γ_r and (rewriting as necessary) decompose it as

$$\Box_R = (\kappa\dot\gamma_d)^2 - \Delta_R\,, \tag{2.3}$$

where κ is the (observer-independent) frequency ratio determined by B. For example: if B is locally a Killing horizon[9] with respect to the local Killing field k, then κ is defined by $\nabla_k k|B = \kappa k|B$. Thus if t is proper time along γ_d, then κt is the geodesic parameter used in Eq. 2.3. (We use the proper time of γ_d to get around the lack of a global time-coordinate, as used by Hawking,[2] in general.) When $\sqrt{\Delta_R}$ is well defined (*e.g.*, when (M, g) is Lorentzian), the operator $\Box_R$ has factors

$$\kappa\dot\gamma_d \pm \sqrt{\Delta_R}\,. \tag{2.4}$$

The inverse Fourier transform with respect to the geodesic parameter τ of γ_d yields first-order pseudodifferential operators

$$\sqrt{\Delta_R} \mp i\frac{\kappa}{2\pi}\xi\,, \tag{2.5}$$

where ξ is the dual variable to τ. To study outgoing radiation, we use the operator $P = \sqrt{\Delta_R} + i\frac{\kappa}{2\pi}\xi$.

Now compute the ζ-function ζ_P of P and, if necessary, remove any degeneracy of the eigenvalue 0 to obtain $\tilde{\zeta}_P$. Fourier-transforming $\tilde{\zeta}_P$ with respect to ξ with ω as the new dual variable to ξ and evaluating at $z = 1$, one obtains an expression for a spectral density at the frequency ω of bosonic radiation.

Fermionic radiation may be handled in a similar way by starting with the Dirac operator obtained from the d'Alembertian operator $\Box$.

3 The Example of a Collapsing Star

As did Hawking,[2] we begin by observing that in the spacetime of a collapsing star, outside of the horizon the d'Alembertian appears as it would in a Schwarzschild spacetime for a suitable choice of coordinates. Let r be such a radial coordinate with $\{r = 0\} = \Sigma$ and let t denote Killing time, so that $k = \partial/\partial t$ is the timelike Killing vector outside the Killing horizon B. We take the d'Alembertian in the form

$$\Box = \frac{\partial^2}{\partial t^2} - \frac{1}{r^2}\left(1 - \frac{2m}{r}\right)\frac{\partial}{\partial r}\left(r^2\left(1 - \frac{2m}{r}\right)\frac{\partial}{\partial r}\right) - \frac{1}{r^2}\left(1 - \frac{2m}{r}\right)L\,, \tag{3.1}$$

where L is an operator which does not involve t or r and which appears the same as it would in a flat spacetime. (Here, one might refer to L as a tangential part; it is the transverse-acting part to which we referred just before Eq. 2.3.) The radial part[8] of this operator is given by

$$\Box_R = \frac{\partial^2}{\partial t^2} - \frac{1}{r^2}\left(1 - \frac{2m}{r}\right)\frac{\partial}{\partial r}\left(r^2\left(1 - \frac{2m}{r}\right)\frac{\partial}{\partial r}\right). \tag{3.2}$$

Letting γ_d be an integral curve of $\partial/\partial t$ and setting $\tau = \kappa t$, we can decompose $\Box_R$ in the form of Eq. 2.3:

$$\begin{aligned}(\kappa\dot{\gamma}_d)^2 - \Delta_R &= (\kappa\dot{\gamma}_d)^2 - \frac{1}{r^2}\left(1 - \frac{2m}{r}\right)\frac{\partial}{\partial r}\left(r^2\left(1 - \frac{2m}{r}\right)\frac{\partial}{\partial r}\right)\\ &= (\kappa\dot{\gamma}_d)^2 - \Delta_R\end{aligned} \tag{3.3}$$

where the dot denotes $\partial/\partial\tau$ as in Eq. 2.3.

Having identified Δ_R, we are concerned with the zeta function of $P = \sqrt{\Delta_R}+i\frac{\kappa}{2\pi}\xi$. Thus we want to study the spectrum of $\sqrt{\Delta_R}$, so we consider the eigenvalue equation $\Delta_R u + \lambda^2 u = 0$ where u is a function of r only. In self-adjoint form, this is

$$\left(r^2\left(1 - \frac{2m}{r}\right)u'\right)' + \lambda^2 r^2\left(1 - \frac{2m}{r}\right)^{-1} u = 0 \tag{3.4}$$

with the prime denoting differentiation with respect to r. By inspection, we see that $u(2m) = 0$ and $u'(2m)$ is finite. For $r \gg 0$, Eq. 3.4 is asymptotic to

$$(r^2u')' + \lambda^2 r^2 u = 0 \tag{3.5}$$

and the spectra are also asymptotic. (Such approximations are implicit in Hawking[2].) Now, the general solution of Eq. 3.5 which is bounded at infinity is

$$u(r) = a\frac{\sin \lambda r}{r} + b\frac{\cos \lambda r}{r} \tag{3.6}$$

for constants a and b. Taking $u(2m) = 0$, then $-b/a = \tan \lambda 2m$ and

$$\lambda_n = \frac{\theta}{2m} + n\frac{\pi}{2m} \tag{3.7}$$

enumerates the asymptotic spectrum of Δ_R, where $\theta = \text{Arctan}(-b/a)$ and $n \in \mathbb{Z}$. Since $\Box\varphi$ is to be smooth on Σ we further require that u be bounded as $r \to 0^+$. Then $b = 0$ whence $\theta = 0$ and $\lambda_n = n(\pi/2m)$. Rescaling r, we may take $\lambda_n = n$.

Thus for $P = \sqrt{\Delta_R} + i\frac{\kappa}{2\pi}\xi$, we have

$$\tilde{\zeta}_P(z) \sim \sum_{n=1}^{\infty} \left(n + i\frac{\kappa}{2\pi}\xi\right)^{-z}, \tag{3.8}$$

where the tilde denotes removal of the degeneracy at the eigenvalue 0. If we rewrite this, we recognize[11] a generalized zeta function of Hurwitz:

$$\begin{aligned} \tilde{\zeta}_P(z) &\sim \sum_{k=0}^{\infty} \left(k + 1 + i\frac{\kappa}{2\pi}\xi\right)^{-z} \\ &= \zeta\left(z, 1 + i\frac{\kappa}{2\pi}\xi\right). \end{aligned} \tag{3.9}$$

Letting ω denote a new dual variable to ξ, it follows from formulas 7.49, p. 174, and 8.58, p. 269, of reference 12 that the Fourier transform with respect to ξ of $\tilde{\zeta}_P(z)$ is approximated by

$$\frac{\left(\frac{2\pi}{\kappa}\right)^z \omega^{z-1}}{\Gamma(z)\left(e^{2\pi\omega/\kappa} - 1\right)}. \tag{3.10}$$

Taking the limit as $z \to 1$ and recalling from reference 1 that $\kappa = 1/4m$ in our spacetime yields

$$\frac{8\pi m}{e^{8\pi m\omega} - 1}. \tag{3.11}$$

The spectral density is now given (*cf.* reference 1) by κ times this as

$$\frac{2\pi}{e^{8\pi m\omega} - 1}, \tag{3.12}$$

in agreement with Hawking[2]. It would be of interest to compare this approximation, here derived from Eq. 3.5, with the true value derived from Eq. 3.4.

Acknowledgements

It is a pleasure to acknowledge the support of E. Binz and Lehrstuhl für Mathematik I of Universität Mannheim, The Wichita State University for a Faculty Summer Research Award, and the Science and Engineering Research Council and London Mathematical Society at the Durham Symposium on Relativity. I also wish to thank J. Barrett and R. Williams for organizing, and the Universities of Cambridge and Newcastle for partially supporting, the Special Meeting on Singular Metrics at which most of this material was presented. Thanks to P. Michor, H. Petry, J. Barrett, and B. Kay for helpful suggestions and encouragement.

References

[1] P. E. Parker, Distributional geometry, *J. Math. Phys.* **20** (1979) 1423–1426.

[2] S. W. Hawking, Particle creation by black holes, *Comm. Math. Phys.* **43** (1975) 199–220.

[3] J. J. Duistermaat and L. Hörmander, Fourier integral operators II, *Acta Math.* **128** (1972) 183–269.

[4] J. K. Beem and P. E. Parker, Pseudoconvexity in general relativity *J. Geom. Phys.* **4** (1987) 71–80.

[5] F. Trèves, *Introduction to Pseudodifferential and Fourier Integral Operators.* New York: Plenum, 1980.

[6] Th. Bröcker and K. Jänich, *Introduction to Differential Topology.* Cambridge: U.P., 1982.

[7] J. Dimock, Scalar quantum field in an external gravitational field, *J. Math. Phys.* **20** (1979) 2549–2555.

[8] S. Helgason, *Analysis on Lie groups and Homogeneous Spaces.* Providence: Amer. Math. Soc., 1972.

[9] R. H. Boyer, Geodesic Killing orbits and bifurcate Killing horizons, *Proc. R. Soc. Lond. A* **311** (1969) 245–252.

[10] S. W. Hawking, Zeta function regularization of path integrals in curved spacetime, *Comm. Math. Phys.* **55** (1977) 133–148.

[11] J. Spanier and K. B. Oldham, *An Atlas of Functions.* New York: Hemisphere, 1987.

[12] V. Ditkin and A. Prudnikov, *Integral Transforms and Operational Calculus.* London: Pergamon, 1965.

Geometry in Partial Differential Equations (pp. 259–315)
eds. A. Pràstaro and Th. M. Rassias

GEOMETRY OF SUPER PDE's [1,2]

AGOSTINO PRÀSTARO
Università degli Studi di Roma "La Sapienza"
Dipartimento di Metodi e Modelli Matematici per le Scienze Applicate
Via A. Scarpa, 10 - 00161 ROMA - Italy

ABSTRACT

Superspaces and supermanifolds are introduced by using the concept of weak differentiability as usually given for locally convex spaces. This allows us to consider in algebraic way superdual spaces and superderivative spaces. In this way we obtain a good generalization of just known superstructures general enough to develop a formal theory for super PDE's that directly extends previous ones for standard manifolds of finite dimension. In particular, we give a Goldschmidt-type criterion of formal superintegrability for super PDE's, and show that a geometric theory of singular supersolutions, with singularities of Thom-Bordman type, can be formulated in the framework of super PDE's too. Conservation superlaws associated to super PDE's are considered and related with some spectral sequences and wholly cohomological character of these equations.

1. Supermanifolds

In the following we shall consider supermanifolds as manifolds modeled on particular locally convex spaces. For such structures we will consider superderivative spaces. Our definition of "supertangent space" allows us to interpret any supervector like the velocity of supercurve. This is profitable in a theory of super PDE's. The price to pay for this is that in general such supertangent space is not a finite dimensional module. However, if one specializes on supermanifolds where the underlying superalgebra A has trivial annihilator of the odd part, ($A_1^{\perp} = 0$), supertangent spaces become finite dimensional also. Such superalgebras are infinite dimensional, as it is required in all physical applications. Thus in this paper we shall almost consider supermanifolds such that the underling superalgebra A has $A_1^{\perp} = 0$. (For other informations on supermanifolds see also the following refs. [2,

---

[1]Some results of this paper were announced at the "International Conference on Differential Geometry and Applications", Dubrovnik (Yugoslavia) - June 26 - July 3, 1988. (See ref. [31]).

[2]Work partially supported by grants CNR/GNFM and MURST 40%,

3, 4, 6, 8, 12, 14, 17, 20, 21, 35, 36, 37]. By using results of this paper we can also extend to SPDE's the quantum geometry of PDE's as formulated in refs. [30, 32, 34]. But this will be the subject for another work in preparation.

1.1. Superspaces

Let $\mathbb{K}$ be the field of real numbers $\mathbb{R}$ or complex numbers $\mathbb{C}$. Denote by $\mathbb{Z}_2 = \{0,1\}$ the additive group with "addition" given by: $0+0=0, 0+1=1, 1+1=0, -1=1$. In the following we shall denote by $|.| : F \longrightarrow \mathbb{Z}_2$ the degree map defined on any set F. A $\mathbb{Z}_2$ **graded commutative locally convex $\mathbb{K}$-algebra** is a l.c $\mathbb{K}$-algebra A (with, or without unit), which is at same time a $\mathbb{Z}_2$-graded l.c.space $A = A_0 \oplus A_1$, such that: (i) $A_r.A_s \subseteq A_{r+s}$; (ii) $a_r a_s = (-1)^{rs} a_r a_s \in A_{r+s}$, $a_r \in A_r$, $a_s \in A_s$. We will denote by $A_1^{\perp}$ the **annihilator** of $A_1 : A_1^{\perp} \equiv \{a \in A | ab = 0 = ba, \forall b \in A_1\}$. ($A_1^{\perp}$ is an ideal of A. Furthermore, if A is a Banach $\mathbb{K}$-algebra, we call A **$\mathbb{Z}_2$-graded commutative Banach $\mathbb{K}$-algebra** if the following condition is satisfied: $\|a_0 + a_1\| = \|a_0\| + \|a_1\|$ $a_r \in A_r$. (Then the canonical involution $J \equiv pr_0 - pr_1$, where $pr_i : A \longrightarrow A_i$, $i = 0, 1$, is an isometric automorphism of $A : \|J(a)\| = \|a\|$).

DEFINITION 1. *A* **superalgebra (resp. Banach superalgebra)** *is a $\mathbb{Z}_2$-graded commutative l.c. $\mathbb{K}$-algebra (resp. $\mathbb{Z}_2$-graded commutative Banach $\mathbb{K}$-algebra) that is also associative with unit, and contains* $\mathbb{K} : A = \mathbb{K} \oplus A_0' \oplus A_1$. *We call* $\mathbb{K}$ *the* **body** *and* $\mathfrak{M} \equiv A_0' \oplus A_1$ *the* **soul** *of* A.

NOTE. The annihilator of A_0, $A_0^{\perp}$ is always trivial: $A_0^{\perp} = 0$. In fact the relation $ab = ba = 0$, for any $a \in A_0$, implies also $a.1 = 1.a = a = 0$).

EXAMPLES. 1) Any commutative Banach algebra, B, over $\mathbb{K}$ (for example the same $\mathbb{K}$), can be considered a particular commutative Banach superalgebra A with $A_0 = B$ and $A_1 = \{0\}$.

2) (**Grassmann algebra**). Every finite-dimensional $\mathbb{Z}_2$-graded commutative algebra $A = A_0 \oplus A_1$ is a Banach superalgebra. In particular if $A = \Lambda\mathbb{K}^q =$ Grassmannian algebra over $\mathbb{K}^q$, $q \in \mathbb{N}$, any element $a \in A$ has a linear representation

$$a = \sum_i a^i e_i, \text{ where } i = (i_1, \dots, i_t), 1 \le i_1 < i_2 < \dots < i_t \le q,$$

$$\text{and } e_i = e_{i_1} \wedge \dots \wedge e_{i_t},$$

being $\{e_1, \dots, e_q\}$ a basis on $\mathbb{K}^q$, and $a_i \in \mathbb{K}$. The subspace A_0 (resp. A_1) consists of the elements a such that each term of its linear representation contains only an even (resp. odd) number of basis elements. One has $\|a\| = \sum_i a^i$. For any Grassmannian algebra $A \equiv \Lambda\mathbb{K}^q$, $A_1^{\perp} = \Lambda^q\mathbb{K}^q$. The non zero elements of $\mathbb{K}$ are the only invertible elements of A. If $q = \infty$ (i.e. $\mathbb{K}^{\infty} \equiv \lim \mathbb{K}^s$), then $A_1^{\perp} = 0$.

A is a projective A_0-module. In fact, it is a free A_0-module: $A = A_0.1 \oplus A_0.e_1 \oplus \dots \oplus A_0.e_q$, with $1 \le q \le \infty$. A_1 is also a free A_0-module, hence projective. A has one maximal ideal $\mathfrak{M}$. So $A/\mathfrak{M} \cong \mathbb{K}$ is its **residue field** [1].

3) (**Banach-Grassmann algebra**). Any Banach-Grassmann algebra in the sense of [12] is a superalgebra. More precisely, a **Banach-Grassmann algebra (BG-algebra)** is a Banach superalgebra A over $\mathbb{K}$ ($\mathbb{K} = \mathbb{R}$ or $\mathbb{C}$) with $A_1^{\perp} = 0$. For such superalgebras one has the following canonical A_0-isomorphisms:

$$L_{A_0}(A_r, A_s) \cong A_{r+s}\,;\, L_A(A,A)_r \cong A_r\,;\, (A_r)^{\oplus}{}_0 \equiv L_{A_0}(A_r, A_0) \cong A_r\,;$$
$$A_r \otimes_{A_0} A_s \cong A_{r+s},\, r = s = 0, \text{ or } r \neq s \in \mathbb{Z}_2;$$
$$L_{A_0}(A; A^{m,n}) \cong L_A\big(L_{A_0}(A^{m,n}; A); A\big) \cong A^{m+n};$$
$$L_{A_0}(A_1 \otimes_{A_0} A_1; A) \cong L_A\big(A_1; L_{A_0}(A_1; A)\big) \cong L^2_{A_0}(A_1; A) \cong A\,.$$

The spectrum, $Sp(a)$, of a every element $a \in A$ consists exactly of one point $Sp(a) \in \mathbb{K}$, and the map $Sp : A \longrightarrow \mathbb{K}$ is a unique non-zero character of A. An element $a \in A$ is invertible iff $Sp(a) \neq 0$. Then

$$a^{-1} = Sp(a)^{-1} \sum_{0 \leq n < \infty} (-1)^n \big(s(a)/Sp(a)\big)^n\,,$$

where $s(a) \equiv a - Sp(a) \in \mathfrak{M} \equiv A_0' \oplus A_1$. For each A_0-linear, map $f : A_1 \longrightarrow A$ one has $Sp\big(f(y)\big) = 0$ and $f(y)z + f(z)y = 0$, for all $y, z \in A_1$. An infinite dimensional Grassmannian algebra is a Banach-Grassmann algebra.

REMARK 1. The **tensor product** of two graded commutative *l.c.* algebras A and B is the $\mathbb{Z}_2$-graded *l.c.* space

$$A \otimes_{\mathbb{K}} B = \oplus_{i \in \mathbb{Z}_2} \big(A \otimes_{\mathbb{K}} B\big)_i, \big(A \otimes_{\mathbb{K}} B\big)_i = \oplus_{i=m+n} A_m \otimes_{\mathbb{K}} B_n,\ m, n, i \in \mathbb{Z}_2\,,$$

together with the structure of $\mathbb{Z}_2$-graded commutative algebra given by

$$(x \otimes y) \cdot (z \otimes t) = (-1)^{|x||z|} xz \otimes yt,\ x, z \in A,\ y, t \in B\,.$$

DEFINITION 2. *Let A be a superalgebra. A* **$\mathbb{Z}_2$-graded *l.c.* A-module,** *is a $\mathbb{Z}_2$-graded l.c. space F that is also a left and right A-module such that:*
(1) $a_r f_s = (-1)^{rs} f_s a_r \in F_{r+s}$, for $a_r \in A_r$, $f_s \in F_s$;
(2) the map $A \times F \longrightarrow F$, $(a, f) \mapsto af$, is continuous.

In table 1 we give some examples of A_0-modules.

Tab. 1 - Examples of A_0-modules and A-modules associated to a superalgebra $A = A_0 \oplus A_1$

A_0-modules:

$$A_0;\ A_1;\ A^{m,n} \equiv (A_0)^m \times (A_1)^n;\ A^{\bar{m},\bar{n}} \equiv (A_1)^m \times (A_0);\ A^{m,n|p,q} \equiv A^{m,n} \times (A/A_1^{\perp})^{p,q}\,.$$

If A is a Banach superalgebra, then $A^{m,n}$ is a Banach space such that the norm of $a = (a^A) \equiv (a^1, \dots, a^{m+n}) \in A^{m,n}$ is $\|a\| = \sum_A \|a^A\|$.

Graded *l.c* A-modules:

$$A^{m+n} \equiv (A^{m+n})_0 \oplus (A^{m+n})_1 \equiv A^{m,n} \oplus A^{\bar{m},\bar{n}};\ A^m \times (A/A_1^{\perp})^n \equiv A^{m,m|n,n}\,.$$

$A^{m,n}$ (resp. $A^{\bar{m},\bar{n}}$) is the subset of A^{m+n} characterized, with respect to the canonical A-basis (e_A) of A^{m+n}, by the following conditions on the components: $|a^A| \in A_{|A|}$, (resp. $|a^A| \in A_{|A|+1}$). $A^{m,m|n,n}$ has a canonical finite set of generators, (but not a basis): $m, n, p, q \in \mathbb{N} \equiv$ set of integers. The group of invertible A_0-linear maps $A^{m,n} \longrightarrow A^{m,n}$ is denoted by $GL(m|n, A)_{A_0}$.

One has the following commutative and exact diagram that relates above modules:

$$\begin{array}{ccccccccc}
 & & & & 0 & & 0 & & \\
 & & & & \downarrow & & \downarrow & & \\
 & & & & A^{\bar{m},n} & & A^{\bar{m},n} & & \\
 & & & & \downarrow & & \downarrow & & \\
0 & \longrightarrow & A^{m,n} & \longrightarrow & A^{m+p,n+q} & \longrightarrow & A^{m,n|p,q} & \longrightarrow & 0 \\
 & & & & \downarrow & & \downarrow & & \\
0 & \longrightarrow & (A_1^{\perp})^{p,q} & \longrightarrow & A^{p,q} & \longrightarrow & (A/A_1^{\perp})^{p,q} & \longrightarrow & 0 \\
 & & & & \downarrow & & \downarrow & & \\
 & & & & 0 & & 0 & &
\end{array}$$

REMARK 2. Let $F_1, \ldots, F_p$, E be $\mathbb{Z}_2$-graded $l.c.A$-modules. The space $L(F_1, \ldots \ldots, F_p; E)$ has a natural $\mathbb{Z}_2$-gradiation. More precisely, if $f \in L(F_1, .., F_p; E)$ one has

$$f_r = \frac{1}{2} f + (-1)^r J' f J \in L(F_1, \ldots, F_p; E)_r, \; r \in \mathbb{Z}_2 .$$

So

$$|f| = r \;\text{ iff }\; f(v^1, \ldots, v^p) \in E_{|v^1| + \cdots + |v^p| + r}, \; v^i \in F_i .$$

Then, we can distinguish two $\mathbb{Z}_2$-graded $l.c.$ subspaces of $L(F_1, \ldots, F_p; E)$ that are graded $l.c.A$-modules:

(a) **space of left $p - A$-linear maps** $\equiv$

$$L_{L,A}(F_1, \ldots, F_p; E) = \{ f \in L(F_1, \ldots, F_p; E) \mid f(v^1, \ldots, a v^k, \ldots, v^p) = (-1)^{|a|(|v^1| + \cdots + |v^{k-1}|)} a f(v^1, \ldots, v^p), a \in A \} .$$

(b) **space of right $p - A$-linear maps** $\equiv$

$$L_{R,A}(F_1, \ldots, F_p; E) = \{ f \in L(F_1, \ldots, F_p; E) | \; f(v^1, \ldots, v^k a, \ldots, v^p) = (-1)^{|a|(|v^{k+1}| + \cdots + |v^p|)} f(v^1, \ldots, v^p) a, a \in A \} .$$

Let $E_1, \ldots, E_p$ be $l.c.A_0$-modules (ungraded), and let F be a $l.c.\mathbb{Z}_2$-graded A-module. Then, $L_{A_0}(E_1, \ldots, E_p; F)$ is the $\mathbb{Z}_2$-graded $l.c.A$-module of all $p - A_0$-linear maps $E_1 \times \ldots \times E_p \longrightarrow F$. The grading is given by $L_{A_0}(E_1, \ldots, E_p; F)_r = L_{A_0}(E_1, \ldots, E_p; F_r)$.

The **tensor product of two graded $l.c.A$-modules** E and F is the $\mathbb{Z}_2$-graded $l.c.$ space $E \otimes_A F \equiv E \otimes_{\mathbb{K}} F / I$, where I is generated by elements of the type $va \otimes u - v \otimes au$, where $v \in E$, $u \in F$, $a \in A$. So on $E \otimes_A F$ one has $va \otimes u = v \otimes au$.

$E \otimes_A F$ becomes an A-module by setting $a(v \otimes u) = av \otimes u$, $(v \otimes u)a = v \otimes ua$. One has

$$a(v \otimes u) = (-1)^{|a|(|v|+|u|)}\,(v \otimes u)a\,.$$

Furthermore, there is a canonical isomorphism $E \otimes_A F \cong F \otimes_A E$ given by $v \otimes u \mapsto (1)^{|v||u|}\, u \otimes v$. The tensor product of graded $l.c.A$-modules results associative. For any A_0-module E we can consider its A-extension $A \otimes_{A_0} E$ by considering A as a A_0-module too. Then $A \otimes_{A_0} E$ becomes a A-module:

$$a'(a \otimes v) = (a'a) \otimes v;\ \ (a \otimes v) \cdot a' = (aa') \otimes v, a, a' \in A, v \in E\,.$$

One has: $a'(a \otimes v) = (1)^{|a'|(|a|+|v|)}(a \otimes v)a'$. If F is another A_0-module we can consider the tensor product of A-modules: $(A \otimes_{A_0} E) \otimes_A (A \otimes_{A_0} F)$ that becomes also a A-module:

$$a\big[(a_1 \otimes v_1) \otimes (a_2 \otimes v_2)\big] = a.(a_1 \otimes v_1) \otimes (a_2 \otimes v_2) = (aa_1 \otimes v_1) \otimes (a_2 \otimes v_2);$$

$$\big[(a_1 \otimes v_1) \otimes (a_2 \otimes v_2)\big]a = (a_1 \otimes v_1) \otimes (a_2 \otimes v_2)a = (a_1 \otimes v_1) \otimes (a_2 a \otimes v_2).$$

One has the following relation:

$$a\big[(a_1 \otimes v_1) \otimes (a_2 \otimes v_2)\big] = (-1)^{|a|(|a_1|+|a_2|)}\,\big[(a_1 \otimes v_1) \otimes (a_2 \otimes v_2)\big]a\,.$$

THEOREM 1. *One has the following commutative diagram where the horizontal line and the vertical line are both exact and $A^{m,n}O_{A_0}A$ is the image of the homomorphism*

$$j : A^{m+n} \longrightarrow (A^{m,n})^{\oplus} \equiv L_{A_0}(A^{m,n}; A),\ \ \textit{defined by: } j(\omega_A) = f,$$

$$f(a) = \sum a^A \omega_A,\ \ \forall a \in A^{m,n}\,.$$

$$\begin{array}{ccccccc}
 & & 0 & & & & \\
 & & \downarrow & & & & \\
 & & (A_1^{\perp})^n & & & & \\
 & & \downarrow & & & & \\
 & & A^{m+n} & & & & \\
 & & j\downarrow & & & & \\
0 & \longrightarrow & (A^{m,n})^{\oplus} & \longrightarrow & A^m \times (A_1^{\oplus})^n & \longrightarrow & 0 \\
 & & \downarrow & & & & \\
 & & (A^{m,n})^{\oplus}/A^{m,n}O_{A_0}A & & & & \\
 & & \downarrow & & & & \\
 & & 0 & & & &
\end{array}$$

(If A is a BG-algebra one has the canonical isomorphism: $(A^{m,n})^{\oplus} \cong A^{m+n} \cong A^{m,n}O_{A_0}A$, given by $\beta \mapsto (\beta^0, \beta^1)$, such that $\beta^r = (\beta^r_A)$ is the unique element of $(A^{m,n})_r$ such that

$$\beta(a)_r = \sum\nolimits_A a^A \beta^r_A \in A_r,\ \ \textit{with } |\beta^r_A| = |A| + r)\,.$$

DEFINITION 3. *By using considerations similar to Theorem 1 ones, we can recognize into* $L_{A_0}(A^{m,n};A^{m',n'})$ *a subspace* $A^{m,n}O_{A_0}A^{m',n'}$ *(***space of superlinear mappings** $A^{m,n}\longrightarrow A^{m',n'}$*) defined by means of the following exact sequence:*

$$0\longrightarrow N\longrightarrow (A^{m',n'})^m\times(A^{\bar{m}',\bar{n}'})^n-j\longrightarrow L_{A_0}(A^{m,m};A^{m',n'})$$

$$\longrightarrow L_{A_0}(A^{m,n};A^{m',n'})/A^{m,n}O_{A_0}A^{m',n'}\longrightarrow 0$$

where j *is the homomorphism defined by* $(\omega_A)\mapsto f$, $f(t^A)^B=t^A\omega_A^B$, $|\omega_A^B|=|A|+|B|$.

$$N=\ker j=\left[(A_1^{\perp})_0^{n'}\times(A_1^{\perp})_1^{m'}\right]^n\equiv(A_1^{\perp})^{n(\bar{m}',\bar{n}')}.$$

REMARK 3. A **superlinear mapping** $A^{m,n}\longrightarrow A^{m',n'}$ is an element of

$$A^{m,n}O_{A_0}A^{m',n'}\cong(A^{m',n'})^m\times\left((A/A_1^{\perp})^{\bar{m}',\bar{n}'}\right)^n=A^{m'm,n'm|n'n,m'n}.$$

More precisely, $(f_A^B)\in A^{m,n}O_{A_0}A^{m',n'}$ iff the following conditions are satisfied:

$$f_A^B\in A \text{ if } |A|=0,\ f_A^B\in A/A_1^{\perp} \text{ if } |A|=1;\ |f_A^B|=|A|+|B|.$$

Note that if

$$f\equiv\left((h_A^B),([\bar{h}_A^B])\right)\in(A^{m',n'})^m\times\left((A/A_1^{\perp})^{\bar{m}',\bar{n}'}\right)^n$$

is a superlinear map $A^{m,n}\longrightarrow A^{m',n'}$, for any $t\equiv(t^A)\in A^{m,n}$ we have

$$f(t)=\sum\nolimits_{1\le A\le m+n}t^Af_A^B\equiv\sum\nolimits_{1\le A\le m}t^Ah_A^B+\sum\nolimits_{m+1\le A\le m+n}t^A\bar{h}_A^B.$$

The last sum is independent on the particular representative used to denote the equivalence class $[\bar{h}_A^B]$.

The group of invertible superlinear maps $A^{m,n}\longrightarrow A^{m,n}$ (**superisomorphisms** of $A^{m,n}$) is denoted by $GL(m|n;A)_s$. Furthermore, we call **space of almost superlinear maps** $A^{m,n}\longrightarrow A^{m',n'}$ the following one: $S(A^{m,n};A^{m',n'})\equiv(A^{m',n'})^m\times(A^{\bar{m}',\bar{n}'})^n$.

It results an extension of $A^{m,n}O_{A_0}A^{m',n'}$ by means of N: the following sequence is exact:

$$0\longrightarrow N\longrightarrow S(A^{m,n};A^{m'n'})\longrightarrow A^{m,n}O_{A_0}A^{m',n'}\longrightarrow 0.$$

The **group of almost superlinear isomorphism** of $A^{m,n}$ is the set

$$GL(m|n;A)\equiv j^{-1}(GL(m|n;A)_s)\subseteq S(A^{m,n};A^{m,n}).$$

If $A\equiv\Lambda V$, $\dim_{\mathbb{R}}V=N$, $A^{m,n}$ is a homeomorphic to $\mathbb{R}^{2^{N-1}(m+n)}$, $GL(m|n;A)$ consists of $(m+n)\times(m+n)$ invertible matrices (M_B^A) with entries in A such that $|M_B^A|=|A|+|B|$, and one has the group homeomorphism: $GL(m|n;A)\cong$

$GL(m,\mathbb{R}) \times GL(n,\mathbb{R}) \times \mathbb{R}^t$, $t \equiv 2^{N-1}(m+n)^2 - m^2 - n^2$. If A is a BG-algebra one has the canonical isomorphisms:

$$L_{A_0}(A^{m,n};A^{m',n'}) \cong A^{m,n}O_{A_0}A^{m',n'} \cong S(A^{m,n};A^{m',n'}) \cong (A^{m',n'})^m \times (A^{\bar{m}',\bar{n}'})^n.$$

DEFINITION 4. *Set* $(A^{m,n})^+ \equiv A^{m,n}O_{A_0}A \cong A^m \times (A/A_1^{\perp})^n$. *We call* $(A^{m,n})^+$ *the* **superdual** *of* $A^{m,n}$.

REMARK 4. $A^{m,n}$ is free, as $\mathbb{K}$-module, and therefore it is projective too. Furthermore, taking into account the monomorphisms $\mathbb{K} \longrightarrow A_0$, $\mathbb{K} \longrightarrow A$, we have also the following: (a) $A_0 \otimes_{\mathbb{K}} A^{m,n}$ is a free, projective, A_0-module; (b) $A \otimes_{\mathbb{K}} A^{m,n}$ is a free, projective, A-module. One has the canonical monomorphisms:

$$A_0 \longrightarrow A_0 \otimes_{\mathbb{K}} A^{m,n}; \quad A \longrightarrow A \otimes_{\mathbb{K}} A^{m,n}.$$

DEFINITION 5. 1) *The* **space of superlinear mappings** *between* $A^{m,n|p,q}$ *and* $A^{m',n'|p',q'}$ *is the subspace*

$$A^{m,n|p,q}O_{A_0}A^{m',n'|p',q'} \subset L_{A_0}(A^{m,n|p,q};\ A^{m',n'|p',q'})$$

defined by means of the following exact commutative diagram:

$$\begin{array}{ccccccc}
0 & \longrightarrow & L_{A_0}(A^{m,n|p,q};A^{m',n'|p',q'}) & \longrightarrow & \begin{array}{c} L_{A_0}(A^{m,n};A^{m',n'})\times \\ \times L_{A_0}\left(A^{m,n};(A/A_1^{\perp})^{p',q'}\right)\times \\ L_{A_0}\left((A/A_1^{\perp})^{p,q};A^{m',n'}\right)\times \\ \times L_{A_0}\left((A/A_1^{\perp})^{p,q};(A/A_1^{\perp})^{p',q'}\right) \end{array} & \longrightarrow & 0 \\
 & & \uparrow & & \uparrow & & \\
0 & \longrightarrow & A^{m,n|p,q}O_{A_0}A^{m',n'|p',q'} & \longrightarrow & \begin{array}{c} A^{m,n}O_{A_0}A^{m',n'}\times \\ \times A^{m,n}O_{A_0}(A/A_1^{\perp})^{p',q'}\times \\ (A/A_1^{\perp})^{p,q}O_{A_0}A^{m',n'}\times \\ \times(A/A_1^{\perp})^{p,q}O_{A_0}(A/A_1^{\perp})^{p',q'} \end{array} & \longrightarrow & 0 \\
 & & \uparrow & & \uparrow & & \\
 & & 0 & & 0 & &
\end{array}$$

2) *In particular we can define also:*

(a) $\widetilde{\overline{A^{m,n,|p,q}}} \equiv \overline{A}O_{A_0}A^{m,n|p,q}$, (**full graded** *of* A_0*-module* $A^{m,n|p,q}$), *where*

$$\overline{A} \equiv A^{1,1|1,1} \cong A \times (A/A_1^{\perp});$$

(b) $(A^{m,n|p,q})^{\dagger} \equiv A^{m,n|p,q}O_{A_0}\overline{A}$, (**full superdual** *of* $A^{m,n|p,q}$);

(c) $(A^{m,n|p,q})^{\smile} \equiv AO_{A_0}A^{m,n|p,q}$, (**graded** *of* A_0*-module* $A^{m,n|p,q}$);

(d) $(A^{m,n|p,q})^+ \equiv A^{m,n|p,q}O_{A_0}A$, (**superdual** *of* $(A^{m,n|p,q})$).

3) *The* **group of superisomorphisms** *of* $A^{m,n|p,q}$ *is denoted by* $GL(m|n;p|q;A)_S$.

Definition 6. *Let A be a superalgebra, and let U be an open subset of $A^{m,n|p,q}$.*

1) *A* **weak superdifferentiable function** $f : U \longrightarrow \overline{A}$ *of class* $\underline{G}^s_w$, $s \geq 0$, *on* $U \subset A^{m,n|p,q}$ *is a weakly s-times differentiable map (in the sense of [13]) (or of class* C^s_w*) such that*

$$Df(x) \in L_{A_0}(A^{m,n}; \overline{A}) \subset L_{\mathbb{K}}(A^{m,n}; \overline{A}), \quad \textit{for all } x \in U .$$

Weak supersmooth *functions (or functions of class* $\underline{G}^s_w$*) are weak superdifferentiable functions of class* $\underline{G}^s_w$, $\forall s \geq 0$. *The set of all such functions is denoted by* $\underline{G}^\infty_w(U; \overline{A})$.

2) *Similarly we can define* **weak superanalytic functions** *(or of class* $\underline{G}^\omega_w$*), functions* $U \longrightarrow \overline{A}$ *of class* C^ω_w *such that their derivates are* A_0*-linear. The set of all such functions is denoted by* $\underline{G}^\omega_w(U; \overline{A})$.

3) *A function* $f : U \longrightarrow A^{m',n'|p',q'}$ *is called* **weak superdifferentiable** *(or of class* $\underline{G}^s_w$, $s \geq 0$*) if all its components* $f^A : U \longrightarrow \overline{A}$*, are of class* $\underline{G}^s_w$*. The set of all such functions is denoted by* $\underline{G}^s_w(U;\ A^{m',n'|p',q'})$ *and is an* $\underline{G}^s_w(U; \overline{A})_0$*-module.*

4) *A* **superdifferentiable function** *of class* G^s_w, $0 \leq s \leq \infty, \omega$, $f : U \subseteq A^{m,n|p,q} \longrightarrow \overline{A}$*, is a weak superdifferentiable function of class* $\underline{G}^s_w$ *such that:*

$$Df(x) \in (A^{m,n|p,q})^\dagger, \quad \forall x \in U .$$

Remark 5. In the following, in order to avoid formal complications, we shall assume that superalgebras are BG-superalgebras. However, this assumption is not too restrictive as superalgebras that are used in all physical applications can be described by considering superalgebras $A = A_0 \oplus A_1$ with trivial A_1-annihilator, i.e., $A_1^\perp = 0$. This implies that A is infinite dimensional.

Theorem 2 (Properties of superdifferentiable functions). 1) *If A is a BG-algebra the weak superdifferentiability coincides with the superdifferentiability. More precisely, for such a function* $f : U \subseteq A^{m,n} \longrightarrow A$*, one has*

$$Df(x) \in (A^{m,n})^+ \cong A^{m+n} .$$

2) *If f and g are two (weak) superdifferentiable functions on U, then $f+g$, fg and $J.f$ are also (weak) superdifferentiable. The set $G^p_w(U;A)$ is a graded commutative algebra. This algebra contains A as the subalgebra of constant functions. Therefore, $G^p_w(U;A)$ is also graded A-module.*

3) *(a) Let* $f : U \longrightarrow A$ *be a weak superdifferentiable function defined on an open subset* $U \subset A^{m,n}$*. Then, for each* $p \in \mathbb{N}$*, the p-derivative of f is a map*

$$D^p f : U \longrightarrow L_{A_0}(A^{m,n} \odot_{A_0} \cdots \odot_{A_0} A^{m,n}; A) .$$

*So we identify the following maps (p-***th partial derivative** *of f):*

$$((\partial_{A_p} \cdots {}_{A_1} \cdot f)) : U \longrightarrow ((A_1 \odot_{A_0} \ldots r \ldots \odot_{A_0} A_1)^{\oplus})^{\binom{p}{r} m^{p-r} n^r}, \; 1 \leq r \leq p,$$

if $|A_1| = \dots |A_{p-r}| = |0,\ |A_{p-r+1}| = \dots = |A_p| = |1;$

$$\big((\partial_{A_p} \cdots {}_{A_1} \cdot f)\big) : U \longrightarrow A^{m^p}, \ \ \textit{if}\ |A_1| = \dots |A_p| = 0\,,$$

such that

$$D^p f(x)(t_1, \dots, t_p) = t_p^{A_p} \dots t_1^{A_1} (\partial_{A_1} \cdots {}_{A_p} f)(x),$$

$$\textit{for all } t_i = (t_i{}^{A_i}) \in A^{m,n},\ (i = 1, \dots, p)\,,$$

$$(\partial_{A_p} \cdots {}_{A_1} \cdot f) = \big(\partial_{A_p} \cdot (\partial_{A_{p-1}} \cdots {}_{A_1} \cdot f)\big),$$

$$(\partial_{A_p} \cdots {}_{A_{k+1}} \cdot {}_{A_k} \therefore {}_{A_1} \cdot f) = (-1)^{|A_k||A_{k+1}|} (\partial_{A_p} \cdots {}_{A_k A_{k+1}} \cdots {}_{A_1} \cdot f)\,.$$

More precisely, these functions are defined by means of the following commutative diagram:

$$D^p f : U \longrightarrow \quad L_{A_0}(A^{m,n} \odot_{A_0} \cdots p \cdots \odot_{A_0} A^{m,n}; A) \cong$$

$$\cong A^{m^p} \times \prod_{1 \le r \le p} \big((A_1 \odot_{A_0} \cdots r \cdots \odot_{A_0} A_1)^{\oplus}\big)^{\binom{p}{r} m^{p-r} n^r}$$

$$\big\|\ (\partial_{A_p} \cdots {}_{A_1} \cdot f) \downarrow \ \text{if } |A_p| = \dots = |A_1| = 0$$

$$U \longrightarrow A$$

$$\big\|\ (\partial_{A_p} \cdots {}_{A_1} \cdot f) \downarrow \ \text{if } |A_p| = 1, |A_{p-1}| = \dots = |A_1| = 0$$

$$U \longrightarrow (A_1)^{\oplus}$$

$$\big\|$$

$$\cdots \qquad \cdots$$

$$\big\|\ (\partial_{A_p} \cdots {}_{A_1} \cdot f) \downarrow \ \text{if } |A_p| = \dots = |A_1| = 1$$

$$U \longrightarrow (A_1 \odot \cdots p \cdots \odot A_1)^{\oplus}$$

(b) Furthermore, as A is a BG-algebra, one has that

$$A_1^{\oplus} \cong A \cong (A_1 \otimes_{A_0} \dots \otimes_{A_0} A_1)^{\oplus} \supset (A_1 \odot_{A_0} \dots \odot_{A_0} A_1)^{\oplus}\,.$$

So the partial derivative of f are all A-valued functions on $U \subseteq A^{m,n}$.

1.2. Supermanifolds and superderivative spaces

DEFINITION 7. *A* **supermanifold of dimension** (m,n) *(over a superalgebra A) (of class G^k_w, $0 \le k \le \infty$, ω) is a locally convex manifold M, modeled on $A^{m,n}$ and with a G^k_w atlas Ξ of local coordinate mappings. So for each open coordinate set $U \subset M$ we have a set of $m+n$ coordinate functions $(x^A) : U \longrightarrow A^{m,n}$ such that $x^A : U \longrightarrow A_0$, $1 \le A \le m$, $x^A : U \longrightarrow A_1$, $m+1 \le A \le m+n$. We denote by $\mathcal{A}_p(M)$ the graded algebra of germs of A-valued weak supersmooth functions at $p \in M$.*

DEFINITION 8. *Let M be a supermanifold of dimension (m,n). The* **tangent space** *at $p \in M$ is the (m,n)-dimensional vector superspace T_pM defined in the following way: $T_pM \equiv$ space of equivalent classes $v = [f]$ of C^1_w (or equivalently C^1) curves $f : I \longrightarrow M$, I-open neighbourhood of $0 \in \mathbb{R}$, $f(0) = p$; two curves f,*

f' are equivalent if for each (equivalently, for some) coordinate system μ around p, the functions $\nu \circ f$, $\mu \circ f' : I \longrightarrow A^{m,n}$, have same derivative at $0 \in \mathbb{R}$.

REMARK 6. T_pM results whether a $\mathbb{K}$-vector space or a topological locally convex $\mathbb{K}$-vector space, so there exists whether the **dual**: $(T_pM)^* \equiv L_{\mathbb{K}}(T_pM;\mathbb{K})$ of T_pM, or the **topological dual** $(T_pM)' \equiv \mathcal{L}_{\mathbb{K}}(T_pM;\mathbb{K})$. On the other hand T_pM is also a (m,n)-dimensional vector superspace, hence an A_0-module. So we can give the following definitions:

DEFINITION 9. *Let M be a supermanifold of dimension (m,n). For any $p \in M$ put:*
*$\tilde{T}_pM \equiv AO_{A_0}T_pM$, (**supertangent space**); $(T_pM)^+ \equiv T_pMO_{A_0}A$, (**superdual of T_pM**).*

REMARK 7. (Relation between supertangent space and space of superderivations). It is useful consider also the following module $\mathrm{Der}(M)_p$ of all the derivations on $\mathcal{A}_p(M) \equiv$ graded algebra of germs of weak supersmooth functions $M \longrightarrow A$ at p. One has the following canonical isomorphism: $\tilde{T}_pM \xrightarrow{j} \mathrm{Der}(M)_p$, where j is given by $j(u) \cdot f = D(f \circ c)(0)$, with c any G^∞_w-curve $c : I \subset A \longrightarrow M$, $c(0) = p$, such that $Dc(0) = u$, and $\forall f \in \mathcal{A}_p(M)$. Set $(\mathrm{Der}(M)_p)^+ \equiv L_A(\mathrm{Der}(M)_p; A)$. One has the following canonical isomorphism: $(T_pM)^+ \xrightarrow{j} \mathrm{Der}(M)_p$, where j is given by $j(u)(v) = v \cdot f$, being f any G^∞_w-function of $\mathcal{A}_p(M)$ such that $Df(p) = u$, $\forall v \in \mathrm{Der}(M)_p$. $\{dx^A(p)\}_{1\leq A\leq m+n}$ is a basis for $\mathrm{Der}(M)^+_p$ and $\{\partial x_A(p)\}_{1\leq A\leq m+n}$ is a basis for $\mathrm{Der}(M)_p$.

NOTATION. We shall denote by $\overline{T}M$ the odd part of $\tilde{T}M = TM \oplus \overline{T}M$. Similar notations will be used for the corresponding tensor extensions. Furthermore, we shall denote by $\dot{T}^r_sM$ the tensor bundle of type (r,s) if each tangent space T_pM is considered as a A_0-module. The corresponding symmetric amd skewsymmetric bundles will be denoted by $\dot{S}^r_sM$ and $\dot{\Lambda}^r_sM$ respectively.

THEOREM 3. 1) *One has the canonical isomorphisms:*

$$(T_pM)^+ \cong AO_{A_0}(T_pM)^+_0 \cong AO_{A_0}(T_pM)^+_1 .$$

2) *One has the canonical isomorphisms:*

$$T_pM \cong (T_pM)^+_0 .$$

3) *One has the canonical pairing:*

$$\langle\ ,\ \rangle : \tilde{T}_pM \times (T_pM)^+ \longrightarrow A.$$

PROOF. 1) In fact one has the following canonical isomorphism:

$$AO_{A_0}(T_pM)^+_0 \cong A_0O_{A_0}(T_pM)^+_0 \times A_1O_{A_0}(T_pM)^+_0 \cong (T_pM)^+_0 \times A_1O_{A_0}(T_pM)^+_0 .$$

On the other hand $A_1O_{A_0}(T_pM)^+_0 \cong \mathrm{Hom}_{A_0}(A_1 \otimes_{A_0} T_pM; A_0)$. Now we can identify $\mathrm{Hom}_{A_0}(A_1 \otimes_{A_0} T_pM; A_0)$ with $(T_pM)^+_1$. In fact, one has the following canonical A_0-homomorphism $\phi : \mathrm{Hom}_{A_0}(T_pM; A_1) \rightarrow \mathrm{Hom}_{A_0}(A_1 \otimes_{A_0} T_pM; A_0)$, given by $\phi(f)(a_1 \otimes$

$v) = f(v) \cdot a_1$. Furthermore, $\ker\phi = \{0\}$, as $A_1^{\perp} = 0$. So taking into account the non-canonical isomorphisms

$$\mathrm{Hom}_{A_0}(T_pM; A_1) \cong A^{\bar{m},\bar{n}} \cong \mathrm{Hom}_{A_0}(A_1 \otimes_{A_0} T_pM; A_0)\,,$$

we can conclude that ϕ is an A_0-isomorphism. So we have

$$AO_{A_0}(T_pM)_0^+ \cong (T_pM)_0^+ \times (T_pM)_1^+ \cong (T_pM)^+\,.$$

The proof of the other part of the theorem follows similarly.

2) Let us fix a chart. Then, we define the following A_0-homomorphism:

$$\phi : T_pM \longrightarrow (T_pM)_0^+,\; v \equiv (v^A) \mapsto \phi(v),\; \phi(v)(u) = \phi(v^A)(u^B) =$$

$$= \sum_{1 \le A \le m+n} \phi_B u^B, \;\text{ with } \phi_B \equiv \delta_{BC} v^C\,.$$

One can see that above definition is indipendent on the chart chosen. In fact, let $(x^A) \longrightarrow (\bar{x}^A)$ be a changement of chart. One has:

$$\phi(v)(u) = \phi(\bar{v}^A)(\bar{u}^B) = \bar{\phi}_B \bar{u}^B = \left[\phi_A(\partial \bar{x}_B \cdot x^A)\right]\left[u^S(\partial x_S \cdot \bar{x}^B)\right]$$

$$= (-1)^{|S|(S|+|B|)+|S|(|S|+|B|)} \phi_A(\partial x_S \cdot \bar{x}^B)(\partial \bar{x}_B \cdot x^A) u^S$$

$$= \phi_A \delta_S^A u^S = \phi_A u^A\,.$$

Furthermore, $\ker\phi = \{0\}$, as $A_1^{\perp} = 0 = A_0^{\perp}$. So taking into account the non-canonical isomorphisms $T_pM \cong A^{m,n} \cong (T_pM)_0^+$, we can conclude the proof.

3) One has the canonical function $f_p : A \longrightarrow A$, given by camposition $f_p \equiv \alpha_p \circ v_p : A \longrightarrow T_pM \longrightarrow A$. So the pairing $\langle\;,\;\rangle_p$ can be defined by $\langle v_p, \alpha_p\rangle = f(1)$, where $1 \in \mathbb{K} \subset A$. □

REMARK 8. (Graded manifolds and supermanifolds). Another important concept, introduced by Berezin [3] and Kostant [14], is that of graded manifold. Bachelor has proved that such supermanifolds can be represented by means of sections of bundles ΛE, for some suitable vector bundle E [2]. The category of supermanifolds is larger than that of graded manifolds. However, for physical applications we can consider a more specialistic category of supermanifolds. More precisely, we will call **regular supermanifold** a supermanifold M of dimension (m,n) such that it admits a projection on a m-dimensional manifold M_B. We will call M_B the **body** of M. This is equivalent to say that for any $p \in M$ we have a system of coordinates adapted to the projection $b : M \longrightarrow M_B$. Then it is possible prove that any Kostant graded manifold (X, A) of dimension (m,n) identifies a regular supermanifold M of dimension (m,n), such that its body M_B coincides with X and A is isomorphic to the sheaf $\mathcal{S} \circ b^{-1}$, where $b : M \longrightarrow M_B$ is the canonical projection and $\mathcal{S}$ is a suitable sheaf of ($A \equiv \Lambda\mathbb{R}^L$)-valued functions on M, $0 \leq L \leq \infty$ [36]. In the following we shall consider only regular supermanifolds.

The following theorem gives a representation of graded p-differential forms as A-valued differential forms on the corresponding body manifolds.

THEOREM 4. *Let M be a supermanifold of dimension (m,n). For any section i of $b : M \longrightarrow M_B$ we have a canonical A-homomorphism*

$$i_* : (T_{i(p)}M)^+ \longrightarrow A \otimes_{\mathbb{K}} (T_pM_B)^* .$$

As a consequence, we have that, for any section i of b we can calculate the pull-back of any graded p-differential form $\alpha : M \longrightarrow \tilde{\Lambda}^0_p M$ on M. We have:

$$i^*\alpha \equiv i_* \circ \alpha \circ i : M_B \longrightarrow A \otimes_{\mathbb{K}} \Lambda^0_p M_B .$$

Any couple (s,i), where s is a G^∞_w section of π and i is a section of b, identifies a mapping $\tilde{\Lambda}^0_p J\underline{D}^k(W) \longrightarrow A \otimes_{\mathbb{K}} \Lambda^0_p M_B$ such that the following diagram is commutative:

$$\begin{array}{ccccc}
A \otimes_{\mathbb{K}} \Lambda^0_p M_B & \xleftarrow{\lambda(i)} & \tilde{\Lambda}^0_p M & \xleftarrow{\lambda(s)} & \tilde{\Lambda}^0_p J\underline{D}^k(W) \\
\downarrow \underline{\pi} & & \overset{0}{\pi} \downarrow & & \downarrow \bar{\pi} \\
M_B & \xrightarrow{\mathrm{i}} & M & \xrightarrow{\mathrm{D^k s}} & \underline{D}^k(W)
\end{array}$$

Then, any section α of $\bar{\pi}$ identifies a section $\big((D^k s) \circ i\big)^*\alpha$ of $\underline{\pi}$ given by

$$\big((D^k s) \circ i\big)^*\alpha = \lambda(s,i) \circ \alpha \circ (D^k s) \circ i,\ \lambda(s,i) \equiv$$
$$\equiv \lambda(i) \circ \lambda(s),\ \lambda(s,i) \equiv I_p \circ \Lambda^p(i_*) \circ \tilde{\Lambda}^0_p(D^k s) ,$$

being I_p the canonical isomorphism $I_p : \Lambda^p\big(A \otimes_{\mathbb{K}} (T_pM_B)^*\big) \cong A \otimes_{\mathbb{K}} \Lambda^p\big((T_pM_B)^*\big)$.

PROOF. In fact, for any i, one has the canonical A-homomorphism

$$h_i : \big(T_{i(p)}M\big)^+ \longrightarrow L_A(A \otimes_{\mathbb{K}} T_pM_B; A) \equiv (A \otimes_{\mathbb{K}} T_pM)^+$$

given by

$$f \mapsto h_i(f)(a \otimes v) = a.f\big(T(i)(v)\big) \in A,\ \forall a \in A,\ v \in T_pM_B,\ f \in (T_{i(p)}M)^+ .$$

Note that h_i is not injective. In fact $K(i) \equiv \ker h_i = \{f \in (T_{i(p)}M)^+ | f | T_{i(p)} i(M_B) = 0\}$. Moreover as T_pM_B is a $\mathbb{K}$-module projective of finite type, we have the A-isomorphism: $(A \otimes_{\mathbb{K}} T_pM_B)^+ \cong A \otimes_{\mathbb{K}} (T_pM_B)^*$. □

REMARK 9. **Lie supergroup** G (modeled on $A^{m,n}$), is defined to be simultaneously a supermanifold of dimension (m,n) such that the mapping $G \times G \longrightarrow G$, $(f,g) \mapsto f \circ g^{-1}$ is superdifferentiable. The set of left-invariant supervector fields on G forms the **Lie superalgebra** $\tilde{A}(G)$ of G. $\tilde{A}(G)$ is just a $(\mathbb{Z}_2)$-graded Lie algebra [35]. In fact $\tilde{A}(G)$ is a graded vector space: $\tilde{A}(G) = A(G) \oplus \overline{A}(G) \equiv \tilde{A}(G)_0 \oplus \tilde{A}(G)_1$, where $A(G)$ (resp. $\overline{A}(G)$) is the set of left-invariant even vector fields (resp. odd vector fields) on G, endowed with the bilinear operation $[X,Y] \equiv L_X Y$. Really, this braket satisfies the following conditions:

$$\text{i)}[X,Y] \in \tilde{A}(G)_{|X|+|Y|}; \quad \text{ii)}[X,Y] = -(-1)^{|X||Y|}[Y,X];$$
$$\text{iii)}(-1)^{|X||Z|}[[X,Y],Z] + (-1)^{|Y||X|}[[Y,Z],X] + (-1)^{|Z||Y|}[[Z,X],Y] = 0.$$

$A(G)$ is an ordinary Lie algebra, with respect to the braket [,]. $A(G)$ and $\tilde{A}(G)$ are A_0-modules. $\tilde{A}(G)$ is a A-module too.

Let M and P be supermanifolds, G a Lie supergroup acting superdifferentially on P, and $\pi : P \longrightarrow M$ a surjective superdifferentiable mapping. The **principal bundle** $(P, M, \pi; G)$ is defined just like on ordinary manifolds, but taking into account that now we are working in the category of supermanifolds.

A graded principal superconnection is a G-compatible superlinear splitting of the following exact sequence of A_0-vector bundles over $P : 0 \longrightarrow v\tilde{T}P \longrightarrow \tilde{T}P \longrightarrow \pi^*\tilde{T}M \longrightarrow 0$. A principal connection gives a splitting $\tilde{T}_pP \cong v\tilde{T}_pP \oplus \tilde{H}_p$, where $\tilde{H}_p$ is the horizontal subspace at $p \in P$ such that the following conditions hold: 1) $\tilde{H}_{pa} = (R_a)^*(\tilde{H}_p)$, for any $p \in P$ and $a \in G$, where R_a is the transformation in P induced by the element $a \in G$, namely $R_a p = pa$; 2) $\tilde{H}_p$ depends superdifferentially on p. Each vector field $X \in \tilde{A}(G)$ induces a fundamental supervector field $\sigma(X)$ on P, and $\sigma : X \longrightarrow \sigma(X)(p)$ is a A_0-isomorphism $\tilde{A}(G) \cong v\tilde{T}_pP$. For any $Y \in \tilde{T}_pP$, let $\tilde{\omega}(Y)$ be the unique $X \in \tilde{A}(G)$ such that $\sigma(X)(p)$ is the vertical component for Y. We call the map $\tilde{\omega} : P \longrightarrow \tilde{T}PO_{A_0}\tilde{A}(G)$ **Ehresmann graded superconnection** of the given principal superconnection on P. Similarly we can define a **principal superconnection** on a principal bundle $(P, M, \pi : G)$ as a G-compatible superlinear-splitting of the following exact sequnce of A_0-vector bundles over P : $0 \longrightarrow vTP \longrightarrow TP \longrightarrow \pi^*TM \longrightarrow 0$. Then, the corresponding **Ehresmann superconnection** ω is a A_0-homomorphism $TP \longrightarrow A(G)$ over $P : \omega : TPO_{A_0}A(G)$. Any graded principal superconnection $\tilde{\omega}$ identifies a principal superconnection ω on P. In fact one has the following commutative diagram where whether the columns or the rows are exact:

$$\begin{array}{ccccc}
0 & \longrightarrow & \tilde{T}PO_{A_0}\tilde{A}(G) & \longrightarrow & L_{A_0}\big(\tilde{T}P; \tilde{A}(G)\big) \\
 & & {\scriptstyle\tilde{\omega}}\,\updownarrows & & \| \\
 & & P & & L_{A_0}\big(TP;\tilde{A}(G)\big) \times L_{A_0}\big(\overline{T}P;\tilde{A}(G)\big) \\
 & & {\scriptstyle\omega}\,\updownarrows & & \downarrow \\
0 & \longrightarrow & TPO_{A_0}A(G) & \longrightarrow & L_{A_0}\big(TP; A(G)\big) \\
 & & & & \downarrow \\
 & & & & 0
\end{array}$$

Furthermore, if $A(G)$ (or T_pP, $\forall p \in P$) is a A_0-module projective of finite type, we have the isomorphism

$$(TP)^{+}{}_{0} \otimes_{A_0} A(G) \cong TPO_{A_0}A(G)\,,$$

and so any Eheresmann superconnection on P can be identified with $A(G)$-valued even differential forms on P.

Definition 10. *The* **superderivative space** $\underline{D}(M,N)$ *for supermanifolds* M *and* N *of dimension* (m,n) *and* (m',n') *respectively is defined by*

$$\underline{D}(M,N) \equiv TMO_{A_0}TN \equiv \bigcup_{(p,q)\in M\times N} T_pMO_{A_0}T_qN\,,$$

where $T_pMO_{A_0}T_qN$ *is the space of superlinear mappings* $T_pM \longrightarrow T_qN$*.*

Remark 10. One has a natural injection over $M\times N : \underline{D}(M,N) \longrightarrow D(M,N)$, where $D(M,N)$ denotes the derivative space considering M and N as locally convex manifolds [13, 15, 38]. $\underline{D}(M;N)$ results a supermanifold of dimension $(m+m'+m\cdot m'+n\cdot n', n+n'+m\cdot n'+n\cdot m')$. If (x^A) is a coordinate system on M and (y^B) is a coordinate system on N, then (x^A, y^B, y^B_A), $1\leq A\leq m+n$, $1\leq B\leq m'+n'$, is a coordinate system on $\underline{D}(M,N)$.

The first **derivative** of a super differentiable mapping $f : M \longrightarrow N$ is a section $Df : M \longrightarrow \underline{D}(M,N)$ of the fiber bundle $\pi_1 : \underline{D}(M,N) \longrightarrow M$. The local expression of Df is as follows: $x^A\circ Df = x^A$, $y^B\circ Df = f^B \equiv y^B\circ f$, $y^B_A\circ Df = (\partial x_A\cdot f^B)$. The **velocity** of a superdifferentiable curve in M, $f : A \longrightarrow M$, is a map $\dot{f} : A \longrightarrow \tilde{T}M$ given by means of the following commutative diagram:

$$\begin{array}{ccc}
\underline{D}(A,M) & \xrightarrow{\;\cong\;} & A\times\tilde{T}M \\
\uparrow{\scriptstyle Df} & & \downarrow \\
A & \xrightarrow{\;\dot{f}\;} & \tilde{T}M
\end{array}$$

The **differential** of a superdifferentiable map $f : M \longrightarrow A$ is a map $df : M \longrightarrow (TM)^+$ given by means of the following commutative diagram:

$$\begin{array}{ccc}
M & \xrightarrow{\;Df\;} & \underline{D}(M,A) \\
{\scriptstyle df}\downarrow & & \|\,\tilde{} \\
(TM)^+ & \longleftarrow & A\times(TM)^+
\end{array}$$

For a superdifferentiable map $f : A\times M \longrightarrow N$ we can define **infinitesimal variation** $\partial f : M \longrightarrow f_0{}^*\tilde{T}M$ as $\partial f = D_1 f\circ j$, being j the inclusion $j : M \longrightarrow A\times M$ given by $p\mapsto (0,p)$. So, if f is a superdeformation of a superdifferentiable section s of the fiber bundle $p : W \longrightarrow M$, such that $f_0 = s$, then $\partial f : M \longrightarrow s^* v\tilde{T}W$, where the symbol v denotes "vertical".

From above considerations we get the following important theorem.

Theorem 5. 1) $\tilde{T}M$ *can be identified with the space of vectors tangent to superdifferentiable curves* $A \longrightarrow M$ *of* M*.*

2) *$((TM)^+)$ can be identified with the space of differentials of superdifferentiable A-valued functions on M.*

DEFINITION 11. 1) *If $\pi : W \longrightarrow M$ is a fiber bundle between supermanifolds, the* **superderivative space of section** *of π, $\underline{D}(W)$, is defined by means of the following commutative diagram:*

$$\begin{array}{ccc} \underline{D}(W) \equiv d^*\underline{D}(M,W) & \longrightarrow & \underline{D}(M,W) \\ & & \downarrow \\ \downarrow & & M \times W \\ & & \downarrow \\ M & \xrightarrow{d} & M \times M \end{array}$$

where d is the diagonal map.

2) **Superderivative spaces of higher orders** *between supermanifolds can be defined putting $\underline{D}^k(M,N) \equiv \underline{D}\big(\underline{D}^{k-1}(M,N)\big)$.*

REMARK 11. We get natural structures of fiber bundles on $\underline{D}^k(M,N)$ as in the case of *l.c.* manifolds: $\pi_{k,h} : \underline{D}^k(M,N) \longrightarrow \underline{D}^h(M,N)$, $0 \leq h \leq k$, $\pi_k : \underline{D}^k(M,N) \longrightarrow M$, $\bar{\pi}_k : \underline{D}^k(M,N) \longrightarrow N$. If $\pi : W \longrightarrow M$ is a fiber bundle between supermanifolds then the k-**superderivative space of sections** of π is a subbundle $\underline{D}^k(W)$ of $\underline{D}^k(M,N)$ over M, and into $\underline{D}^k(W)$ we can distinguish some further subbundles homonymous to classic case ones. For example the **holonomic k-jet superderivative space**:

$$\begin{aligned} J\underline{D}^k(W) \equiv \big\{u \in \underline{D}^k(W) | \exists \text{ superdifferentiable sections } s \text{ of } \pi : D^k s(p) = \\ = u,\ p = \pi_k(u)\big\}. \end{aligned}$$

A system of coordinates on a k-jet superderivative space can be written:

$$\{x^A, y^B, y^B{}_A, y^B{}_{A_1 A_2}, \dots, y^B{}_{A_1 \cdots A_k}\},\ 1 \leq A \leq m+n,\ 1 \leq B \leq m' + n'$$

where

$$y^B{}_{A_1 \cdots A_s A_{s+1} \cdots A_q} = (-1)^{|A_s||A_{s+1}|} y^B{}_{A_1 \cdots A_{s+1} A_s \cdots A_q}\,.$$

So $J\underline{D}^k(W)$ is a supermanifolds of dimension $\big(m' + (m'-m)P + (n'-n)Q, n' + (m'-m)Q + (n'-n)P\big)$, with

$$\begin{aligned} P &= \sum\nolimits_{0 \leq 2j \leq \min(n,k)} \binom{n}{2j} \left[\binom{m+k-2j}{k-2j} - 1\right]; \\ Q &= \sum\nolimits_{0 \leq 2j+1 \leq \min(n,k)} \binom{n}{2j+1} \left[\binom{m+k-2j-1}{k-2j} - 1\right]. \end{aligned}$$

One has:

$$|y^B{}_{A_1 \cdots A_p}| = |B| + |A_1| + \ldots + |A_p|\,.$$

PROPOSITION 1. 1) *$J\underline{D}^k(W)$ is an affine bundle over $J\underline{D}^{k-1}(W)$ with associated vector bundle $\pi^*_{k,0}(\dot{S}^k_0 MO_{A_0} vTW)$. So we have the following exact sequence of vector bundles over $J\underline{D}^k(W)$:*

$$0 \longrightarrow \pi^*_{k,0}(\dot{S}^k_0 MO_{A_0} vTW) \longrightarrow vTJ\underline{D}^k(W) \longrightarrow \pi^*_{k,k-1} vT\underline{D}^{k-1}(W) \longrightarrow 0 .$$

2) The relation with k-jet derivative spaces is given by the following exact commutative diagram:

$$\begin{array}{ccccccccc} & & 0 & & 0 & & 0 & & \\ & & \downarrow & & \downarrow & & \downarrow & & \\ 0 & \longrightarrow & \pi^*_{k,0}(\dot{S}^k_0 MO_{A_0} vTW & \longrightarrow & vTJ\underline{D}^k(W) & \longrightarrow & \pi^*_{k,k-1} vT\underline{D}^{k-1}(W) & \longrightarrow & 0 \\ & & \downarrow & & \downarrow & & \downarrow & & \\ 0 & \longrightarrow & \pi^*_{k,0}(S^k_0 MO_{\mathbb{K}} vTW) & \longrightarrow & vTJD^k(W) & \longrightarrow & \pi^*_{k,k-1} vTJD^{k-1}(W) & \longrightarrow & 0 . \end{array}$$

PROPOSITION 2. *Let M and N be supermanifolds of dimension (m,n) and (m',n') respectively, with respect to a superalgebra A. Then, for the superderivative space $\underline{D}(M,N)$ one has the following commutative diagrams of fiber bundles over $M \times N$, where the lines are exact:*

(A)

$$\begin{array}{ccccc} 0 & \longrightarrow & \underline{D}(M,N) & \longrightarrow & L_{A_0}(TM;TN) \\ & & \uparrow & & \uparrow \\ 0 & \longrightarrow & (TM)^+_0 \otimes_{A_0} TN & \longrightarrow & (TM)^{\oplus}_0 \otimes_{A_0} TN \end{array}$$

(B) (TN=fiber bundle of projective A_0-modules)

$$\begin{array}{ccccc} & & \underline{D}(M,N) & & \\ & & \uparrow & & \\ 0 & \longrightarrow & (TM)^+_0 \otimes_{A_0} TN & \longrightarrow & L_{A_0}(TM;TN) \\ & & \uparrow & & \\ & & 0 & & \end{array}$$

(C) (TN (or TM)=fiber bundle of projective A_0-modules of finite type)

$$0 \longrightarrow \underline{D}(M,N) \cong (TM)^+_0 \otimes_{A_0} TN \longrightarrow 0 .$$

2. Superchomology of supermanifolds

In order to develop a formal theory of super PDE's we shall use a (co)homological theory for supermanifolds. So scope of this chapter is to explicitly characterize the (co)homological structures of supermanifolds.

2.1. The de Rham supercohomology

Set $\widetilde{\Omega}^\bullet(M) \equiv \bigoplus_{p\geq 0}\widetilde{\Omega}^p(M)$, where $\widetilde{\Omega}^p(M) \equiv G^\infty_w(\Lambda^p T^+ M) = G^\infty_w(\widetilde{\Lambda}^0_p M)$. $\widetilde{\Omega}^\bullet(M)$ has a natural structure of A-module. The **exterior superdifferential** is a A-linear first order differential operator $\tilde{d}$: $\widetilde{\Omega}^\bullet(M) \longrightarrow \widetilde{\Omega}^\bullet(M)$, defined by means of the following commutative diagram, for any superdifferential form $\alpha \in \widetilde{\Omega}^p(M)$

$$\begin{array}{ccccc}
M & \xrightarrow{D\alpha} & J\underline{D}(\widetilde{\Lambda}^0_p M) & \longrightarrow & JD(\widetilde{\Lambda}^0_p M) \\
 & & \|\wr & & \|\wr \\
\tilde{d}\alpha\downarrow & & \widetilde{\Sigma}^0_p \oplus \widetilde{\Lambda}^0_{p+1}M & \longrightarrow & \Sigma^p \oplus \widetilde{\Lambda}^0_{p+1}M \\
 & & \downarrow & & \downarrow \\
\widetilde{\Lambda}^0_{p+1}M & = & \widetilde{\Lambda}^0_{p+1}M & = & \widetilde{\Lambda}^0_{p+1}M
\end{array}$$

where $J\underline{D}(\widetilde{\Lambda}^0_p M)$ is the first jet-superderivative space of the bundle $\widetilde{\Lambda}^0_p M \longrightarrow M$ and $\widetilde{\Sigma}^p$ (resp. Σ^p) is a $\mathbb{K}$-vector subbundle over M of $J\underline{D}(\widetilde{\Lambda}^0_p M)$ (resp. $JD(\widetilde{\Lambda}^0_p M)$). One case see that $\tilde{d}^2 \equiv \tilde{d}\circ\tilde{d} = 0$, so that $\{\widetilde{\Omega}^\bullet(M), \tilde{d}\}$ is a cochain complex (**de Rham supercomplex** of M). (For abuse of notation we shall write d for $\tilde{d}$).

DEFINITION 1. *The* **de Rham supercohomology** *of the supermanifold M is the homology $\widetilde{H}^\bullet(M)$ of the de Rham supercomplex on M.*

In the following table we also define some further useful cochain complexes and the corresponding cohomology spaces.

THEOREM 0. *One has the following canonical A-isomorphisms:*

$$\begin{aligned}
\widetilde{\Omega}^\bullet(M) &\cong AO_{A_0}\dot{\Omega}^\bullet(M) \cong \dot{\Omega}^\bullet(M) \oplus A_1 O_{A_0}\dot{\Omega}^\bullet(M) \\
&\cong AO_{A_0}\overline{\Omega}^\bullet(M) \cong \overline{\Omega}^\bullet(M) \oplus A_1 O_{A_0}\overline{\Omega}^\bullet(M)\,.
\end{aligned}$$

PROOF. In order to prove this theorem we shall use some lemmas.

LEMMA 1. *One has the following canonical isomorphisms:*

$$\begin{aligned}
\widetilde{T}^0_p M &\equiv \bigotimes\nolimits^p_A T^+ M \cong (\bigotimes\nolimits^p_{A_0} TM) O_{A_0} A \\
&\equiv \mathrm{Hom}_{A_0}(\bigotimes\nolimits^p_{A_0} TM; A) \equiv \mathrm{Hom}_{A_0}(\dot{T}^p_0 M; A)\,.
\end{aligned}$$

PROOF OF LEMMA 1. In fact one has the following canonical A-homomorphism:

$$\phi : \bigotimes\nolimits^p_A T^+ M \longrightarrow \mathrm{Hom}_{A_0}(\bigotimes\nolimits^p_{A_0} TM; A\bigotimes\nolimits_A \cdots \bigotimes\nolimits_A A) \cong \mathrm{Hom}_{A_0}(\bigotimes\nolimits^p_{A_0} TM; A)\,,$$

given by $\phi(\alpha_1 \otimes \ldots \otimes \alpha_p)(v_1 \otimes \ldots \otimes v_p) = \alpha_1(v_1)\ldots\alpha_p(v_p) \in A$. The mapping ϕ is injective. Infact, in any chart, one has:

$$\alpha_1(v_1)\ldots\alpha_p(v_p) = \alpha_{A_1\ldots A_p} z^{A_1\ldots A_p}\,, \quad \text{with} \quad \alpha_{A_1\ldots A_p} \in A, z^{A_1\ldots A_p} \in A\,.$$

So in order to have $\phi(\alpha_1 \otimes \ldots \otimes \alpha_p) = 0$, one should have $\alpha_{A_1\ldots A_p} z^{A_1\ldots A_p} = 0$, for any $z^{A_1\ldots A_p} \in A$; in particular for $z^{A_1\ldots A_p} = 1$; this implies that $\alpha_{A_1\ldots A_p} = 0$. Now, taking into account the following non-canonical isomorphisms:

$$\bigotimes_A^p T^+M \cong A^{(m+n)^p} \cong \mathrm{Hom}_{A_0}(\bigotimes_{A_0}^p TM; A)$$

we conclude that ϕ is a canonical isomorphism. □

Tab. 2

Cochain complex	Cohomology spaces	Name
$\{\widetilde{\Omega}^\bullet(M), \tilde{d}\}$	$\widetilde{H}^\bullet(M)$	de Rham supercohomology
$\{\dot{\Omega}^\bullet(M), \dot{d}\}$	$\dot{H}^\bullet(M)$	de Rham even supercohomology
$\{A_1 O_{A_0} \dot{\Omega}^\bullet(M), \bar{d}\}$	$\overline{H}^\bullet(M)$	de Rham odd supercohomology
$\{\Omega^\bullet(M), d\}$	$H^\bullet(M)$	de Rham cohomoloy
$\{A \bigotimes_{\mathbb{K}} \Omega^\bullet(M), d\}$	$H^\bullet(M; A)$	de Rham cohomology with coefficients in A
$\{A_0 \bigotimes_{\mathbb{K}} \Omega^\bullet(M), d\}$	$H^\bullet(M; A_0)$	de Rham cohomology with coefficients in A_0
$\{A_1 \bigotimes_{\mathbb{K}} \Omega^\bullet(M), d\}$	$H^\bullet(M; A_1)$	de Rham cohomology with coefficients in A_1

Note. $\dot{\Lambda}_p^0 M = (\Lambda^p T^+M)_0^+$, $\overline{\Lambda}_p^0 M = (\Lambda^p T^+ M)_1^+$, $\Omega^p(M) \equiv C_w^\infty(\Lambda_p^0 M)$, $\dot{\Omega}^p(M) \equiv G_w^\infty(\dot{\Lambda}_p^0 M)$, $\overline{\Omega}^p(M) \equiv G_w^\infty(\overline{\Lambda}_p^0 M)$, $\widetilde{\Omega}^p(M) \equiv G_w^\infty(\widetilde{\Lambda}_p^0 M)$; $\widetilde{\Lambda}_0^0 M \equiv M \times A$; $\dot{\Lambda}_0^0 M \equiv M \times A_0$; $\overline{\Lambda}_0^0 M \equiv M \times A_1$; $\widetilde{\Omega}^0(M) = G_w^\infty(M, A)$; $\dot{\Omega}^0(M) = G_w^\infty(M, A_0)$; $\overline{\Omega}^0(M) = G_w^\infty(M, A_1)$. Furthermore, we put $\bigotimes_{A_0}^0 T_p M \equiv A_0$.

Lemma 2. *One has the following canonical splitting:* $\widetilde{T}_p^0 M \cong (\bigotimes_{A_0}^p TM)_0^+ \oplus (\bigotimes_{A_0}^p TM)_1^+$.

Lemma 3. *One has the following canonical A-isomorphism:*

$$\widetilde{T}_p^0 M \cong AO_{A_0}(\bigotimes_{A_0}^p TM)_0^+ \cong AO_{A_0} TM)_1^+$$
$$\cong AO_{A_0}(\dot{T}_0^p M)_0^+ \cong AO_{A_0}(\dot{T}_0^p M)_1^+ .$$

Proof of Lemma 3. We have: $AO_{A_0}(\bigotimes_{A_0}^p TM)_0^+ \cong (\bigotimes_{A_O}^p TM)_0^+ \oplus A_1 O_{A_0} (\bigotimes_{A_0}^p TM)_0^+$.

On the other hand

$$\begin{aligned} A_1 O_{A_0}(\bigotimes_{A_0}^p TM)_0^+ &\cong \mathrm{Hom}_{A_0}(A_1; \mathrm{Hom}_{A_0}(\bigotimes_{A_0}^p TM; A_0)) \\ &\cong \mathrm{Hom}_{A_0}(A_1 \bigotimes_{A_0} (\bigotimes_{A_0}^p TM); A_0) \\ &\cong \mathrm{Hom}_{A_0}((\bigotimes_{A_0}^p TM) \bigotimes_{A_0} A_1; A_0) \\ &\cong \mathrm{Hom}_{A_0}((\bigotimes_{A_0}^p TM); \mathrm{Hom}_{A_0}(A_1; A_0)) \\ &\cong \mathrm{Hom}_{A_0}(\bigotimes_{A_0}^p TM; A_1) \cong (\bigotimes_{A_0}^p TM)_1^+ . \end{aligned}$$

So we have: $AO_{A_0}(\bigotimes_{A_0}^p TM)_0^+ \cong (\bigotimes_{A_0}^p TM)_0^+ \oplus (\bigotimes_{A_0}^p TM)_1^+ \cong \widetilde{T}_p^0 M$. The other part of the lemma can be proved in a similar way. □

LEMMA 4. *One has the following isomorphisms:* $\widetilde{\Lambda}^0_p M \cong AO_{A_0}(\Lambda^p TM)^+_0 \cong AO_{A_0}(\Lambda^p TM)^+_1$.

PROOF OF LEMMA 4. This is obtained by means of the following exact commutative diagram:

$$\begin{array}{ccccccc}
0 & \longrightarrow & \widetilde{T}^0_p M & \longrightarrow & AO_{A_0}(\dot{T}^p_0 M)^+_i & \longrightarrow & 0 \\
 & & \uparrow & & \downarrow & & \\
0 & \longrightarrow & \widetilde{\Lambda}^0_p M & \longrightarrow & AO_{A_0}(\dot{\Lambda}^p_0 M)^+_i & \longrightarrow & 0, i \in \mathbb{Z}_2 . \\
 & & \uparrow & & & & \\
 & & 0 & & & &
\end{array}$$

□

LEMMA 5. *One has the following isomorphisms:*

$$\widetilde{\Lambda}^0_p M \cong \dot{\Lambda}^0_p M \oplus A_1 O_{A_0} \dot{\Lambda}^0_p M \cong \overline{\Lambda}^0_p M \oplus A_1 O_{A_0} \overline{\Lambda}^0_p M .$$

Then, from above Lemma 5, taking into account that

$$G^\infty_w(A_1, O_{A_0} \dot{\Lambda}^0_p M) \cong A_1 O_{A_0} G^\infty_w(\dot{\Lambda}^0_p M) = A_1 O_{A_0} \dot{\Omega}^p(M)$$

we get the proof of the first two steps in the theorem. Finally in order to get the two last ones we must observe the following:

$$\begin{aligned}
&\mathrm{Hom}_{A_0}(A; \dot{\Omega}^p(M)) \cong \mathrm{Hom}_{A_0}(A; \mathrm{Hom}_{A_0}(\widetilde{\Omega}^p(M); A_0)) \\
\cong\, &\mathrm{Hom}_{A_0}(\widetilde{\Omega}^p(M); A) = \dot{\Omega}^p(M) \oplus \overline{\Omega}^p(M); \\
&\mathrm{Hom}_{A_0}(A; \overline{\Omega}^p(M)) \cong \mathrm{Hom}_{A_0}(A : \mathrm{Hom}_{A_0}(\widetilde{\Omega}^p(M); A_1)) \\
\cong\, &\mathrm{Hom}_{A_0}(\widetilde{\Omega}^p(M); A) = \dot{\Omega}^p(M) \oplus \overline{\Omega}^p(M) .
\end{aligned}$$

□

Note. The superdifferential $\tilde{d}$ splits into two mappings: $\tilde{d} = (\dot{d}, \bar{d})$. For abuse of notation we shall also write all these symbols simply by d.

THEOREM 1. *One has the following canonical exact commutative diagram of* $\mathbb{K}$*-modules:*

$$\begin{array}{ccccc}
 & & 0 & & 0 \\
 & & \uparrow & & \uparrow \\
0 & \longrightarrow & A_1 O_{A_0} \dot{\Omega}^\bullet(M) & \longrightarrow & \mathrm{Hom}_{\mathbb{K}}(A_1, \dot{\Omega}^\bullet(M) \\
 & & \uparrow & & \uparrow \\
0 & \longrightarrow & \widetilde{\Omega}^\bullet(M) & \longrightarrow & \mathrm{Hom}_{\mathbb{K}}(A, \dot{\Omega}^\bullet(M)) \\
 & & \uparrow & & \uparrow \\
0 & \longrightarrow & \dot{\Omega}^\bullet(M) & \longrightarrow & \mathrm{Hom}_{\mathbb{K}}(A_0, \dot{\Omega}^\bullet(M)) \\
 & & \uparrow & & \uparrow \\
 & & 0 & & 0
\end{array}$$

The first vertical line and the horizontal lines are exact lines of A_0-modules homomorphisms also. Furthermore, one has the canonical isomorphisms of $\mathbb{K}$*-modules:*

$$\mathrm{Hom}_{\mathbb{K}}(A, \dot{\Omega}^\bullet(M) \cong \mathrm{Hom}_{\mathbb{K}}(A_0, \dot{\Omega}^\bullet(M)) \oplus \mathrm{Hom}_{\mathbb{K}}(A_1, \dot{\Omega}^\bullet(M)).$$

PROOF. It follows directly from above results. □

THEOREM 2. 1) *One has the following long exact sequences:*
(i) *(A_0-modules):*

$$\dot{H}^0(M) \longrightarrow \widetilde{H}^0(M) \longrightarrow \overline{H}^0(M) \longrightarrow \dot{H}^1(M) \longrightarrow \dots$$
$$\dots \longrightarrow \dot{H}^q(M) \longrightarrow \widetilde{H}^q(M) \longrightarrow \overline{H}^q(M) \longrightarrow \dot{H}^{q+1}(M) \longrightarrow \dots$$

(ii) *($\mathbb{K}$-modules):*

$$\mathrm{Hom}_{\mathbb{K}}(A_0, \dot{H}^0(M)) \longrightarrow \mathrm{Hom}_{\mathbb{K}}(A, \dot{H}^0(M)) \longrightarrow$$
$$\longrightarrow \mathrm{Hom}_{\mathbb{K}}(A_1, \dot{H}^0(M)) \longrightarrow \mathrm{Hom}_{\mathbb{K}}(A_0, \dot{H}^1(M)) \longrightarrow \dots$$
$$\dots \longrightarrow \mathrm{Hom}_{\mathbb{K}}(A_0, \dot{H}^q(M)) \longrightarrow \mathrm{Hom}_{\mathbb{K}}(A, \dot{H}^q(M)) \longrightarrow$$
$$\longrightarrow \mathrm{Hom}_{\mathbb{K}}(A_1, \dot{H}^q(M)) \longrightarrow \mathrm{Hom}_{\mathbb{K}}(A_0, \dot{H}^{q+1}(M)) \longrightarrow$$

2) *One has the following isomorphisms*
(A_0-modules):

$$\widetilde{H}^q(M) \cong \dot{H}^q(M) \times \overline{H}^q(M);$$

($\mathbb{K}$-modules):

$$\mathrm{Hom}_{\mathbb{K}}(A, \dot{H}^q(M)) \cong \mathrm{Hom}_{\mathbb{K}}(A_0, \dot{H}^q(M)) \oplus \mathrm{Hom}_{\mathbb{K}}(A_1, \dot{H}^q(M)).$$

Furthermore, $\widetilde{H}^q(M) = 0$ *iff* $\dot{H}^q(M) = 0$.

PROOF. It is a direct consequence of Theorem 1 and of a well known result of homological algebra concerning the cohomology of shorty exact sequences $0 \longrightarrow A \longrightarrow B \longrightarrow C \longrightarrow 0$ of cochain complexes [11, 26]. Furthermore, we have used the dual universal coefficient theorem of homological algebra [11], and the fact that A is a projective A_0-module. □

THEOREM 3. *One has the canonical isomorphisms of* $\mathbb{K}$*-modules:*

$$H^q(M; A) \cong A \bigotimes_{\mathbb{K}} H^q(M); \qquad H^q(M; A_0) \cong A_0 \bigotimes_{\mathbb{K}} H^q(M);$$
$$H^q(M; A_1) \cong A_1 \bigotimes_{\mathbb{K}} H^q(M); \quad H^q(M; A) \cong H^q(M; A_0) \times H^q(M; A_1).$$

PROOF. Note that the cochain complex $\{\Omega^\bullet(M), d\}$ is a flat cochain complex of $\mathbb{K}$-modules and that A is a $\mathbb{K}$-module. So we can apply the universal coefficient theorem of homological algebra to have the following canonical exact sequence of $\mathbb{K}$-modules:

$$0 \longrightarrow A \bigotimes_{\mathbb{K}} H^q(M) \longrightarrow H^q(M; A) \longrightarrow \mathrm{Tor}_1^{\mathbb{K}}(A, H^{q-1}(M)) \longrightarrow 0$$

and the following non-canonical isomorphsm of $\mathbb{K}$-modules:

$$H^q(M;A) \cong A \bigotimes_{\mathbb{K}} H^q(M) \oplus \mathrm{Tor}_1^{\mathbb{K}}(A, H^{q+1}(M)).$$

One the other hand, A, as $\mathbb{K}$-module is free, and it is also projective. Then we have

$$\mathrm{Tor}_1^{\mathbb{K}}(A, H^{q+1}(M)) = 0.$$

□

THEOREM 4. *If the supermanifold M is the product of two supermanifolds M_i, $i = 1, 2, M = M_1 \times M_2$ then $\{\Omega^\bullet(M), d\}$ splits into a bicomplex $\{\Omega^{\bullet,\bullet}(M), d_1, d_2\}$, with $\Omega^{r,s}(M) \equiv \Omega^r(M_1) \otimes_{\mathbb{K}} \Omega^s(M_2)$, and for the de Rham cohomology $H^\bullet(M)$ we have the following canonical isomorphism of $\mathbb{K}$-modules:*

$$H^\bullet(M) \cong H^\bullet(M_1) \bigotimes_{\mathbb{K}} H^\bullet(M_2), \quad e.g.\ H^p(M) \cong \bigoplus_{r+s=p} H^r(M_1) \bigotimes_{\mathbb{K}} H^s(M_2).$$

PROOF. In fact, as $\{\Omega^\bullet(M_i), d\}$ are flat cochain complexes we can apply the Künneth theorem of homological algebra and obtain the non-canonical isomorphism

$$H^p(M) \cong \oplus_{r+s=p} H^r(M_1) \bigotimes_{\mathbb{K}} H^s(M_2) \oplus \bigoplus_{r+s=p-1} \mathrm{Tor}_1^{\mathbb{K}}(H^r(M_1), H^s(M_2))$$

On the other hand, $H^r(M_i)$ are projective and so $\mathrm{Tor}_1^{\mathbb{K}}(H^r(M_1), H^s(M_2)) = 0$. □

THEOREM 5. *One has the following isomorphism of A-modules:*

$$H^p(M_1 \times M_2; A) \cong \bigoplus_{r+s=p} H^r(M_1; A) \bigotimes_A H^s(M_2; A).$$

PROOF. It is a direct consequence of Theorem 3 and Theorem 4. Infact one has the following isomorphisms:

$$\begin{aligned} H^p(M_1 \times M_2; A) &\cong A \bigotimes_{\mathbb{K}} H^p(M_1 \times M_2; A) \\ &\cong A \bigotimes_{\mathbb{K}} \Big(\bigoplus_{r+s=p} H^r(M_1; A) \bigotimes_{\mathbb{K}} H^s(M_2; A) \Big) \\ &\cong \bigoplus_{r+s=p} \Big((A \bigotimes_{\mathbb{K}} H^r(M_1; A) \Big) \bigotimes_A (A \bigotimes_{\mathbb{K}} H^s(M_2; A) \Big) \\ &\cong \bigoplus_{r+s=p} H^r(M_1; A) \bigotimes_A H^s(M_2; A). \end{aligned}$$

□

THEOREM 6. *Let M be a regular (m, n)-dimensional supermanifold. Then one has the canonical $\mathbb{K}$-isomorphism:*

$$H^p(M) \cong H^p(M_B).$$

PROOF. Let M be modeled on $A^{m,n}$, where A is a superalgebra. The fiber bundle $b \colon M \longrightarrow M_B$ has fiber $F \equiv A_0'^m \oplus A_1^n$ that is a $\mathbb{K}$-module. So we can apply a well known result of algebraic topology [24, 39] to prove that holds above isomorphism. □

COROLLARY 1. *If M is a regular (m,n)-dimensional supermanifold one has the following canonical $\mathbb{K}$-isomorphism:*

$$H^p(M;A) \cong A \bigotimes_{\mathbb{K}} H^p(M_B)\,.$$

2.2. The singular supercohomology

In the following we shall assume $\mathbb{K} = \mathbb{R}$. The standard p-simplex is the convex set $\Delta^p \subset \mathbb{R}^{p+1}$ consisting of all $(p+1)$-tuples $(a_0, \ldots, a_n)$ of real numbers given by

$$\Delta^p \equiv \{(a_0, \ldots, a_p) \in \mathbb{R}^{p+1} \Big| \sum_{0 \le i \le p} a_i \le 1\,, \quad a_i \ge 0\}\,.$$

For each $p \ge 0$, we define **boundary map** $\partial_i^p \colon \Delta^p \longrightarrow \Delta^{p+1}$, $0 \le i \le p+1$, given by

$$p=0: \left| \begin{array}{l} \partial_0^0(0) = 1 \\ \\ \partial_1^0(0) = 0 \end{array} \right. \quad ; p \ge 1: \left| \begin{array}{l} \partial_0^p(a_1, \ldots, a_p) = \Big(1 - \sum_{1 \le i \le p} a_i, a_1, \ldots, a_p\Big) \\ \\ \partial_i^p(a_1, \ldots, a_p) = (a_1, \ldots, a_{i-1}, 0, a_i, \ldots, a_p)\,, \\ \qquad 1 \le i \le p+1\,. \end{array} \right.$$

A **differentiable singular p-simplex** u in a supermanifold X is a map $u : \Delta^p \longrightarrow X$ which extends to be C_w^∞ (or equivalently C^∞) map of a neighborhood of Δ^p in $\mathbb{K}^{p+1}$ into X. If $p \ge 1$ we define **i-th face** of u the differentiable $(p-1)$-singular simplex $\partial_i u \equiv u \circ \partial_i^p$. We call **support** of u the image $u(\Delta^p) \subset X$. A **p-chain** c in X, (with coefficients in $\mathbb{K}$), is a finite linear combination $c = \sum_i c^i u_i$ of differentiable p-simplex u_i in X where $c^i \in \mathbb{K}$.

DEFINITION 2. *Let X be a supermanifold modeled on $A^{m,n}$. The* **group of p-chains** *of X, with coefficients in $\mathbb{K}$, $p \in \mathbb{Z}$, is the set $C_p(X, \mathbb{K})$ of p chains in X. $C_p(X;\mathbb{K})$ is a free $\mathbb{K}$-module. A basis of $C_p(X;\mathbb{K})$ is just the set of differentiable singular p-simplex in X. For $p < 0$ $C_p(X;\mathbb{K})$ is defined to be zero. Let $\partial : C_p(X;\mathbb{K}) \longrightarrow C_{p-1}(X;\mathbb{K})$ be the homomorphism defined by $\partial u = \sum_{0 \le i \le p} (-1)^i \partial_i u\colon \Delta^{p-1} \longrightarrow X$; ∂ is called the* **boundary homomorphism**. *As $\partial\partial = 0$, one has that $\{C_p(X;\mathbb{K}), \partial\}$ is a chain complex, (for 0-chains one has $\partial \equiv 0$). So $C_p(X;\mathbb{K})$ identifies a covariant functor from the category of supermanifolds to the category $\mathbb{K}$-modules.*

DEFINITION 3. *The* **singular homology** *of X, with coefficients in $\mathbb{K}$, is the homology $H_\bullet(X;\mathbb{K})$ of the chain complex $\{C_p(X;\mathbb{K}), \partial\}$. ($H_p(X;\mathbb{K})$ is a $\mathbb{K}$-module).*

THEOREM 7. *Let X and Y be supermanifolds. Then one has the following isomorphism of $\mathbb{K}$-modules:*

$$H_p(X \times Y; \mathbb{K}) \cong \bigoplus_{r+s=p} H_r(X;\mathbb{K}) \bigotimes_{\mathbb{K}} H_s(Y;\mathbb{K})\,.$$

PROOF. The singular simplex $C_\bullet(X \times Y; \mathbb{K})$ is homotopy equivalent to the bicomplex ([44])

$$\left\{C_{\bullet,\bullet}(X \times Y; \mathbb{K}), \partial_1, \partial_2\right\} \equiv \left\{ \bigoplus_{r+s=p} C_r(X; \mathbb{K}) \bigotimes\nolimits_{\mathbb{K}} C_s(Y; \mathbb{K}), \partial_1, \partial_2 \right\}.$$

So, we can apply the Künneth theorm of homological algebra to obtain the isomorphism

$$H_p(X \times Y; \mathbb{K}) \cong \bigoplus_{r+s=p} H_r(X; \mathbb{K}) \bigotimes\nolimits_{\mathbb{K}} H_s(Y; \mathbb{K}) \oplus \bigoplus_{r+s=p-1} \mathrm{Tor}_1^{\mathbb{K}}(H_r(X; \mathbb{K}), H_s(Y; \mathbb{K}))$$

One the other hand $H_p(X; \mathbb{K})$ are free-modules, therefore projective. So

$$\mathrm{Tor}_1^{\mathbb{K}} \Big(H_r(X; \mathbb{K}), H_s(Y; \mathbb{K})\Big) = 0 .$$

□

DEFINITION 4. *The p-**cochain group** is the dual $C^p(X; \mathbb{K})$ of $C_p(X; \mathbb{K})$:*

$$C^p(X; \mathbb{K}) \equiv \mathrm{Hom}_{\mathbb{K}} \Big(C_p(X; \mathbb{K}); \mathbb{K}\Big) \equiv C_p(X; \mathbb{K})^* .$$

REMARK 1. $C^p(X; \mathbb{K})$ is a $\mathbb{K}$-module. One has the pairing: $\langle , \rangle$: $C^p(X; \mathbb{K}) \times C_p(X; \mathbb{K}) \longrightarrow \mathbb{K}$. Then we can define the cochain complex $\{C^p(X; \mathbb{K}), \delta\}$ considering the **coboundary** of a p-cochain $\alpha \in C^p(X; \mathbb{K})$ as the p-cochain $\delta\alpha \in C^{p+1}(X; \mathbb{K})$ such that

$$\langle \delta\alpha, c \rangle + (-1)^p \langle \alpha, \partial c \rangle = 0 .$$

DEFINITION 5. *The **singular cohomology** of X (with coefficients in $\mathbb{K}$), is the cohomology of the cochain complex $\{C^p(X; \mathbb{K}), \delta\}$.*

THEOREM 8 (Relation between singular cohomology and singular homology). *One has the canonical isomorphism:*

$$H^p(X; \mathbb{K}) \cong \mathrm{Hom}_{\mathbb{K}}(H_p(X; \mathbb{K}); \mathbb{K}) .$$

PROOF. As $C_p(X; \mathbb{K})$ is a free chain complex over $\mathbb{K}$ we can use the universal coefficients theorem in cohomology [11] and say that there is the following isomorphism of $\mathbb{K}$-modules:

$$H^p(X; \mathbb{K}) \cong \mathrm{Hom}_{\mathbb{K}}(H_p(X; \mathbb{K}); \mathbb{K}) \oplus \mathrm{Ext}^1_{\mathbb{K}}(H_{p-1}(X; \mathbb{K}), \mathbb{K}) .$$

As $\mathbb{K}$ is injective one has that $\mathrm{Ext}^1_{\mathbb{K}}(H_{p-1}(X; \mathbb{K}), \mathbb{K}\Big) = 0$. □

THEOREM 9 (Relation between singular cohomology and de Rham cohomology). *One has the canonical $\mathbb{K}$-isomorphism:*

$$H^p(X; \mathbb{K}) \cong H^p(X) .$$

PROOF. Let $\omega \in \Omega^p(X)$ be a differential p-form defined on a neighborhood of the image of a p-chain $c = \sum_i a^i u_i$ in X. We define **integral of ω over c** by the following formula:

$$\int_c \omega \equiv \sum_i a^i \left(\int_{u_i} \omega \right) \equiv \sum_i a^i \left(\int_{\Delta^p} u_i^* \omega \right) ,$$

where $\int_{\Delta^p} u_i^* \omega$ are the standard Riemannian integrals.

Then one has a natural $\mathbb{K}$-homomorphism $\zeta^* \colon \Omega^p(X) \longrightarrow C^p(X; \mathbb{K})$, given by $\omega \mapsto \zeta^*(\omega)$, $\zeta^*(\omega)(c) = \int_c \omega$. The map ζ^* passes to the quotient giving the above isomorphism. □

THEOREM 10. *Let X and Y be supermanifolds. Then one has the following canonical $\mathbb{K}$-isomorphism:*

$$H^p(X \times Y; \mathbb{K}) \cong \bigoplus_{r+s=p} H^r(X; \mathbb{K}) \otimes_{\mathbb{K}} H^s(Y; \mathbb{K}) .$$

PROOF. The proof is similar to Theorem 5 one. □

DEFINITION 6. *Let $A \equiv \mathbb{K} \oplus A_0' \oplus A_1$ be a* BG *algebra and let X be a supermanifold modeled on $A^{m,n}$. The* **group of p-chains**, *of the supermanifold X, with coefficients in A, $p \in \mathbb{Z}_2$, is the set $C_p(X; A) \equiv A \bigotimes_{\mathbb{K}} C_p(X; \mathbb{K})$.*

REMARK 2. $C_p(X; A)$ is a $\mathbb{Z}_2$-graded free A-module: $C_p(X; A)_r = \{c = \sum_i c^i u_i | c^i \in A_r\}$, $r \in \mathbb{Z}_2$. One has $a_r c_s = (-1)^{rs} c_s a_r \in C_p(X; A)_{r+s}$. For $p < 0$ the group $C_p(X; A)$ is defined to be zero.
Let $\partial \colon C_p(X; A) \longrightarrow C_{p-1}(X; A)$ be the homomorphism defined by

$$\partial u = \sum_{0 \le i \le p} (-1)^i \partial_i u \colon \Delta^{p-1} \longrightarrow X ;$$

∂ is called the **boundary homomorphism**. As $\partial \circ \partial = 0$, one has that $\{C_p(X;A), \partial\}$ is a chain complex. (For 0-chains one has $\partial \equiv 0$). So $C_p(X; A)$ identifies a covariant functor from the category of superdifferentiable manifolds to the category of $\mathbb{Z}_2$-graded free A-modules.

DEFINITION 7. *The* **singular homology** *of X,* **with coefficients** *in A, is the homology $H_\bullet(X; A)$ of the chain complex $\{C_p(X; A), \partial\}$.*

REMARK 3 $H_p(X; A)$ is a $\mathbb{Z}_2$-graded A-module, for $p \geq 0$.

THEOREM 11. *One has the following canonical isomorphism of $\mathbb{K}$-modules:*

(a) $$H_p(X; A) \cong A \otimes_{\mathbb{K}} H_p(X; \mathbb{K}) ;$$

(b) $$H_p(X \times Y; A) \cong \bigoplus_{r+s=p} H_r(X; A) \bigotimes_A H_s(X; A)$$

PROOF.
(a) $H_p(X;A) \cong H_p(A \otimes_{\mathbb{K}} C_\bullet(X;\mathbb{K})) \cong A \otimes_{\mathbb{K}} H_p(X;\mathbb{K}) \oplus \mathrm{Tor}_1^{\mathbb{K}}(H_{p-1}(X;\mathbb{K}),A)$.
One the other hand

$$\mathrm{Tor}_1^{\mathbb{K}}(H_{p-1}(X;\mathbb{K}),A) = 0\,,$$

(A is a projective $\mathbb{K}$-module).

(b)

$$\begin{aligned} H_p(X\times Y;A) &\cong A \bigotimes_{\mathbb{K}} H_p(X\times Y;\mathbb{K}) \cong A \bigotimes_{\mathbb{K}} \bigoplus_{r+s=p} H_r(X;\mathbb{K}) \bigotimes_{\mathbb{K}} H_s(Y;\mathbb{K}) \\ &\cong \bigoplus_{r+s=p} (A \bigotimes_{\mathbb{K}} H_r(X;A)) \bigotimes_A (A \bigotimes_{\mathbb{K}} H_s(Y;\mathbb{K})) \\ &\cong \bigoplus_{r+s=p} H_r(X;A) \bigotimes_A H_s(Y;A). \end{aligned}$$

□

DEFINITION 8. *The p-***supercochain groups** *is the superdual $\widetilde{C}^p(X;A)$ of $C_p(X;A)$:*

$$\widetilde{C}^p(X;A) \equiv \mathrm{Hom}_A(C_p(X;A),A)\,.$$

REMARK 4 $\widetilde{C}^p(X;A)$ is a $\mathbb{Z}_2$-graded A-module:

$$\widetilde{C}^p(X;A)_r = \left\{\alpha \in \widetilde{C}^p(X;A) | \langle \alpha, c_s\rangle \in A_{r+s}; c_s \in C_p(X;A)_s\right\}.$$

$\langle\ ,\ \rangle$ denotes the following pairing: $\langle\ ,\ \rangle : \widetilde{C}^p(X;A) \times C_p(X;A) \longrightarrow A$. The corresponding cochain complex $\{\widetilde{C}^p(X;A),\delta)$ is defined by considering the **coboundary** of a p-supercochain $\alpha \in \widetilde{C}^p(X;A)$ as the p-supercochain $\delta\alpha \in \widetilde{C}^{p+1}(X;A)$ such that

$$\langle \delta\alpha, c\rangle + (-1)^p \langle \alpha, \delta c\rangle = 0\,.$$

DEFINITION 9. *The* **singular supercohomology** *of X with coefficients in A, is the cohomology $\widetilde{H}^\bullet(X;A)$ of the cochain complex $\{\widetilde{C}^p(X;A),\delta\}$.*

In the following table we resume some further useful cochain complexes and the corresponding cohomology spaces.

Tab. 3

Cochain complex	Cohomology spaces	Name
$\{\widetilde{C}^\bullet(X,A),\delta\}$	$\widetilde{H}^\bullet(X;A)$	Singular supercohomology
$\{C^p(X;A) \equiv \mathrm{Hom}_{\mathbb{K}}(C_p(X;\mathbb{K});A),\delta\}$	$H_s^\bullet(X;A)$	Singular cohomology with coeff. in A

THEOREM 12 (Relationship between singular cohomology with coefficients in A, singular homology and de Rham cohomology). *One has the canonical isomor-*

phisms of $\mathbb{K}$*-modules:*

$$H^p_s(X;A) \cong \mathrm{Hom}_{\mathbb{K}}(H_p(X;\mathbb{K});A) \cong H^p(X) \oplus \check{H}(X;A)\,, \quad \textit{where}$$
$$\check{H}(X;A) \equiv \mathrm{Hom}_{\mathbb{K}}(H_p(X;\mathbb{K});\mathfrak{M})\,.$$

PROOF. We can apply the dual universal coefficient theorem of homological algebra [11] to the free cochain complex of $\mathbb{K}$-modules $C^p(X;A)$ and obtain the following isomorphism of $\mathbb{K}$-modules:

$$H^p_s(X;A) \cong \mathrm{Hom}_{\mathbb{K}}(H_p(X;\mathbb{K});A) \oplus \mathrm{Ext}^1_{\mathbb{K}}(H_{p-1}(X;\mathbb{K}),A)\,.$$

But $\mathrm{Ext}^1_{\mathbb{K}}(H_{p-1}(X;\mathbb{K}),A) = 0$ being A a $\mathbb{K}$-module. □

COROLLARY 2. $H^p_s(X;A) \cong H^p(X)$ *iff* $\mathfrak{M} = 0$.

THEOREM 13 (Relationship between singular supercohomology and singular cohomology with coefficients in A). 1) *One has the canonical isomorphisms:*

$$\text{(a)} \quad \widetilde{H}^p(X;A) \cong \mathrm{Hom}_A(H_p(X;A);A)\,; \qquad \text{(b)} \quad \widetilde{H}^p(X;A) \cong H^p_s(X;A)\,.$$

PROOF. As $C_p(X;\mathbb{K})$ is a free $\mathbb{K}$-module one has also that $A \bigotimes_{\mathbb{K}} C_p(X;\mathbb{K})$ is a free A-module, so we can apply the dual universal coefficient theorem of homological algebra to the free cochain complex of A-modules $\{\mathrm{Hom}_A(A \bigotimes_{\mathbb{K}} C_\bullet(X;\mathbb{K});A) \equiv \widetilde{C}^\bullet(X;A), \delta\}$ and obtain the isomorphism:

$$\begin{aligned}
\widetilde{H}^p(X;A) \cong \mathrm{Hom}_A(H_p(A \textstyle\bigotimes_{\mathbb{K}} C_\bullet(X;\mathbb{K}));A) \textstyle\bigoplus& \\
\textstyle\bigoplus \mathrm{Ext}^1_A(H_{p-1}(A \bigotimes_{\mathbb{K}} C_\bullet(X;\mathbb{K});A)& \\
\cong \mathrm{Hom}_A(H_p(X;A);A) \textstyle\bigoplus \mathrm{Ext}^1_A(H_{p-1}(X;A),A)&\,.
\end{aligned}$$

But $\mathrm{Ext}^1_A(H_{p-1}(X;A),A) = 0$ being $H_{p-1}(X;A) \cong A \bigotimes_{\mathbb{K}} H_{p-1}(X;\mathbb{K})$ a projective A-module. So we get the canonical isomorphism of A-modules $\widetilde{H}^p(X;A) \cong \mathrm{Hom}_A(H_p(X;A);A)$. On the other hand one also

$$\begin{aligned}
\widetilde{H}^p(X;A) &\cong \mathrm{Hom}_A(H_p(A \textstyle\bigotimes_{\mathbb{K}} C_\bullet(X;\mathbb{K}));A) \\
&\cong \mathrm{Hom}_A(A \textstyle\bigotimes_{\mathbb{K}} H_p(X;\mathbb{K});A) \cong \mathrm{Hom}_{\mathbb{K}}(H_p(X;\mathbb{K});\mathrm{Hom}_A(A;A)) \\
&\cong \mathrm{Hom}_{\mathbb{K}}(H_p(X;\mathbb{K});A) \cong H^p_s(X;A)\,.
\end{aligned}$$

□

THEOREM 14. *One has the following canonical isomorphisms:*

1) *(A-modules)* $\widetilde{H}^p(X;A) \cong AO_{A_0}\widetilde{H}^p(X;A)_r$, $r \in \mathbb{Z}_2$;
2) *(A_0-modules)* $\widetilde{H}^p(X;A)_r \cong \mathrm{Hom}_{\mathbb{K}}(H_p(X;\mathbb{K});A_r)$, $r \in \mathbb{Z}_2$;
3) *(A-modules)* $H^p_s(X;A) \cong AO_{A_0}H^p_s(X;A)_r$, $r \in \mathbb{Z}_2$.

PROOF. 1) Let us consider the following canonical splitting:

$$\widetilde{H}^p(X;A) \cong \widetilde{H}^p(X;A)_0 \oplus \widetilde{H}^p(X;A)_1 .$$

On the other hand $AO_{A_0}\widetilde{H}^p(X;A)_0 \cong \widetilde{H}^p(X;A)_0 \oplus A_1O_{A_0}\widetilde{H}^p(X;A)_0$. We can identify $A_1O_{A_0}\widetilde{H}^p(X;A)_0$ with $\widetilde{H}^p(X;A)_1$. In fact, let us first consider the following A_0-isomorphisms:

$$\begin{aligned}\widetilde{H}^p(X;A)_1 &= \left\{f \in \widetilde{H}^p(X;A) \equiv \mathrm{Hom}_A(H_p(X;A);A) | f(v) \in A_{|v|+1}\right\} \\ &\cong \mathrm{Hom}_{A_0}(H_p(X;A_0);A_0) \times \mathrm{Hom}_{A_0}(H_p(X;A_1);A_1) .\end{aligned}$$

So we get:

$$\begin{aligned}A_1O_{A_0}\widetilde{H}^p(X;A)_0 &\cong A_1O_{A_0}\,\mathrm{Hom}_{A_0}(H_p(X;A_0);A_0)\times \\ &\qquad \times A_1O_{A_0}\,\mathrm{Hom}_{A_0}(H_p(X;A_1);A_1) \\ &\cong \mathrm{Hom}_{A_0}(H_p(X;A_0);A_1) \times \mathrm{Hom}_{A_0}(H_p(X;A_1);A_0) \\ &\cong \widetilde{H}^p(X;A)_1 .\end{aligned}$$

The other part of the theorem 1) follows similarly.
2) It is enough to consider the canonical isomorphism $\widetilde{H}^p(X;A) \cong H^p_s(X;A)$.
3) It is a direct consequence of above theorems. □

THEOREM 15 (Relation between singular (super)-cohomology and de Rham cohomology). *If X is a regular supermanifold one has the following canonical isomorphisms of* $\mathbb{K}$*-modules:* $\widetilde{H}^p(X;A) \cong A \bigotimes_{\mathbb{K}} H^p(X) \cong H^p(X;A)$.

PROOF. Let us consider the following lemma.

LEMMA 6. *One has the canonical isomorphism of* $\mathbb{K}$*-vector spaces:*$H_p(X;\mathbb{K}) \cong H_p(X_B;\mathbb{K})$.

PROOF OF LEMMA 6. From $H^p(X) \cong H^p(X_B)$ it follows the following exact sequence

$$0 \longrightarrow \mathrm{Hom}_{\mathbb{K}}(H_p(X;\mathbb{K});\mathbb{K}) \longrightarrow \mathrm{Hom}_{\mathbb{K}}(H_p(X_B;\mathbb{K});\mathbb{K}) \longrightarrow 0 .$$

Then taking into account that we are working with $\mathbb{K}$-vector spaces, we have also the following exact sequence:

$$0 \longrightarrow H_p(X;\mathbb{K}) \longrightarrow H_p(X_B;\mathbb{K}) \longrightarrow 0 .$$ □

Now, we have

$$\begin{aligned}A \bigotimes_{\mathbb{K}} H^p(X) &\cong A \bigotimes_{\mathbb{K}} H^p(X_B) \cong A \bigotimes_{\mathbb{K}} \mathrm{Hom}_{\mathbb{K}}(H_p(X_B;\mathbb{K});\mathbb{K}) \\ &\cong \mathrm{Hom}_A(A \bigotimes_{\mathbb{K}} H_p(X_B;\mathbb{K});A) \cong \mathrm{Hom}_A(A \bigotimes_{\mathbb{K}} H_p(X;\mathbb{K});A) \\ &\cong \mathrm{Hom}_{\mathbb{K}}(H_p(X;\mathbb{K});A) \cong H^p_s(X;A) .\end{aligned}$$

THEOREM 16 (Relation between singular supercohomology and the Rham superchomology)*One has the canonical A-monomorphism:* $\tilde{\zeta}^\bullet : \widetilde{H}^p(X) \longrightarrow H^p_s(X;A)$.

PROOF. Let $\omega \in \tilde{\Omega}^p(X)$ be a differential p-form defined on a neighborhood of the image of a p-chain $c = \sum_i a^i u_i$ in X. We define **integral of ω over c** the following formula:

$$\int_c \omega \equiv \sum_i a^i \Big(\int_{u_i} \omega \Big) \equiv \sum_i a^i \Big(\int_{\Delta^p} u_i^* \omega \Big),$$

where $u_i^* \omega \Delta^p \longrightarrow A \otimes_{\mathbb{K}} \Lambda_p^0(\Delta^p)$, and $\int_{\Delta^p} u_i^* \omega$ are the standard Riemannian integrals.

Then one has a natural A-homomorphism $\tilde{\zeta}^\bullet : \tilde{\Omega}^p(X) \longrightarrow C^p(X;A)$ given by $\omega \mapsto \tilde{\zeta}^\bullet(\omega)$, $\tilde{\zeta}^\bullet(\omega)(c) = \int_c \omega$. The map passes to the quotient giving an A-homomorphism $\tilde{\zeta}^\bullet : \tilde{H}^p(X) \longrightarrow H_s^p(X;A)$.

One can easily see that $\tilde{\zeta}^\bullet$ is injective. □

In the following table we resume the isomorphic cohomology spaces as it results from above theorems.

Tab. 4 - Important isomorphic (super) cohomology spaces

Name	(Super) Cohomology spaces
de Rham cohomology with coefficients in A	$H^\bullet(X;A) \cong A \bigotimes_{\mathbb{K}} H^\bullet(X) \cong A \bigotimes_{\mathbb{K}} H^\bullet(X_B)$
Singular supercohomology	$\tilde{H}^\bullet(X;A) \cong \mathrm{Hom}_A(H^\bullet(X;A);A)$
Singular cohomology with coefficients in A	$H_s^\bullet(X;A) \cong \mathrm{Hom}_{\mathbb{K}}(H^\bullet(X;A);\mathbb{K})$

THEOREM 17. *If X is a contractible supermanifold, then*

$$H_s^p(X;A) \cong \tilde{H}^p(X;A) \cong H^p(X;A) \cong \tilde{H}^p(X) = 0, \quad \textit{for} \ \ p \geq 1 .$$

PROOF. We know that for contractible manifolds $H^p(X)$=0, for $p \geq 1$. Taking into account the canonical isomorphism $H^p(X) \cong \mathrm{Hom}_{\mathbb{K}}\ (H_p(X;\mathbb{K});\mathbb{K})$ we have that if $H^p(X) = 0 \implies H_p(X;\mathbb{K}) = 0$. Then by using Theorem 12 we get also $H_s^p(X;A) = 0$, for $p \geq 1$. Finally, by using Theorem 13 and Theorem 15 we conclude the proof. □

THEOREM 18 (Stoke's Theorem).

$$\langle d\phi, c\rangle + (-1)^{\dim \phi} \langle \phi, \partial c\rangle = 0, \ \textit{for any} \ \phi \in \tilde{\Omega}^{p-1}(X), \ \textit{and} \ c \in C_p(X;A) .$$

3. Super partial differential equations

The principal result of this chapter is theorem 1 that gives an extension to supermanifolds of a previous result by H. Goldschmidt for even manifolds [9]. Other interesting results are theorems 2-4. These prove that also the behaviour of super PDE's with respect to singular supersolutions is similar to classic case one. In particular, in theorem 3 and 4 we are able to extend to SPDE's some well known and useful properties of PDE's, founded on their prolongations. This has been made by requiring that the underling superalgebra A is the inductive limit of Noetherian superalgebras. Such algebras are not, in general Noetherian rings, but allow the same to obtain behaviours for SPDE's similar to the classic cases ones, where the hypothesis of Nowtherianity playes an essential role. This is a very remarkable result as one can easily find important examples of such superalgebras that are BG-superalgebras too, (see below). Moreover, we characterize singular supersolutions by means of characteristic superclasses and show that in order to solve Cauchy problems all the characteristic supernumbers of Cauchy hypersurfaces data must be zero. (see theorems 5-7). Other important results are also reported in theorems 10-12, that give a criterion to recognize superconservation laws related to symmetries properties of super PDE's and criteria of wholly supercohomologicity. In particular, we introduce, and study, a generalized version of Vinogradov spectral sequence for super PDE's (theorem 9), and utilize it in order to characterize superconservation laws and wholly supercohomology.

3.1. Formal theory of super PDE's

DEFINITION 1. *A super PDE (SPDE) of order k in the fiber bundle $\pi: W \longrightarrow M$, defined in the category of supermanifolds, is a subbundle $E_k \subset J\underline{D}^k(W)$ of the jet superderivative space $J\underline{D}^k(W)$ over M.*

EXAMPLES. 1) (**Holonomic k-superconnections**). A holonomic k-superconnection on $\pi: W \longrightarrow M$ is a SPDE $C_k \subset J\underline{D}^k(W)$ given as image of a section $\rceil: J\underline{D}^{k-1}(W) \longrightarrow J\underline{D}^k(W)$ of $\pi_{k,k-1}$: $C_k \equiv \rceil(J\underline{D}^{k-1}(W)) \subset J\underline{D}^k(W)$. C_k identifies (and it is identified by) an affine morphism

$$\lceil: J\underline{D}^k(W) \longrightarrow {\pi_{k-1,0}}^*(S_0^k M O v T W)$$

over $J\underline{D}^{k-1}(W)$, whose fibre derivative is $vD\lceil = 1$. One has $\lceil \circ \rceil = 0$, $\lceil = 1 - \rceil \circ \pi_{k,k-1}$. Let $\{x^A, y^B, \dots, y^B_{A_1\dots A_{k-1}}, \bar{y}^B_{A_1\dots A_k}\}$ be fibered coordinates on ${\pi_{k-1,0}}^*(S_0^k M O v T W)$. Then one has the following local representation of $\lceil$:

$$x^A \circ \lceil = x^A, \dots, y^B_{A_1\dots A_{k-1}} \circ \lceil = y^B_{A_1\dots A_{k-1}}, \bar{y}^B_{A_1\dots A_k} \circ \lceil = y^B_{A_1\dots A_k},$$

where

$$-\lceil^B_{A_1\dots A_k} = y^B_{A_1\dots A_k} \circ \rceil: J\underline{D}^{k-1}(W) \longrightarrow A.$$

So the local expression of C_k is: $y^B_{A_1\dots A_k} + \lceil^B_{A_1\dots A_k} = 0$.

2) (**Holonomic superconnections**). In the particular case that $k = 1$, a holonomic 1-superconnection ($\equiv$ holonomic superconnection) on W identifies a A_0-supersplitting of the following exact sequence of bundles of A_0-modules over W: $0 \longrightarrow vTW \longrightarrow TW \longrightarrow \pi^* TM \longrightarrow 0$.

3) (*Principal superconnections*). Furthermore, if $W \equiv P$ is a G-principal bundle in the category of supermanifolds, a principal superconnection on P can be characterized as a G-superinvariant superconnection. More precisely, the SPDE C_1 is

superinvariant under the natural action of G on $J\underline{D}(P)$, eg. the following diagram

$$\begin{array}{ccccc} C_1 \subset & J\underline{D}(P) & \xrightarrow{\pi_{1,0}} & P \\ \mathrm{J}\underline{\mathrm{D}}(\phi_\alpha)|\mathrm{C}_1 \downarrow & & \downarrow \mathrm{J}\underline{\mathrm{D}}(\phi_\alpha) & \downarrow \phi_\alpha \\ C_1 \subset & J\underline{D}(P) & \xrightarrow{\pi_{1,0}} & P \end{array}$$

is commutative for any $a \in G$, being $\phi\colon G \times P \longrightarrow P$ the superdifferentiable action map of G on P. This is equivalent to say that the section $\rceil\colon P \longrightarrow J\underline{D}(P)$ is G-superinvariant, that is $\phi^*_\alpha \rceil \equiv J\underline{D}(\phi_\alpha) \circ \rceil \circ \phi^{-1}_\alpha = \rceil$, or that the splitting $\lceil\colon TP \longrightarrow vTP$ gives the following commutative diagram:

$$\begin{array}{ccc} TP & \xrightarrow{\lceil} & vTP \\ \mathrm{T}(\phi_\alpha) \downarrow & & \downarrow \mathrm{vT}(\phi_\alpha) \\ TP & \xrightarrow{\lceil} & vTP \end{array}$$

for any $a \in G$. So, we recover the definition of principal superconnection given in section 1.3. Furthermore, there exists a unique section $\rceil/G\colon P/G \cong M \longrightarrow J\underline{D}(P)/G$ such that the following diagram

$$\begin{array}{ccc} J\underline{D}(P) & \xrightarrow{\underline{\pi}_G} & J\underline{D}(P)/G \\ \rceil \uparrow & & \uparrow \rceil/G \\ P & \xrightarrow{\pi_G} & P/G \cong M \end{array}$$

is commutative, where $\underline{\pi}_G$ and π_G are the canonical projections.

DEFINITION 2. 1) *The r-**superprolongation** of $\underline{E}_k \subset J\underline{D}^k(W)$ is*

$$\underline{E}_{k+r} \equiv J\underline{D}^r(E_k) \cap J\underline{D}^k(W) \subset J\underline{D}^{k+r}(W)\,.$$

2) *The **supersymbol** of $\underline{E}_{k+r}$, $r \geq 0$, is a family of A_0-modules over $\underline{E}_k$ given by*

$$\dot{g}_{k+r} = vT\underline{E}_{k+r} \cap \pi^*_{k+r,0}(\dot{S}^{k+r}_0 MO_{A_0} vTW)\,.$$

3) *A* SPDE *$\underline{E}_k \subset J\underline{D}^k(W)$ is **formally superintegrable** if $\forall r \geq 0$ one has that $\dot{g}_{k+r+1} \longrightarrow \underline{E}_k$ is a bundle of A_0-modules and $\underline{E}_{k+r+1} \longrightarrow \underline{E}_{k+r}$ is surjective.*

PROPOSITION 1. *If $\underline{E}_k$ is a formally superintegrable* SPDE, *then one has the following sequence of A_0-bundles over $\underline{E}_k$:*

$$0 \longrightarrow \pi^*_{k+r,k}\dot{g}_{k+r} \longrightarrow vT\underline{E}_{k+r} \longrightarrow \pi^*_{k+r,k+r-1} vT\underline{E}_{k+r-1}$$

that means that $\underline{E}_{k+r}$ is an affine bundle over $\underline{E}_{k+r-1}$ with associated vector bundle

$$\pi^*_{k+r-1,k}\dot{g}_{k+r} \longrightarrow \underline{E}_{k+r-1}\,.$$

PROPOSITION 2. *Associated to $\underline{E}_k$ one has the following complexes of A_0-modules over $\underline{E}_k$ (δ-**supersequence**):*

$$\begin{array}{l} 0 \longrightarrow \dot{g}_m \longrightarrow TMO_{A_0}\dot{g}_{m-1} \longrightarrow \dot{\Lambda}^2_0 MO_{A_0}\dot{g}_{m-2} \longrightarrow \ldots \longrightarrow \\ \longrightarrow \dot{\Lambda}^{m-k}_0 MO_{A_0}\dot{g}_k \longrightarrow \dot{\Lambda}^{m-k+1}_0 MO_{A_0}(\dot{S}^{k-1}_0 MO_{A_0} vTW) \longrightarrow \ldots \end{array}$$

DEFINITION 3. *We call the corresponding cohomology* $\{H_q^{m-j,j}\}_{q\in \underline{E}_k}$ *at* $(\dot\Lambda_0^j M O_{A_0} \dot g_{m-j})_q$ *the* **Spencer supercohomology** *of* $\underline{E}_k$.
We say that $\underline{E}_k$ *is* r-**superacyclic** *if* $H^{m,j} = 0$, $m \geq k$, $0 \leq j \leq r$. *We say that* E_k *is* **superinvolutive** *if* $H^{m,j} = 0$, $m \geq k$, $j \geq 0$ *(e.g. the* δ*-supersequences are exact for* $m \geq k$*).*

Then we have the following important criterion of formal superintegrability that extends provious one given by H. GOLDSCHMIDT for PDE's [9].

THEOREM 1 (Criterion of formal superintegrability). *Let* $\underline{E}_k \subset J\underline{D}^k(W)$ *be a SPDE in the category of supermanifolds. Then, there is an integer* $k_0 = k + h \geq k$, *depending only on* k*, the dimension of* M *and the dimension of* W *such that* $\underline{E}_k$ *is superinvolutive and such that if* $\dot g_{k+r+1}$ *is a bundle of* A_0*-modules over* $\underline{E}_k$*, and* $\underline{E}_{k+r+1} \longrightarrow E_{k+r}$ *is surjective for* $0 \leq r \leq h$*, then* $\underline{E}_k$ *is formally superintegrable.*

PROOF. Let $\underline{E}_k \subset JD^k(W)$ be a k-order SPDE on $\pi\colon W \longrightarrow M$. The symbol g_k of $\underline{E}_k$ is the family of A_0-modules over $\underline{E}_k$ characterized by the following exact sequence:

$$0 \longrightarrow \dot g_k \longrightarrow \dot S_0^k M O_{A_0} vTW \xrightarrow{\ \sigma\ } F_0 \equiv vTJ\underline{D}^k(W)/vT\underline{E}_k\,.$$

So $\dot g_k = \ker\sigma$ (Δ). Similarly for the superprolongation $\dot g_{k+r}$ we have $\dot g_{k+r} = \ker\sigma_r$, where σ_r is defined by means of the following commutative diagram of A_0-modules:

$$\begin{array}{ccc}
\dot S_0^{k+r} M O_{A_0} vTW & \xrightarrow{\ \sigma_r\ } & \dot S_0^r M O_{A_0} F_0 \\
\downarrow & & \psi\downarrow \\
\dot S_0^r M \otimes_{A_0} \dot S_0^k M P_{A_0} vTW & \longrightarrow & \mathrm{Hom}_{A_0}(\dot S_0^r M, \mathrm{Hom}_{A_0}(\dot S_0^k M, vTW))
\end{array}$$

where the map ψ is induced by the natural projection

$$\dot S_0^k M P_{A_0} vTW \longrightarrow \dot S_0^k M O_{A_0} vTW/\dot g_k\,.$$

Let us consider the δ-complex:

$$\begin{aligned}
&0 \leftarrow \dot S_0^m M \leftarrow TM \otimes_{A_0} \dot S_0^{m-1} M \leftarrow \dot\Lambda_0^2 M \otimes_{A_0} \dot S_0^{m-2} M \leftarrow \dots \leftarrow \\
&\leftarrow \dot\Lambda_0^p M \otimes_{A_0} \dot S_0^{m-p} M \leftarrow \dots \leftarrow 0
\end{aligned}$$

where $\dot S_0^l M = 0$ for $l < 0$. This is similar to the classic one, and similarly we can prove that this complex is exact. (For the classic case see e.g. refs [9,15]). By applying the functor $\mathrm{Hom}_{A_0}(-, vTW)$ to the δ-complex, we obtain the following sequences: (□)

$$\begin{array}{ccc} 0 \longrightarrow \mathrm{Hom}_{A_0}(\dot{S}_0^m M, vTW) \longrightarrow & \mathrm{Hom}_{A_0}(TM \otimes_{A_0} \dot{S}_0^{m-1} M, vTW) \longrightarrow \dots \\ \| & \| \\ 0 \longrightarrow \mathrm{Hom}_{A_0}(\dot{S}_0^m M, vTW) \longrightarrow & \mathrm{Hom}_{A_0}(TM, \mathrm{Hom}_{A_0}(\dot{S}_0^{m-1} M, vTW)) \longrightarrow \dots \end{array}$$

$$\begin{array}{c} \dots \longrightarrow \mathrm{Hom}_{A_0}(\dot{\Lambda}_0^p M \otimes_{A_0} \dot{S}_0^{m-p} M, vTW) \longrightarrow \dots \\ \| \\ \dots \longrightarrow \mathrm{Hom}_{A_0}(\dot{\Lambda}_0^p M, \mathrm{Hom}_{A_0}(\dot{S}_0^{m-p} M, vTW)) \longrightarrow \dots \end{array}$$

As Hom is an additive left exact functor [11,26], above sequence is also exact. Now, since the diagram

$$\begin{array}{ccc} \dot{S}_0^{k+l} M & \longrightarrow & \dot{S}_0^l M \otimes_{A_0} \dot{S}_0^k M \\ \delta \uparrow & & \delta \uparrow \\ TM \otimes_{A_0} \dot{S}_0^{k+l-1} M & \longrightarrow & TM \otimes_{A_0} \dot{S}_0^{l-1} M \otimes_{A_0} \dot{S}_0^k M \end{array}$$

commutes, we can see that $\delta(\dot{g}_{k+1}) \subseteq TMO_{A_0}\dot{g}_k$, and hence δ induces a map

$$\delta \colon \dot{\Lambda}_0^j MO_{A_0}\dot{g}_{k+l} \longrightarrow \dot{\Lambda}_0^{j+1} MO_{A_0}\dot{g}_{k+l-1}\,.$$

In fact, by considering equation (Δ) we have the following commutative and exact diagram of families of A_0-modules over $\underline{E}_k$:

$$\begin{array}{ccccc} 0 \longrightarrow \dot{\Lambda}_0^p MO_{A_0}\dot{g}_{k+r+1} & \longrightarrow & \dot{\Lambda}_0^p MO_{A_0}(\dot{S}_0^{k+r+1} MO_{A_0} vTW) & \xrightarrow{\sigma_{r+1}} & \dot{\Lambda}_0^p MO_{A_0}(\dot{S}_0^{r+1} MO_{A_0} F_0) \\ \downarrow \delta & & \downarrow \delta & & \downarrow \delta \\ 0 \longrightarrow \dot{\Lambda}_0^{p+1} MO_{A_0}\dot{g}_{k+r} & \longrightarrow & \dot{\Lambda}_0^{p+1} MO_{A_0}(\dot{S}_0^{k+r} MO_{A_0} vTW) & \xrightarrow{\sigma_r} & \dot{\Lambda}_0^{p+1} MO_{A_0}(\dot{S}_0^r MO_{A_0} F_0) \end{array}$$

The induced map δ on the left is still such that $\delta^2 = 0$. Therefore the complex (□) induces the following one

$$\begin{array}{c} 0 \longrightarrow \dot{g}_m \longrightarrow \dots \longrightarrow TMO_{A_0}\dot{g}_{m-1} \longrightarrow \dot{\Lambda}_0^2 MO_{A_0}\dot{g}_k \longrightarrow \dots \\ \dots \longrightarrow \dot{\Lambda}_0^{m-k} MO_{A_0}\dot{g}_k \longrightarrow \dot{\Lambda}_0^p MO_{A_0}\dot{g}_{m-p} \longrightarrow \dots \end{array}$$

with $\dot{g}_{k+r-p} \equiv \dot{S}_0^{k+r-p} MO_{A_0} vTW$ if $r < p$ and $\dot{g}_{k+r-p} = 0$ if $k + r - p < 0$.
It is useful to consider also the graded counterpart of $\dot{g}_{k+r} : AO_{A_0}\dot{g}_{k+r} \equiv \tilde{g}_{k+r}$. Of course one has:

(a)
$$\tilde{g}_{k+r} = v\widetilde{T}\underline{E}_{k+r} \cap \pi_{k+r,0}{}^*(\widetilde{S}_0^{k+r} MO_A v\widetilde{T}W);$$

(b) exact sequence of A-modules:

$$0 \longrightarrow \tilde{g}_{k+r} \longrightarrow \widetilde{S}_0^{k+r} MO_A v\widetilde{T}W \longrightarrow \widetilde{S}_0^r MO_A \widetilde{F}_0 \equiv v\widetilde{T}J\underline{D}^k(W)/v\widetilde{T}\underline{E}_k\,;$$

(c) One has also a **graded δ-sequence**. In the following commutative diagram this is related to the δ-sequence:

$$
\begin{array}{ccccccccc}
0 \longrightarrow \tilde{g}_m & \longrightarrow & (TM)^+ \otimes_A \tilde{g}_{m-1} & \longrightarrow & \tilde{\Lambda} M_2^0 \otimes_A \tilde{g}_{m-2} & \longrightarrow \dots \longrightarrow & \tilde{\Lambda}_0^{m-k+1} M O_{A_0}(\dot{S}_0^{k+1} O_{A_0} vTW) \\
\| & & \|\tilde{} & & \|\tilde{} & & \|\tilde{} \\
0 \longrightarrow \tilde{g}_m & \longrightarrow & \widetilde{TM} O_{A_0} \dot{g}_{m-1} & \longrightarrow & \tilde{\Lambda}_0^2 M O_{A_0} \dot{g}_{m-2} & \longrightarrow \dots \longrightarrow & \tilde{\Lambda}_0^{m-k+1} M O_{A_0}(\dot{S}_0^{k+1} O_{A_0} vTW) \\
\|\tilde{} & & \|\tilde{} & & & & \|\tilde{} \\
0 \longrightarrow A O_{A_0} \dot{g}_m & \longrightarrow & (A O_{A_0} TM) O_{A_0} \dot{g}_{m-1} & & & \longrightarrow \dots \longrightarrow & (A O_{A_0} \dot{\Lambda}_0^{m-k+1} M) O_{A_0}(\dot{S}_0^{k+1} M O_{A_0} vTW) \\
\|\tilde{} & & \|\tilde{} & & & & \|\tilde{} \\
0 \longrightarrow A O_{A_0} \dot{g}_m & \longrightarrow & A O_{A_0}(TM O_{A_0} \dot{g}_{m-1}) & & & \longrightarrow \dots \longrightarrow & A O_{A_0}(\dot{\Lambda}_0^{m-k+1} M) O_{A_0}(\dot{S}_0^{k+1} M O_{A_0} vTW)
\end{array}
$$

In other words if we denote by $\dot{\Delta}$ and $\tilde{\Delta}$ the δ-**supersequence** and **graded** δ-**superseuqence** respectively, then one has the following isomorphism:

$$(\star\star\star) \qquad \tilde{\Delta} \cong A O_{A_0} \dot{\Delta} \equiv \mathrm{Hom}_{A_0}(A, \dot{\Delta}).$$

Note also that, as we are working with complexes of A_0-modules, where A is a BG-algebra, one has the isomorphism $\mathrm{Hom}_{A_0}(A, \dot{\Delta}) = \dot{\Delta} \oplus \overline{\Delta}$, where $\overline{\Delta}$ is the odd complement of Δ: $\overline{\Delta} \equiv \mathrm{Hom}_{A_0}(A_1, \dot{\Delta})$. Let us denote the corresponding **graded Spencer supercohomology** of $\underline{E}_k$ by $(\widetilde{H}_q^{m-j,j})_{q \in \underline{E}_k}$.
This allows us to define for $\underline{E}_k$ also the following:
(i) **graded** r-**superaciclicity**: $\widetilde{H}_q^{m,j} = 0$, $m \geq k$, $0 \leq j \leq r$; (ii) **graded supere-involutivness**: $\widetilde{H}_q^{m,j} = 0$, $m \geq k$, $j \geq 0$. One has the following

LEMMA 1. *1) $\tilde{g}_{k+r}$ is a bundle of A-modules over $\underline{E}_k$ iff $\dot{g}_{k+r}$ is a bundle of A_0-modules over $\underline{E}_k$.*
2) $\tilde{g}_{k+r}$ is graded r-superacyclic or graded r-superinvolutive iff $\dot{g}_{k+r}$ is r-superacyclic or superinvolutive;
3) $\tilde{g}_k$ as well $\dot{g}_k$ are always 1-superacyclic.

PROOF OF LEMMA 1. 1) Really $\tilde{g}_{k+r} \cong A O_{A_0} \dot{g}_{k+r} \equiv \mathrm{Hom}_{A_0}(A, \dot{g}_{k+r})$.
2) Let assume that $H^q(\dot{\Delta}) = 0$, $0 \leq q \leq r$. As $\mathrm{Hom}_{A_0}(A, -)$ is an additive left exact functor we can conclude that also $H^q(\mathrm{Hom}_{A_0}(A, \dot{\Delta})) = 0$, for $0 \leq q \leq r$. Conversely, let $H^q(\tilde{\Delta}) = 0$, $0 \leq q \leq r$. By using the isomorphism $(\star\star\star)$ one has also the bijection $H(\tilde{\Delta}) \cong H(\dot{\Delta}) \oplus H(\overline{\Delta})$. Therefore, when $H(\tilde{\Delta}) = 0$ must be also $H(\dot{\Delta})$.
3) The proof can be conduced by considerations similar to the classic case ones. See ref. [9]. □

REMARK 1. As $\tilde{g}_{k+r}$ is a family of free A-modules of finite type, we can prove for it all the classic considerations made in formal theory of PDE [9] and induce them on the even part $\dot{g}_{k+r}$ by using Lemma 1. So we have the following statements:
4) If $\dot{g}_k$ is s-superacyclic or superinvolutive, $\dot{g}_{k+r}$ is also s-superacyclic or superinvolutive;

5) $(\dot{g}_{k+r})_{+s} = (\dot{g}_k)_{+(r+s)} = \dot{g}_{k+r+s}$;
6) If $\dot{g}_k$ is 2-superacyclic and if $\dot{g}_{k+1}$ is a bundle of A_0-modules over $\underline{E}_k$, then $\dot{g}_{k+r}$ is a bundle of A_0-modules over $\underline{E}_k$, $\forall\, r \geq 1$;
7) If $\dot{g}_k$ is involutive and if $\dot{g}_k$ is a bundle of A_0-modules over $\underline{E}_k$, then $\dot{g}_{k+r}$ is a bundle of A_0-modules over $\underline{E}_k$, $\forall\, r \geq 0$;
8) Let $\underline{E}_k \subset J\underline{D}^k(W)$ be a SPDE of order k on $\pi\colon W \longrightarrow M$ and let $\hat{\underline{E}}_1 \subset J\underline{D}(J\underline{D}^{k-1}(W))$ be a SPDE of first order on $\widehat{W} \equiv J\underline{D}^{k-1}(W)$ image of $\underline{E}_k$ by the monomorphism $J\underline{D}^k(W) \longrightarrow J\underline{D}(J\underline{D}^{k-1}(W))$. The map $D^{k-1}\colon \underline{W} \longrightarrow \underline{\widehat{W}}$ establishes a one to one correspondence between the set of solutions of $\underline{E}_k$ and that of $\widehat{\underline{E}}_1$. Here $\underline{W}$ (resp. $\underline{\widehat{W}}$) denotes shave of sections of W (resp. $\widehat{W}$). Moreover, we have the identification: $\underline{E}_{k+r} = \widehat{\underline{E}}_{1+r}$, $\dot{g}_{k+r} = \dot{\hat{g}}_{1+r}$, $\forall\, r \geq 0$;
9) Let $\underline{E}_1 \subset J\underline{D}(W)$ be a SPDE of first order on $\pi\colon W \longrightarrow M$. Then there exists an integer $k_0(m,n|m',n') \geq 1$, $(m,n) = \dim M$, $(m',n') = \dim W$, such that $\dot{g}_{k_0}$ is involutive;
10) Let $\underline{E}_k \subset J\underline{D}^k(W)$ be a SPDE of order k on $\pi\colon W \longrightarrow M$. Then there exists an integer $k_0(m,n|m',n';k) \geq k$ such that $\dot{g}_{k_0}$ is involutive.
From above considerations the proof of Theorem 1 follows directly. □

REMARK 2. In the superanalytic case, under the hypotheses of theorem 1, we have that, for any initial condition $q \in \underline{E}_k \subset J\underline{D}^k(W)$, there exists a G^ω_w solution of $\underline{E}_k$. Really, we can choice points $q_{k+t} \in \underline{E}_{k+t}$, $t \geq -k$, where for $t < 0$, $\underline{E}_{k+t} = \pi_{k,k+t}(\underline{E}_k)$, such that $\pi_{k+t,k+t-1}(q_{k+t}) = q_{k+t-1}$. Then a local solution can be represented into a suitable neighborhood U of $p \equiv \pi_k(q)$, by the following functions:

$$s^c(x,\theta) = \sum_{0 \leq r+s \leq \infty, 0 \leq s \leq n} \frac{1}{r!}(x-b)^{A_1}\dots(x-b)^{A_r}(\theta-\beta)^{B_1}\dots$$
$$\dots(\theta-\beta)^{B_s}(\partial_{A_1\dots A_r B_1 \dots B_s} s^c)(b,\theta)$$

$1 \leq c \leq m'' + n''$, (fiber dimension of $W = (m'', n'')$), where $(x,\theta) \in U$, $(b,\beta) = p$, and

$$(\partial_{a_1\dots A_r B_1\dots B_s} s^c)(b,\theta) = y^c_{A_1\dots A_r B_1 \dots B_s}(q_{r+s}),$$

where $y^c_{A_1\dots A_r B_1\dots B_s}$ are vertical coordinates on $J\underline{D}^{r+s}(W)$.

REMARK 3. It is a simply corollary of Theorem 1 that in the superanalytic case, any superintegrable SPDE is **completely superintegrable** also: it admits local solution of class G^ω_w for any initial condition.

3.2. Singular supersolutions for SPDE's

Similarly to the ordinary case [15, 16, 19, 34, 40, 41, 42, 43], we can develop a geometric theory of singular supersolutions for SPDE's. In fact, one recognizes on $J\underline{D}^k(W)$ a canonical distribution, **Cartan superdistribution**, $\mathbb{E}_k \subset TJ\underline{D}^k(W)$, that is the distrbution on $J\underline{D}^k(W)$ spanned by tangent spaces to graphs of k-superderivatives of sections of the bundle $\pi : W \longrightarrow M$. If $\dim M = (m,n)$, we call **Vinogradov-supermanifolds** (**V-supermanifolds**) the (m,n)-dimensional integral supermanifolds of $\mathbb{E}_k$. We call also **Ehresmann-Vinogradov supermanifolds** (EV-**supermanifolds**) the (m,n)-dimensional integral supermanifolds of $\mathbb{E}_k$ that locally can be represented as the image of the k-superderivative of some section of π. Then a **singular supersolution** of $\underline{E}_k \subset J\underline{D}^k(W)$ is a closed sub-supermanifold $V \subset J\underline{D}^k(W)$ with the property that for each $q \in V$ there exists a supersmooth V-supermanifold $\overline{V} \subset J\underline{D}^{k+r}(W)$, $r \geq 0$, such that V coincides with $\pi_{k+r,k}(\overline{V})$ in some neighborhood of q and $\overline{V} \cap U \subset \underline{E}_{k+r}$, for some neighborhood U of some point $\overline{q} \in \pi^{-1}_{k+r,k}(q) \cap \overline{V}$. By using the "blow up" $\overline{V}$ at each point it is possible to construct a smooth supermanifold $\overline{V}$, called the **blow-up** of V. The points $q \in V$ for which $q \in \Sigma(\overline{V}) \equiv$ set of singular points of the map $\pi_{k+r}|\overline{V} : \overline{V} \longrightarrow M$, will be called the **geometric singularities** of the supersolution V. A **non singular** supersolution of $\underline{E}_k$ is a EV-supermanifold $V \subset J\underline{D}^k(W)$ contained into $\underline{E}_k$ also. We define a super PDE $\underline{E}_k \subset J\underline{D}^k(W)$ **V-formally superintegrable at** $q \in \underline{E}_k$ if there exists a V-supermanifold L tangent to $\underline{E}_k$ at this point with infinite order. $\underline{E}_k$ is **formally superintegrable at** $q \in \underline{E}_k$ if it is V-formally superintegrable at q and L is diffeomorphically projected onto M. A super PDE $\underline{E}_k \subset J\underline{D}^k(W)$ is formally superintegrable iff the projections $\pi_{k+s+1,k+s} : (\underline{E}_k)_{+(s+1)} \longrightarrow (\underline{E}_k)_{+s}$ are supersmooth bundles, $s \geq 0$. An equation $\underline{E}_s \subset J\underline{D}^s(W)$ is called an **intermediate superintegral** of the super PDE $\underline{E}_k$ if its $(k-s)$-superprolongation is contained in $\underline{E}_k$: $(\underline{E}_s)_{+(k-s)} \subset \underline{E}_k$. The number s is the **order** of the intermediate superintegral. An equation $\underline{E}_k \subset J\underline{D}^k(W)$ is called **V-superinvolutive** if: (a) its first superprolongation $\underline{E}_k$ is **superregular**, at each point $q \in (\underline{E}_k)_{+1}$ the subspaces $T_q((\underline{E}_k)_{+1})$ and $\dot{S}_0^{k+1}(T_xM)O_{A_0}vT_pW$, $p \equiv \pi_{k+1,0}(q)$, intersect transversally; (b) the map $\pi_{k+1,k} : (\underline{E}_k)_{+1} \longrightarrow \underline{E}_k$ is surjective; (c) $\underline{E}_k$ is superacyclic. Then one can directly obtain results that generalize previous ones obtained for PDE's on ordinary manifolds. In particular, we have the following important theorems.

THEOREM 2. *Let $\underline{E}_k \subset J\underline{D}^k(W)$ be a superregular system of order k on $\pi : W \longrightarrow M$. If $\dot{g}_{k+1}$ is bundle of A_0-modules over $\underline{E}_k$ and if $\dot{g}_k$ is a 2-superacyclic, then*

$$\underline{E}^{(1)}_{k+r} = (\underline{E}^{(1)}_k)_{+k} \quad where \quad \underline{E}^{(1)}_{k+r} \equiv \pi_{k+r+1,k+r}(\underline{E}_{k+r+1}) \subset J\underline{D}^{k+r}(W)\,.$$

PROOF. We shall use an induction on r. Start with $r = 1$. As $\dot{g}_{k+1}$ is a bundle of A_0-modules over $\underline{E}_k$, it follows that $\pi_{k+1,k} : \underline{E}_{k+1} \longrightarrow \underline{E}_k$ is a morphism of constant superrank and that $\underline{E}_{k+1}$ is an affine bundle over $\underline{E}^{(1)}_k \equiv \pi_{k+1,k}(\underline{E}_{k+1}) \subset \underline{E}_k$, modeled on the restriction of $\dot{g}_{k+1}$ to $\underline{E}^{(1)}_k$. Moreover, we have $\dim \dot{g}_{k+1} = \dim \underline{E}_{k+1} - \dim \underline{E}^{(1)}_k$.

Furthermore, under the hypotheses of the theorem, we have the exact sequence of bundles of A_0-modules over $\underline{E}_k$:

$$0 \longrightarrow \dot{g}_{k+2} \longrightarrow \dot{S}_0^{k+2} MO_{A_0} vTW \longrightarrow \dot{S}_0^2 MO_{A_0} F_0 \longrightarrow TMO_{A_0} F_1 ,$$

where F_1 is the cokernel of the homomorphism $\dot{S}_0^{k+1} MO_{A_0} vTW \to TMO_{A_0} F_0$ of bundles of A_0-modules over $\underline{E}_k$. In fact, we have the following commutative diagram which has exact columns because $\dot{g}_k$ is 2-acyclic and exact rows, except parhaps the upper one

$$\begin{array}{ccccccc}
 & 0 & & 0 & & 0 & \\
 & \downarrow & & \downarrow & & \downarrow & \\
0\to & \dot{g}_{k+2} & \to & \dot{S}_0^{k+2} MO_{A_0} vTW & \to & \dot{S}_0^2 MO_{A_0} F_0 & \to TMO_{A_0} F_1 \\
 & \downarrow \delta & & \downarrow \delta & & \downarrow \delta & \| \\
0\to & TMO_{A_0} \dot{g}_{k+1} & \to & TMO_{A_0} \dot{S}_0^{k+1} MO_{A_0} vTW & \to & TMO_{A_0} TMO_{A_0} F_0 & \to TMO_{A_0} F_1 \\
 & \downarrow \delta & & \downarrow \delta & & \downarrow \delta & \\
0\to & \dot{\Lambda}_0^2 MO_{A_0} \dot{g}_k & \to & \dot{\Lambda}_0^2 MO_{A_0} \dot{S}_0^k MO_{A_0} vTW & \to & \Lambda_0^2 MO_{A_0} F_0 & \\
 & \downarrow \delta & & \downarrow \delta & & \downarrow & \\
0\to & \dot{\Lambda}_0^3 MO_{A_0} \dot{S}_0^{k-1} MO_{A_0} vTW & = & \dot{\Lambda}_0^3 MO_{A_0} \dot{S}_0^{k-1} MO_{A_0} vTW & \to & 0 &
\end{array}$$

Now, as F_1 is a bundle of A_0-modules over $\underline{E}_k$, then $J\underline{D}(F_1)$ is a bundle of A_0-modules over $J\underline{D}(\underline{E})$ and we have exact sequnce of bundles of A_0-modules over $\underline{E}_{k+1}$:

$$0 \longrightarrow TMO_{A_0} F_1 \longrightarrow J\underline{D}(F_1)|\underline{E}_{k+1} \longrightarrow \pi^*_{k+1,k} F_1 \longrightarrow 0 .$$

As the morphism $\pi_{k+1,k} : \underline{E}_{k+1} \longrightarrow \underline{E}_k$ has constant rank, we have:

$$\begin{aligned}
\underline{E}_{k+1}^{(1)} &\equiv \pi_{k+2,k+1}(\underline{E}_{k+2}) \\
&= \pi_{k+2,k+1}\big(J\underline{D}(\underline{E}_{k+1}) \cap J\underline{D}^{k+2})\big) \subseteq J\underline{D}(\pi_{k+1,k})\big(J\underline{D}(\underline{E}_{k+1})\big)\cap \\
&\quad \cap J\underline{D}^{k+1}(W) \\
&= J\underline{D}(\underline{E}_k^{(1)}) \cap J\underline{D}^{k+1}(W) = (\underline{E}_k^{(1)})_{+1} .
\end{aligned}$$

On the other hand we have the following commutative and exact diagram:

$$(\star =) \qquad \begin{array}{ccccccc}
0 & \longrightarrow & \underline{E}_{k+1}^{(1)} & \longrightarrow & \underline{E}_{k+1} & \overset{k_1}{\underset{0}{\rightrightarrows}} & TMO_{A_0} F_1 \\
 & & \downarrow & & \| & & \downarrow \varepsilon_1 \\
0 & \longrightarrow & (\underline{E}_k^{(1)})_{+1} & \longrightarrow & \underline{E}_{k+1} & \overset{J\underline{D}(k)}{\underset{0}{\rightrightarrows}} & J\underline{D}(F_1)|\underline{E}_{k+1} .
\end{array}$$

It follows that $\underline{E}_{k+1}^{(1)} = (\underline{E}_k^{(1)})_{+1}$ bacause $\varepsilon_1 : TMO_{A_0} F_1 \longrightarrow J\underline{D}(F_1)|\underline{E}_{k+1}$ is a monomorphism of bundles of A_0-modules over $\underline{E}_{k+1}$. (Note that diagram $(\star =)$ is a direct consequence of the following exact sequnces

$$\underline{E}_{k+2} \longrightarrow \underline{E}_{k+1} \overset{k_1}{\underset{0}{\rightrightarrows}} TMO_{A_0} F_1 ; \quad \underline{E}_{k+1} \longrightarrow \underline{E}_k \overset{k}{\underset{0}{\rightrightarrows}} F_1 .$$

We call k the **supercurvature map** of $(\underline{E}_k)$. Therefore, the case $r = 1$ is proved. Suppose, now, that $\underline{E}^{(1)}_{k+r} = (\underline{E}^{(1)}_k)_{+r}$. As $\dot{g}_{k+1}$ is a bundle of A_0-modules over $\underline{E}_k$ and $\dot{g}_k$ is a 2-acyclic, then $\dot{g}_{k+r}$ is a bundle of A_0-modules over $\underline{E}_k$, $\forall r \geq 1$. It follows that $\pi_{k+r+1,k+r} : \underline{E}_{k+r+1} \longrightarrow \underline{E}_{k+r}$ is a morphism of constant super rank and $\underline{E}^{(1)}_{k+r}$ a fibered subsupermanifold of $J\underline{D}^{k+r}(W)$. So we have

$$(\underline{E}^{(1)}_k)_{+(r+1)} = ((\underline{E}^{(1)}_k)_{+r})_{+1} = (\underline{E}^{(1)}_{k+r})_{+1}\,.$$

But $\dot{g}_{k+r+1}$ is a bundle of A_0-modules over $\underline{E}_k$ and its pull-back $\pi^*_{k+r,k}\dot{g}_{k+r+1}$ is a bundle of A_0-modules over $\underline{E}_{k+r}$ which also 2-superacyclic, because $\dot{g}_k$ is 2-superacyclic. Then by using theorem for $r = 1$ we have

$$(\underline{E}^{(1)}_k)_{+(r+1)} = \big(\underline{E}^{(1)}_{k+r}\big)_{+1} = \underline{E}^{(1)}_{k+r+1}\,.$$

THEOREM 3. *Let $\underline{E}_k \subset J\underline{D}^k(W)$ be a sufficiently superregular SPDE of order k on $\pi : W \longrightarrow M$, i.e., $\underline{E}_k$ is a superregular system such that the morphisms $\pi_{k+r+s,k+r} : \underline{E}_{k+r+s} \longrightarrow \underline{E}_{k+r}$ have constant superrank, $\forall r, s \geq 0$. If A is the inductive limit of Noetherian rings A_λ,*

$$A = \lim_{\lambda \to \infty} \quad A_\lambda\,,$$

then there exist integer $s \geq 0$ and $k_s \geq k$ such that $\underline{E}^{(s)}_{k_s} \subset J\underline{D}^{k_s}(W)$ is a superinvolutive formally superintegrable system with the same supersolutions as $\underline{E}_k$ and such that $(\underline{E}^{(s)}_{k_s})_{+r} = \underline{E}^{(s)}_{k_s+r}$.

PROOF. As the system is sufficiently superregular it follows that $\underline{E}^{(s)}_{k+r} \equiv \pi_{k+r+s,k+r}(\underline{E}_{k+r+s})$ is a fibered subsupermanifold of $J\underline{D}^{k+r}(W)$, $\forall r, s \geq 0$. For r fixed we have a chain of subsets:

$$J\underline{D}^{k+r}(W) \supset \underline{E}_{k+r} \supset \underline{E}^{(1)}_{k+r} \supset \ldots \supset \underline{E}^{(s)}_{k+r} \supset \ldots$$

such that $\dim J\underline{D}^{k+r}(W) \geq \dim \underline{E}_{k+r} \geq \dim \underline{E}^{(1)}_{k+r} \geq \ldots \geq \dim \underline{E}^{(s)}_{k+r} \geq \ldots$ and there must exist an integer $s(r)$ such that the chain becomes stationary, i.e. to say $\underline{E}^{(s(r))}_{k+r} = \underline{E}^{(s(r)+1)}_{k+r} = \ldots = \overline{\underline{E}}^{(s)}_{k+r}$. We know that $\dot{g}_{k+r}$ becomes superinvolutive when r is big enough. So without loss of generality let us suppose from now on that $\dot{g}_k$ is superinvolutive. Then, we have the following commutative diagram

$$\begin{array}{ccccc}
0 & \longrightarrow & \underline{E}^{(s)}_{k+r+1} & \longrightarrow & J\underline{D}(\underline{E}^{(s)}_{k+r}) \\
 & & \downarrow & & \downarrow \\
0 & \longrightarrow & \underline{E}^{(s+1)}_{k+r} & \longrightarrow & \underline{E}^{(s)}_{k+r} \\
 & & \downarrow & & \downarrow \\
 & & 0 & & 0
\end{array}$$

Let $\dot{g}^{(s)}_{k+r}$ be the supersymbol of $\underline{E}^{(s)}_{k+r}$. Then we have the following monomorphism

$$0 \longrightarrow \dot{g}^{(s)}_{k+r+s} \longrightarrow TMO_{A_0}\dot{g}^{(s)}_{k+r}$$

that is induced by the following one

$$(\mathrm{id}_{J\underline{D}^{k+r}(W)})^{(1)} : J\underline{D}^{k+r+1}(W) \longrightarrow J\underline{D}(J\underline{D}^{k+r}(W)) .$$

Furthermore, as we have the following exact sequence

$$0 \longrightarrow \dot{g}^{(s)}_{k+r+1} \longrightarrow vT\underline{E}^{(s)}_{k+r+1} \longrightarrow vT\underline{E}^{(s+1)}_{k+r} \longrightarrow 0$$

we get that $\dim \dot{g}^{(s)}_{k+r+1} = \dim \underline{E}^{(s)}_{k+r+1} - \dim \underline{E}^{(s+1)}_{k+r}$, so $\dot{g}^{(s)}_{k+r+1}$ is a bundle of A_0-modules over $\underline{E}^{(s)}_{k+r+1}$. Then we have that

$$\delta\big(\dot{g}^{(s)}_{k+r+1}\big) \subset TMO_{A_0}\dot{g}^{(s)}_{k+r} .$$

Now, $\underline{E}_k$ and $\underline{E}^{(s)}_{k+r}$ have the same supersolutions, $\forall r,\ s \geq 0$. In fact, every supersolution of $\underline{E}_k$ is a supersolution of $\underline{E}_{k+r+s}$ and thus of $\underline{E}^{(s)}_{k+r}$. Conversely, every supersolution of $\underline{E}^{(s)}_{k+r}$ is a supersolution of $\underline{E}_{k+r}$ because $\underline{E}^{(s)}_{k+r} \subset \underline{E}_{k+r}$ and thus is a supersolution of $\underline{E}_k$. Now, we can construct inductively a chain of supersystems $\underline{E}_k$, $\underline{E}^{(1)}_k, \ldots, \underline{E}^{(s)}_{k_s}, \ldots$ with superinvolutive symbols and such that $\big(\underline{E}^{(s)}_{k_s}\big)_{+r} = \underline{E}^{(s)}_{k_s+r}$ $\forall r,\ s \geq 0$. Then, we can prove that there exists an integer s_0 such that

$$\dot{g}^{(s)}_{k+r} = \dot{g}^{(s_0)}_{k+r},\ \forall s \geq s_0,\ r \geq 0 .$$

In fact $\dot{g}^{(s)}_{k+r}$ is a subsuperbundle of $\dot{S}^{k+r}_0 MO_{A_0}vTW$, $\forall r,\ s \geq 0$. Furthermore, we have the following commutative diagram:

$$\begin{array}{ccc} \dot{g}^{(s+1)}_{k+r+1} & \xrightarrow{\ \delta\ } & TMO_{A_0}\dot{g}^{(s+1)}_{k+r} \\ \downarrow & & \downarrow \\ \dot{g}^{(s)}_{k+r+1} & \xrightarrow{\ \delta\ } & TMO_{A_0}\dot{g}^{(s)}_{k+r} . \end{array}$$

Let $p \in M$, U be a neighborhood of p in M trivializing W and $\{x^A, y^B, y^B_A, \ldots, y^B_{A_1\cdots A_k}\}$ be local coordinates for $J\underline{D}^k(W)$ over U. We define locally $\dot{g}^{(s)}_{k+r}$ by linear system such as

$$A^C_B(p)\, y^B + A^{CA}_B(p) y^B_A + \ldots + A^{CA_1\ldots A_{k+r}}_B(p) y^B_{A_1\ldots A_{k+r}} = 0 .$$

For p fixed we can consider the A_0-module generated by the polynomials in $\{\chi_A\}$ with value in $A^{m'',n''}$, $((m'',n'')$ = fiber dimension of $(W))$, and indexed by C:

$$\begin{aligned}&\{A^C_B(p)\chi_B + A^{CA}_B(p)\chi^B_A + \ldots + A^{CA_1\ldots A_{k+r}}_B(p)\chi^B_{A_1\ldots A_{k+r}}\}\\ &\in A^{m'',n''}[\chi] = A^{m'',n''}[\chi_1, \ldots, \chi_{m+n}] .\end{aligned}$$

$\dot{g}^{(s)}_{k+r+1}$ is defined locally by the linear system

$$A^C_B(p)y^B_D + A^{CA}_B(p)y^B_{AD} + \ldots + A_B^{CA_1\ldots A_{k+r}}(p)y^B_{A_1\ldots A_{k+r}D} = 0\,.$$

It follows that the A_0-module associated to $(\dot{g}^{(s)}_{k+r})_{+1}$ is generated by the products by χ_A of the polynomials associated to $\dot{g}^{(s)}_{k+r}$. Furthermore, from the last diagram it follows that

$$\delta\big(\dot{g}^{(s)}_{k+r+1}\big) \subset TMO_{A_0}\dot{g}^{(s)}_{k+r} \quad \text{and thus} \quad \big(\dot{g}^{(s)}_{k+r}\big)_{+1} \supset \dot{g}^{(s)}_{k+r+1}\,.$$

Then we can see that the A_0-module associated to $(\dot{g}^{(s)}_{k+r})_{+1}$ is a A_0-submodule of the A_0-module associated to $\dot{g}^{(s)}_{k+r+1}$. Now we let us consider the graded symbol:

$$(=\star) \qquad \tilde{g}^{(s)}_{k+r} \equiv \mathrm{Hom}_{A_0}(A, \dot{g}^{(s)}_{k+r})\,.$$

To $\tilde{g}^{(s)}_{k+r}$ we can associate the A-module

$$A^{m''+n''}[\chi] = \mathrm{Hom}_{A_0}\left(A, A^{m'',n''}[\chi]\right).$$

We call $\dot{M}^{(s)}_{k+r}$ (resp. $\widetilde{M}^{(s)}_{k+r}$) the $A_0[\chi]$-submodule (resp. $A[\chi]$-submodule) of the $A_0[\chi]$-module (resp. $A[\chi]$-module) $A^{m'',n''}[\chi]$ (resp. $A^{m''+n''}[\chi]$) generated by the polynomials associated as above to $\dot{g}^{(s)}_k, \ldots, \dot{g}^{(s)}_{k+r}$ (resp. $\tilde{g}^{(s)}_k, \ldots, \tilde{g}^{(s)}_{k+r}$). Moreover,

$$\dot{M}^{(s)}_{k+r_1} \subseteq \dot{M}^{(s)}_{k+r_2} \ (\text{ resp. } \ \widetilde{M}^{(s)}_{k+r_1} \subseteq \widetilde{M}^{(s)}_{k+r_2})$$

whenever $r_1 \leq r_2$. We shall use the following.

LEMMA 2. *Let (A_λ) be an inductive system of rings. Let $A \equiv \lim_{\lambda\to\infty} A_\lambda$. Let M be a A-module of finite presentation, i.e., one has an exact sequence*

$$A^m \longrightarrow A^n \xrightarrow{\quad f \quad} M \longrightarrow 0\,,$$

for some $m < n \in \mathbb{N}$. (This is equivalent to say that $M \cong A^n/\ker f \cong A^n/A^m$). Then, there exists a λ and a A_λ-module M_λ of finite presentation such that

$$M \cong M_\lambda \otimes_{A_\lambda} A\,.$$

PROOF OF LEMMA 2 See e.g. ref. [7].

□

Now, as in our case A is the inductive limit of Noetherian rings, we have that also $A[\chi] = \lim_{\lambda\to\infty} A_\lambda[\chi]$, where each $A_\lambda[\chi]$ are Noetherian rings. On the other hand $A^{m'',n''}[\chi]$ is a free $A[\chi]$-module of finite type, hence a $A[\chi]$-module of finite presentation. Then, from above lemma 2 we get that

$$A^{m'',n''}[\chi] \cong A_\lambda^{m'',n''}[\chi] \otimes_{A_\lambda[\chi]} A[\chi]$$

for some λ. Note that each $A_\lambda^{m'',n''}[\chi]$ are Noetherian $A_\lambda[\chi]$-modules [1, 10]. Moreover, each $\widetilde{M}_{k+r}^{(s)}$ is also a $A[\chi]$-module of finite presentation. So we can write

$$\widetilde{M}_{k+r}^{(s)} \cong \left(\widetilde{M}_{k+r}^{(s)}\right)_{\bar{\lambda}} \otimes_{A_{\bar{\lambda}}[\chi]} A[\chi]$$

for some $\bar{\lambda}$. As $\widetilde{M}_{k+r}^{(s)} \subset A^{m'',n''}[\chi] \implies \left(\widetilde{M}_{k+r}^{(s)}\right)_{\bar{\lambda}} \subset A_\lambda^{m'',n''}[\chi]$. As $A_\lambda^{m'',n''}[\chi]$ is a Noetherian it follows that the increasing chain

$$\widetilde{M}_k^{(s)} \subseteq \widetilde{M}_{k+1}^{(s)} \subseteq \dots$$

is stationary and we call $\widetilde{M}^{(s)}$ its maximum element which is a finitely generated submodule of $A^{m''+n''}[\chi]$. Now, as $\tilde{g}_{k+r}^{(s)} \supseteq \tilde{g}_{k+r}^{(s+1)}$, we have $\widetilde{M}_{k+r}^{(s)} \subseteq \widetilde{M}_{k+r}^{(s+1)}$, $\forall s \geq s_0$, and the increasing chain

$$\widetilde{M} = \widetilde{M}^{(0)} \subseteq \widetilde{M}^{(1)} \subseteq \dots \subseteq \widetilde{M}^{(s)} \subseteq \dots$$

is stationary, i.e., there exists an integer s_0 such that $\widetilde{M}^{(s)} = \widetilde{M}^{(s_0)}$, $s \geq s_0$. This implies that $\tilde{g}_{k+r}^{(s)} = \tilde{g}_{k+r}^{(s_0)}$, $\forall s \geq s_0$, and $r \geq 0$. Note that this equality is true $\forall p \in M$, because the $\tilde{g}_{k+r}^{(s)}$ are, by hypothesis, bundles of A-modules over M, $\forall s$, $r \geq 0$. Finally, taking into account relation $(=\star)$ between $\tilde{g}_{k+r}^{(s)}$ and $\dot{g}_{k+r}^{(s)}$ we get also that $\dot{g}_{k+r}^{(s)} = \dot{g}_{k+r}^{(s_0)}$, $\forall s \geq s_0$, and $r \geq 0$. □

EXAMPLES OF SUPERALGEBRAS WITH $A_1^{\perp} = 0$ GIVEN AS INDUCTIVE LIMIT OF NOETHERIAN RINGS. If B is a commutative $\mathbb{K}$-algebra, $A_\lambda \equiv B \otimes_{\mathbb{K}} \Lambda(\mathbb{K}^\lambda)$ is a $\mathbb{K}$-superalgebra that is a Noetherian ring. For example we can take $B \equiv \mathbb{K}[\![x^1, \dots, x^m]\!]$. Then, $A \equiv \lim_{\lambda\to\infty} A_\lambda \cong B \otimes_{\mathbb{K}} \Lambda(\mathbb{K}^\infty)$ is just a $\mathbb{K}$-superalgebra with $A_1^{\perp} = 0$ obtained as inductive limit of Noetherian $\mathbb{K}$-superlagebras.

REMARK 4. We call **EV-superplane** at point $q \in J\underline{D}^k(W)$ the linear subspace in $T_q J\underline{D}^k(W)$ of the form, $T_q D^k s(M)$, for some G^∞-section s of π such that $D^k s(x) = q$, $x = \pi_k(q)$. Any point $q \in J\underline{D}^k(W)$ identifies a unique EV-superplane L_q at $q' \equiv \pi_{k,k-1}(q) \in J\underline{D}^{k-1}(W)$: $L_q \subset T_{q'} J\underline{D}^{k-1}(W)$. If $P \subset T_{\bar{q}} J\underline{D}^{k-1}(W)$ is some subspace, we set

$$S_{(i,j)}(P) \equiv \left\{q \in J\underline{D}^k(W) | \bar{q} \equiv \pi_{k,k-1}(q),\ \dim(L_q \cap P) = (i,j)\right\}.$$

We call $S_{(i,j)}(q) \equiv S_{(i,j)}(L_q)$ **characteristic supercones**. These are affine subspaces of $\pi_{k,k-1}^{-1}(\bar{q})$. Furthermore, as $S_{(i,j)}(q)$ can be considered pointed at q, it can be identified with its tangent space $T_q S_{(i,j)}(q)$ that is subspace of $T_q \pi_{k,k-1}^{-1}(\bar{q}) \cong S_0^k(T_x M) O_{A_0} v T_p W$, $p \equiv \pi_{k,0}(q)$.

The **characteristic supermanifold**, $\mathrm{Char}_{(i,j)}(\underline{E}_k)_q$, of all (i,j)-dimensional characteristic subspaces for a SPDE $\underline{E}_k \subset J\underline{D}^k(W)$ at $q \in \underline{E}_k$ is defined by:

$$\mathrm{Char}_{(i,j)}(\underline{E}_k)_q \equiv \bigcup_{q' \in \dot{g}_{k,q} \cap S_{(m-i,n-j)}(q)} \mathrm{Char}_{(i,j)}(\underline{E}_k)_{q,q'},$$

where $\mathrm{Char}_{(i,j)}(\underline{E}_k)_{q,q'} \equiv \{\xi \in (T_xM)_0^+ - \{0\}| < \xi, T(\pi_{k-1})(L_q \cap L_{q'}) >= 0\}$, $x = \pi_k(q) = \pi_k(q')$. $T(\pi_{k-1})(L_q \cap L_{q'})$ is the $(m-i, n-j)$-dimensional subspace of T_xM identified by q'. A vector ξ contained in a (1,1)-dimensional characteristic subspace is called a **characteristic supercovector** of $\underline{E}_k$ at q.

The following theorem relates the concept of characteristic supercovector with a criterion of existence of singular supersolutions. This generalizes previous one by V. Lychagin given for ordinary manifolds [18].

THEOREM 4. *Let $\underline{E}_k \subset J\underline{D}^k(W)$ be a SPDE of order k on $\pi\colon W \longrightarrow M$, $\dim M = (m,n)$, $\dim W = (m+m', n+n')$, such that $\underline{E}_k = \ker K$, being $K\colon J\underline{D}^k(W) \longrightarrow F$ a differential operator of class G^∞, where F is a bundle of vector superspaces of dimension (p,q) over M. Let $L \subset \underline{E}_k$ be a supersolution. Let us denote by $\Sigma^{(i,j)}(L)$ the* **set of singular points of type** *(i,j), i.e.*

$$\Sigma^{(i,j)}(L) \equiv \{q \in L | \dim\ker(T\pi_k|T_qL) = (i,j)\},\ 1 \le j \le m,\ 1 \le i \le n.$$

Let Ξ be the subspace

$$\Xi \equiv T(\pi_k)(T_qL) \subset T_xM,\ x \equiv \pi_k(q).$$

Let

$$\widetilde{\Xi} \equiv \mathrm{Hom}_{A_0}(A, \Xi) \subset \widetilde{T}(\pi_k)(\widetilde{T}_qL) \subset \widetilde{T}_xM$$

be the graded extension of Ξ. Set

$$\tilde{g}_{k+s,q}(\widetilde{\Xi}) = \tilde{g}_{k+s,q} \cap \widetilde{\Xi}^{(k+s)}(W)_p \subset S_0^{k+s}(\widetilde{T}_xM) O_A v\widetilde{T}_pW\,,$$

being $\widetilde{\Xi}^{(k+s)}(W)_p$ the subspace of $S_0^{k+s}(\widetilde{T}_xM)O_Av\widetilde{T}_pW$ of all A-linear $(k+s)$-symmetric applications $\widetilde{T}_xM \times \dots_{k+s} \dots \times \widetilde{T}_xM \longrightarrow v\widetilde{T}_pW$ that annihilate on $\widetilde{\Xi}$, $p = \pi_{k,0}(q)$. Then there exists a number s_0 such that $(\underline{E}_k)_{+s}$, $s \ge s_0$, has no more singular supersolution of type $\Sigma^{(i,j)}(L)$ for any point $\bar{q} \in (\underline{E}_k)_{+s}$ such that $\pi_{k+s,k}(\bar{q}) = q$, if the following conditions are satisfied: (a) *A is the inductive limit of Noetherian rings: $A = \lim_{\lambda\to\infty} A_\lambda$;* (b) *The complexified projective spaces $P(\widetilde{\Xi}^{\mathbf{C}})$ and $P(\widetilde{\mathrm{Char}}_{(i,j)}(\underline{E}_k)_q^{\mathbf{C}})$ intersect in a finitely number of points, $\lambda_1, \dots, \lambda_r$, and*

$$\sum_{1 \le i \le r} \dim_{A^{\mathbf{C}}} \tilde{g}_{k,q}(\lambda_i)^{\mathbf{C}} < \mathrm{codim}\,\widetilde{\Xi} = i + j\,.$$

Furthermore, if above conditions are satisfied for all subspaces $\Xi \subset T_xM$, $\mathrm{codim}\,\Xi = (i,j)$, then the system $\underline{E}_{k+s} \subset J\underline{D}^{k+s}(W)$, $s \ge s_0$, do not have solutions passing trough points at which there is a singularity of type $\Sigma^{(i,j)}$.

PROOF. Note that $q \in \Sigma^{(i,j)}(L)$ iff $\dot{g}_{k,q}(L) = T_qL \cap \dot{g}_{k,q}$, where

$$\dot{g}_{k,q}(L) \equiv \ker\big(T(\pi_k)|T_qL\big) = \ker\big(T(\pi_{k,k-1})|T_qL\big)$$

and $\dim \dot{g}_{k,g}(L) = (i,j)$. In fact, we know that $q \in \Sigma^{(i,j)}(L)$ iff $\dim\dot{g}_{k,q}(L) = (i,j)$. One the other hand taking into account the following exact sequences of families of vector spaces over $\underline{E}_k$:

$$\begin{array}{ccccccccc}
0 & \longrightarrow & \dot{S}_0^k MO_{A_0}vTW & \longrightarrow & vTJ\underline{D}^k(W) & \longrightarrow & vTJ\underline{D}^{k-1}(W) & \longrightarrow & 0 \\
 & & \uparrow & & \uparrow & & & & \\
0 & \longrightarrow & \dot{g}k & \longrightarrow & vT\underline{E}_k & & & & \\
 & & \uparrow & & \uparrow & & & & \\
 & & 0 & & 0 & & & &
\end{array}$$

we get that $\dot{g}_{k,q}(L) = T_qL \cap \dot{g}_{k,q}$ and $\dot{g}_{k,q} \subset \dot{S}_0^k(T_xM)O_{A_0}vT_pW$. Now, we can have the following two possibilities: A) L is a supersmooth supermanifold in some neighborhood of the point q, but the projection $\pi_k|L : L \longrightarrow M$ has a singularity: $\ker\big(T(\pi_k)|T_qL\big) \neq 0$; B) L is not a supersmooth supermanifold at $q \in \underline{E}_k$, but this singularity resolved by prolongation, i.e., there exists a supersmooth V-manifold $L^{(s)} \subset (\underline{E}_k)_{+s} \subset J\underline{D}^{k+s}(W)$, passing through the point $\bar{q}$, $q = \pi_{k+s,k}(\bar{q})$, such that at some neighborhood U of q, $L \cap U = \pi_{k+s,k}(L^{(s)})$. On the other hand, case (B) can be reduced to case (A), as we can pass from the equation $\underline{E}_k$ to the equation $(\underline{E}_k)_{+s} \subset J\underline{D}^{k+s}(W)$ at the point $\bar{q}$. Hence, in the sequel we shall consider case (A) only.

Let $\Sigma(L)$ be the set of singular points of the projection $\pi_k|L$. $\Sigma(L)$ has a natural stratification, namely:

$$\Sigma(L) = \bigcup_{1\leq i\leq m, 1\leq j\leq n} \Sigma^{(i,j)}(L)\,.$$

Let us describe the tangent plane T_qL to the V-manifold L at the point $q \in \Sigma^{(i,j)}(L)$. Set $\Xi \equiv T(\pi_k)(T_qL) \subset T_xM, x \equiv \pi_k(q)$. Ξ is a vector superspace of dimension $(m-i, n-j)$. Let us denote by $\Xi^{(k)}(W)_p, p \equiv \pi_{k,0}(q)$, the subspace of $S_0^k(T_xM)O_{A_0}vT_pW$ of all A_0-linear k-symmetric applications $T_xM \times \ldots_k \ldots \times T_xM \longrightarrow vT_pW$ that annihilate on Ξ. Let us denote by $\dot{g}_{k,q}(\Xi)$ the subspace of $\dot{S}_0^k(T_xM)O_{A_0}vT_pW$ defined by

$$\dot{g}_{k,q}(\Xi) \equiv \dot{g}_{k,q} \cap \Xi^{(k)}(W)_p\,.$$

Thus, T_qL, $q \in \Sigma^{(i,j)}(L)$ determines $(m-i, n-j)$-dimensional subspace $\Xi \subset T_xM$ such that

$$\dot{g}_{k,q}(L) \subset \dot{g}_{k,q}(\Xi) \subset \dot{S}_0^k(T_xM)O_{A_0}vT_pW\,.$$

If L is not supersmooth at q, but becomes supersmooth after taking the sth prolongation (case (B)), then the tangent space $T_{\bar{q}}L^{(s)}$, under the condition $\bar{q} \in \Sigma^{(i,j)}(L^{(s)})$ also defines a $(m-i, n-j)$-dimensional vector superspace

$$\Xi \equiv T(\pi_{k+s})(T_{\bar{q}}L^{(s)}) \subset T_xM\,,$$

and a (i,j)-dimensional superspace

$$\dot{g}_{k+s,q}(L) \equiv T_{\bar{q}}L^{(s)} \cap \dot{S}_0^{k+s}(T_xM)O_{A_0}vT_pW\,.$$

We have:

$$\dot{g}_{k+s,q}(L) \subset \dot{g}_{k+s,q}(\Xi) = \dot{g}_{k+s,q} \cap \Xi^{(k+s)}(W)_p \subset \dot{S}_0^{k+s}(T_xM)O_{A_0}vT_pW\,,$$

where $\Xi^{(k+s)}(W)_p$ is the subspace of $\dot{S}_0^{k+s}(T_xM)O_{A_0}vT_pW$ of all $(k+s)$-symmetric A_0-linear maps $T_xM \times \ldots_{k+s} \ldots \times T_xM \longrightarrow vT_pW$ that annihilate on Ξ. Of course the necessary condition for the existence of such a supersolution is the following:

$$(\odot\,\square) \qquad \text{(case (A))} \;\; \dim \dot{g}_{k,q}(\Xi) \geq (i,j); \;\; \text{(case (B))} \;\; \dim \dot{g}_{k+s,q}(\Xi) \geq (i,j)\,.$$

In order to investigate under which conditions $(\odot\,\square)$ fail, let us consider the graded situation, i.e. let us consider the supertangent space $\widetilde{T}_qL = \mathrm{Hom}_A(A, T_qL) = AO_{A_0}T_qL$, to the V-supermanifold L at the point $q \in \Sigma^{(i,j)}(L)$. Of course, one has $\dim_A \ker\big(\widetilde{T}(\pi_k)|\widetilde{T}_qL\big) = i+j$ and $\dim_A \widetilde{\Xi} = m+n-(i+j)$. Thus $\widetilde{T}_qL$ determines $(m+n)-(i+j)$-dimensional subspace $\widetilde{\Xi} \subset \widetilde{T}_xM$ such that

$$\tilde{g}_{k,q}(L) \equiv \ker\big(\widetilde{T}(\pi_k)|\widetilde{T}_qL\big) \subset \tilde{g}_{k,q}(\widetilde{\Xi}) \subset S_0^k(\widetilde{T}_xM)O_Av\widetilde{T}_pW\,.$$

Similarly, in the case (B) we have

$$\tilde{g}_{k+s,q}(L) \equiv \widetilde{T}_{\bar{q}}L^{(s)} \cap S_0^{k+s}(\widetilde{T}_xM)O_Av\widetilde{T}_pW\,;$$
$$\tilde{g}_{k+s,q}(L) \subset \tilde{g}_{k+s,q}(\widetilde{\Xi}) = \tilde{g}_{k+s,q} \cap \widetilde{\Xi}^{(k+s)}(W)_p \subset S_0^{k+s}(\widetilde{T}_xM)O_Av\widetilde{T}_pW\,.$$

Then, the necessary conditions for the existance of such a supersolution are the following:
$(\odot\,\star)$ $\dim \tilde{g}_{k,q}(\widetilde{\Xi}) \geq i+j$; $\dim \tilde{g}_{k+s,q}(\widetilde{\Xi}) \geq i+j$. Now, conditions $(\odot\,\star)$ are satisfied iff conditions $(\odot\,\square)$ are. So, let us investigate under which conditions $(\odot\,\star)$ fail. This is more simple as we can work with finitely generated A-modules and therefore we can more easily reproduce the classical situation where one works with finitely generated $\mathbb{R}$-modules. Set $\dim_A \tilde{g}_{k+s,q}(\widetilde{\Xi}) \equiv d(q,k,s,\Xi)$. We shall use the following lemmas.

LEMMA 3. *Let $u \in L(E,F)$, where E and F are superspaces of dimension (m,n) and (m',n') repsectively. Then, one has the following canonical isomorphism:*

$$\text{(a)}\;\; \ker(\tilde{u}) \cong AO_{A_0}\ker(u) \equiv \widetilde{\ker}(u)\,; \qquad \text{(b)}\;\; \mathrm{Coker}(\tilde{u})^+ \cong \ker(\tilde{u}^+)\,;$$
$$\text{(c)}\;\; \mathrm{Im}(\tilde{u})^+ \cong \mathrm{Im}(\tilde{u}^+)\,; \qquad \text{(d)}\;\; \ker(\tilde{u})^+ \cong \mathrm{Coker}(\tilde{u}^+)\,.$$

PROOF OF LEMMA 3.

$$\text{(a)} \qquad \ker(\tilde{u}) \subset AO_{A_0}E \equiv \big\{\alpha \in L_{A_0}(A;E) | u\circ\alpha : A \longrightarrow F, u\circ\alpha = 0\big\}\,.$$

This means that $u\big(\alpha(a)\big) = 0$, $\forall a \in A$, i.e. $\alpha \in \ker(\tilde{u})$ iff $\alpha : A \longrightarrow \ker(u) \subset E$. Propositions (b), (c) and (d) follow from the fact that $\widetilde{E}$ and $\widetilde{F}$ are free A-modules of finite type, and $E^+ \cong L_A(\widetilde{E};A)$. So, from the first of the following exact commutative diagrams we get the second one

$$
\begin{array}{ccccccccc}
 & & & & & & 0 & & \\
 & & & & & & \downarrow & & \\
0 & \longrightarrow & \ker(\tilde{u}) & \longrightarrow & \widetilde{E} & \longrightarrow & \mathrm{Im}(\tilde{u}) & \longrightarrow & 0 \\
 & & & \tilde{u} & \downarrow & & \downarrow & & \\
 & & & & \widetilde{F} & \longrightarrow & \widetilde{F} & & \\
 & & & & & & \downarrow & & \\
 & & & & & & \mathrm{Coker}(\tilde{u}) & & \\
 & & & & & & \downarrow & & \\
 & & & & & & 0 & &
\end{array}
$$

$$
\begin{array}{ccccccccc}
 & & & & & & 0 & & \\
 & & & & & & \uparrow & & \\
0 & \longleftarrow & \ker(\tilde{u})^+ & \longleftarrow & \widetilde{E}^+ & \longleftarrow & \mathrm{Im}(\tilde{u})^+ & \longleftarrow & 0 \\
 & & & \tilde{u}^+ & \uparrow & & \uparrow j^+ & & \\
 & & & & \widetilde{F}^+ & = & \widetilde{F}^+ & & \\
 & & & & & & \uparrow & & \\
 & & & & & & \mathrm{Coker}(\tilde{u})^+ & & \\
 & & & & & & \uparrow & & \\
 & & & & & & 0 & &
\end{array}
$$

This is enough to prove above isomorphisms (b), (c) and (d). □

LEMMA 4. *Let $\underline{E}_k \subset J\underline{D}^k(W)$ be a SPDE of order k on π: $W \longrightarrow M$, $\dim M = (m,n)$, $\dim W = (m+m', n+n')$, such that $\underline{E}_k = \ker K$, being K: $J\underline{D}^k(W) \longrightarrow F$ a differential operator of class G^∞, where F is a bundle of vector superspaces of dimension (p,q) over M. Then one has $\dot{g}_{k+s} = \ker\left(\dot{\sigma}_{k+s}(K)\right)$, (resp. $\tilde{g}_{k+s} = \ker\left(\tilde{\sigma}_{k+s}(K)\right)$ where $\dot{\sigma}_{k+s}(K)$ (resp. $\tilde{\sigma}_{k+s}(K)$) is the s-prolongation of the supersymbol (resp. graded supersymbol) of the differential operator K.*

PROOF OF LEMMA 4. The proof of this lemma follows directly by using above lemma and the following commutative diagrams with exact lines:

$$
\begin{array}{ccccccc}
 & & & & 0 & & 0 \\
 & & & & \downarrow & & \downarrow \\
0 & \longrightarrow & \dot{g}_{k+s} & \longrightarrow & \dot{S}_0^{k+s} M O_{A_0} \upsilon TW & \xrightarrow{\dot{\sigma}_{k+s}(K)} & \dot{S}_0^{s} M O_{A_0} \upsilon TF \\
 & & & & \downarrow & & \downarrow \\
 & & & & \upsilon TJ\underline{D}^{k+s}(W) & \xrightarrow{\upsilon T(K^{(s)})} & \upsilon TJ\underline{D}^{s}(F) \\
 & & & & \downarrow & & \downarrow \\
 & & & & \upsilon TJ\underline{D}^{k+s-1}(W) & \xrightarrow{\upsilon T(K^{(s-1)})} & \upsilon TJ\underline{D}^{s-1}(W) \\
 & & & & \downarrow & & \downarrow \\
 & & & & 0 & & 0
\end{array}
$$

$$\begin{array}{ccccccc}
 & & & 0 & & 0 & \\
 & & & \downarrow & & \downarrow & \\
0 \longrightarrow & \tilde{g}_{k+s} & \longrightarrow & \widetilde{S}_0^{k+s} M O_{A_0} v\widetilde{T}W & \xrightarrow{\tilde{\sigma}_{k+s}(K)} & \widetilde{S}_0^{s} M O_{A_0} v\widetilde{T}F & \\
 & & & \downarrow & & \downarrow & \\
 & & & v\widetilde{T}J\underline{D}^{k+s}(W) & \xrightarrow{v\widetilde{T}(K^{(s)})} & v\widetilde{T}J\underline{D}^{s}(F) & \\
 & & & \downarrow & & \downarrow & \\
 & & & v\widetilde{T}J\underline{D}^{k+s-1}(W) & \xrightarrow{v\widetilde{T}(K^{(s-1)})} & v\widetilde{T}J\underline{D}^{s-1}(W) & \\
 & & & \downarrow & & \downarrow & \\
 & & & 0 & & 0. & \square
\end{array}$$

LEMMA 5. *Let $A \equiv \bigoplus_{n=0}^{\infty} A_n$ be a graded ring obtained as inductive limit of Noetherian graded rings A_λ:*

$$A = \lim_{\lambda \to \infty} A_\lambda \,, \quad A_\lambda = \bigoplus_{n=0}^{\infty} A_{\lambda,n} \,.$$

Let $\bigoplus_{n=0}^{\infty} M_n$ be a graded A-module of finite presentation. Then Hilbert-Serre theorem [1, 7] can be extended to M.

PROOF OF LEMMA 5. In fact, from Lemma 2 we can write $M \cong M_\lambda \otimes_{A_\lambda} A$, for some λ, where $M_\lambda = \bigoplus_{n=0}^{\infty} M_{\lambda,n}$ is a graded A_λ-module of finite presentation and A_λ is a Noetherian graded ring. $M_{\lambda,n}$ is an A_0-module of finite presentation. Then, for any $\mathbb{Z}$-valued additive function μ defined on the class of all finitely generated $A_{\lambda,0}$-modules we can define the Poincare' series of M like the Poincare' series of M_λ:

$$P(M,t) = P(M_\lambda, t) = \sum_{0 \le n \le \infty} \mu(M_{\lambda,n}) t^n \in \mathbb{Z}[\![t]\!] \,.$$

As M_λ is a finitely generated graded A_λ-module, with A_λ a graded Noetherian ring, we can apply Hilbert-Serre theorem [1, 7]. □

LEMMA 6. *Let A be a ring and let E and F be A-modules. Let E be projective and F of finite presentation. Then one has that $L_A(E;F)$ is a E^+-module of finite presentation.*

PROOF OF LEMMA 6. Let us first note that on E^+ we can induce the ring structure from A. Furthermore, one has the natural mapping ϕ: $E^+ \times L_A(E;F) \longrightarrow L_A(E;F)$, given by, $\phi(\alpha, f) : E \longrightarrow F, \phi(\alpha, f)(v) = \alpha(v) f(v)$. We can easily see that above mapping satisfies the (E^+)-module-axioms. Now, as F is a A-module of finite presentation, we have an exact sequence $A^m \longrightarrow A^n \longrightarrow F \longrightarrow 0$, for some $m, n \in \mathbb{N}$. As a consequence, we have, taking into account that E is projective, also the following exact commutative diagram:

$$\begin{array}{ccccccc}
\mathrm{Hom}_A(E, A^m) & \longrightarrow & \mathrm{Hom}_A(E, A^n) & \longrightarrow & \mathrm{Hom}_A(E, F) & \longrightarrow & 0 \\
\| & & \| & & \| & & \\
(E^+)^m & \longrightarrow & (E^+)^n & \longrightarrow & L_A(E;F) & \longrightarrow & 0
\end{array}$$

that proves that $L_A(E;F)$ is a (E^+)-module of finite presentation. □

Now, taking into account that $\tilde{g}_{k+s,q}(\widetilde{\Xi})$ is a free A-module of finite type, hence of finite presentation, from above lemmas we can see that

$$M(\widetilde{\Xi}) \equiv \sum_{s\geq 0} \tilde{g}_{k+s,q}(\widetilde{\Xi})^+$$

is a $(\widetilde{\Xi}^\bullet(A)_x)^+$-module of finite presentation, where

$$(\widetilde{\Xi}^\bullet(A)_x)^+ \equiv \sum_{s\geq 0} (\widetilde{\Xi}^{(k+s)}(A)_x)^+ \,,$$

being $\widetilde{\Xi}^{(k+s)}(A)_x$ the space of all A-linear $(k+s)$-symmetric applications

$$\widetilde{T}_x M \times \dots {}_{k+s} \dots \times \widetilde{T}_x M \longrightarrow A$$

that annihilate on $\widetilde{\Xi}$; $\widetilde{\Xi}^{(k+s)}(A)_x$ is a projective A-module. Then, we can use Lemma 5, and state that for sufficiently large s the function $d(q,k,s,\Xi)$ is polynomial in s of degree $\leq (i+j)-1$ (the Hilbert polynomial). Hence, if we assume that $(\odot\, \star)$ breaks down beginning from some s_0, then $d(q,k,s,\Xi)$ =constant for all $s \geq s_0$. On the other hand the degree of the Hilbert polynomial coincides with the dimension of the support of the module $M(\widetilde{\Xi})_\lambda$, for some λ, as

$$M(\widetilde{\Xi}) \cong M(\widetilde{\Xi})_\lambda \bigotimes_{\left(\widetilde{\Xi}^\bullet(A)_x\right)^+_\lambda} \left(\widetilde{\Xi}^\bullet(A)_x\right)^+ .$$

Thus the failure of the inequality $(\odot\star)$ is equivalent to the zero dimensionality of the algebraic manifold $\mathrm{Supp}(M(\widetilde{\Xi}))_\lambda$. To describe $\mathrm{Supp}(M(\widetilde{\Xi}))_\lambda$ let us complexify the objects under considerations. $\widetilde{\mathrm{Char}}_{(i,j)}(\underline{E}_k)^{\mathbf{C}}_q$ is conical, hence in complex projective space $P(T^+_x M^{\mathbf{C}})$, a subset $P(\widetilde{\mathrm{Char}}_{(i,j)}(\underline{E}_k)^{\mathbf{C}}_q)$, whose elements are the lines $\{t\nu | t \in \mathbf{C}, \nu \in \widetilde{\mathrm{Char}}_{(i,j)}(\underline{E}_k)^{\mathbf{C}}_q\}$ is defined. Then the set

$$\widetilde{\mathrm{Char}}_{(i,j)}(\underline{E}_k)^{\mathbf{C}}_q(\widetilde{\Xi})_\lambda \equiv \left\{\text{characteristic covectors } \nu \in \widetilde{\Xi}^{(1)}(A)^{\mathbf{C}}_{x\lambda}\right\}$$

is obtained by the following intersection:

$$\widetilde{\mathrm{Char}}_{(i,j)}(\underline{E}_k)^{\mathbf{C}}_q(\widetilde{\Xi})_\lambda = \widetilde{\mathrm{Char}}_{(i,j)}(\underline{E}_k)^{\mathbf{C}}_q \cap \widetilde{\Xi}^{(1)}(A)^{\mathbf{C}}_{x\lambda} .$$

Similarly for the corresponding projective spaces one has:

$$P\Big(\widetilde{\mathrm{Char}}_{(i,j)}(\underline{E}_k)^{\mathbf{C}}_q(\widetilde{\Xi})_\lambda\Big) = P\Big(\widetilde{\mathrm{Char}}_{(i,j)}(\underline{E}_k)^{\mathbf{C}}_q\Big) \cap P\Big(\widetilde{\Xi}^{(1)}(A)^{\mathbf{C}}_{x\lambda}\Big) .$$

Furthermore, by generalizing a well known result by H. Goldschmidt, we can prove that the vector $\nu \in \widetilde{\Xi}^{(1)}(A)^{\mathbf{C}}_{x\lambda}, \nu \neq 0$, is characteristic iff the localization $M_\nu(\widetilde{\Xi}^{\mathbf{C}})_\lambda$ of the module $M(\widetilde{\Xi}^{\mathbf{C}})_\lambda$ with respect to the maximal ideal

$$\tilde{\mu}_{\nu,\lambda} \equiv \left\{p \in (\widetilde{\Xi}^\bullet(A)^{\mathbf{C}}_x)^+_\lambda | p(\nu) = 0\right\}$$

does not vanish.

Then, we get that $\mathrm{Supp}(M(\widetilde{\Xi}^{\mathbf{C}})_\lambda) = P(\widetilde{\mathrm{C}}\mathrm{har}_{(i,j)}(\underline{E}_k)_q^{\mathbf{C}}(\widetilde{\Xi})_\lambda)$. □

Let us now characterize singularities by means of characteristic superclasses. Let $\pi : W \longrightarrow M$ be a fiber bundle over a (m,n)-dimensional supermanifold M. Set:

$$\underline{I}_k(W) \equiv \bigcup_{u \varepsilon J\underline{D}^k(W)} \underline{I}_k(W)_u \,,$$

$\underline{I}_k(W)_u \equiv$Grassmannian of integral superplanes at u. An **integral superplane** at a point $u \in J\underline{D}^k(W)$ is defined to be an (m,n)-dimensional subspaces of $(\underline{\mathbb{E}_k})_u$ tangent to some integral supermanifold of the Cartan distribution $(\underline{\mathbb{E}_k} \subset TJ\underline{D}^k(W))$. It is also useful to consider the graded counterpart. More precisely we call **integral graded superplane** at $u \in J\underline{D}^k(W)$ the $(m+n)-A-$dimensional subsuperspaces L lying in the supertangent to some integral supermanifold, which pass through u. Let us denote by $\tilde{\underline{I}}_k(W)_u$ Grassmannian of integral graded superplanes at u and by $\tilde{\underline{I}}_k(W)$ the corresponding bundle over $J\underline{D}^k(W)$. Note that the graded counterpart of the superstructures considered allow us to work with free A-modules of finite type and so we can more easily reproduce well known results obtained for ordinary manifolds [18, 19, 34]. Let $\tilde{\underline{I}}(\underline{E}_{k+s})$ be the fiber bundle of Grassmannian $(m+n)$-dimensional integral superplanes of the Cartan graded superdistribution $\widetilde{\underline{\mathbb{E}}}(\underline{E}_{k+s})$ on $\underline{E}_{k+s}$ being $\underline{E}_k \subset J\underline{D}^k(W)$ a SPDE. If $\underline{E}_{k+s} = J\underline{D}^{k+s}(W)$ one has

$$\tilde{\underline{I}}(J\underline{D}^{k+s}(E)) = \tilde{\underline{I}}_{k+s}(W) \,.$$

DEFINITION 4. *Each (m,n)-dimensional integral supermanifold $V \subset J\underline{D}^k(W)$ determines a* **tangential** map $i_V : V \longrightarrow \tilde{\underline{I}}_k(W)$ *given by* $i_V(q) = \widetilde{T}_q V \in \tilde{\underline{I}}_k(W)_q$.

Then, any integral V-supermanifold $V \subset J\underline{D}^k(W)$ determines a map

$$\widetilde{H}^i(\tilde{\underline{I}}_k(W); A) \longrightarrow \widetilde{H}(V; A)$$

given by $\omega \longrightarrow i_V^* \omega \equiv$ **characteristic superclass** *on V corresponding to* $\omega \in \widetilde{H}^i(\tilde{\underline{I}}_k(W); A)$.

Then we have the following important theorem that extends previous one obtained by Lychagin for ordinary manifolds [18].

THEOREM 5. 1) *Let $\underline{E}_k \subset J\underline{D}^k(W)$ be a SPDE. Each cohomology superclass $\omega \in \widetilde{H}^i(\tilde{\underline{I}}(\underline{E}_{k+s}); A)$ defines* **characteristic superclass** $\omega_v \equiv i_v^* \omega$ *on the singular solutions $V \subset \underline{E}_{k+s}$ of $\underline{E}_k$.*

2) $\widetilde{H}^\bullet(\tilde{\underline{I}}(\underline{E}_{k+s}); A)$ *is an algebra over* $\widetilde{H}^\bullet(\underline{E}_{k+s}; A)$.

3) *Let $\underline{E}_k$ be formally superintegrable SPDE then the fibre bundles*

$$\pi_{k+s,k+s-1} \colon \underline{E}_{k+s} \longrightarrow \underline{E}_{k+s-1}, s \geq 1 \,,$$

are affine subbundles in $\pi_{k+s,k+s-1} \colon J\underline{D}^{k+s}(W) \longrightarrow J\underline{D}^{k+s-1}(W)$ *and, hence*

$$\widetilde{H}^\bullet(\underline{E}_{k+s}; A) \cong \widetilde{H}^\bullet(\underline{E}_k; A) \,.$$

4) *If V is a regular integral supermanifold, then all its supercharecteristics classes are zero in dimension* ≥ 1.

PROOF. Let us explicitly prove only point 4).

We shall prove that if V is a non-singular integral supermanifold, then the mapping $i_V^*\colon \widetilde{H}^i(\tilde{\underline{I}}_k(W); A) \longrightarrow \widetilde{H}^i(V; A)$ is the zero map. In fact, one has the following commutative and exact diagram:

$$\begin{array}{ccccccc} & & & & 0 & & \\ & & & & \downarrow & & \\ & & \widetilde{H}^i(\tilde{\underline{I}}_k(W); A) & \xrightarrow{i_V^*} & \widetilde{H}^i(V; A) & & \\ & & i_V^* \downarrow & & \downarrow & & \\ 0 & \longleftarrow & A \otimes_{\mathbb{K}} H^\bullet(V_B) & \longleftarrow & A \otimes_{\mathbb{K}} H^\bullet(V) & \longleftarrow & 0 \\ & & & & \downarrow & & \\ & & & & 0 & & \end{array}$$

So any characteristic superclass $i_V^*\omega$ on V can be written as

$$(\star\star\star) \qquad i_V^*\omega = \sum_i a^i \otimes \beta_i\,,$$

where $a^i \in A$, and $\beta_i \in H^\bullet(V_B)$ are characteristic superclasses on V_B. On the other hand if V is non-singular, must be also non-singular v_B; hence all β_i must be zero [18].

Then, from $(\star\star\star)$ it follows that $i_V^*\omega = 0$, $\forall \omega \in \widetilde{H}^i(\tilde{\underline{I}}_k(W); A)$. □

THEOREM 6. 1) **Lie supertransformations**, *that are G_w^∞-diffeomorphisms $\phi\colon J\underline{D}^k(W) \longrightarrow J\underline{D}^k(W)$ which preserves the Cartan's graded superdistribution, $T(\phi)(\widetilde{\mathbb{E}}_k)_u = (\widetilde{\mathbb{E}}_k)_{\phi(u)}$, $\forall u \in J\underline{D}^k(W)$, preserve cohomology space $\widetilde{H}^i(\tilde{I}_k(W); A)$.*
2) **Characteristic Lie supertransformations**, *that are Lie supertransformations that transform each integral supermanifold in intself, preserve the space of characteristic superclasses of any integral manifold.*
(3) *In particular, any characteristic Lie supertransformation that preserves also the classes of $\widetilde{H}^i(\tilde{\underline{I}}_k(W); A)$ preserves characteristic superclasses of any integral supermanifold.*

PROOF. Every Lie supertransformation ϕ defines a lift $\tilde{\phi}\colon \tilde{\underline{I}}_k(W) \longrightarrow \tilde{\underline{I}}_k(W)$ such that the following diagram

$$\begin{array}{ccc} \tilde{\underline{I}}_k(W) & \xrightarrow{\tilde{\phi}} & \tilde{\underline{I}}_k(W) \\ \downarrow & & \downarrow \\ J\underline{D}^k(W) & \xrightarrow{\phi} & J\underline{D}^k(W) \end{array}$$

is commutative. More precisely one has $\tilde{\phi}(L_u) = \widetilde{T}(\phi)(L_u)_{\phi(u)}$. □

REMARK 5. Boundary value problem, for analytic regular supersolutions of SPDE's can be solved by means of revisited version of Cartan-Khaler theorem. So we have the following.

THEOREM 7 (Generalized Cartan-Khäler Theorem). *Let $\underline{E}_k \subset J\underline{D}^k(W)$ be superanalytic and superinvolutive formally superintegrable SPDE on the fiber bundle $\pi : W \longrightarrow M$, $(\dim M = (m,n))$. Then there exists a unique solution $s \in G^\omega_w(W)$ such that it satisfies the following boundary conditions:*

1) $D^{k-1}s(x_0) = q$, where q is a fixed point on $\pi_{k,k-1}(E_k) \subset J\underline{D}^{k-1}(W)$ and $x_0 = \pi_{k-1}(q) \in M$;

2) For the α^i_k parametric derivatives $(\partial x_{\alpha_1} \dots \partial x_{\alpha_x} \cdot s^j)(x)$ of class (i,j), $(i = 1, \dots, m, j = 1, \dots, n)$ (the components of class i (resp. j) are that where do not exist indexes i_α (resp. j_α), with $1 \le i_\alpha < i$, (resp. $1 \le j_\alpha < j$), but exist indexes $i_\alpha = i$ (resp. $j_\alpha = j$), and eventually exist indexes $i_\alpha > i$ (resp. $j_\alpha > j$), one as $(\partial x_{\alpha_1} \dots \partial x_{\alpha_k} \cdot s^j) = f_{\alpha_1 \dots \alpha_k}(x)$ =fixed, $\forall x \in X$, where X is a subsupermanifold of M passing through x_0, locally characterized by means of the equations: $x^{i+1} = x_0^{i+1}, \dots, x^m = x_0^m, x_0^{j+1}, \dots, x^n = x_0^n$.

REMARK 6. The Cauchy problem for (singular) supersolutions of SPDE can be formulated in the following way. A **Cauchy data** for system $\underline{E}_k \subset J\underline{D}^k(W)$ is defined by a $(m-1, n-1)$-dimensional integral supermanifold $\sigma \subset \underline{E}_{k+r}$ of the Cartan superdistribution $\mathbb{E}_{k+r}$, beside a section h of the bundle $\underline{\tilde{I}}(\underline{E}_{k+r})|\sigma \longrightarrow \sigma$, such that $h(\tilde{q}) \supset T_{\tilde{q}}\sigma$, for all $\tilde{q} \in \sigma$. A **supersolution of the Cauchy problem** with initial data, (σ, h), is defined to be an (m,n)-dimensional integral supermanifold V, with boundary such that $V \subset \underline{E}_{k+r}$, $\partial V = \sigma$ and $T_{\tilde{q}}V = h(\tilde{q})$, $\forall \tilde{q} \in \sigma$. With each supercohomology class $\omega \in \widetilde{H}^{m-1}(\underline{\tilde{I}}(\underline{E}_{k+r}))$ and each Cauchy data (σ, h) we associate the **characterisitc supernumber** $\chi_\omega(\sigma, h) = \langle h^*\omega, z_\sigma \rangle$, where $z_\sigma \in \widetilde{H}_{m-1}(\sigma; A)$ is a fundamental cycle of σ.

Then we have the following important theorem.

THEOREM 8 (Cauchy problem for SPDE's and characteristic supernumbers). *For solvability of the Cauchy problem (σ, h) for SPDE's it is necessary that all the characteristic supernumbers $\chi_\omega(\sigma, h)$ be equal to zero.*

PROOF. As supermanifolds involved are regulare ones, we have the following canonical isomorphisms, (see theorems of section 2);

$$\widetilde{H}^{m-1}(\sigma; A) \cong A \otimes_{\mathbb{K}} H^{m-1}(\sigma_B); \widetilde{H}_{m-1}(\sigma; A) \cong A \otimes_{\mathbb{K}} H_{m-1}(\sigma_B; \mathbb{K}),$$

where σ_B is the body of σ. (In this way has sense to talk of fundamental cycle of σ). Hence, the characteristic supernumbers can be written as follows:

$$(\star\star) \qquad \chi_\omega(\sigma, h) = \langle h^*\omega, z_\sigma \rangle = \langle a^i \otimes \beta_i, 1 \otimes z_{\sigma_B} \rangle = a^i \langle \beta_i, z_{\sigma_B} \rangle,$$

where $a^i \in A$, $\beta_i \in H^{m-1}(\sigma_B)$, and z_{σ_B} is a fundamental cycle of σ_B. On the other hand, if $V \subset \underline{E}_{k+r}$ is an integral supermanifold such that $\partial V = \sigma$, we get also $\partial V_B = \sigma_B$. Then, taking into account that for ordinary manifolds, a necessary condition for solvability of Cauchy problem is that all the characteristic numbers $\langle \beta_i, z_{\sigma_B}\rangle$ must be zero, from $(\star\star)$ we get that, in order to solve the Cauchy problem for super PDE's all the characteristic supernumbers $\chi_\omega(\sigma, h)$ must be zero too. □

3.3. Conservation laws for super PDE's

In this section we shall consider a spectral sequence that can be associated to super PDE's and that are useful to characterize the space of conservation laws of super PDE's. The wholly cohomological character of super PDE's will be considered and a general criterion to recognize conservation laws for super PDE's will be given. These results generalize previous ones formulated for PDE's; (see refs. [22, 23, 28, 29, 30, 31, 43]). (For informations on spectral sequences see e.g. [24, 39]).

REMARK 7. $J\underline{D}^\infty(W) \equiv \lim \underline{D}^k(W)$ has a canonical connection, that is a canonical splitting $TJ\underline{D}^\infty(W) \cong vTJ\underline{D}^\infty(W) \oplus H_W$, where H_W is the canonical horizontal subbundle $H_W \subset TJ\underline{D}^\infty(W)$ given by

$$H_W \equiv \bigcup_{q\in J\underline{D}^\infty(W)} H_q\,,$$

where $H_q \equiv \mathrm{im}(D(D^\infty s)(q)$, $p = \pi_\infty(q)$, $D^\infty s(p) = q$, and $D(D^\infty s)(p)\colon\ T_pM \longrightarrow T_qJ\underline{D}^\infty(W)$. This map does not depend on the particular section s used but only on the point q. Associated to H_W we have the graded bundle

$$H_W^\perp \equiv \bigcup_{q\in J\underline{D}^\infty(W)} H_q^\perp \subset T^+J\underline{D}^\infty(W)\,,$$

where

$$H_q^\perp \equiv \left\{\beta \in T_q^+J\underline{D}^\infty(W)|\beta|H_q = 0\right\}.$$

H_W is **superflat** in the sense that $dG_w^\infty(H_W^\perp)|H_W = 0$.

A section $u\colon M \longrightarrow J\underline{D}^\infty(W)$ is called **superflat** if $Tu(M) \subset H_W$. The requirement of superflatness for a section $u\colon M \longrightarrow J\underline{D}^\infty(W)$ of π_∞ is equivalent to the following equivalent conditions:

(a) $u^*G_w^\infty(H_W^\perp) = 0$; (b) $u = D^\infty s$, $s = \pi_{\infty,0}\circ u\colon M \longrightarrow W$.

Let $\underline{E}_k \subset J\underline{D}^k(W)$ be a formally superintegrable SPDE on a fiber bundle $\pi\colon W \longrightarrow M$. Then, the canonical connection on $J\underline{D}^\infty(W)$ identifies a canonical connection on $\underline{E}_\infty \equiv \lim \underline{E}_{k+r}$. More precisely, we have a canonical splitting

$$T\underline{E}_\infty \cong vT\underline{E}_\infty \oplus H_{\underline{E}_k}, \quad \text{where} \quad H_{\underline{E}_k} \equiv H_W|\underline{E}_\infty \subset T\underline{E}_\infty\,.$$

$H_{\underline{E}_k}$ is **superflat** in the sense that $dG_w^\infty(H_{\underline{E}_k}^\perp)|H_{\underline{E}_k} = i^*dG_w^\infty(H_W^\perp)|H_{\underline{E}_k} = 0$, where i is the canonical embedding $i\colon \underline{E}_\infty \subset J\underline{D}^\infty(W)$. A section $u \in G_w^\infty(\underline{E}_k)$ of $\pi_\infty|\underline{E}_\infty\colon \underline{E}_\infty \longrightarrow M$ is a solution of $\underline{E}_k$ iff one of the following equivalent conditions holds: (a) u is superflat; (b) $u^*G^\infty_w(H^\perp_{\underline{E}_k})=0$. So the pair $(\underline{E}_\infty, H_{\underline{E}_k})$ is a good-substitute for $\underline{\mathrm{Sol}}(\underline{E}_k)\equiv$ set of non-singular supersmooth solutions of $\underline{E}_k$, and it can express important formal aspects of $\underline{E}_k$.

REMARK 8. Set $\widetilde{\Omega}^p(J\underline{D}^\infty(W))\equiv \lim\{\pi^*_{\infty,k}G^\infty_w(\Lambda^pT^+J\underline{D}^k(W))\}$. Similarly we can define $\widetilde{\Omega}^p(\underline{E}_\infty)$ for a SPDE $\underline{E}_k\subset J\underline{D}^k(W)$. The the canonical connection on $\underline{E}_\infty$, identifies a bigradiation on $\widetilde{\Omega}^r(\underline{E}_\infty)$: $\widetilde{\Omega}^r(\underline{E}_\infty)\equiv\oplus_{p+q}\widetilde{\Omega}^{p,q}\underline{E}_\infty$, where $\widetilde{\Omega}^{p,q}\underline{E}_\infty\equiv G^\infty_w(\Lambda^p vT^+\underline{E}_\infty\otimes_A\Lambda^qH^+_{\underline{E}_k})$.

Since $H_{\underline{E}_k}$ is superflat, d has only $(1,0)$ and $(0,1)$ components:

$$\left.\begin{array}{lll}(0,1)\ \text{component}: & \delta\colon\widetilde{\Omega}^{p,q}\underline{E}_\infty & \longrightarrow\widetilde{\Omega}^{p+1,q}\underline{E}_\infty\\ (0,1)\ \text{component}: & (-1)^p\partial\colon\widetilde{\Omega}^{p,q}\underline{E}_\infty & \longrightarrow\widetilde{\Omega}^{p,q+1}\underline{E}_\infty\end{array}\right|(\mathrm{d}^2=0)\implies\delta^2=\partial^2=\delta\partial-\partial\delta=0\,.$$

We call $\{\widetilde{\Omega}^{\bullet,\bullet}\underline{E}_\infty;\delta,\partial\}$ the **variational superbicomplex** of $\underline{E}_k$.

The superflatness of $u\in G^\infty_w(\underline{E}_\infty)$ can be expressed using the variation bicomplex of $\underline{E}_k$ by one of the following equivalent conditions:

1) $u^*\widetilde{\Omega}^{+,\bullet}\underline{E}_\infty=0$, $\widetilde{\Omega}^{+,\bullet}\underline{E}_\infty\equiv\oplus_{p\geq0}\widetilde{\Omega}^{p,\bullet}\underline{E}_\infty$;
2) $u^*\widetilde{\Omega}^{1,q}\underline{E}_\infty=0$, for some $q\in\mathbb{Z}(0,n-1)\equiv$ set of integers in $[0,n-1]$;
3) $u^*\widetilde{\Omega}^{1,0}\underline{E}_\infty=0$;
4) $du^*f=u^*\partial f$, for all $f\in G^\infty_w(\underline{E}_\infty;A)$.

DEFINITION 5. *The filtration $F^p\widetilde{\Omega}^\bullet(\underline{E}_\infty)\equiv\oplus_{p'\geq p}\widetilde{\Omega}^{p',\bullet}\underline{E}_\infty$, is compatible with d and induces a spectral sequence $\{\widetilde{E}^{p,q}_r,d_r\}$ that we call* **Vinogradov graded superspectral sequence** *of $\underline{E}_k\subset J\underline{D}^k(W)$.* (This is a natural extension to SPDE's of one introduced by VINOGRADOV for PDE's [42]).

THEOREM 9.

1) *The Vinogradov graded superspectral sequence converges to $\widetilde{H}^\bullet(\underline{E}_\infty)$.*
2) *If $\underline{E}_k$ is a path-connected formally superintegrable SPDE one has:*

$$\widetilde{H}^\bullet(\underline{E}_k)\cong\widetilde{H}^\bullet(\underline{E}_\infty)\,.$$

PROOF.

1) Note that the filtration in Definition 5 is: (a) **exhaustive**, i.e.,

$$\widetilde{\Omega}^\bullet(\underline{E}_\infty)=\cup_pF^p\widetilde{\Omega}^\bullet(\underline{E}_\infty)\,;$$

and (b) **bounded above**, i.e., for each n there exists a value $s(n)$ with $F^{s(n)}\widetilde{\Omega}^\bullet(\underline{E}_\infty)=\{0\}$. This $s(n)$ is just -1 (realy $\widetilde{\Omega}^\bullet(\underline{E}_\infty)$ is positive). Now, from the second property it follows that $\widetilde{E}^{\bullet,\bullet}_r$ is weakly convergent [11]. Then, taking into account also the property (a) it follows [11] that $\widetilde{E}^{\bullet,\bullet}_r$ converges to $H(\widetilde{\Omega}^\bullet(\underline{E}_\infty))=\widetilde{H}^\bullet(\underline{E}_\infty)$.

2) If $\underline{E}_k$ is formally superintegrable $(\underline{E}_k)_{+s}\longrightarrow(\underline{E}_k)_{+(s-1)}$ is an affine bundle, $\forall s\geq1$, with fiber $F_{k,s}$ contractible. So we get $H^q(F_{k,s})=0$, $q>0$, and $H^0(F_{k,s})=$

$\mathbb{K}$. On the other hand, from the following commutative and exact diagram

$$\begin{array}{ccccccc} 0 & \longrightarrow & \widetilde{H}^\bullet((\underline{E}_k)_{+s}) & \longrightarrow & \widetilde{H}^\bullet((\underline{E}_k)_{+s};A) & \cong & A\otimes_{\mathbb{K}} H^\bullet((\underline{E}_k)_{+s}) \\ & & \uparrow b & & \| & & \| \\ 0 & \longrightarrow & \widetilde{H}^\bullet((\underline{E}_k)_{+(s-1)}) & \longrightarrow & \widetilde{H}^\bullet((\underline{E}_k)_{+(s-1)};A) & \cong & A\otimes_{\mathbb{K}} H^\bullet((\underline{E}_k)_{+(s-1)}) \end{array}$$

we get that the mapping b is an isomorphism. □

REMARK 9. In order to intrepret the meaning of such a structure, let us define a **solution supermanifold** of $\underline{E}_k$ as a map $\beta\colon X \longrightarrow \underline{\mathrm{Sol}}(\underline{E}_k)$, being X a (d_1, d_2)-dimensional supermanifold ($d_i \geq 0, i = 1, 2$) such that the map $\underline{\beta}\colon X \times M \longrightarrow E$, $\underline{\beta}(x, y) = D^\infty(\beta(x))(y)$, $x \in X, y \in M$, is supersmooth. Let $\beta\colon \widetilde{\Omega}^\bullet(\underline{E}_\infty) \longrightarrow \widetilde{\Omega}^\bullet(X \times M)$ be the induced map. $\widetilde{\Omega}^\bullet(X \times M)$ has an obvious underlying bicomplex structure $\{\widetilde{\Omega}^{\bullet,\bullet}(X \times M), d_X, d_M\}$ and $\beta^\bullet$ is a bicomplex map (it preserves filtration). Let $\{\widetilde{E}_r^{p,q}, d_r\}$ be the spectral sequence induced by the filtration $'F^p \equiv \oplus_{p'\geq p}\widetilde{\Omega}^{p',\bullet}(X \times M)$. One has:

$$\text{(a)}\ \ '\widetilde{E}_1^{p,q} \cong \widetilde{\Omega}^p(X) \otimes_A \widetilde{H}^q(M); \quad \text{(b)}\ \ '\widetilde{E}_2^{p,q} \cong \widetilde{H}^p(X) \otimes_A \widetilde{H}^q(M).$$

So $\{'E_1^{\bullet,q}, 'd_1\}$ is isomorphic to the $\widetilde{H}^q(M)$-valued **de Rham supercomplex** on X; (see section 2.1). Furthermore, $\beta^\bullet$ identifies a spectral sequence homomorphism $\{\beta_r\}\colon \{\widetilde{E}_r^{p,q}\} \longrightarrow \{'\widetilde{E}_r^{p,q}\}$ that gives a representation of the superspectral sequence of $\underline{E}_k$, associated to any solution supermanifold β. In particular one has the following homomorphisms:

$$\beta_1\colon \widetilde{E}_1^{p,q} \longrightarrow \widetilde{\Omega}^p(X) \otimes_A \widetilde{H}^q(M); \quad \beta_2\colon \widetilde{E}_2^{p,q} \longrightarrow \widetilde{H}^p(X) \otimes_A \widetilde{H}^q(M).$$

If the dimension of X is $(0, 0)$ then a solution supermanifold is identified with a solution $s \in \underline{\mathrm{Sol}}(\underline{E}_k)$, and we have the homomorphisms:

$$s_r\colon \widetilde{E}_r^{0,q} \longrightarrow \widetilde{H}^q(M), [\mu] \mapsto s_r([\mu]) = [D^\infty s^*\mu], r = 1, 2.$$

EXAMPLES

1. If $\underline{E}_k$ is an ordinary superdifferential equation $\widetilde{E}_1^{0,0}$ is the space of **superconstants of motions**

2. $\widetilde{E}_1^{0,m-1}$ is the space of **conserved superlaws** of $\underline{E}_k$. (For the classic case see also refs. [22, 29, 48, 49, 50, 51]). Note that physical interpretation of any conservation superlaw $\mu \in \widetilde{E}_1^{0,m-1}$ is made by considering the graded $(m-1)$-differential form $(D^\infty S)^*\mu\colon M \longrightarrow \widetilde{\Lambda}_{m-1}^0 M$, where s is a G_w^∞-solution of $\underline{E}_k$, restricted to any m-dimensional submanifold N of M obtained as the image of a section i of $b\colon M \longrightarrow M_B$:

$$i^*(D^\infty s)^*\mu\colon M_B \longrightarrow A \otimes_{\mathbb{K}} \Lambda_{m-1}^0 M_B.$$

(See Theorem 1.4). Really, $\alpha \equiv i^*(D^\infty S)^*\mu$ results a closed A-valued $(m-1)$-differential form on M_B so that for any two cobording $(m-1)$-chains B, B' on M_B: $B = B' + \partial C$, we have

$$\int_B \alpha = \int_{B'} \alpha + \int_{\partial C} \alpha = \int_{B'} \alpha + \int_C d\alpha = \int_{B'} \alpha\,.$$

Note that we have used the Stokes' formula (see Theorem 2.18).

3. $\widetilde{E}_1^{0,m}$ is the space of **Lagrangian superdensities**. Then $d_1^{0,m}[\mu] = 0$ is the **Euler Lagrange superequation** of the Lagrangian superdensity $[\mu] \in \widetilde{E}_1^{0,m}$ and $d_1^{0,m}$ is the **Euler-Lagrange superoperator**. Note that in order to consider variational calculus on supermanifolds it should be necessary to use an extended version of usual integral calculus on ordinary manifolds. However, from the physical point of view the action integrals to use in any classical (non-quantized) description of supergravity, necessitates the integration on ordinary submanifolds of the base supermanifold M, obtained as sections of the fiber bundle $b\colon M \longrightarrow M_B$. Now, a Lagrangian function on the fiber bundle $\pi\colon W \longrightarrow M$, in the category of supermanifolds, is a function $f\colon J\underline{D}^k(W) \longrightarrow A$. If $\dim M = (m, n)$, a corresponding Lagrangian superdensity $\Omega(f)$ is the horizontal graded m-differential form

$$(\square\star) \qquad \Omega[f] = f.\pi_k^*\mu\colon J\underline{D}^k(W) \longrightarrow \widetilde{\Lambda}_m^0 J\underline{D}^k(W)\,,$$

where μ is a m-form on M. Notice that as A is a l.c.s. and M_B is an ordinary m-dimensional manifold, the action integral

$$I_B[s;i] = \int_B i^*(D^k s)^*\Omega[f] = \int_B (f \circ D^k s \circ i).i^*\mu$$

is well defined for any compact domain $B \subset M_B$.

The effective construction of conservation laws for PDE's defined in the category of ordinary manifolds can be made by a fully covariant method given in refs. [28, 29] and further developed in refs. [22, 23, 31, 31]. That method can be directly extended to the category of supermanifolds. More precisely we have the following.

THEOREM 10 (Conservation super laws for super PDE's).

1) *Let $\underline{E}_k \subset J\underline{D}^k(W)$ be a k-order SPDE on a fiber bundle $\pi\colon\ W \longrightarrow M$ between supermanifolds W and M respectively. Let μ be a m-form on the (m,n)-dimensional supermanifold M. Let $\mathbb{E}_H$ be the involutive distribution generated by the infinitesimal generators of the supersymmetry group H of $\underline{E}_k$. Then for any differential invariant $f\colon J\underline{D}^k(W) \longrightarrow A$ of $\mathbb{E}_H$ we have some configuration $s\colon M \longrightarrow W$ that leads a conservation superlaw if* $\underline{\mathrm{Sol}}(\underline{E}_k) \cap \mathrm{Ext}(\beta) \neq \emptyset$*, where* $\mathrm{Ext}(\beta)$ *denotes the set of extremals of the following Lagrangian superdensity: $\beta \equiv f.\pi_k^*\mu\colon\ J\underline{D}^k(W) \longrightarrow \widetilde{\Lambda}_m^0 J\underline{D}^k(W)$.*

2) *Furthermore, if $\underline{E}_k$ is formally superintegrable and $\underline{E}_{2k} \equiv \underline{E}_{k+k}$ is a subsupermanifold of $(EL)_f \subset J\underline{D}^{2k}(W)$, where $(EL)_f$ is the Euler-Lagrange superequa-*

tion of f, then any supersolution s of $\underline{E}_k$ brings a dynamic conservation laws of class G^∞_w associated to f.

PROOF. The proof can be conduced in a way similar to the case for ordinary manifolds [28, 30]. □

REMARK 10. From above considerations we get that for any C^∞_w section i of b we have the following map:

$$J_i\colon \underline{\mathrm{Sol}}(\underline{E}_k) \times \widetilde{E}_1^{0,m-1} \longrightarrow H^{m-1}(A \otimes_{\mathbb{K}} \Omega^\bullet(M_B)) \cong A \otimes_{\mathbb{K}} H^{m-1}(M_B)\,,$$

given by

$$J_i(s,\beta) = [((D^\infty s)\circ i)^*\beta]\,.$$

In general the partial map $J_{i,s}\colon \widetilde{E}_1^{0,m-1} \longrightarrow A \otimes_{\mathbb{K}} H^{m-1}(M_B)$ is not surjective for any s belonging to $\underline{\mathrm{Sol}}(\underline{E}_k)$. This suggests to give the following.

DEFINITION 6. *For any C^∞_w section $i\colon M_B \longrightarrow M$ of b let $[A\otimes_{\mathbb{K}} H^{m-1}(M_B)]_i$ denote the image of the A-homomorphism*

$$(a_i)^*\colon \widetilde{H}^{m-1}(M) \longrightarrow A \otimes_{\mathbb{K}} H^{m-1}(M_B)\,.$$

$[A \otimes_{\mathbb{K}} H^{m-1}(M_B)]_i$ represents the space of conservation superlaws leaded by the even m-dimensional manifold $i(M_B)$. A SPDE $\underline{E}_k \subset J\underline{D}^k(W)$ is **wholly supercohomologic** *if in any cohomological class $[\mu] \in [A \otimes_{\mathbb{K}} H^{m-1}(M_B)]_i$ there exists a $(m-1)$-differential form (non trivial) obtained by means of above Theorem 10, for suitable G^∞_w solution s of $\underline{E}_k$.*

REMARK 11 Of course if $H^{m-1}(M_B) = 0$, any SPDE that admits conservation laws obtained by means of Theorem 10 is wholly supercohomologic. However, in less simple situations ($H^{m-1}(M_B) \neq 0$) the wholly supercohomological character of a SPDE is not assured and so it is useful the following criteria.

THEOREM 11 (Criteria of wholly supercohomologicity for SPDE's). *Let $\underline{E}_k$ be a k-order formally superintegrable SPDE on a fiber bundle $\pi\colon W \longrightarrow M$, in the category of supermanifolds, where M is a (m,n)-dimensional supermanifold. Let $\underline{E}_k$ be complete with respect to a conservation law $\beta\colon J\underline{D}^k(W) \longrightarrow \widetilde{\Lambda}^0_{m-1} J\underline{D}^k(W)$, which is of the type considered in above Theorem 10. Then we have the following criteria of wholly supercohomologicity:*
1) $\underline{E}_k$ is wholly supercohomologic if $\widetilde{H}^{m-1}(M) \cong \pi_0(\underline{\mathrm{Sol}}(\underline{E}_h))$.
2) If $\underline{E}_k$ is also a conic SPDE, it is wholly supercohomologic if $\widetilde{E}_2^{0,m-1} \cong \pi_0(\underline{\mathrm{Sol}}(\underline{E}_k))$, where π_0 denotes the 0-th Hurewicz homotopy group functor.

PROOF.
1) As we have the natural A-epimorphism $(a_i)^*\colon \widetilde{H}^{m-1}(M) \longrightarrow [A\otimes_{\mathbb{K}} H^{m-1}(M_B)]_i$, induced by the section $i\colon M_B \longrightarrow M$, it follows that $\underline{E}_k$ will be surely wholly supercohomologic if in any class of $\widetilde{H}^{m-1}(M)$ there is a differential form of the type $(D^k s)^*\beta$ for some supersolution s of $\underline{E}_k$. But this is equivalent to say that $\widetilde{H}^{m-1}(M) \cong \pi_0(G^\infty_{adm}(\underline{E}_k))$, where $G^\infty_{adm}(\underline{E}_k)$ is the set of G^∞_w-sections u of $\underline{E}_k$

that can be represented as $u = D^k s$ for some section s of π. For functorial property we have also that $\pi_0(G^{\infty}_{adm}(\underline{E}_k)) \cong \pi_0(\underline{\mathrm{Sol}}(\underline{E}_k))$.
2) We know that for any section $s \in \underline{\mathrm{Sol}}(\underline{E}_k)$, there exists the following homomorphism $s_2 \colon \widetilde{E}_2^{0,q} \longrightarrow \widetilde{H}^q(M)$. On the other hand if $\underline{E}_k$ is a conic SPDE, (that is W is a vector bundle, $\underline{E}_k$ admits the zero section of W as a solution and the gauge group of homoteties of W belongs to the group of symmetries of $\underline{E}_k$), above homomorphism s_2 becomes an isomorphism. (The proofs of these statements can be conduced in a way similar to one for finite dimensional manifolds as given in ref. [23]). So, from point 1, we have that if

$$\widetilde{E}_2^{0,m-1} \cong \widetilde{H}^{m-1}(M) \cong \pi_0(\underline{\mathrm{Sol}}(\underline{E}_k)) .$$

$\underline{E}_k$ is surely wholly supercohomologic. □

THEOREM 12. *Let $\underline{E}_k \subset J\underline{D}^k(W)$ be a linear SPDE, in the category of supermanifolds, that is complete with respect to a conservation law, then $\underline{E}_k$ is wholly supercohomologic if $\widetilde{H}^{m-1}(M) = 0$.*

EXAMPLES.

All the linear superequations of the field theory on the Minkowsky superspace-time (that is a supermanifold of dimension (4,4)), are wholly supercohomologic.

REFERENCES

[1] M.F. ATIYAH - I.G. MACDONALD: *Introduction to commutative algebra*, Addison-Wesley, Publishing Company, London 1969.

[2] M. BACHELOR: Trans. Amer. Math. Soc., **253**(1979), 329-338; Trans. Amer. Math. Soc. 258(1)(1980), 257-270.

[3] F.A. BEREZIN: *Introduction to superanalysis*, D. Reidel Pulishing, Dordrecht, Holland 1987.

[4] F.A. BEREZIN - D.A. LEITES: Soviet Math. Dokl., **16**(1975), 1218-1221.

[5] N. BOURBAKI: *Algebra*, Herman, Paris 1970.

[6] B. DE WITT: *Supermanifolds*, Cambridge University Press, Cambridge 1984.

[7] J. DIEUDONNE': Advances Math., **3**(1960),322-413.

[8] K. GAWEDZKI: Ann. Inst. Henri Poincaré, **27**(4)(1977), 335-366.

[9] H. GOLDSCHMIDT: J. Differential Geom., **1**(1967), 269-307.

[10] A. GROTHENDIECK - J. DIEUDONNÉ: J. Publ. Math. I.H.E.S., **4**(1960), 1-228; **8**(1961), 1-222; **17**(1963), 137-223; **32**(1967), 1-361.

[11] J. HILTON - V. STANMBACH: *A course in homological algebra*, Springer-Verlag, Berlin 1971.

[12] A. JADCZYK - K. PILCH: Commun. Math. Phys. **78**(1981), 373-390.

[13] H.H. KELLER: Lect. Notes Math. **417**(1974), 1-142, Springer-Verlag, Berlin.

[14] B. KOSTANT: Lect. Notes Math., **570**(1977), 127-306. Springer-Verlag, Berlin.

[15] I.S. KRASILISHCHIK - V.V. LYCHAGIN - A.M. VINOGRADOV: *Geometry of Jet Spaces and Nonlinear Partial Differential Equations*, Gordon & Breach, N.Y. 1986.

[16] LANG: *Introduction to differential manifolds*, Interscience, N.Y., 1962.

[17] D.A. LEITES: Russian Math. Surveys **35**(1)(1980), 1-64.

[18] V.V. LYCHAGIN: Soviet Math. Dokl., **24**(3)(1981), 680-685; **25** (3)(1982); 685-689;**28**(1)(1983),275-279.

[19] V.V. LYCHAGIN: Soviet. Math., **51(6)**(1990), 2735-2757.

[20] YU.I. MANIN: J. Soviet. Math., **30**(2)(1985), 1927-1975.

[21] YU.I. MANIN: Proceedings Steklov Inst. Math., **3**(1985), 107-124; **3**(1986), 113-136.

[22] V. MARINO - A. PRASTARO: Lect. Notes Math., **1209**(1986), 222-234, Springer-Verlag, Berlin.

[23] V. MARINO - A. PRASTARO: Rep. Math. Phys., **26(2)**(1987/8), 211-225.

[24] J. MCCLEARY: *User's guide to spectral sequences*, Publish or Perish in., USA 1985.

[25] J. MILNOR - J.D. STASHETF: *Characteristic classes*, Annals Math. Studies, 76, Princeton 1974.

[26] D.G. NORHCOTT: *An introduction to homological algebra*, Cambridge 1962.

[27] A. PRASTARO: Boll. Un. Mat. Ital., **5**(17-B)(1980),704-726; **5**(FM-S)(1981), 69-106; **5**(FM-S)(1981),107-129.

[28] A. PRASTARO: Riv. Nuovo Cimento, **5**(4)(1982),1-22.

[29] A. PRASTARO: Rend. Sem. Mat., Torino, **40**(2)(1982), 89-117; **43**(1)(1985), 199-216.

[30] A. PRASTARO: *Dynamic conservation laws*, Geometrodynamics Proceedings 1985, World Scientific, Singapore (1985), 283-420.

[31] A. PRASTARO: *Wholly cohomological PDE's*, International Conference on Differential Geometry and Applications, Dubrovnik 1988, Univ. Beograd & Univ. Novi Sad (1989), 305-314.

[32] A. PRASTARO: *On the singular solutions of PDE's*, Atti X Congresso AIMETA, Pisa 1990, AIMETA (1990), 17-20.

[33] A. PRASTARO: Boll. Un. Mat. Ital., **(7)5-B**(1991),977-1001.

[34] A. PRASTARO: Rep. Math. Phys., **30(3)**(1991), 273-352.

[35] A. PRASTARO - T. REGGE: Ann. Inst. H. Poincare, **44(1)**, (1986), 39-89.

[36] A. ROGERS: J. Math. Phys., **21**(6)(1980), 1352-1365; J. Math. Phys., **22** (3) (1981), 443-444.

[37] M.J. ROTHSTEIN: Trans. Amer. Math. Soc., **297**(1)(1986), 159-180.

[38] A. ROUX: Publications du Department de Mathematiiques, Lyon **7-4**(1970), 1-43.

[39] R. SWITZER: *Algebraic topology-homotopy and homology*, Springer-Verlag, Berlin 1975.

[40] A.M. VINOGRADOV: J. Soviet Math., **17**(1)(1981), 1624-1649.
[41] A.M. VINOGRADOV: Soviet Math. Dokl., **14**(3)(1973), 661-665; **19** (1)(1978), 144-148.
[42] A.M. VINOGRADOV: Lect. Notes Math., 1108(1984) 77-102, Springer-Verlag, Berlin.
[43] A.M. VINOGRADOV: J. Math. Anal. Appl., **100**(1)(1981), 1-129.

Geometry in Partial Differential Equations (pp. 316–322)
eds. A. Pràstaro and Th. M. Rassias

ON A GEOMETRIC APPROACH TO AN EQUATION OF J. D'ALEMBERT

AGOSTINO PRÀSTARO
Universitá degli Studi di Roma "La Sapienza", Dipartimento di Metodi e Modelli Matematici per le Scienze Applicate, Via A. Scarpa, 4-00161 Roma, Italy

and

THEMISTOCLES M. RASSIAS
University of LaVerne, Department of Mathematics, P. O. Box 51105, Kifissia, Athens, Greece 14510

By using a geometric framework for PDE's we prove that the set of solutions of the J. D'Alembert equation

$$\left(\frac{\partial^2}{\partial x \partial y} \log f\right) = 0$$

is larger than the set of smooth functions of two variables $f(x, y)$ of the form $f(x,y) = h(x) \cdot g(y)$. This agrees with a previous counter example by Th. M. Rassias given to a statement by C. Stéphanos. A generalization of this result to functions of more than two variables is sketched also.

Questions of representation of functions in several variables by means of functions of a lesser number of variables have occupied the interest of mathematicians for centuries. One of these questions is closely connected with the thirteenth problem of D. Hilbert (1862–1943) and concerns the solvability of algebraic equations [7]. Functions of certain special forms have been investigated by several authors, including J. d'Alembert (1717–1783), who as early as 1747 proved that each sufficiently smooth scalar function h of the form

$$f(x,y) = h(x) \cdot g(y) \tag{1}$$

has to satisfy the partial differential equation

$$\left(\frac{\partial^2}{\partial x \partial y} \log f\right) = 0 \tag{2}$$

(see [3]). A generalization of the form (1) to a *finite* sum of products of functions in single variables

$$f(x,y) = \sum_{1 \le i \le n} h_i(x) \cdot g_i(y) \tag{3}$$

was considered since the beginning of this century and it forms the *fundamental problem* in the subject. The functions of the *tensor product* (3) play a significant role in many areas of both pure and applied mathematics, for example in the theory of integral equations, ordinary and partial differential equations, and the approximation theory. In the year 1904 in the section *Arithmetics and Algebra* at the 3rd International Congress of Mathematicians in Heidelberg, C. Stéphanos announced the following result: *The functions of the type (1) form the space of all solutions of the partial differential equation with the "Wronskian" of order* $(k+1)$

$$\det W_{n+1}(f) = \begin{vmatrix} f & f_y & \cdots & f_{y^n} \\ f_x & f_{xy} & \cdots & f_{xy^n} \\ \vdots & \vdots & & \vdots \\ f_{x^n} & f_{x^n y} & \cdots & f_{x^n y^n} \end{vmatrix} = 0\ . \tag{4}$$

His talk was published in [22] (see also [23]) where some further applications and consequences were discussed. However, no proof of the above result was given and no smoothness condition on the given function h was mentioned. It is our belief that C. Stéphanos was thinking of analytic functions only. In fact, 80 years later, Th. M. Rassias gave in [16] a counter-example of a function in two real variables satisfying the differential equation (2), which is not of the form (1). (Eq. (2) can be written in the form (4) for $n = 1$). For the development of the subject see [13],[18]. The new interest in the fundamental problem was revived in the beginning of the 1980's, when F. Neumann proved the basic theorem involving the equation (4) for functions of class C^n. This important result was published in [10] (see also [11]), where also a general criterion for a function $f\colon X \times Y \to \mathbb{R}$ (or

$\mathbb{C}$, respectively), of the form (3) was given in the form of the functional equation

$$\begin{vmatrix} f(x_1, y_1) & f(x_1, y_2) & \cdots & f(x_1, y_{n+1}) \\ f(x_2, y_1) & f(x_2, y_2) & \cdots & f(x_2, y_{n+1}) \\ \vdots & \vdots & & \vdots \\ f(x_{n+1}, y_1) & f(x_{n+1}, y_2) & \cdots & f(x_{n+1}, y_{n+1}) \end{vmatrix} = 0$$

for all $x_i \in X$ and $y_j \in Y$. For the proof of this criterion no structure on the sets X and Y is required. (Most of the theorems here hold for both real- and complex-valued functions). The problem of representing a function f in several (more than two) variables by

$$f(x_1, \ldots, x_k) = \sum_{1 \le i \le n} f_{i_1}(x_1) \cdots f_{i_k}(x_k) \tag{5}$$

was proposed by Th. M. Rassias in [17]. H. Gauchman and L. A. Rubel [4] obtained some new results on finite sums expansions of the form (3), especially for real analytic functions. Moreover, they developed an interesting approach to finding special solutions of the classical partial differential equations, the so-called *extended separation of variables*. The first existence theorem on the decomposition (5) was discovered by F. Neuman [12]. Later, M. Čadek and J. Šimša [1] found an effective criterion for the existence of the decomposition (5) by making use of a system of functional equations, which does not require any assumption on the function f. Furthermore, in [2] Čadek and Šimša outlined a way to find systems of partial differential equations whose solution spaces form the family of all sufficiently smooth functions f of type (5). J. Šimša [19] has introduced some new functional expressions for functions of the form (3) using the so-called Casorati determinants. For a given function f that cannot be expressed in the form (3) with a prescribed value of f, it is an interesting problem to find the *best* approximation of the type

$$f(x, y) = \sum_{1 \le i \le n} f_i(x) \cdot f_i(y) \tag{6}$$

in a given metric functional space. For the case of L^2 spaces, this problem was recently solved by J. Šimša [20]. However, for spaces without an inner product structure, the problem seems to be difficult and open. A weaker version of the above approximation problem, namely to find a *good* approximation (6) and also an estimation of the error, was treated in [21] (see the forthcoming book by Th. M. Rassias & J. Šimša [18]).

In this paper we announce the following theorem.

Theorem. *The set, Sol*(d'A)*, of solutions of d'Alembert equation* (*2*) *is larger than the set of functions which can be represented by* (*1*).

Proof. (In the following we will give a sketch of the proof. For a more detailed version see [15]). By using the modern geometric theory of PDE's ([6,8,9,14,24]) we can prove that Eq. (2) can be identified with a submanifold (d'A) of the second jet-derivative space $JD^2(\mathbb{R}^2, \mathbb{R})$ on the trivial fiber bundle π: $\mathbb{R}^3 \to \mathbb{R}^2$, and that (d'A) is a formally integrable PDE. In the analytic case this implies complete integrability, i.e., the effective construction of analytic solutions in suitable neighbourhoods of any point $p \in \mathbb{R}^2$, with $p = \pi_2(q)$, where π_2: $JD^2(\mathbb{R}^2, \mathbb{R}) \to \mathbb{R}^2$ is the canonical projection. Now, we can see that (d'A) can be identified with the set of points of $JD^2(\mathbb{R}^2, \mathbb{R})$ that can be represented in the form $(D^2 f)(x, y) = D^2(h \cdot g)(x, y)$, where $h = h(x)$, and $g = g(y)$. (This, of course, will imply that the set of functions of the form (1) are solutions of Eq. (2)). In fact, (d'A) is diffeomorphic to the image P^2 of the canonical mapping

$$j : JD^2(\mathbb{R}, \mathbb{R}) \times JD^2(\mathbb{R}, \mathbb{R}) \to JD^2(\mathbb{R}^2, \mathbb{R}), \quad \text{defined by}$$
$$j(D^2(h(x), D^2 g(y)) = D^2(h \cdot g)(x, y) \; .$$

More precisely the following diagram

$$\begin{array}{ccccc}
JD^2(\mathbb{R},\mathbb{R}) \times JD^2(\mathbb{R},\mathbb{R}) & = & JD^2(\mathbb{R},\mathbb{R}) \times JD^2(\mathbb{R},\mathbb{R}) \to \eta \to & & JD^2(\mathbb{R},\mathbb{R}) \otimes JD^2(\mathbb{R},\mathbb{R}) \\
\downarrow & & \downarrow j & & \downarrow \rho \\
(\text{d'A}) & \subset & JD^2(\mathbb{R}^2,\mathbb{R}) & = & JD^2(\mathbb{R}^2,\mathbb{R})
\end{array}$$

is commutative, where: (i) η is the canonical morphism of vector bundles over $\mathbb{R}^2(x, y)$, and (ii) ρ is defined by the following equations:

$$x \circ \rho = x, \quad y \circ \rho = y, \quad u \circ \rho = \bar{u}\bar{\bar{u}}, \quad u_x \circ \rho = \bar{u}\bar{\bar{u}}, \quad u_x \circ \rho = \bar{u}_x \bar{\bar{u}} \; ,$$
$$u_y \circ \rho = \bar{u}\bar{\bar{u}}_y, \quad u_{xx} \circ \rho = \bar{u}_{xx}\bar{\bar{u}}, \quad u_{yy} \circ \rho = \bar{u}\bar{\bar{u}}_{yy}, \quad u_{xy} \circ \rho = \bar{u}_x \bar{\bar{u}}_y \; ,$$

where $(x, \bar{u}, \bar{u}_x, \bar{u}_{xx}, y, \bar{\bar{u}}, \bar{\bar{u}}_y, \bar{\bar{u}}_{yy})$ are coordinates on $JD^2(\mathbb{R}, \mathbb{R}) \times JD^2(\mathbb{R}, \mathbb{R})$. The image of P^2 by means of ρ is the set of points $D^2 f(x, y) \in JD^2(\mathbb{R}^2, \mathbb{R})$ that can be written in the following way:

$$D^2 f(x, y) = D^2(hg)(x, y) \tag{7}$$

where $h = h(x)$ and $g = g(y)$ are C^∞-functions. Furthermore, P^2 is a 7-dimensional submanifold of $JD^2(\mathbb{R}^2, \mathbb{R})$ as ρ has constant rank 7. Moreover, $P^2 \subseteq$ (d'A) as all points of $JD^2(\mathbb{R}^2, \mathbb{R})$ that satisfy (7) belong to (d'A) also. On the other hand the mapping j is onto (d'A). In fact for any point $q \in$ (d'A) we can find a point $u \equiv ((D^2h)(x), (D^2g)(y)) \in JD^2(\mathbb{R}, \mathbb{R}) \times JD^2(\mathbb{R}, \mathbb{R})$ such that $j(q) = u$. Indeed the point u is given by

$$u = (D^2 f_y(x), D^2 f_x(y)) ,$$

where f_x and f_y are the partial mappings $\mathbb{R} \to \mathbb{R}$ at the points x and y, respectively. So the proof that P is diffeomorphic to (d'A) is complete. On the other hand, any solution of (d'A) can be identified with 2-dimensional integral manifolds of the Cartan distribution ([8,9,14,24]) $\mathbb{E}_2 \subseteq T$(d'A) of (d'A). Observe that the image C of $C^\infty(\mathbb{R}, \mathbb{R}) \times C^\infty(\mathbb{R}, \mathbb{R})$ into $C^\infty(\mathbb{R}^2, \mathbb{R})$ by means of the canonical mapping

$$\begin{aligned} c : C^\infty(\mathbb{R}, \mathbb{R}) \times C^\infty(\mathbb{R}, \mathbb{R}) &\to C^\infty(\mathbb{R}, \mathbb{R}) \otimes_{\mathbb{R}} C^\infty(\mathbb{R}, \mathbb{R}) \\ &\cong C^\infty(\mathbb{R}^2, \mathbb{R} \otimes_{\mathbb{R}} \mathbb{R}) \cong C^\infty(\mathbb{R}^2, \mathbb{R}) \end{aligned}$$

represents the set of solutions of (d'A) that can be represented in the form (1). In fact, $f \in C^\infty(\mathbb{R}^2, \mathbb{R})$ can be expressed in the form (1) iff $f \in C$. On the other hand, taking into account that $P \cong$ (d'A), it follows that if $f \in C$ the graph of $D^2 f$ is contained in (d'A) $\subset JD^2(\mathbb{R}^2, \mathbb{R})$. Therefore, C can be identified with the subset of Sol(d'A) defined by those integral manifolds of (d'A) that can be obtained as graphs of 2-jet-derivatives of functions $f(x, y)$ expressed in the form (1).

Let us now prove that Sol(d'A) properly contains C. In fact, one can see that the Cartan distribution $\mathbb{E}_2 \subset T$(d'A) on (d'A), that characterizes the solutions of (d'A) is the annihilator of the ideal generated by the following differential 1-forms on $JD^2(\mathbb{R}^2, \mathbb{R})$:

$$\omega_\alpha = \left| \begin{array}{l} \omega_0 \equiv dF = du\; u_{xy} + u\; du_{xy} - du_x\; u_y - u_x\; du_y \\ \omega_1 \equiv du - u_x\; dx - u_y\; dy \\ \omega_2 \equiv du_x - u_{xx}\; dx - u_{xy}\; dy \\ \omega_3 \equiv du_y - u_{yx}\; dx - u_{yy}\; dy \end{array} \right.$$

Let us consider two 1-dimensional integral manifolds V_1 and V_2 of A^2 and B^2 respectively, where A^2 (resp. B^2) is the submanifold of (d'A) that represents second derivatives of functions $f(x, y)$ that are constant with respect to the variable y (resp. x). Furthermore, $V_i, i = 1,2$, are also 1-dimensional integral submanifolds of (d'A). Let $v_i, i = 1,2$, be characteristic vector fields of (d'A) that are not contained in the tangent space TV_i. More precisely we can take

$$v_1 \equiv v^x = u[\partial y + u_y \partial u + u_{yx}\partial u_x + u_{yy}\partial u_y] + u_{xx}u_y\partial u_{xx} + u_{yy}u_x\partial u_{yx}$$
$$v_2 \equiv v^y = u[\partial x + u_x \partial u + u_{xx}\partial u_x + u_{xy}\partial u_y] + u_{xx}u_y\partial u_{xy} + u_{yy}u_x\partial u_{yy} \ .$$

Let $\phi_{i,t} : (\text{d'A}) \to (\text{d'A})$ be the 1-parameter groups of diffeomorphisms generated by $v_i : \partial\phi_i = v_i$. Then, $N_i \equiv \bigcup_t \phi_{i,t}(V_i), i = 1,2$, are 2-dimensional integral manifolds of (d'A). Such manifolds cannot, in general, be represented as the graphs of the second-order jet-derivatives of functions $f(x,y)$ of the type (1). □

Remark. This result can be directly generalized to functions of more than two variables by considering the following generalized D'Alembert equation $(\text{d'A})_n \subset JD^n(\mathbb{R}^n, \mathbb{R})$:

$$(\text{d'A})_n \qquad \left(\frac{\partial^n}{\partial x_1 \cdots \partial x_n} \log f\right) = 0 \ .$$

In fact, the set of solutions of $(\text{d'A})_n$ properly contains graphos of the n-jet-derivatives of all functions $f \in C^\infty(\mathbb{R}^n, \mathbb{R})$ expressed in the form

$$f(x^1, \ldots, x^n) = f_1(x^1) \cdots f_n(x^n) \ .$$

References

1. M. Čadek and J. Šimša, *Decomposable functions of several variables*, Aequat. Math. **40** (1990), 8–25.
2. M. Čadek and J. Šimša, *Deccmposition of smooth functions of two multidimensional variables*, Czechoslovak Math. J. 116 **41** (1991), 342–358.
3. J. D'Alembert, *Recherches sur la curbe que forme une corde tendue mise en vibration I–II*, Hist. Acad. Berlin, 1747, pp. 214–249.
4. H. Gauchaman and L. A. Rubel, *Sums of products of functions of x times functions of y*. Linear Algebra Appl. **125** (1989), 19–63.
5. V. Guillemin and S. Sternberg, *An algebraic model of transitive differential geometry*, Bull. Am. Math. Soc. **70** (1964), 16–28.
6. H. Goldschmidt, *Integrability criteria for systems of nonlinear partial differential equations*, J. Diff. Geometry **1** (1967), 269–307.

7. D. Hilbert, *Mathematical problems* (Lecture delivered before the International Congress of Mathematicians at Paris in 1900), Bull. Amer. Soc. **8** (1902), 437–479.
8. I. S. Krasil'shchik, V. Lychagin, and A. M. Vinogradov, *Geometry of Jet Spaces and Nonlinear Partial Differential Equations*, Gordon & Breach, London, 1986.
9. V. Lychagin, *Geometry theory of singularities of solutions of nonlinear differential equations*, J. Soviet Math. **51**(6) (1990), 2735–2757.
10. F. Neuman, *Functions of two variables and matrices involving factorizations*, C. R. Math. Rep. Acad. Sci., Canada **3** (1981), 7–11.
11. F. Neuman, *Factorizations of matrices and functions of two variables*, Czechoslovak Math. J. **32**(107) (1982), 582–588.
12. F. Neuman, *Finite sums of products of functions in single variables*, Linear Algebra Appl. **134** (1990), 153–164.
13. F. Neuman and Th. M. Rassias, *Functions decomposable into finite sums of products* Constantin Carathéodory: An International Tribute, vol. II, World Scientific, Singapore, 1991, pp. 956–963.
14. A. Pràstaro, *Quantum geometry of PDE's*, Rep. Math. Phys. **30**(3) (1991), 273–352.
15. A. Pràstaro and Th. M. Rassias, *A geometric approach of an equation of J. d'Alembert*, Proc. Amer. Math. Soc., (to appear).
16. Th. M. Rassias, *A criterion for a function to be represented as a sum of products of factors*, Bull. Math. Acad. Sinica **14** (1986), 377-382.
17. Th. M. Rassias, *Problem P 286*, Aequat. Math. **38** (1989), 111–112.
18. Th. M. Rassias and J. Šimša, *Finite Sums Decompositions in Mathematical Analysis*, John Wiley & Sons (to appear).
19. J. Šimša, *A note on certain functional determinants*, Aequat. Math. (1992) (to appear).
20. J. Šimša, *The best L^2-approximation by finite sums of functions with separable variables*, Aequat. Math. (1992) (to appear).
21. J. Šimša, *Stability of a differential criterion for decomposable functions of two variables*, (to appear).
22. C. M. Stéphanos, *Sur une catégorie d'équations fonctionnelles*, Rend. Circ. Mat., Palermo **18** (1904), 360–362.
23. C. M. Stéphanos, *Sur une catégorie d'équations fonctionnelles*, Math. Kongress, Heidelberg, 1904, pp. 200–201; 1095, pp. 10–194.
24. A. M. Vinogradov, *Geometry of nonlinear differential equations*, J. Soviet. Math. (1) **17** (1981), 1624–1649.

Geometry in Partial Differential Equations (pp. 323–328)
eds. A. Pràstaro and Th. M. Rassias

GEOMETRIC PREQUANTIZATION OF THE EINSTEIN'S VACUUM FIELD EQUATIONS

MIRCEA PUTA
University of Timisoara, Department of Mathematics, Blvd V Parvan 4, 1900 Timisoara, Romania

The geometric prequantization of the Einstein's vacuum field equations is described and its connection with the reduction under the symmetry group of all space time diffeomorphisms is pointed out.

1. Introduction

In [2], Arnowitt, Deser and Misner have realized classical Einstein's vacuum field equations of general relativity as Hamiltonian equations on an appropriate phase space. In this paper using their formalism, usually called the ADM-formalism, and the methods and conventions of Fischer and Marsden [5] the geometric prequantization of the Einstein's vacuum field equations is described and its connection with the reduction under the symmetry group of all space-time diffeomorphisms is pointed out.

2. ADM — Formalism

Let V_4 be a 4-dimensional manifold with a Lorentzian metric ${}^{(4)}g$, M a compact, 3-dimensional oriented manifold and i: $M \hookrightarrow V_4$ an embedding of M such that the embedded manifold $\sum = i(M)$ is space like, that is, the pullback $i^*({}^{(4)}g) = g$ is a Riemannian metric on M. The Einstein's vacuum field equations state that the Ricci tensor vanishes:

$$\mathrm{Ric}({}^{(4)}g) = 0 \ .$$

These are the Euler-Lagrange equations for the Lagrange density:

$$\mathcal{L}(g) = R({}^{(4)}g)\mu({}^{(4)}g) ,$$

where $R({}^{(4)}g)$ denotes the scalar curvature of ${}^{(4)}g$ and $\mu({}^{(4)}g) = \sqrt{-\det({}^{(4)}g)}d^4x$, is its volume element.

In the pioneering work of Dirac [3], Wheeler [7] and Arnowitt, Deser and Misner [2], it was shown how to write these equations as Hamiltonian evolution equations. In the sequel we shall sketch their formalism, usually called ADM-formalism, following the methods and conventions of Fischer and Marsden [1], [5].

Let $\mathcal{M}$ be the space of C^∞ Riemannian metrics on M, $\mathcal{D}$ the C^∞ diffeomorphism group of M, $T\mathcal{M} \simeq M \times S_2$ the tangent bundle of $\mathcal{M}$, where S_2 is the space of C^∞ 2-covariant symmetric tensor fields on M and S_d^2 the space of C^∞ 2-contravariant symmetric tensor densities on M, i.e. $S_d^2 \approx S^2 \otimes \Omega$, where Ω is the bundle of densities (3-forms) on M.

Define $T^*\mathcal{M} \approx \mathcal{M} \times S_d^2 = \{(g,\pi) |\, g \in \mathcal{M}, \pi \in S_d^2\}$. It is a subbundle of the "true" cotangent bundle of $\mathcal{M}$. Since $T^*\mathcal{M}$ is open in $S_2 \times S_d^2$, the tangent space of $T^*\mathcal{M}$ at $(g,\pi) \in T^*\mathcal{M}$ is $T_{(g,\pi)}(T^*\mathcal{M}) \approx S_2 \times S_d^2$ and then the canonical symplectic structure ω on $T^*\mathcal{M}$ is given by:

$$\omega_{(g,\pi)}((h_1,\pi_1),(h_2,\pi_2)) = \int_M \pi_2 h_1 - \pi_1 h_2 \ .$$

Let $C^\infty = C^\infty(M,R)$ denote the smooth real valued functions on $M, C_d^\infty =$ the space of smooth scalar densities on $M, \mathfrak{X} =$ the space of smooth vector fields on M, $\wedge_d^1 =$ the space of smooth 1-form densities on M and k the second fundamental form of the embedding i of M in V_4.

Consider the functions:

$$\mathcal{G} : T^*\mathcal{M} \to C_d^\infty;\ (g,\pi) \mapsto \left[\pi' \cdot \pi' - \frac{1}{2}(\operatorname{tr}\pi')^2 - R(g)\right] d\mu(g)$$

$$\mathcal{F} = 2\delta : T^*\mathcal{M} \to \wedge_d^1;\ (g,\pi) \mapsto 2\delta_g \pi = -2\pi^j_{i|j} ,$$

where $\pi = \pi' d\mu(g), \pi' = [(\operatorname{tr} k)g - k]^\#$, $\#$ indicating the contravariant form of a covariant tensor with indices raised by $g, \pi' \cdot \pi' = (\pi')^{ij}(\pi')_{ij}, \delta_g \pi$ is the divergence of π with respect to the Riemannian metric g and $R(g)$ is the scalar curvature of g.

With these notations we have:

Theorem 1. ([2]) *The Einstein's vacuum field equations are Hamilton's equations with the Hamiltonian:*

$$H = N\mathcal{G} + X \cdot \mathcal{F} ,$$

i.e.

$$\begin{cases} \dfrac{\partial g}{\partial \lambda} = \dfrac{\delta}{\delta \pi}(N\mathcal{G} + X \cdot \mathcal{F}) \\ \dfrac{\partial \pi}{\partial \lambda} = -\dfrac{\delta}{\delta g}(N\mathcal{G} + X \cdot \mathcal{F}) \end{cases}$$

where N and X are respectively the lapse function and the shift vector fields of Arnowitt, Deser and Misner.

3. Prequantization

Let $(T^*\mathcal{M}, \omega)$ be the phase space of the Einstein's vacuum field equations. Since $\omega = d\theta$ is an exact form, $T^*\mathcal{M}$ is a quantizable manifold. Indeed, the line bundle $(L^\omega, \pi, T^*\mathcal{M})$ is simply the trivial bundle, $L^\omega \simeq T^*\mathcal{M} \times \mathbb{C}$, the Hermitian structure on L^ω is defined by:

$$((g, \pi, z_1), (g, \pi, z_2)) = \bar{z}_1 z_2 ,$$

and the space of smooth sections of the line bundle L^ω can be identified with $C^\infty(T^*\mathcal{M}, \mathbb{C})$.

Let $C_b(T^*\mathcal{M}, \mathbb{C})$ be the space of smooth, complex valued function f on $T^*\mathcal{M}$ such that $\int_{T^*\mathcal{M}} f \cdot \bar{f} < \infty$. It is a pre-Hilbert space under the inner product given by:

$$\langle f, g \rangle = \int_{T^*\mathcal{M}} f \cdot \bar{g}$$

and let $\mathcal{H}$ be its completion. Then for each $f \in C^\infty(T^*\mathcal{M}, \mathbb{R})$ the prequantum operator δ_f corresponding to f is given by:

$$\begin{cases} \delta_f : \mathcal{H} \to \mathcal{H} \\ \delta_f = -i\hbar X_f - X_f \lrcorner \theta + f , \end{cases}$$

where $\hbar$ is Planck's constant divided by 2π.

Remark. In the above construction the topology on the space of Riemannian metrics, tensor fields, density-tensor fields is the Sobolev topology $W^{s,p}, s > 3/p$.

4. Reduction

It is easy to see that the group $\mathcal{D}$ acts in a canonical way on $\mathcal{M}$ by pull-back:

$$\phi_\mu : g \in \mathcal{M} \to \phi_\mu(g) \overset{\text{def}}{=} (\mu^{-1})^*(g) \in \mathcal{M} ,$$

and the infinitesimal generator of this action is given by:

$$X \in T_e\mathcal{D} = \mathfrak{X} \to \left(\frac{d}{dt}\right)((\mu_t^{-1})^* g)|_{t=0} = -L_X g \in \mathcal{M} ,$$

where L_X is the Lie derivative and $\mathfrak{X}$ is the Lie algebra of $\mathcal{D}$.

The action ϕ of $\mathcal{D}$ on $\mathcal{M}$ can be canonically lifted to a symplectic action of $\mathcal{D}$ on the cotangent bundle $T^*\mathcal{M}$ and its momentum mapping is:

$$J : T^*\mathcal{M} \to \mathfrak{X}^*$$
$$J(g,\pi)(X) = \int_M \langle \pi, L_X g \rangle .$$

Of particular interest is the case $J^{-1}(0)$ referred to as the divergence constraint in general relativity. The isotropy group is all of $\mathcal{D}$ so the reduced space in the sense of Marsden–Weinstein [6] is:

$$(T^*\mathcal{M})_0 = J^{-1}(0)/\mathcal{D} .$$

If we work near a metric with no isometries, then under some technical assumptions (see the above Remark) $(T^*\mathcal{M})_0$ is a manifold by Ebin–Palais–Slice theorem [4]. We can conclude by the general reduction technique that $((T^*\mathcal{M})_0, \omega_0)$ is a weak, exact symplectic manifold, $\omega_0 = d\theta_0$ and moreover the following equality also holds:

$$\pi_0^* \theta_0 = i_0^* \theta ,$$

where $\pi_0 : J^{-1}(0) \to (T^*\mathcal{M})_0$ is the projection and $i_0 : J^{-1}(0) \to T^*\mathcal{M}$ is the inclusion.

5. Prequantization of the Reduced Phase-Space

In this section we shall prove that the geometric prequantization and the Marsden–Weinstein reduction of the Einstein's vacuum field equations are interchangeable processes.

Let $((T^*\mathcal{M})_0, \omega_0)$ be the reduced phase space described in the above section. Since $\omega_0 = d\theta_0$ is an exact form, $(T^*\mathcal{M})_0$ is a quantizable manifold and let $(\mathcal{H}^0, \delta^0)$ be the Hilbert representation space and the quantum operator given by the Kostant–Souriau geometric quantization (see Sec. 4). Then we have:

Theorem 2. *Let $\mathcal{H}_\mathcal{D}$ be the Hilbert space of $\mathcal{D}$-invariant states of the system, i.e. $\mathcal{H}_\mathcal{D} = \{f \in \mathcal{H}|_f$ is $\mathcal{D}$-invariant$\}$ and $C^\infty_\mathcal{D}(T^*\mathcal{M}, \mathbb{R})$ the space of $\mathcal{D}$-invariant smooth real valued functions on $T^*\mathcal{M}$. Then for each $f \in C^\infty(T^*\mathcal{M}, \mathbb{R})$, $g \in \mathcal{H}_\mathcal{D}$ and for their induced functions $f_0 \in C^\infty_\mathcal{D}((T^*\mathcal{M})_0, \mathbb{R})$, $g_0 \in \mathcal{H}^0_\mathcal{D}$, i.e. $f \circ \pi_0 = f \circ i_0$, $g \circ \pi_0 = g \circ i_0$, we have:*

(i) $\delta_f(g) \in \mathcal{H}_\mathcal{D}$

(ii) $[\delta_f(g)]_0 = \delta^0_{f_0}(g_0)$ *or in other words, prequantization and the reduction are interchangeable processes.*

Proof. (i) This assertion is a straightforward computation and we shall omit it.

(ii) We can write successively:

$$\begin{aligned}
X_{f_0}(g_0)(\pi_0(x)) &= (X_{f_0})_{\pi_0(x)}(g_0) \\
&= T_x(X_f)_x(g_0) \\
&= (X_f)_x(g_0 \circ \pi_0) \\
&= (X_f)_x(g \circ i_0) \\
&= (X_f(g))(i_0(x)) \ ;
\end{aligned}$$

$$\begin{aligned}
[\theta(X_f)(\pi_0(x))] \cdot [g_0(\pi_0(x))] &= [(\theta_0)_{\pi_0(x)}(X_{f_0})_{\pi_0(x)}] \cdot [g \circ i_0(x)] \\
&= [(\theta_0)_{\pi_0(x)}(T_x\pi_0(X_f)_x][g(i_0(x))] \\
&= [\pi_0^*(\theta_0)_{\pi_0(x)}(X_f)_x][g(i_0(x))] \\
&= [\theta_{i_0(x)}(X_f)_{i_0(x)}][g(i_0(x))] \\
&= [\theta(X_f)(i_0(x)][g(i_0(x))] \\
&= [\theta(X_f) \cdot g](i_0(x)) \ ;
\end{aligned}$$

$$\begin{aligned}
[f_0 \cdot g_0](\pi_0(x)) &= f_0(\pi_0(x)) \cdot g_0(\pi_0(x)) \\
&= f(i_0(x)) \cdot g(i_0(x)) \\
&= (f \cdot g)(i_0(x))
\end{aligned}$$

and so we obtain the desired result. □

Remark. Using similar arguments we can also prove that the projections onto $T^*\mathcal{M}_D$ of the characteristics of the prequantum operator δ_f are the characteristics of the prequantum operator $\delta^0_{f_0}$.

6. Conclusions

In this paper we have discussed the geometric prequantization of the Einstein's vacuum field equations and its connection with the Marsden–Weinstein reduction with respect to the symmetry group of all spacetime diffeomorphisms. It is an open problem to try to extend the above results to the general case of geometric quantization. Of course the problem is to find a polarization P on $T^*\mathcal{M}$ such that the Hamiltonian H to be constant along P, or to extend in infinite dimensional picture the concept of BKS-kernel. Unfortunately these notions are not yet clear in infinite dimensional context.

References

1. R. Abraham and J. Marsden, *Foundations of Mechanics*, Benjamin Cummings, Reading, Mass., 1978.
2. R. Arnowitt, S. Deser, and C. W. Misner, *Dynamical structure and definition of energy in general relativity*, Phys. Rev. **116** (1959), 1322–1330.
3. P. A. M. Dirac, *The theory of gravitation in Hamiltonian form*, Proc. R. Soc. London A. **246** (1959), 333–343.
4. D. Ebin, *On the space of Riemannian metrics*, Symp. Pure Math. Amer. Math. Soc. **15** (1970), 11–40.
5. A. E. Fischer and J. Marsden, *The initial value problem and the dynamical formulation of general relativity*, General Relativity, An Einstein Centenary Volume, eds. S. Hawking and W. Israel, Cambridge Univ. Press, 1979, 138–221.
6. J. Marsden and A. Weinstein, *Reduction of symplectic manifolds with symmetry*, Rep. Math. Phys. **5** (1974), 121–130.
7. J. A. Wheeler, *Geometrodynamics and the issue of the final state*, Relativity, Groups and Topology, eds. C. M. Dewitt and B. S. Dewitt, Gordon and Breach, New York, 1964.

Geometry in Partial Differential Equations (pp. 329–344)
eds. A. Pràstaro and Th. M. Rassias

ON DIFFERENTIAL EQUATIONS AND CARTAN'S PROJECTIVE CONNECTIONS

YOURY R. ROMANOVSKY
Program Systems Institute of the Russian Academy of Sciences
Pereslavl-Zalessky 152140, Russia

ABSTRACT

We describe a method for constructing a normal projective connection associated with a second order ordinary differential equation. Our construction shows that differential equations generate Cartan's connections in the same way as Riemannian metrics generate Levi-Civita's connections. We use this consruction to give a classification, in terms of projective curvature, of differential equations admitting Lie groups of point symmetries.

1. Introduction

This work is devoted to Elie Cartan's theory of *normal projective connections*[1] and its applications to differential equations. Projective connections themselves provide a geometric theory of ordinary differential equations. However, this theory gave rise to the notion of *Cartan connection*[2] which has a wider range of applications. For example, the Cartan connection associated with a second order partial differential equation of Monge-Ampère type was discovered in the paper by T. Morimoto[3].

In §2, we begin with a second order ordinary differential equation and describe a natural way leading to the notion of normal projective connection associated with a *cubic differential equation* (see Eq. 8). An important intermediate step is the construction of a *projective structure*[2]. A projective structure is a principal bundle in which the normal projective connection appears in an intrisic manner. Cubic differential equations and normal projective connections are in one-to-one correspondence. The differential 1-form of a normal projective connection is an absolute parallelism on a projective structure. Hence, the equivalence problem for cubic differential equations under the pseudo-group of coordinate transformations reduces to the equivalence problem for absolute parallelisms. The latter problem has a well-known solution[4].

In §3, we explain how one can use the connection form to solve the equivalence problem for a pair of cubic differential equations. Here we discuss the notion of *projective curvature* and prove an elegant Cartan's criterion for an equation to be locally equivalent to the differential equation of straight lines (see Eq. 1). The other interesting result is Cartan's *theorema egregium* stating that the curves on a surface in Euclidean space that are orthogonal to the level lines of Gaussian curvature can be computed in terms of the differential equation of geodesics (see Eq. 3).

In §4, we use the connection form to investigate the *symmetry group* of a cubic differential equation. Here we deduce the generalized *Bianchi identities.* As a consequence, we give a necessary and sufficient condition, in terms of projective curvature, for an equation to admit a symmetry group of dimension $n \geq 3$. It should be emphasized that such an equation is locally equivalent either to the differential equation of straight lines or to the differential equation of quadrics (see Eq. 2).

Finally, we note that an analogous classification can be obtained for cubic differential equations with a symmetry group of dimension $n = 2$. Moreover, in this case the machinery of normal projective connections allows us to find *by quadratures* the infinitesimal generators of a symmetry group, *i.e.* the generators of its Lie algebra. Thus, if an equation has a symmetry algebra of dimension two, then the generators of this algebra can be found by quadratures. This means that the solutions of the equation can be found by quadratures too[5]. We will study the problem of integration by quadratures for cubic differential equations in a forthcoming paper.

2. Projective Structures

We begin by recalling that the general solution of a second order ordinary differential equation $y'' = f(x, y, y')$ is, by definition, the two-parameter family of its integral curves on the plane $\mathbf{R}^2_{(x,y)}$.

Example 1. The differential equation of straight lines:

$$y'' = 0. \tag{1}$$

The general solution: $x \cos t_1 + y \sin t_1 + t_2 = 0$.

Example 2. The differential equation of quadrics:

$$y'' = \pm(xy' - y)^3. \tag{2}$$

The general solution: $XY(X^2 \pm Y^2 - 1) = 0$ where $X = (x \cos t_1 + y \sin t_1)t_2$, $Y = (y \cos t_1 - x \sin t_1)/t_2$.

Example 3. The differential equation of geodesics:

$$y'' = a_0 + a_1 y' + a_2 y'^2 + a_3 y'^3, \tag{3}$$

where

$$\begin{aligned}
a_0 &= \quad ((EE_y + FE_x)/2 - EF_x)/(EG - F^2),\\
a_1 &= \quad (3FE_y/2 - FF_x - EG_x + E_xG/2)/(EG - F^2),\\
a_2 &= -(3FG_x/2 - FF_y - GE_y + G_yE/2)/(EG - F^2),\\
a_3 &= -((GG_x + FG_y)/2 - GF_y)/(EG - F^2).
\end{aligned}$$

The general solution: geodesics of the Riemannian metric $Edx^2 + 2Fdx\,dy + Gdy^2$. Here we note that Eq. 1 and Eq. 2 are particular cases of Eq. 3.

Let us consider a special equivalence problem: determine all local diffeomorphisms $\varphi\colon \mathbf{R}^2_{(x,y)} \to \mathbf{R}^2_{(x,y)}$ which take the general solution of Eq. 1 into the general solution of the target equation $y'' = f(x, y, y')$. We say that the target equation is locally equivalent to the differential equation of straight lines, if the set of such diffeomorphisms is not empty. The first obstacle to construct φ arises at the level of a one-parameter pencil of integral curves:

$$y - y_0 = p(x - x_0) + \frac{f(x_0, y_0, p)}{2}(x - x_0)^2 + O((x - x_0)^3), \quad p = y'(x_0). \tag{4}$$

Example[6]**.** The pencil of circles

$$y = px + a(p)(x^2 + y^2)$$

can be rectified by a diffeomorphism in a neigbourhood of the point $(0, 0)$, if and only if the function $a(p)$ is a linear function. In other words, the circles should have one more common point. Otherwise the rectifying transformations are not smooth at the origin of $\mathbf{R}^2_{(x,y)}$. This is an immediate consequence of the following proposition.

Proposition 1. *Pencil (4) transforms by a diffeomorphism*

$$x = \varphi^1(z^1, z^2)$$
$$y = \varphi^2(z^1, z^2)$$

into the form

$$z^2 = qz^1 + O(|z|^3), \quad p = \frac{u^2_2 q + u^2_1}{u^1_2 q + u^1_1}, \quad u^i_j = \frac{\partial \varphi^i}{\partial z^j}(0), \tag{5}$$

if and only if $f(p)$ *is a cubic polynomial:*

$$f(p) = A + 3Bp + 3Cp^2 + Dp^3. \tag{6}$$

Suppose φ, ψ *is a pair of diffeomorphisms transforming pencil (4) to pencil (5). Then there is a unique projective transformation*

$$h\colon \mathbf{R}^2_z \to \mathbf{R}^2_z, \quad h^i = \frac{a^i_j z^j}{1 + a_j z^j}, \tag{7}$$

such that

$$j^2_0(\varphi^{-1} \circ \psi) = j^2_0(h),$$

where $j^2_0(\varphi)$ *is the 2-jet of* φ *at* $z = 0$.

Proof. Follows from a straightforward calculation.

Proposition 1 has three important corollaries.

Corollary 1. *The condition* $\partial^4 f/\partial y'^4 = 0$ *is a necessary condition for the equation* $y'' = f(x,y,y')$ *to be locally equivalent to the differential equation of straight lines. The set of "suspicious equations"*

$$y'' = A(x,y) + 3B(x,y)y' + 3C(x,y)y'^2 + D(x,y)y'^3 \tag{8}$$

is invariant under the pseudo-group of coordinate transformations.

We call Eq. 8 a *cubic differentual equation.* From now on we shall be only interested in cubic differentual equations. For the reader whose interest is in geometry of general second order ordinary differential equations, we would recommend the paper by E. Cartan[7].

Corollary 2. *A cubic differential equation generates a principal fibre bundle*

$$\pi: P \to \mathbf{R}^2_{(x,y)} \tag{9}$$

whose fibre $\pi^{-1}(x_0, y_0)$ *consists of 2-jets of diffeomorphisms*

$$\varphi: \mathbf{R}^2_z \to \mathbf{R}^2_{(x,y)}, \quad \varphi(0) = (x_0, y_0), \tag{10}$$

which take pencil (5) to pencil (4) where $f(p)$ *is cubic polynomial (6). The structure group* H *of bundle (9) coincides with the group of projective transformations (7).*

We recall[2] that the r-jet $j^r_0(\varphi)$ of diffeomorphism (10) at 0 is, by definition, an *r-frame* of $\mathbf{R}^2_{(x,y)}$ at (x_0, y_0). The set of r-frames of $\mathbf{R}^2_{(x,y)}$ will be denoted by P^r. It is a smooth manifold. For example, each 3-frame $u \in P^3$ has a unique polynomial representation $u = j^3_0(\varphi)$ of the form

$$\varphi^i(z) = u^i + u^i_j z^j + u^i_{jk} z^j z^k + u^i_{jkl} z^j z^k z^l.$$

The coefficients u^i_{jk}, u^i_{jkl} are symmetric in the lower indices. We take $(u^i, u^i_j, u^i_{jk}, u^i_{jkl})$ as the natural coordinate system in P^3. The bundle

$$\pi_r: P^r \to \mathbf{R}^2_{(x,y)}, \quad \pi_r(j^r_0(\varphi)) = \varphi(0),$$

is a principal fibre bundle. Its structure group G^r consists of r-jets of the diffeomorphisms $g: \mathbf{R}^2_z \to \mathbf{R}^2_z$, $g(0) = 0$. The group G^r acts on P^r by $j^r_0(\varphi) \cdot j^r_0(g) = j^r_0(\varphi \circ g)$. For example, if (a^i_j, a^i_{jk}) are the natural coordinates of $j^2_0(g) \in G^2$, then we have

$$(u^i, u^i_j, u^i_{jk}) \cdot (a^i_j, a^i_{jk}) = (u^i, u^i_p a^p_j, u^i_p a^p_{jk} + u^i_{qr} a^q_j a^r_k).$$

Let $H^2 = \{j^2_0(h) \in G^2 \mid h \in H\}$. Then H^2 is isomorphic to H. This follows from the formula

$$j^2_0(h) = (a^i_j, -(a^i_j a_k + a^i_k a_j)/2).$$

Definition[2]. A subbundle P of P^2 with structure group H^2 is called a *projective structure* on $\mathbf{R}^2_{(x,y)}$.

Corollary 3. *Bundle (9) is a projective structure on* $\mathbf{R}^2_{(x,y)}$. *Conversely, any projective structure on* $\mathbf{R}^2_{(x,y)}$ *is generated by a cubic differential equation. Cubic differential equations and projective structures are in one-to-one correspondence.*

Proof. Let P be a projective structure. Then there is a unique cross section

$$s\colon \mathbf{R}^2_{(x,y)} \to P, \quad s(x,y) = (x,y,u^i_j(x,y),u^i_{jk}(x,y)),$$

such that

$$u^i_j = \delta^i_j, \quad u^i_{jk} = -\frac{1}{2}\Pi^i_{jk},$$
$$\Pi^1_{11} + \Pi^2_{12} = \Pi^2_{22} + \Pi^1_{21} = 0,$$

where δ^i_j is Kronecker's delta. This cross section is called the *natural frame* (*le repère naturel* in Cartan[1]). Thus, each element of P has a unique representation of the form

$$s \cdot j^2_0(h) = (x,y,a^i_j, -(a^i_j a_k + a^i_k a_j + \Pi^i_{qr} a^q_j a^r_k)/2). \tag{11}$$

In the sequel, we shall use

$$(x,y,a), \quad a = \begin{pmatrix} 1 & a_1 & a_2 \\ 0 & a^1_1 & a^1_2 \\ 0 & a^2_1 & a^2_2 \end{pmatrix}, \tag{12}$$

as the natural coordinate system in P. A simple calculation shows that diffeomorphisms (11) take pencil (5) to pencil (4) where $f(p)$ is polynomial (6) with the coefficients

$$\begin{cases} A = -\Pi^2_{11}, \\ B = \Pi^1_{11} = -\Pi^2_{12} = -\Pi^2_{21}, \\ C = -\Pi^2_{22} = \Pi^1_{12} = \Pi^1_{21}, \\ D = \Pi^1_{22}. \end{cases} \tag{13}$$

The proof is complete.

In order to solve the special equivalence problem for Eq. 8, we should give now the answer to the following question.

Does there exist the prolongation of the projective structure P

$$P_\infty = \{\varphi\colon \mathbf{R}^2_z \to \mathbf{R}^2_{(x,y)} \mid j^2_0(\varphi) \in P\}$$

such that all composite maps

$$\varphi^{-1} \circ \psi : \mathbf{R}^2_z \to \mathbf{R}^2_z \qquad (\varphi, \psi \in P_\infty)$$

are collineations of $\mathbf{R}^2_z$ *?*

If the answer to this question is affirmative, then P_∞ is the solution of the problem. It is clear that P_∞ is a principal fibre bundle over $\mathbf{R}^2_{(x,y)}$ with structure group H. Let φ be an arbitrary element of P_∞ and $\varphi_t \in P_\infty$ be a one-parameter family such that $\varphi_0 = \varphi$. Then the composite map $\chi_t = \varphi^{-1} \circ \varphi_t$ is a collineation

$$\chi_t^i = \frac{b^i(t) + b_j^i(t) z^j}{1 + b_j(t) z^j}, \tag{14}$$

where $b^i(0) = b_j(0) = 0$, $b_j^i(0) = \delta_j^i$. Hence we have

$$d\chi_t^i|_{t=0} = db^i + db_j^i\, z^j - \frac{1}{2}(\delta_j^i\, db_k + \delta_k^i\, db_j) z^j z^k + O(|z|^3). \tag{15}$$

On the other hand, we obtain

$$d\chi_t^i|_{t=0} = \theta^i + \theta_j^i\, z^j + \theta_{jk}^i\, z^j z^k + O(|z|^3), \tag{16}$$

where $\theta = (\theta^i, \theta_j^i, \theta_{jk}^i)$ is the so-called *canonical (differential) 1-form*[8] on P^3 at the point $j_0^3(\varphi)$. The values of θ can be viewed as the 2-jets of infinitesimal composite maps. A straightforward computation gives the following formulæ:

$$\begin{aligned}
\theta^i &= v_p^i\, du^p, \\
\theta_j^i &= v_p^i\, du_j^p - 2 v_p^i u_{qj}^p\, \theta^q, \\
\theta_{jk}^i &= v_p^i\, du_{jk}^p - 2 v_p^i u_{qj}^p\, \theta_k^q - 3 v_p^i u_{qjk}^p\, \theta^q,
\end{aligned}$$

where (v_j^i) is the inverse matrix of (u_j^i); $(u^i, u_j^i, u_{jk}^i, u_{jkl}^i) = j_0^3(\varphi)$. Now comparing expressions (15) and (16), we see that the restriction of the canonical form θ to

$$P_3 = \{ j_0^3(\varphi) \mid \varphi \in P_\infty \}$$

satisfies the condition

$$\varepsilon_{jk}^i \overset{\text{def}}{=} \theta_{jk}^i - \frac{1}{2}(\delta_j^i\, \theta_{kk}^k + \delta_k^i\, \theta_{jj}^j) \;=\; 0.$$

In other words, P_3 is a subbundle of P^3 annihilating the Pfaffian system $\{\varepsilon_{jk}^i\}$.

Now, in answer to the preceding question, we have the following fundamental observation: the prolongation P_∞ does not generally exist, while the prolongation P_3 can be constructed for any projective structure P.

Proposition 2. *For any projective structure P there exists a unique map*

$$\sigma : P \to P^3 \qquad (j_0^2 \circ \sigma = id)$$

such that

$$\sigma^*(\varepsilon_{jk}^i) \;=\; 0.$$

The bundle

$$\pi_3|_{P_3} : \ P_3 \stackrel{\text{def}}{=} \sigma(P) \ \rightarrow \ \mathbf{R}^2_{(x,y)}$$

is a principal subbundle of P^3 *with structure group* $H^3 = \{j^3_0(h) \mid h \in H\} \simeq H^2 \simeq H$.

Proof. Follows from a straightforward computation in the natural coordinates.

Proposition 2 enables us to take the canonical form θ on P^3 back to a form on the projective structure P. We set

$$\omega = \begin{pmatrix} 0 & \omega_1 & \omega_2 \\ \omega^1 & \omega^1_1 & \omega^1_2 \\ \omega^2 & \omega^2_1 & \omega^2_2 \end{pmatrix},$$

where

$$\omega^i = \sigma^*(\theta^i), \quad \omega^i_j = \sigma^*(\theta^i_j), \quad \omega_j = -\sigma^*(\theta^j_{jj}). \tag{17}$$

Definition. The differential 1-form ω on P is called the *normal projective connection* associated with the projective structure P.

Remark. The motivation behind this definition is the following simple observation. Each tangent vector $\dot{u}(t) \in T(P)$ generates an infinitesimal collineation of the form (14) where

$$\begin{pmatrix} 1 & b_1 & b_2 \\ b^1 & b^1_1 & b^1_2 \\ b^2 & b^2_1 & b^2_2 \end{pmatrix} = I - \omega(\dot{u})\,dt,$$

I is the unit matrix. This infinitesimal shift is interpreted as the infinitesimal "*parallel displacement*" of the *tangent projective plane* to $\mathbf{R}^2_{(x,y)}$ at the point $\pi(u)$ in the direction $\pi_*(\dot{u})$. As usual the infinitesimal picture can be integrated up to give the global picture of parallel displacement along curves.

The correspondence between normal projective connections and cubic differential equations satisfies the following condition.

Proposition 3. *Let* (P,ω) *and* $(\overline{P},\overline{\omega})$ *be normal projective connections associated with Eq. 8 and Eq.* $\bar{8}$, *respectively. Then the map* $\Psi: P \rightarrow \overline{P}$ *takes the form* $\overline{\omega}$ *back into the form* ω, *i.e.* $\Psi^*(\overline{\omega}) = \omega$, *if and only if* $\Psi = \Phi|_P$ *where* $\Phi: P^2 \rightarrow P^2$ *is the natural prolongation of the map* $\varphi: \mathbf{R}^2_{(x,y)} \rightarrow \mathbf{R}^2_{(x,y)}$ *which takes the general solution of Eq.* $\bar{8}$ *back into the general solution of Eq. 8.*

Proof. Follows from Corollary 3 and the properties of the canonical form[7] θ.

Finally, to conclude this section, we shall express the connection form ω in terms of natural coordinate system (12) on a projective structure P. For this goal, one should construct the prolongation $S: \mathbf{R}^2_{(x,y)} \rightarrow P_3$ of the natural frame s such that $S^*(\varepsilon^i_{jk}) = 0$. Then the map σ is given by

$$\sigma\left(s \cdot j^2_0(h)\right) = S \cdot j^3_0(h).$$

Now, from definition (17) it follows that ω is given by the "gauge transformation"

$$\omega = a^{-1}wa + a^{-1}da + I\,a_i\omega^i, \quad w = \begin{pmatrix} 0 & w_1 & w_2 \\ w^1 & w_1^1 & w_2^1 \\ w^2 & w_1^2 & w_2^2 \end{pmatrix}, \tag{18}$$

where

$$\begin{aligned} w^i &= \quad s^*(\theta^i) \quad = \quad \delta_1^i\,dx + \quad \delta_2^i\,dy, \\ w_j^i &= \quad s^*(\theta_j^i) \quad = \Pi_{j1}^i\,dx + \Pi_{j2}^i\,dy, \\ w_j &= -S^*(\theta_{jj}^j) = \Pi_{j1}\,dx + \Pi_{j2}\,dy, \end{aligned}$$

$$\begin{aligned} \Pi_{12} = \Pi_{21} = \quad \Pi_{22}^1\Pi_{11}^2 - \Pi_{11}^1\Pi_{22}^2 &+ \frac{\partial \Pi_{22}^2}{\partial x} + \frac{\partial \Pi_{11}^1}{\partial y}, \\ \Pi_{11} = 2(\Pi_{11}^1\Pi_{11}^1 - \Pi_{11}^2\Pi_{22}^2) &- \frac{\partial \Pi_{11}^1}{\partial x} - \frac{\partial \Pi_{11}^2}{\partial y}, \\ \Pi_{22} = 2(\Pi_{22}^2\Pi_{22}^2 - \Pi_{11}^1\Pi_{22}^1) &- \frac{\partial \Pi_{22}^1}{\partial x} - \frac{\partial \Pi_{22}^2}{\partial y}. \end{aligned}$$

The coefficients Π_{jk}^i are given by formula (13).

3. The Equivalence Problem

The equivalence problem is the following question. Given a pair of cubic differential equations: Eq. 8 and Eq. $\overline{8}$, one should determine all diffeomorphisms $\varphi\colon \mathbf{R}^2_{(x,y)} \to \mathbf{R}^2_{(x,y)}$ which take the general solution of Eq. 8 into the general solution of Eq. $\overline{8}$. If the set of such diffeomorphisms is not empty, then the equations are said to be locally equivalent.

From proposition 3 it follows that the normal projective connection is a good tool to solve the equivalence problem. Actually, the components ω^i, ω_j^i, ω_j of Cartan's connection (18) form a basis of the cotangent space $T_u^*(E)$ at each point $u = (x, y, a)$ of the direct product

$$E = \mathbf{R}^2_{(x,y)} \times H.$$

This means that ω is an absolute parallelism on E. Proposition 3 asserts that the equivalences φ of Eq. 8 and Eq. $\overline{8}$ are in one-to-one correspondence with diffeomorphisms

$$\Psi\colon E \to E \tag{19}$$

such that

$$\Psi^*(\overline{\omega}) = \omega. \tag{20}$$

This correspondence is given by

$$\Psi(x, y, a) = (\,\varphi^1,\, \varphi^2,\, \varphi_\star \cdot a\,), \tag{21}$$

where $\varphi_\star$ is the *projective differential* of φ:

$$\varphi_\star = \begin{pmatrix} 1 & J_x & J_y \\ 0 & \varphi_x^1 & \varphi_y^1 \\ 0 & \varphi_x^2 & \varphi_y^2 \end{pmatrix}, \qquad J = -\frac{1}{3}\log(\varphi_x^1\varphi_y^2 - \varphi_y^1\varphi_x^2),$$

which maps the tangent projective plane to $\mathbf{R}^2_{(x,y)}$ at p to the tangent projective plane to $\mathbf{R}^2_{(x,y)}$ at $\varphi(p)$.

Thus, the equivalence problem for cubic differential equations reduces to the equivalence problem for absolute parallelisms which can be solved by the standard method[4]. Following this method, we describe the finite system of invariants recognizing a cubic differential equation up to local equivalence. The algorithm that will find such invariants rests on the notion of *structure equations.* For the problem under consideration the structure equations are nothing else but the structure equations of Cartan's connection ω:

$$d\omega = -\omega \wedge \omega + I\,\omega_i \wedge \omega^i + \Omega\,\omega^1 \wedge \omega^2, \tag{22}$$

where

$$\Omega = \begin{pmatrix} 0 & \Omega_1 & \Omega_2 \\ 0 & 0 & 0 \\ 0 & 0 & 0 \end{pmatrix}, \quad \Omega_j = L_i a_j^i |a|, \quad |a| = a_1^1 a_2^2 - a_2^1 a_1^2,$$

$$L_1 = \frac{\partial \Pi_{12}}{\partial x} - \frac{\partial \Pi_{11}}{\partial y} + \Pi_{11}\Pi_{12}^1 - \Pi_{12}\Pi_{11}^1 + \Pi_{12}\Pi_{12}^2 - \Pi_{22}\Pi_{11}^2,$$

$$L_2 = \frac{\partial \Pi_{22}}{\partial x} - \frac{\partial \Pi_{12}}{\partial y} + \Pi_{11}\Pi_{22}^1 - \Pi_{12}\Pi_{12}^1 + \Pi_{12}\Pi_{22}^2 - \Pi_{22}\Pi_{12}^2.$$

The differential 2-form $\Omega\,\omega^1 \wedge \omega^2$ is called the *projective curvature* of ω. It follows from structure equations (22) that the projective curvature vanishes, *i.e.* $\Omega = 0$, if and only if the "parallel displacement" along each contractible closed path is the identical transformation.

We are now in position to describe the method for solving a special equivalence problem.

Theorem 1 (Elie Cartan). *A cubic differential equation is locally equivalent to the differential equation of straight lines, if and only if its projective curvature vanishes.*

Proof. Let ω and $\overline{\omega}$ be the normal projective connections associated with Eq. 8 and Eq. 1 respectively. It is clear that

$$d\overline{\omega} = -\overline{\omega} \wedge \overline{\omega} + I\,\overline{\omega}_i \wedge \overline{\omega}^i. \tag{23}$$

Since the exterior derivative is an intrinsic operation, from (20) and (22) we obtain

$$\Psi^*(\overline{\Omega}) = \Omega. \tag{24}$$

Hence we have

$$d\omega = -\omega \wedge \omega + I\,\omega_i \wedge \omega^i \tag{25}$$

provided Eq. 8 is locally equivalent to Eq. 1. Now subtracting Eq. (25) from Eq. (23), we find

$$d(\omega - \overline{\omega}) = -(\omega - \overline{\omega}) \wedge \omega - \overline{\omega} \wedge (\omega - \overline{\omega}) + I\,(\omega_i - \overline{\omega}_i) \wedge \overline{\omega}^i + I\,\overline{\omega}_i \wedge (\omega_i - \overline{\omega}^i).$$

Thus, the differential system $\omega - \overline{\omega} = 0$ is a Frobenius system on the direct product $E \times E$. The 8-parameter family of its 8-dimensional integral surfaces is nothing else but the family of graphs of diffeomorphisms (19) satisfying condition (20). The proof is complete.

In the general case ($\Omega \neq 0$, $\overline{\Omega} \neq 0$), condition (24) is a first (but not a unique!) obstacle to construct an equivalence φ. Let us study the geometric meaning of this obstacle. For this aim, we set

$$\gamma = \Omega_1 \omega^1 + \Omega_2 \omega^2 = (L_1\,dx + L_2\,dy)|a|.$$

The general solution of the first order equation

$$\gamma = 0 \tag{26}$$

is a one-parameter family of curves on the plane $\mathbf{R}^2_{(x,y)}$. Since condition (24) is equivalent to $\Psi^*(\overline{\gamma}) = \gamma$, we see that any equivalence φ of Eq. 8 and Eq. $\overline{8}$ must take the general solution of Eq. 26 into the general solution of Eq. $\overline{26}$.

Theorem 2 (Elie Cartan). *Suppose ω is the normal projective connection associated with differential equation of geodesics (3). Then the general solution of Eq. 26 consists of the curves orthogonal to the level lines of Gaussian curvature.*

Proof. See Chapitre V in the book by Cartan[1].

Remark. In other words, the theorem states that the curves on a surface in Euclidean space that are orthogonal to the level lines of Gaussian curvature can be computed in terms of the differential equation of geodesics. In this sense, Cartan's theorem is analogous to Gauss's *theorema egregium* stating that Gaussian curvature of a surface can be computed in terms of the Riemannian metric.

Now, we shall describe the total system of obstacles to construct an equivalence φ. Consider the family

$$\mathcal{F}_r = \{\,\Omega,\ \nabla(\Omega), \ldots, \underbrace{\nabla \cdots \nabla}_{r \text{ times}}(\Omega)\,\}, \qquad \nabla = (\nabla_i, \nabla_i^j, \nabla^j),$$

where $\nabla_i, \nabla_i^j, \nabla^j$ are the *covariant derivatives* with respect to the absolute parallelism ω, *i.e.* the vector fields on E such that

$$df = \nabla_i(f)\,\omega^i + \nabla_i^j(f)\,\omega_j^i + \nabla^j(f)\,\omega_j$$

for any smooth function $f \in C^\infty(E)$; $r = \operatorname{ord}(\omega)$ is the *order* of the parallelism ω, that is the smallest $r \geq 0$ such that $\operatorname{rank}(\mathcal{F}_r) = \operatorname{rank}(\mathcal{F}_{r+1})$; $\operatorname{rank}(\mathcal{F})$ equals the number of functionally independent elements in $\mathcal{F}$. Recall that the $\operatorname{rank}(\omega)$ is, by definition, the number $\operatorname{rank}(\mathcal{F}_r)$. Let us fix the system

$$\{\, x_1 = \Omega_1,\ x_2 = \Omega_2,\ \ldots,\ x_k \,\} \subset \mathcal{F}_r, \quad r = \operatorname{ord}(\omega), \quad k = \operatorname{rank}(\omega),$$

of functionally independent elements. Then any other element $f \in \mathcal{F}_{r+1}$ generates a unique function $F_f(t_1, \ldots, t_k)$ defined on an open subset $\mathcal{D}$ of $\mathbf{R}^k$ such that

$$f = F_f(x_1, \ldots, x_k).$$

It is clear that the domain $\mathcal{D}$ does not depend on f.

Theorem 3. *Let ω and $\overline{\omega}$ be normal projective connections associated with Eq. 8 and Eq. $\bar{8}$, respectively. Then the equations are locally equivalent, if and only if the following conditions hold:*

(a) $\operatorname{ord}(\omega) = \operatorname{ord}(\overline{\omega})$, $\operatorname{rank}(\omega) = \operatorname{rank}(\overline{\omega})$;

(b) $F_f(t_1, \ldots, t_k) = \overline{F}_{\overline{f}}(t_1, \ldots, t_k)$ *for each* $(t_1, \ldots, t_k) \in \mathcal{D} \cap \overline{\mathcal{D}} \neq \emptyset$ *and any pair* $(f, \overline{f}) \in \mathcal{F}_{r+1} \times \overline{\mathcal{F}}_{r+1}$ *of covariant derivatives of the same names.*

Proof. The necessity is evident. To prove sufficiency, we observe that the restriction of the differential system $\omega - \overline{\omega} = 0$ to the submanifold

$$x_i - \overline{x}_i = 0 \quad (1 \leq i \leq k)$$

of the direct product $E \times E$ is a Frobenius system whenever conditions (a) and (b) hold. The $(8 - k)$-parameter family of its 8-dimensional integral surfaces is nothing else but the family of graphs of diffeomorphisms (19) satisfying condition (20). See the book by S. Sternberg[4] for details.

See also the paper by C. Grissom *et al.*[9] for an alternative presentation of Cartan's method for solving the equivalence problem for second order equations.

4. Group Analysis

There exists an extensive literature on the symmetry groups of differential equation. Here, we just mention the paper by L. Hsu and N. Kamran[10] which is sufficiently close to the material of the present section.

By a (point) symmetry of a differential equation, we understand a local diffeomorphism $\varphi\colon \mathbf{R}^2_{(x,y)} \to \mathbf{R}^2_{(x,y)}$ which takes the general solution of the equation into itself. From proposition 3 it follows that the symmetry group of a cubic differential equation is isomorphic to the group

$$\operatorname{Sym}(\omega) = \{\, \Psi\colon E \to E \mid \Psi^*(\omega) = \omega \,\}$$

of symmetries of the normal projective connection associated with the equation. The isomorphism is given by relation (21). In order to investigate the group $\mathrm{Sym}(\omega)$, we shall use the machinery developed in §3. In particular, we shall determine the dimension of the group by the formula

$$\dim \mathrm{Sym}(\omega) = 8 - \mathrm{rank}(\omega)$$

which follows from Theorem 3. Now, we supply this formula with three lemmas that allow one to simplify the computation of the number $\mathrm{rank}(\omega)$.

Lemma 1. *Let $\mathcal{F}^s$ be the following family of functions*

$$\mathcal{F}^s = \{\,\Omega,\ \partial(\Omega),\ \ldots,\ \underbrace{\partial\cdots\partial(\Omega)}_{s \text{ times}}\,\},\quad \partial = (\nabla_1, \nabla_2),$$

and $\langle\mathcal{F}^s\rangle$ be the linear space spanned by $\mathcal{F}^s$. Then

$$\nabla^i(\mathcal{F}^s) \subset \langle\mathcal{F}^s\rangle,\quad \nabla^i_j(\mathcal{F}^s) \subset \langle\mathcal{F}^s\rangle$$

for all $s \geq 0$. In particular,

$$\mathrm{rank}(\mathcal{F}_s) = \mathrm{rank}(\mathcal{F}^s).$$

Proof. Applying the exterior derivative d to the structure equation (22) and using a result from exterior algebra known as Cartan's lemma[4], we find

$$\nabla^i(\Omega_j) = \nabla^2_1(\Omega_1) = \nabla^1_2(\Omega_2) = 0,$$

$$\nabla^1_1(\Omega_1) = 2\Omega_1,\ \nabla^1_2(\Omega_1) = \Omega_2,\ \nabla^2_2(\Omega_1) = \Omega_1,$$
$$\nabla^2_2(\Omega_2) = 2\Omega_2,\ \nabla^2_1(\Omega_2) = \Omega_1,\ \nabla^1_1(\Omega_2) = \Omega_2.$$

This proves the statement of the lemma in the case $s = 0$. To prove the lemma in general, we proceed by induction using the commutation relations

$$[\nabla^1, \nabla_1] = -2\nabla^1_1 - \nabla^2_2 \qquad [\nabla^1, \nabla_2] = -\nabla^1_2$$
$$[\nabla^2, \nabla_1] = -\nabla^2_1 \qquad [\nabla^2, \nabla_2] = -\nabla^1_1 - 2\nabla^2_2$$
$$[\nabla^1_1, \nabla_1] = \nabla_1 \qquad [\nabla^1_1, \nabla_2] = 0$$
$$[\nabla^2_2, \nabla_1] = 0 \qquad [\nabla^2_2, \nabla_2] = \nabla_2$$
$$[\nabla^1_2, \nabla_1] = \nabla_2 \qquad [\nabla^1_2, \nabla_2] = 0$$
$$[\nabla^2_1, \nabla_1] = 0 \qquad [\nabla^2_1, \nabla_2] = \nabla_1$$

which are nothing else but the same structure equations (22) written in terms of covariant derivatives.

Corollary. *Theorem 3 will remain valid, if we replace (everywhere) the families $\mathcal{F}_s$ by the families $\mathcal{F}^s$.*

Lemma 2. *If* $\Omega \neq 0$, *then the functions*

$$x_1 = \Omega_1, \quad x_2 = \Omega_2, \quad y_1 = \nabla_1(\Omega_1), \quad y_2 = \nabla_2(\Omega_2) \tag{27}$$

are functionally independent. In particular,

$$\mathrm{rank}(\mathcal{F}^1) \geq 4.$$

Proof. Follows from the relation

$$\begin{pmatrix} \dfrac{\partial y_1}{\partial a_1} & \dfrac{\partial y_1}{\partial a_2} \\ \dfrac{\partial y_2}{\partial a_1} & \dfrac{\partial y_2}{\partial a_2} \end{pmatrix} = -5 \begin{pmatrix} x_1 & 0 \\ 0 & x_2 \end{pmatrix},$$

which one can readily verify using the explicit formula

$$\nabla_i = X_i - (a_q^p w_p(X_i) + a_i a_q)\frac{\partial}{\partial a_q} - (a_q^p w_p^r(X_i) + a_q a_i^r + a_i a_q^r)\frac{\partial}{\partial a_q^r},$$

where $X_i = a_i^1 \dfrac{\partial}{\partial x} + a_i^2 \dfrac{\partial}{\partial y}$.

Lemma 3. *If* $\Omega \neq 0$, *then*

$$\mathrm{rank}(\omega) \geq 5, \tag{28}$$

i.e. the dimension of the symmetry group is not greater than three.

Proof. Lemmas 1 and 2 imply that $\mathrm{rank}(\mathcal{F}_1) = \mathrm{rank}(\mathcal{F}^1) \geq 4$. The equality is attained at the equation

$$y'' = y^2. \tag{29}$$

Hence we need to look at the family $\mathcal{F}^2$. Let us denote

$$\Omega'_{i,j} = \nabla_j(\Omega_i), \quad \Omega''_{i,jk} = \nabla_k \nabla_j(\Omega_i)$$

and show that

$$\mathrm{rank}(\mathcal{F}^2) \geq 5.$$

Suppose, on the contrary, that $\mathrm{rank}(\mathcal{F}^2) < 5$, *i.e.*

$$\Omega'_{i,j} = \Sigma_{i,j}(x_1, x_2, y_1, y_2), \quad \Omega''_{i,jk} = \Sigma_{i,jk}(x_1, x_2, y_1, y_2).$$

Then we immediately find

$$\Sigma_{i,j} = \frac{1}{5}\frac{y_i x_j}{x_i} + \frac{4}{5}\frac{x_i y_j}{x_j}, \qquad \Sigma_{i,ii} = \frac{6}{5}\frac{y_i^2}{x_i},$$

$$\Sigma_{i,ij} = \frac{1}{5}\frac{y_i^2 x_j}{x_i^2} + \frac{y_i y_j}{x_j}, \qquad \Sigma_{i,jj} = \frac{4}{5}\frac{x_i y_j^2}{x_j^2} + \frac{2}{5}\frac{y_i y_j}{x_i}.$$

Indeed, the sought functions $\Sigma_{i,j}$ must satisfy the generalized *Bianchi identities*

$$\nabla^1(\Omega'_{i,j}) = -(\delta_i^1 + 4\delta_i^2)x_2, \qquad \nabla_1^1(\Omega'_{i,j}) = 2\Omega'_{i,j}, \qquad \nabla_2^1(\Omega'_{i,j}) = y_2,$$
$$\nabla^2(\Omega'_{i,j}) = -(\delta_j^1 + 4\delta_j^2)x_1, \qquad \nabla_2^2(\Omega'_{i,j}) = 2\Omega'_{i,j}, \qquad \nabla_1^2(\Omega'_{i,j}) = y_1,$$

which can be derived from the commutation relations of Lemma 1. We integrate these identities and determine the functions $\Sigma_{i,j}$ explicitly, using the fact that the differential operators ∇^i and ∇_j^i take the following form

$$\nabla^i = -5x_i\frac{\partial}{\partial y_i},$$
$$\nabla_1^1 = 2x_1\frac{\partial}{\partial x_1} + x_2\frac{\partial}{\partial x_2} + 3y_1\frac{\partial}{\partial y_1} + y_2\frac{\partial}{\partial y_2},$$
$$\nabla_2^2 = 2x_2\frac{\partial}{\partial x_2} + x_1\frac{\partial}{\partial x_1} + 3y_2\frac{\partial}{\partial y_2} + y_1\frac{\partial}{\partial y_1},$$
$$\nabla_j^i = x_j\frac{\partial}{\partial x_i} + (\Omega'_{1,2} + \Omega'_{2,1})\frac{\partial}{\partial y_i}$$

within the coordinate system (27). The sought functions $\Sigma_{i,jk}$ must satisfy the similar "Bianchi identities" which also can be derived from the commutation relations of Lemma 1. Therefore, these functions can be found explicitly too.

Using now the commutation relation

$$[\nabla_2, \nabla_1] = \Omega_1\nabla^1 + \Omega_2\nabla^2,$$

we derive two more "Bianchi identities"

$$\nabla_j(\Omega''_{i,ii}) - \nabla_i(\Omega''_{i,ij}) = (-1)^i 5x_i^2. \tag{30}$$

Substituting the functions $\Sigma_{i,j}$ and $\Sigma_{i,jk}$ into Eq. 30 and taking into account the fact that the differential operators ∇_1 and ∇_2 take the form

$$\nabla_1 = y_1\frac{\partial}{\partial x_1} + \Omega'_{2,1}\frac{\partial}{\partial x_2} + \Omega''_{1,11}\frac{\partial}{\partial y_1} + \Omega''_{2,21}\frac{\partial}{\partial y_2},$$
$$\nabla_2 = y_2\frac{\partial}{\partial x_2} + \Omega'_{1,2}\frac{\partial}{\partial x_1} + \Omega''_{2,22}\frac{\partial}{\partial y_2} + \Omega''_{1,12}\frac{\partial}{\partial y_1}$$

within the coordinate system (27), we see that the left-hand side of Eq. 30 equals zero. Hence the assumption $\operatorname{rank}(\mathcal{F}^2) < 5$ contradicts the condition $\Omega \neq 0$, and the lemma is proven.

Note that the equality in (28) is attained at the differential equation of quadrics (2). The symmetry group of Eq. 2 is isomorphic to $SL(2, \mathbf{R})$. It consists of linear unimodular transformations of the plane $\mathbf{R}^2_{(x,y)}$.

Set

$$\Omega_{i,i} = \Omega'_{i,i}, \quad \Omega_{i,j} = \Omega''_{i,j} - \Sigma_{i,j}, \quad \Omega_{i,jk} = \Omega''_{i,jk} - \Sigma_{i,jk},$$

where $\Omega'_{i,j}$, $\Omega''_{i,jk}$, $\Sigma_{i,j}$, $\Sigma_{i,jk}$ are defined in the proof of Lemma 3.

Theorem. *The following conditions are equivalent:*
(i) *Eq. 8 admits a three-dimensional group of symmetries,*
(ii) *Eq. 8 is locally equivalent to Eq. 2,*
(iii) *the projective curvature of Eq. 8 satisfies the relations*

$$\begin{aligned} \Omega_{1,2} + \Omega_{2,1} &= 0, \\ 3\Omega_{i,ii}\Omega_{1,2} + 5\Omega_i^3 &= 0, \\ \Omega_{i,ij}\Omega_i + \Omega_{j,i}\Omega_{i,i} - \Omega_{i,ii}\Omega_j &= 0, \\ \Omega_{i,jj}\Omega_j + 2\Omega_{j,i}\Omega_{j,j} - \Omega_{j,jj}\Omega_i &= 0. \end{aligned} \tag{31}$$

Remark. Relaton (31) is satisfied identically for an equation of the form

$$y'' = A(x,y). \tag{32}$$

This means that relation (31) is a necessary condition for a cubic differential equation to be locally equivalent to Eq. 32. Relation (31) is also a necessary and sufficient condition for a normal projective connection to coincide with the induced connection on a surface in three dimensional projective space[11].

Proof. Suppose Eq. 8 admits a three-dimensional group of symmetries. Then Lemmas 1–3 give $\text{rank}(\mathcal{F}^1) \geq 4$, $\text{rank}(\mathcal{F}^2) = 5$. Using the technique of Lemma 1, one can readily verify that the equality $\text{rank}(\mathcal{F}^1) = 4$ contradicts the condition (i). For example, in the case of Eq. 29 we get $\text{rank}(\mathcal{F}^1) = 4$, $\text{rank}(\mathcal{F}^2) = 5$, $\text{rank}(\mathcal{F}^3) = 6$. Therefore, the condition (i) is equivalent to the constraint

$$\text{rank}(\mathcal{F}^1) = \text{rank}(\mathcal{F}^2) = 5.$$

In other words, the covariant derivatives $\Omega_{1,2}$, $\Omega_{i,jk}$ are functions of x_1, x_2, y_1, y_2 and $z = \Omega_{2,1}$. Using the fact that these functions are subjected to the generalized Bianchi identities, we determine them explicitly and, thereby, obtain the condition (iii) which defines (Theorem 3) a unique differential equation up to local equivalence. This differential equation coincides with Eq. 2.

Acknowledgements

It is a pleasure to thank V.V. Lychagin for many valuable discussions on this research. Also, I thank S.V. Duzhin for having the time to read the manuscript, and for many helpful criticisms and corrections.

References

1. E. Cartan, *Leçons sur la Théorie des Espaces à Connexion Projective*, (Gauthier-Villars, Paris, 1937).
2. S. Kobayashi and T. Nagano, On Projective Connections, *J. Math. Mech.* **13** (1964) 215-235.
3. T. Morimoto, La géométrie des équations de Monge-Ampère, *C. R. Acad. Sc. Paris (Sér. A)* **289** (1979) 25-28.
4. S. Sternberg, *Lectures on Differential Geometry*, (Prentice-Hall, Englwood Cliffs, NJ, 1964).
5. S.V. Duzhin and V.V. Lychagin, Symmetries of Distributions and Quadrature of Ordinary Differential Equations, *Acta Appl. Math.* **24** (1991) 29-57.
6. A.G. Khovansky, On Rectification of Circles, *Siberian Mathematical Journal* **XXI** (1980) 221-226 [Russian].
7. E. Cartan, Sur les Variétés à Connexion Projective, *Bull. Soc. Math. France* **52** (1924) 205-241.
8. S. Kobayashi, Canonical Forms on Frame Bundles of Higher Order Contact, *Proc. Symp. Pure Math.* **III** (1961) 186-193.
9. C. Grissom, G. Thompson, and G. Wilkens, Linearization of Second Order Ordinafy Differential Equations via Cartan's Equivalence Method, *J. Differential Equations* **77** (1989) 1-15.
10. L. Hsu and N. Kamran, Classification of Second Order Ordinary Differential Equations Admitting Lie Groups of Fibre-Preserving Point Symmetries, *Proc. London Math. Soc.* **58** (1989) 387-416.
11. V. Prokofiev, Embedding of Two-Dimensional Spaces of Normal Projective Connection in Three-Dimensional Projective Space, *Dokl. Acad. Nauk SSSR* **XXXVI** (1942) 95-97 [Russian].

Geometry in Partial Differential Equations (pp. 345–375)
eds. A. Pràstaro and Th. M. Rassias

SMOOTH MARGINAL ANALYSIS OF BIFURCATION OF EXTREMALS

YURI I. SAPRONOV
Department of Mathematics Voronezh State University 394693, Voronezh, Russia

It is well known that many nonlinear problems of mathematical physics admit a variational statement

$$V(x) \to \inf, \quad x \in X .$$

The solution of this problem is often accompanied by the construction and analysis of a marginal function (the term is borrowed from [1])

$$W(y) = \inf_{p(x)=y} V(x), \quad y \in Y \tag{0.1}$$

and a marginal mapping φ: $Y \to X$

$$\varphi(y) = \{x \in X \,|\, p(x) = y, \quad V(x) = W(y)\} \tag{0.2}$$

where p is some (fibering) mapping from X and Y.

In the present paper the scheme of a marginal analysis is applied under the following additional requirements: 1) X, Y are smooth Banach manifolds, 2) a marginal mapping is single-valued, 3) functional V and mappings p, φ are smooth. These requirements make it possible to reduce the problem under consideration to the problem

$$W(y) \to \inf, \quad y \in Y .$$

Functions of the form (0.1) naturally appear in problems of mechanics, geometry and optimal control. The situations in which Y is a finite-dimensional manifold are of particular interest. The local method of Lyapunov-Schmidt, modified for variational problems [2–6], gives an example of such a situation. Another example is Morse theory, in which the functional of action is studied via its restriction on a finite-dimensional manifold of piece-wise geodesic curves [7,8]. The theory of local reductions is more advanced than that of the global ones. Papers [9,10], in which W and φ are constructed on the basis of the contracting mapping principle, can serve as an example of the use of nonlocal reductions. These papers consider the problems, in which $p^{-1}(y)$ is an affine space and this fact is essentially made use of.

Finite-dimensional reductions produce an important effect: the problem with an infinite-dimensional number of degrees of freedom is reduced to the problem with a finite and *a priori* bounded number of degrees of freedom, whereas in generally accepted approximations of the Rietz approximation type, the number of degrees of freedom grows with the growth of the order of accuracy.

The method of accurate finite-dimensional reductions explains the phenomenal success of the analysis of supercritical states of elastic hulls via approximated reductions on the basis of Rietz approximations with respect to several fixed modes of deflections (see, for example, [11]).

The method of finite-dimensional reductions can also be considered as the development of the method of elastic parameters [12–16] in the mechanics of elastic constructions, according to which an elastic construction is attached to the finite set of key geometric and mechanical parameters, adequately representing the qualitative state of an elastic system.

For illustration let us consider the simplest example: a rectilinear Euler rod in the state of a supercritical equilibrium. The elasticity of a lengthwise compressed rod, rigidly fixed at the ends, is described by a boundary value problem [17]

$$-\ddot{x} - \lambda \sin x = 0 \tag{0.3}$$

$$x(0) = x(1) = 0 \ . \tag{0.4}$$

Let X be a space of functions of class C^2 on [0,1], satisfying condition (0.4). Then (0.3) is an Euler-Lagrange equation of extremals of the functional

$$V(x, \lambda) = \int_0^1 \left(\frac{\dot{x}^2}{2} + \lambda(\cos x - 1) \right) ds \tag{0.5}$$

in this space. Functional (0.5) represents a full energy of the deformed rod.

The supercritical elasticity of the rod can be studied with the help of the following marginal functions

$$W_j(\xi,\lambda) = \inf_{p_j(x)=\xi} V(x,\lambda), \quad \xi \in \mathbb{R}, \; j = 1,\dots,6 \tag{0.6}$$

where

$$p_1(x) = \int_0^1 x(s)\sin \pi s\, ds, \quad p_2(x) = \int_0^1 \sin x(s) ds \;,$$

$$p_3(x) = \left(\int_0^1 x^2(s) ds\right)^{\frac{1}{2}}, \quad p_4(x) = \left(\int_0^1 (1-\cos x(s)) ds\right)^{\frac{1}{2}},$$

$$p_5(x) = \left(\int_0^1 \frac{\dot{x}^2(s)}{2} ds\right)^{\frac{1}{2}}, \quad p_6(x) = x\left(\frac{1}{2}\right).$$

This list could be continued. The value of each functional p_j is a natural parameter, characterizing the rod's deflection. For $\lambda < 4\pi^2$ each function (0.6) is smooth. For sufficiently small ξ and $\lambda - \pi^2$ all these functions by a local change of coordinates are reduced to the form

$$\text{const} + \alpha(\lambda)\xi^2 + \xi^4$$

where $\alpha(\lambda) = 0(\lambda - \pi^2)$, $\alpha'(\pi^2) < 0$. This statement is substantiated in the results given below.

The author presented the contents of the present paper at the conference dedicated to M. A. Krasnosel'skii's 70th birthday (Voronezh, 1990) and at Moscow P. S. Alexandrov topological conference (1990).

The author is grateful to V. I. Arnol'd, Yu. G. Borisovich, Yu. E. Gliklikh and A. T. Fomenko for helpful discussions and comments.

1. Condition of Geodesic Convexity

Many principles of the existence and construction of extremals in nonlinear analysis are based on the property of convexity of a functional [18–22]. Let us give some sufficiently typical statements.

In what follows the term "smooth" denotes the membership of class C^∞.

Let $f\colon E \to F$ be a smooth Fredholm mapping of Banach spaces with a zero index [23]. Let f be proper (pre-image of a compact is compact) and potential

$$\langle f(x), h\rangle = \frac{\partial V}{\partial x}(x)h \tag{1.1}$$

where V is some smooth functional on E (a potential of mapping f), $\langle\cdot,\cdot\rangle$ is a scalar product in some Hilbert space H, in which F is continuously embedded with the conditiion of density of E in H. It is assumed that E is continuously embedded in F.

If the condition of positivity

$$\left\langle \frac{\partial f}{\partial x}(x)h, h \right\rangle > 0, \quad x \in E, \ h \in E\backslash 0 \ , \tag{1.2}$$

is fulfilled, then in E there is a point a which is a unique solution to equation

$$f(x) = 0 \tag{1.3}$$

and a point of a global minimum V on E.

If U is a star domain centred at point a (any point of this domain is connected with a by a segment of a straight line in U) and a is a solution to Eq. (1.3), then under the fulfillment of condition (1.2) at point a functional V takes the value of the global minimum.

If f_t: $E \to F$ is a smooth family of smooth potential Fredholm mappings of a zero index with potential $V_t, t \in [0,1]$, and U is a bounded closed domain in E, on which (1.2) is fulfilled for all t, $a \in \operatorname{int} U$ is a unique solution to equation $f_0(x) = 0$, $f_t(x) \neq 0$, $t \in [0,1]$, $x \in \partial U$, then in the case of properness of the mapping $U \times [0,1] \to F$, $(x,t) \to f_t(x)$, there is a point a from which goes out a smooth curve $x = x(t)$, $t \in [0,1]$, $x(0) = a$, for which $f_t(x(t)) = 0$, $t \in [0,1]$, and $x(t)$ is a point of a global minimum V_t on U.

Analogous statements can be obtained on a smooth submanifold, using Lagrange multipliers [21,24]. However, there is also another approach based on the geodesic convexity of a functional.

Let μ be a smooth submanifold of E. Let us call μ complemented if the following two conditions are fulfilled: 1) for any $a \in \mu$ and every $l \in E^*$, for which $l/_{T_a(\mu)} = 0$, there is $b \in E$, for which

$$\langle b, x\rangle = l(x) \quad \forall\, x \in E$$

(the set of such vectors b at point a is denoted by $N_a(\mu)$); 2) $N_a(\mu)$ smoothly depends on $a \in \mu$ with respect to a metric of space E (the set $\bigcup_{a\in\mu} \{a, N_a(\mu)\}$ is called a smooth subbundle of a trivial bundle $\mu \times E$ over μ).

These conditions are satisfied if μ is a finite-defect submanifold in E given by the equations

$$u_1(x) = \cdots = u_m(x) = 0 \tag{1.4}$$

where $u_j(x)$ are smooth potentials (in the sense of (1.1)) of some smooth mappings g_j: $E \to E$.

For a complemented submanifold μ a smooth family of projectors $\Pi(x)$: $E \to T_x(\mu)$ orthogonal in metric $\langle\cdot,\cdot\rangle$ is defined. For a finite-defect μ the family $\Pi(x)$ is extended to the smooth family of the projectors $\hat{\Pi}(x)$: $F \to \hat{T}_x(\mu)$, where $\hat{T}_x(\mu) = \{y \in F | \langle y, u\rangle = 0 \ \forall\ u \in N_x(\mu)\}$ is a direct complement to $N_x(\mu)$ in F. The set $T_x(\mu)$ smoothly depends on $x \in \mu$ (it forms a subbundle of a trivial bundle $\mu \times F \to \mu$).

If μ is a subset of a unit defect ($m = 1$ in (1.4)), then

$$\hat{\Pi}(x)y = y - |g_1(x)|^{-2}\langle y, g_1(x)\rangle g_1(x) \ . \tag{1.5}$$

The field $X(x) = \hat{\Pi}(x)f(x)$ is called a codifferential of V on μ (at point x). Operator

$$\nabla^2(V)(a) := \hat{\Pi}(a)\frac{\partial}{\partial x}(X)(a) : T_a(\mu) \to \hat{T}_a(\mu) \tag{1.6}$$

is called the second codifferential of functional V on μ at point a. The quadratic form

$$\langle\nabla^2(V)(a)h, h\rangle = \left\langle \frac{\partial X}{\partial x}(a)h, h\right\rangle, \quad h \in T_a(\mu) \tag{1.7}$$

is naturally connected with (1.6).

Functional V is called geodesically convex on the domain $O \in \mu$, if for any point $a \in O$ the form (1.7) is positive

$$\left\langle \frac{\partial X}{\partial x}(a)h, h\right\rangle > 0 \quad \forall\ h \neq 0 \ .$$

The curve $x(t)$ on μ, $x(\alpha) = a$, $x(\beta) = b$, is called a geodesic, connecting points a and b on μ, if this curve belongs to class C^2 and the following equality holds

$$\Pi(x(t))(\ddot{x}(t)) = 0 \ . \tag{1.8}$$

The domain $O \subset \mu$ is called a geodesic star domain centred at point $a \in O$, if a is connected with any other point of this domain by a unique geodesic. If O is a star domain centred at any of its points, then O is called a geodesically convex domain (see [18]).

Under a regular parametrization $x = x(u)$, $u \in M$ the curve $u(t)$ can become a geodesic curve $x(t) = x(u(t))$ iff the classical equation of a geodesic [25]

$$\ddot{u} + \Gamma_u(\dot{u}, \dot{u}) = 0 \tag{1.9}$$

is fulfilled, where $\Gamma_u \colon M \times M \to M$ is a smooth, with respect to u, family of symmetrical bilinear mappings (connection) obtained by the composition of a bilinear mapping

$$\Pi(x)\frac{\partial^2 X}{\partial u^2} : \ M \times M \to T_x(\mu)$$

and a linear mapping inverse to the mapping

$$\frac{\partial X}{\partial u} : \ M \to T_x(\mu), \quad x = x(u) \ .$$

The complemented manifold is a local geodesic star manifold, i.e. each point of this manifold has a geodesic star neighborhood centred at this point. The proof follows from the theorem of solvability of the Cauchy problem for a differential equation in a Banach space [25].

Theorem 1.1. *Let O be a geodesic star domain in μ with point $\bar{a}$ as centre. Let $X(\bar{a}) = 0$ and let the condition of positivity of the form (1.7) $\forall\ a \in O$ be fulfilled. Then $\bar{a}$ is a point of a global minimum of V on O.*

Proof. Let $x(t)$ be a geodesic connecting $\bar{a}$ with an arbitrary point a: $x(\alpha) = \bar{a}$, $x(\beta) = a$. Since $\frac{d}{dt}V(x(t)) = \langle X(x(t)), \dot{x}(t)\rangle$ and $\frac{d}{dt^2}V(x(t)) = \langle \frac{\partial X}{\partial x}(x(t))\dot{x}(t), \dot{x}(t)\rangle + \langle X(x(t)), \ddot{x}(t)\rangle$ then, with due account of the equality $\langle X(x(t)), \ddot{x}(t)\rangle = 0$ (see (1.8)) we have $\frac{d^2}{dt^2}V(x(t)) > 0$ and also $\frac{d}{dt}V(x(t)) > 0$, $t > \alpha$, or $V(x(\alpha)) > V(x(\beta))$.

Definition. Potential V is called Fredholm on μ if its second codifferential (1.6) is Fredholm at every point $a \in \mu$. Index (Fredholm) of $\nabla^2(V)$ is called index of V.

Theorem 1.2. *Let imbedding $E \subset F$ be compact and let a Fredholm mapping $f: E \to F$ have potential V. Let μ be a finite-defect submanifold given by relations (1.4). Then operator (1.6) is a Fredholm one and its iindex coincides with index f.*

Proof follows from the following remark: If A: $E \to F$ is a Fredholm operator, then operator PA: $E_0 \to F_0$ is also a Fredholm one, where $E_0 \subset E$, $F_0 \subset F$, codim $E_0 < \infty$, codim $F_0 < \infty$, P is a projector F on F_0. In the case codim $F_0 =$ codim E_0 we have Ind $A =$ Ind $P(A|_{E_0})$. Compact perturbations of Fredholm operators preserve a Fredholm property and index [18,23].

Theorem 1.3. *Let O be a closed geodesic convex domain in μ, V_t be a smooth family of smooth geodesic convex Fredholm potentials of a zero index on O. Let the family of codifferentials X_t give a proper mapping $\mu \times [0,1] \to F$ and let $X_0(a) = 0$, $a \in O$, $X_t(x) \neq 0$ $\forall$ $x \in \partial O$. Then in O there is a smooth curve $x = x(t)$; $x(0) = a$, $t \in [0,1]$ such that $x(t)$ is a point of a global minimum for V_t on O for any $t \in [0,1]$.*

Proof. From the condition of properness it follows that the set S of solutions to the equation

$$X_t(x) = 0, \quad x \in O, \quad t \in [0,1]$$

is compact in $O \times [0,1]$. From the geodesic convexity it follows that the intersection of S and the hyperplane $t =$ const includes not more than a single-point set. The Fredholm condition with a zero index, the condition of convexity and the relation $S \cap (\partial O \times [0,1]) = \emptyset$ on the basis of the theorem of an implicit function lead to the openness of the image of the mapping $S \to [0,1]$, $(x,t) \to t$. From the compactness of S the closure of this image follows. Consequently, the projection of S coincides with [0,1] and the equation $X_t(x) = 0$ has a solution for every $t \in [0,1]$.

Remark. The possibility of applying the above scheme may appear to be limited due to the requirement of complementation of a submanifold μ. In this respect it should be noted that many submanifolds become complemented after the replacement of H by H^1 with a stronger metric.

2. Regular Finite-Dimensional Reductions

Let the extremals of a smooth family of smooth functionals $V(\cdot,\lambda)$ be defined by the equation $f(x,\lambda) = 0$, where $f(\cdot,\lambda)$: $E \to F$ is a smooth family of smooth Fredholm mappings, $f(\cdot,\lambda) = \text{grad}_H V(\cdot,\lambda)$ and E is compactly imbedded in F, Ind $\frac{\partial f}{\partial x}(x,\lambda) = 0$. Let in E a smooth family of smooth submanifolds μ_ξ be given by the equations

$$u_1(x) = \xi_1, \dots, u_n(x) = \xi_n;\ u_j(0) = 0\ \ \forall\, j$$

under the condition that $\forall\, k$ the gradient $g_k = \text{grad}_H u_k$ is defined and the field of these gradients gives a smooth mapping $E \to E$. Assume also that the condition of stationary state

$$f(0,0) \in \text{Span}\{g_1(0), \dots, g_n(0)\}$$

holds and let the second codifferential V be positive on μ_ξ for sufficiently small ξ and in a sufficiently small neighborhood of zero O in E. In the subsequent sections the neighborhood of zero in $\mathbb{R}^n$ is denoted by O^n and the neighborhood of zero in some subspace of codimension n in E is denoted by $O^{\infty-n}$.

Theorem 2.1. *For sufficiently small x, λ, ξ, when the above conditions are fulfilled, the function*

$$W(\xi,\lambda) = \inf_{x\in\mu_\xi} V(x,\lambda) \tag{2.1}$$

is smooth. Moreover, there exists a smooth family of the C^∞-diffeomorphisms

$$\varphi(\cdot,\cdot,\lambda):\ O^n \times O^{\infty-n} \to O,\ \ \varphi(0,0,0) = 0\ ,$$

such that $W(\xi,\lambda) = V(\varphi(\xi,0,\lambda),\lambda)$, $u_j(\varphi(\xi,\nu,\lambda)) = \xi_j$, $\xi = (\xi_1,\dots,\xi_n) \in O^n, \nu \in O^{\infty-n}$.

Proof. Let O be an open set in E whose intersection with μ_ξ for sufficiently small ξ gives a star neighborhood centred at point $a(\xi,\lambda)$, $a(0,0) = 0$, where $a(\xi,\lambda)$ is a solution to the equation

$$X(x,\lambda) = 0,\ \ x \in \mu_\xi\ . \tag{2.2}$$

Here $X(x,\lambda)$ is the first codifferential of V on μ_ξ. From the compactness of the imbedding $E \subset F$ it follows that the second codifferential at every point $a \in \mu_\xi$ is a Fredholm one of zero index. From its positivity it follows that $\nabla^2(V)(0,0)$: $T_0(\mu_0) \to \hat{T}_0(\mu_0)$ is an isomorphism. By the theorem of an implicit function in O there exists a unique solution $a(\xi, \lambda)$ to Eq. (2.2) (O is sufficiently small), which is at the same time a point of a local minimum of V on μ_ξ (Theorem 1.3). It is easy to establish that $a(\xi, \lambda) = \varphi(\xi, 0, \lambda)$ where $\varphi(\cdot, \cdot, \lambda)$ is a mapping $O^n \times O^{\infty-n} \to O$, inverse to the mapping ψ: $O \to O^n \times O^{\infty-n}$, $\psi(x) = (u_1(x), \dots, u_n(x), A(x)X(x, \lambda))$. Here $A(x)$ is an arbitrary smooth family of reversible operators $\hat{T}_x(\mu_\xi) \to T_0(\mu_0)$. It is easy to verify the reversibility of Frechet derivative of the mapping ψ in zero.

Remark. The compactness of the imbedding $E \subset F$ was needed for establishing a Fredholm property of the second codifferential. This condition can be replaced by any other guaranteeing the Fredholm property of the second codifferential in the neighborhood of zero. For example, the following equality can be required

$$f(0,0) = 0 \ .$$

The compactness of the imbedding $E \subset F$ makes it possible to establish also the properness of the mapping [23]. The property of properness is at the basis of many principles of solvability of equations, in particular, the principles of the existence of non-local reductions [54].

The following statement is the consequence of Theorem 1.3.

Theorem 2.2. *Let the conditions described before the statement of Theorem 2.1 be fulfilled and let the mapping*

$$X(\cdot, \lambda) : O \to F$$

where $X(x, \lambda) = \hat{\Pi}(x) f(x, \lambda)$ *is the first codifferential and* $\hat{\Pi}(x)$ *is a projector on* $\hat{T}_x(\mu_\xi)$, *be proper and have no zeros on* ∂O *for* $|\xi| \leq C$. *Then the function*

$$W(\xi, \lambda) = \inf_{x \in \mu_\xi \cap O} V(x, \lambda), \quad |\xi| \leq C$$

is well-defined and it is smooth.

The finite-dimensional reduction allows one to reduce the problem of existence of an extremal to the corresponding statement on a finite-dimensional manifold. For a more detailed investigation of extremals a more detailed information about the structure of the function (2.1) is required. It is possible to construct Taylor approximations of the function

W and a marginal mapping. This allows one to make a qualitative analysis of the behaviour of the function W in the small and, consequently, the behaviour of functional V and also to calculate asymptotic approximations to bifurcating extremals (with respect to the supercritical increase of parameter λ).

Theorem 2.3. *The conclusion of Theorem 2.1 is preserved if the condition of compactness of the imbedding $E \subset F$ is replaced by the condition $f(0,0) = 0$. In this case for sufficiently small x, ξ and λ the function (2.1) can be represented in the form*

$$W(\xi,\lambda) = V(0,\lambda) + \tilde{W}(\xi,\lambda) + o(|\xi|^2 + |\lambda|^2)$$

where

$$\tilde{W}(\xi,\lambda) = \inf_{\substack{\langle l_j, x\rangle = \xi_j \\ j=1,\dots,n}} \left\{\frac{1}{2}\langle Ax, x\rangle + \langle b, x\rangle\right\}$$

$$A = \frac{\partial f}{\partial x}(0,0), \quad b = f(0,\lambda), \quad l_j = \operatorname{grad}_H u_j(0) \ .$$

Proof. Let $x = x(\xi,\lambda)$ be a function, for which $x(0,0) = 0$, $X(x(\xi,\lambda), \lambda) \equiv 0$, $u(x(\xi,\lambda)) \equiv \xi$. Here $X(x,\lambda) = \hat{\Pi}(x) f(x,\lambda)$. From the condition of positivity of the second codifferential it follows that $X(x,0)$ has in zero a non-degenerate singularity.

Searching for the main linear part for X, we obtain

$$x(\xi,\lambda) = \sum_{j=1}^{m} d_j\lambda_j + \sum_{k=1}^{n} c_k\xi_k + o(|\lambda| + |\xi|) \tag{2.3}$$

where

$$d_j \in T_0(\mu_0), \quad \hat{\Pi}(0)\frac{\partial f}{\partial x}(0,0)d_j + \hat{\Pi}(0)\frac{\partial f}{\partial \lambda_j}(0,0) = 0 \ ,$$

$$\hat{\Pi}(0)\frac{\partial f}{\partial x}(0,0)c_k = 0, \quad \langle c_k, l_j\rangle = \delta_{k,j} \ . \tag{2.4}$$

In accordance with the notations introduced in the statement of the theorem, we can write

$$V(x,\lambda) = V(0,\lambda) + \langle b(\lambda), x\rangle + \frac{1}{2}\langle Ax, x\rangle + \cdots \tag{2.5}$$

Since $W(\xi,\lambda) = V(x(\xi,\lambda),\lambda)$, then Taylor summands of the function $W(\xi,\lambda)$ up to the second order are determined by the linear summands of the expansion (2.3). But these summands are identically defined by the relations (2.4), which in their turn are completely defined by the quadratic part of the expansion (2.5). Since $T_0(\mu_0)$ is given by the relations $\langle l_j, x\rangle = 0$, then from (2.4) we obtain the variational problem

$$\frac{1}{2}\langle Ax, x\rangle + \langle b(\lambda), x\rangle \to \inf$$
$$\langle l_j, x\rangle = \xi_j, \quad j = 1, \dots, n$$

defining the linear part of (2.3).

Theorem 2.3 is useful in the cases where the behaviour of extremals is quite accurately defined by a quadratic part of W (see [26–28] for such problems). If it is necessary to apply Taylor summands of a higher order, one can turn to general algorithms of the theory of branching of solutions of nonlinear equations [29,30]. These algorithms in many cases are realized only with the help of a computer (see, for example, [31–33]). In some cases, connected with the presence of symmetry and other circumstances simplifying calculations, the latter can be done manually [34–36].

The theorem given below substantiates the approximated methods for calculating supercritical deflections of elastic hulls, based on Rietz approximations with respect to the fixed set of modes of deflection.

Assume that $f(0,\lambda) = 0$ and on some neighbourhood $\tilde{O}$ of zero in $\mathbb{R}^m$ the set of functions (the leading modes of bifurcation) $e_1(\lambda), \dots, e_n(\lambda)$ which is smooth and normed in H is given, for which

$$\frac{\partial f}{\partial x}(0,\lambda)e_j(\lambda) = \alpha_j(\lambda)e_j(\lambda) \quad \forall j \,. \tag{2.6}$$

And let

$$\left\langle \frac{\partial f}{\partial x}(x,\lambda)\delta, \delta \right\rangle > 0 \quad \forall \nu \perp e_j(\lambda), \quad \nu \neq 0\,; \tag{2.7}$$
$$j = 1, \dots, n; \quad \lambda \in \tilde{O}, \quad x \in O$$

(here O is a neighborhood of zero in E). Then for a sufficiently small neighbourhood O the function

$$W(\xi,\lambda) = \inf_{\nu} V\left(\sum \xi_j e_j(\lambda) + \nu, \lambda\right), \quad \nu \in O, \quad \nu \perp e_j(\lambda)\,,$$

is well-defined and it is sufficiently smooth (for sufficiently small ξ and λ). Moreover, there exists a smooth function $\nu = \phi(\xi, \lambda)$ with values in an orthogonal complement (with respect to metrics H) to $L = \mathrm{Span}(\{e_j(\lambda)\})$ such that $\phi(0, \lambda) = 0$ and $\phi(x, 0) = o(\|x\|_E^2)$ and

$$W(\xi, \lambda) = V(\sum \xi_j e_j(\lambda) + \phi(\xi, \lambda), \lambda) .$$

Thus, here a reducing diffeomorphism is given in the form

$$\varphi(\xi, u, \lambda) = \sum \xi_j e_j(\lambda) + \phi(\xi, \lambda) + u .$$

In the case of the invariance of $V(\cdot, \lambda)$ with respect to an orthogonal (in H) group action $G \times H \to H$, mapping E, F, L into itself, the function $W(\cdot, \lambda)$ is invariant with respect to action G, obtained by the restriction on L of the given action [36]. In particular, if on H the set of such involutions $\{J_k\}$ that $J_k(E) \subset E$, $J_k(F) \subset F$, $k = 1, \ldots, n$ and

$$V(J_k x, \lambda) = V(x, \lambda), J_k e_k(\lambda) = -e_k(\lambda), \quad J_k e_r(\lambda) = e_r(\lambda) , \quad r \neq k \tag{2.8}$$

then

$$W(\xi, \lambda) = \widehat{W}(\eta, \lambda), \quad \eta = (\xi_1^2, \ldots, \xi_n^2)$$

where $\widehat{W}$ is some smooth function.

Theorem 2.4. *Let (2.6)–(2.8) be fulfilled and let*

$$V(x, \lambda) = V_\lambda^{(0)} + V_\lambda^{(2)}(x) + V_\lambda^{(3)}(x) + V_\lambda^{(4)}(x) + o(\|x\|_E^4) ,$$

$V_\lambda^{(k)}(x) = \mathcal{V}_\lambda^{(k)}(x, \ldots, x)$, *where* $\mathcal{V}_\lambda^{(k)}$ *is a symmetric form of order* k. *Let*

$$\mathrm{grad}_H V_0^{(3)}(u) = 0 \quad \forall\, u \in L .$$

Then $W(\xi, \lambda) = \widehat{W}(\eta, \lambda)$, $\eta = (\xi_1^2, \ldots, \xi_r^2)$, *where*

$$\widehat{W}(\eta, \lambda) = V_\lambda^{(0)} + \frac{1}{2} \sum \alpha_j(\lambda)\eta_j + (B\eta, \eta) + O(\lambda)o(\xi) + o(|\eta|^2) ,$$
$$B = (a_{i,j}), \quad a_{i,i} = V_0^{(4)}(e_i(0)) ,$$
$$a_{i,j} = 3\mathcal{V}_0^{(4)}(e_i(0), e_i(0), e_j(0), e_j(0)) .$$

The proof follows from an easily verifiable general relation

$$W_0^{(4)}(\xi) = V_0^{(4)}(u) - \frac{1}{2}\langle AC(u), C(u)\rangle \tag{2.9}$$

where

$$u = \sum_{j=1}^{n} \xi_j e_j(0), \quad A = \frac{\partial f}{\partial x}(0,0), \quad AC = \mathrm{grad}_H V_0^{(3)} .$$

3. Equivalence of Regular Reductions

The following question is natural: are marginal functions in the same boundary value problem, obtained by different methods, equivalent? For regular reductions the answer is positive if some natural conditions are fulfilled.

The system of smooth functionals $u_1(x), \dots, u_n(x)$ (without a parameter) on E is called reducing for a smooth functional V at point $a \in E$ if, firstly, the system of differentials $\frac{\partial u_j}{\partial x}$ is linearly independent and, secondly, there exists such smooth mapping φ: $O^n \to E$, $\varphi(0) = a$, where O^n is a neigbourhood of zero in $\mathbb{R}^n$, that $u_j(\varphi(\xi)) = \xi_j \ \forall\, j$ and $\varphi(\xi)$ is a unique point of minimum for V on $\mu_\xi = \cap_{j=1}^n u_j^{-1}(\xi)$; this point is non-degenerate conditional extremal (the second codifferential V on μ_ξ is positive). The function $W(\xi) = V(\varphi(\xi))$ is called a key function for V in the system $u_1, \dots, u_n$.

The point $\bar{x}$ from a sufficiently small neighbourhood of point a is critical for V iff $\bar{x} = \varphi(\bar{\xi})$, where $\bar{\xi}$ is some critical point for W. The non-degeneracy of $\bar{x}$ for V is equivalent to the non-degeneracy of $\bar{\xi}$ for W.

Let for V on E two reducing systems $u' = \{u_1', \dots, u_n'\}$ and $u'' = \{u_1'', \dots, u_m''\}$ be given under the condition that u'' is a part of u'. Then we say that the first system is an expansion of the second one (or the second system is a contraction of the first one).

Theorem 3.1. *If a pair of key functions W and $\widehat{W}$ of functional V on E is constructed at point $a \in E$ on the basis of the reducing systems u and $\hat{u}$, having a common expansion, then W and $\widehat{W}$ are positively stably smoothly equivalent (in zeros), i.e. the germs of these functions become smoothly equivalent in zeros after the positively defined quadratic forms of additional variables are added to them.*

Proof. It is sufficient to consider the case when one system is an expansion of the other. Let

$$u = \{u_1, \dots, u_n\}, \quad \hat{u} = \{\hat{u}_1, \dots, \hat{u}_{n+p}\} ,$$
$$u_j = \hat{u}_j , \quad j = 1, \dots, n ,$$

and let $\varphi(\xi)$ and $\hat{\varphi}(\hat{\xi})$ be critical points in fibres μ_ξ and $\hat{\mu}_{\hat{\xi}}$,

$$\xi = (\xi_1, \dots, \xi_n) \in O^n, \quad \hat{\xi} = (\hat{\xi}_1, \dots, \hat{\xi}_{n+p}) \in O^{n+p} .$$

Since $u_j(\varphi(\xi)) = \xi_j$, $\hat{u}_k(\hat{\varphi}(\hat{\xi})) = \hat{\xi}_k$, $\hat{\mu}_{\hat{\xi}} \subset \mu_\xi$, then the mapping φ: $O^n \to \mathbb{R}^{n+p}$ obtained by the composition of the mappings $\xi \to \varphi(\xi)$ and $x \to (\hat{u}_1(x), \dots, \hat{u}_{n+p}(x))$ is represented in the form

$$\varphi(\xi) = (\xi_1, \dots, \xi_n, \quad \hat{u}_{n+1}(\varphi(\xi)), \dots, \hat{u}_{n+p}(\varphi(\xi)))$$

and, by construction, $\varphi(\xi) = \hat{\varphi}(\varphi(\xi))$. Consequently,

$$W(\xi) = V(\varphi(\xi)) = V(\hat{\varphi}(\psi(\xi))) = \widehat{W}(\psi(\xi)) . \tag{3.1}$$

Since $x = \varphi(\xi) = \hat{\varphi}(\psi(\xi))$ is a critical point for V/μ_ξ and $V/\hat{\mu}_{\varphi(\xi)}$ simultaneously, then

$$\left(\frac{\partial \widehat{W}}{\partial \hat{\xi}_{n+j}}\right)(\varphi(\xi)) = \left(\frac{\partial V}{\partial x}(\hat{\varphi}(\hat{\xi}))\frac{\partial \hat{\varphi}}{\partial \hat{\xi}_{n+j}}\right)(\varphi(\xi)) = 0, \quad j = 1, \dots, p .$$

Consequently, $\varphi(\xi)$ is a critical point for the restriction $\widehat{W}$ on fibre $N_\xi = \{\xi \in \hat{O}^{n+p}\, \hat{\xi}_j = \xi_j, j \le p\}$. From $u_j(\hat{\varphi}(\hat{\xi})) = \xi_j$ we obtain $\frac{\partial u_j}{\partial x} \cdot \frac{\partial \hat{\varphi}}{\partial \xi_k} = \delta_{j,k}$. Consequently, each tangent vector u to fibre N_ξ is represented in the form

$$u = \sum_{j=1}^{p} u_j \frac{\partial \hat{\varphi}}{\partial \hat{\xi}_{n+j}}(\hat{\xi}) . \tag{3.2}$$

If $h = u + \nu$, where u is presented in the form (3.2) and ν is an arbitrary tangent vector to $\hat{\mu}_{\hat{\xi}}$ at point $\varphi(\xi)$, then

$$\begin{aligned}\frac{\partial^2 V}{\partial x^2}(\varphi(\xi))(h,h) &= \frac{\partial^2 V}{\partial x^2}(\varphi(\xi))(u,u) + 2\frac{\partial^2 V}{\partial x^2}(\varphi(\xi))(u,\nu) \\ &\quad + \frac{\partial^2 V}{\partial x^2}(\varphi(\xi))(\nu,\nu) \\ &= \sum \frac{\partial^2 \widehat{W}}{\partial \xi_{n+j}\partial \xi_{n+i}}(\psi(\xi))u_j u_i + \frac{\partial^2 V}{\partial x^2}(\varphi(\xi))(\nu,\nu)\end{aligned} \tag{3.3}$$

(here an obvious equality $\frac{\partial^2 V}{\partial x^2}(\varphi(\xi))(u,\nu) = 0$ is used). Since, by definition, the forms $\frac{\partial^2 V}{\partial x^2}(\varphi(\xi))(h,h)$ and $\frac{\partial^2 V}{\partial x^2}(\varphi(\xi))(\nu,\nu)$ are positively defined, then from representation (3.3) also follows a positive definiteness of the form $\sum \frac{\partial^2 \widehat{W}}{\partial \xi_{n+j} \partial \xi_{n+i}}(\varphi(\xi)) u_j u_i$. Consequently (see (3.1)) $W(\xi)$ is obtained by the reduction from $\widehat{W}(\hat{\xi})$ with the help of the reducing system of the functionals $\hat{\xi} \to \xi_j$, $1 \le j \le n$. The statement of the theorem follows now from the lemma of splitting [37].

4. Intermediate Reduction

In some cases a marginal function is conveniently calculated via an intermediate reduction with respect to the function of the greater number of variables.

Let for the equation $f(x) = 0$ with potential $V(x)$ a set of reducing functionals be given and let

$$W(\xi) = \inf_{p(x)=\xi} V(x), \quad x \in o, \quad \xi \in \mathbb{R}^n$$

be the corresponding smooth marginal function, for which it is required to calculate the initial segment of Taylor series with respect to ξ:

$$W(\xi) = \sum_{|k| \le r} W^{(k)}(\xi) + o(|\xi|^r) \, .$$

Given the function $\widehat{W}(\hat{\xi})$ (or its segment of Taylor series), obtained by the reduction with respect to the expanded system of functionals $\{\hat{p}_j(x)\}$, $\hat{\xi} \in \mathbb{R}^{n+m}$

$$\widehat{W}(\hat{\xi}) = \inf_{\hat{p}(x)=\hat{\xi}} V(x), \quad x \in O, \quad \hat{\xi} \in \mathbb{R}^{n+m} \, .$$

Let $\hat{\xi} = (\xi, \eta)$, $\xi \in \mathbb{R}^n$, $\eta \in \mathbb{R}^m$. It is easy to see that

$$W(\xi) = \inf_{\eta \in U} \widehat{W}(\xi, \eta), \quad \xi \in \mathbb{R}^n \, , \tag{4.1}$$

where U is some neighbourhood of zero in $\mathbb{R}^m$. Taylor coefficients W can be calculated starting from $\widehat{W}$.

Theorem 4.1. *Let a smooth function W be obtained on the basis of the relation (4.1), where $\widehat{W}(\xi, \eta)$ is a semi-quasihomogeneous [37] function with* *weights* $\hat{\nu} = (\nu_1, \ldots, \nu_n, \frac{1}{2}, \ldots, \frac{1}{2})$. *Then* W *is a semi-quasihomogeneous function with weights* $\nu = (\nu_1, \ldots, \nu_n)$.

Proof. From (4.1) it follows that $W(\xi) = \widehat{W}(\xi, \hat{\phi}(\xi))$ where $\hat{\phi}(\xi)$ is an implicit function, defined by the equation

$$\mathrm{grad}_\eta \widehat{W}(\xi, \eta) = 0 .$$

Hence it follows that the order in zero of the function $\hat{\phi}$ with respect to quasihomogeneous filtration with weights $\nu_1, \ldots, \nu_n$ is no less than $\frac{1}{2}$:

$$\mathrm{ord}_\nu \hat{\phi} \geq \frac{1}{2} .$$

Consequently (see [37])

$$\mathrm{ord}_\nu \widehat{W}(\xi, \hat{\phi}(\xi)) \geq 1, \quad \mathrm{ord}_\nu \frac{\partial W}{\partial \xi_j} \geq 1 - \nu_j, \quad j = 1, \ldots, n .$$

If we assume that ν-quasihomogeneous part of the first order of the function $W(\xi)$ is infinite-to-one in zero, then the system of equations

$$\left[\frac{\partial W}{\partial \xi_1}\right]^\nu_{1-\nu_1} = \cdots = \left[\frac{\partial W}{\partial \xi_n}\right]^\nu_{1-\nu_n} = 0$$

($[\cdot]^\nu_\alpha$ is a sum of monomials, having order α in ν-filtration) has a continuous branch of solutions $\xi(t) \in \mathbb{C}^n$, $t \in [0,1]$, passing through zero: $\xi(0) = 0$. Since $\frac{\partial \widehat{W}}{\partial \eta}(\xi, \hat{\phi}(\xi)) \equiv 0$ then $\frac{\partial W}{\partial \xi}(\xi) = \frac{\partial \widehat{W}}{\partial \xi}(\xi, \hat{\phi}(\xi))$ and, consequently, the system of equations

$$\left[\frac{\partial \widehat{W}}{\partial \xi_1}\right]^{\hat{\nu}}_{1-\nu_1} = \cdots = \left[\frac{\partial \widehat{W}}{\partial \xi_n}\right]^{\hat{\nu}}_{1-\nu_n} = \left[\frac{\partial \widehat{W}}{\partial \eta_1}\right]^{\hat{\nu}}_{\frac{1}{2}} = \cdots = \left[\frac{\partial \widehat{W}}{\partial \eta_m}\right]^{\hat{\nu}}_{\frac{1}{2}} = 0$$

has a solution $\hat{\xi}(t) = (\xi(t),\ [\hat{\phi}]^\nu_{\frac{1}{2}}\ (\xi(t)))$. But this contradicts a semi-quasihomogeneity of the function $\widehat{W}(\xi, \eta)$.

Remark. From a semi-quasihomogeneity of $\widehat{W}(\xi, \eta)$ in $\hat{\nu}$-filtration, $\hat{\nu} = (\nu_1, \ldots, \nu_n, \frac{1}{2}, \ldots, \frac{1}{2})$, it follows that

$$\widehat{W}(\xi, \eta) = \alpha(\xi) + (\beta(\xi), \eta) + \frac{1}{2}(A\eta, \eta) + \cdots$$

where α is ν-quasihomogeneous first order function and β is ν--quasihomogeneous vector field of order $\frac{1}{2}$, A is a reversible matrix. Consequently, $A[\hat{\phi}]^{\nu}_{\frac{1}{2}}(\xi) = \beta(\xi)$ and (see (2.9))

$$\begin{aligned}[W]^{\nu}_{1}(\xi) &= \alpha(\xi) + (\beta(\xi), [\hat{\phi}]^{\nu}_{\frac{1}{2}}(\xi)) + \frac{1}{2}(A[\hat{\phi}]^{\nu}_{\frac{1}{2}}(\xi), [\hat{\phi}]^{\nu}_{\frac{1}{2}}(\xi)) \\ &= \alpha(\xi) - \frac{1}{2}(A^{-1}\beta(\xi), \beta(\xi)) \,. \end{aligned} \tag{4.2}$$

5. Special One-dimensional Reduction for Even Functionals

In non-linear boundary value problems of mathematical physics the symmetry of evenness $V(-x, \lambda) = V(x, \lambda)$ is often present. And

$$V(x, \lambda) = V_1(x) + \lambda V_2(x) \tag{5.1}$$

where

$$\begin{aligned} V_1(x) = \frac{1}{2}\langle Ax, x\rangle + \cdots, \quad V_2(x) = -\frac{1}{2}\langle Bx, x\rangle + \cdots, \\ \langle Ax, x\rangle > 0 \quad \text{for} \quad x \neq 0, \quad \langle Bx, x\rangle \geq 0 \,. \end{aligned} \tag{5.2}$$

The investigation of the branching of the extremals of functional (5.1) is naturally connected with the question of the distribution of the extremals with respect to the manifolds of the levels of functional V_2 (or V_1). Thus an even marginal function arises

$$W(\xi, \lambda) = \inf_{-V_2(x) = \xi^2} V(x, \lambda) \,.$$

The reduction considered differs from those considered above in that here the fibre $V_2(x) = 0$ is singular. This singularity can be done away with by the scaling change of the variable $x = \xi y$ (the first step of σ-process [38]). Consequently, we obtain

$$W(\xi, \lambda) = \xi^2(-\lambda + \widehat{W}(\xi)) \tag{5.3}$$

where

$$\widehat{W}(\xi) = \inf_{\frac{1}{2}\langle By, y\rangle + \xi^2 o(\|y\|_E^2) = 1} \left\{ \frac{1}{2}\langle Ay, y\rangle + \xi^2 V_1^{(4)}(y) + o(\xi^4) \right\} \,.$$

For $\xi = 0$ we have

$$\widehat{W}(0) = \inf_{y\in\mu}\frac{1}{2}\langle Ay, y\rangle, \quad \mu = \left\{y \,\middle|\, \frac{1}{2}\langle By, y\rangle = 1\right\}.$$

Let $\tilde{V}_j(y,\xi) = \xi^{-2}V_j(\xi y)$. The first and the second codifferentials of the quadratic form $T(y) = \frac{1}{2}\langle Ay, y\rangle$ at an arbitrary point y on the surface $\frac{1}{2}\langle By, y\rangle = 1$ are written in the form

$$f(y) = Ay - \frac{\langle Ay, By\rangle}{|By|^2}By \tag{5.4}$$

$$\nabla^2(T)(y)h = \Pi(y)\left(Ah - \frac{\langle Ay, By\rangle}{|By|^2}Bh\right), \quad h \in T_y(\mu). \tag{5.5}$$

Conditions (1.2) and (1.3) have the following form here

$$Ay = \lambda_1 By, \quad \lambda_1 = \frac{\langle Ay, By\rangle}{|By|^2}, \tag{5.6}$$

$$\langle Ah, h\rangle > \lambda_1\langle Bh, h\rangle, \quad y \in U, \quad h \in T_y(\mu). \tag{5.7}$$

These conditions are fulfilled if, for example, A is an isomorphism from E into F and B is completely continuous and a minimal eigenvalue $\lambda_1 = \min\{\lambda \mid \dim\operatorname{Ker}(A - \lambda B) > 0\}$ is single. Let also e_1 be a basic vector in $\operatorname{Ker}(A - \lambda_1 B)$, for which $\frac{1}{2}\langle Be_1, e_1\rangle = 1$ (i.e. $e_1 \in \mu$). The existence of the neighbourhood $V \subset \mu$, in which e_1 is a unique solution to (5.6) and $\forall\, x \in V$ (5.7) is fulfilled, is easily established on the basis of the theory of Riesz-Schauder [39]. Let $\operatorname{grad}_H V_2$ be a completely continuous mapping from E into F. And let

$$\operatorname{grad}_H V_1(x) = Ax + \nu_1(x)$$

where ν_1 is a completely continuous mapping from E into F. Then a unique (in the small) solution $y = \varphi(\xi)$, for which

$$\varphi(\xi) = e_1 + d\xi^2$$

of the equation

$$\Pi(y)\operatorname{grad}_H \tilde{V}_1(y,\xi) = 0$$

is simultaneously a point of a local minimum for $\tilde{V}_1(\cdot,\xi)$ on $\mu_\xi = \{y \mid \tilde{V}_2(y,\xi) = 1\}$ (for sufficiently small ξ), where $\Pi(y)$ is an orthoprojector (in H) on the tangent subspace with respect to μ_ξ at point y.

Since $V_2(\varphi(\xi),\xi) = 1$, then

$$2\xi\langle \operatorname{grad}_H \tilde{V}_2(\varphi(\xi),\xi), d\rangle + \frac{\partial \tilde{V}_2}{\partial \xi}(\varphi(\xi),\xi) = 0$$

and hence we obtain the equality

$$-\langle Bd, e_1\rangle + V_2^{(4)}(e_1) = 0$$

from which, with due account of (5.6), it follows that

$$\begin{aligned}\widehat{W}(\xi) &= \frac{1}{2}\langle Ae_1, e_1\rangle + (V_1^{(4)}(e_1) + \langle Ae_1, d\rangle)\xi^2 + O(\xi^4)\\ &= \frac{1}{2}\langle Ae_1, e_1\rangle + (V_1^{(4)}(e_1) + \lambda_1 V_2^{(4)}(e_1))\xi^2 + O(\xi^4)\ ,\\ \lambda_1 &= \frac{\langle Ae_1, Be_1\rangle}{|Be_1|^2}\ .\end{aligned}$$

Hence for $W(\xi,\lambda)$ we have the representation (see (5.3)):

$$W(\xi,\lambda) = \Big(-\lambda + \frac{1}{2}\langle Ae_1, e_1\rangle\Big)\xi^2 + (V_1^{(4)}(e_1) + \lambda_1 V_2^{(4)}(e_1))\xi^4 + O(\xi^6)\ .$$

Consequently (see Theorem 2.4) we established the following identity

$$\inf_{\langle x,e_1\rangle=\xi} V(x,\lambda) = \inf_{-V_2(x)=\xi^2} V(x,\lambda) \pmod{\xi^6}\ . \tag{5.8}$$

From (5.8) it is easy to show the coincidence (with respect to mod ξ^6) of the functions from the series (0.6): W_1, W_3, W_4, W_5.

6. Reduction in the Problem of Equilibrium of Elastic Plate

Let us consider a scaled Kármán equation [11,40]

$$\Delta^2\omega - [\omega,\varphi] + \lambda\omega_{xx} = \Delta^2\varphi + \frac{1}{2}[\omega,\omega] = 0 \tag{6.1}$$

under boundary value conditions

$$\omega = \Delta\omega = \varphi = \Delta\varphi = 0|_{\partial\Omega_a}\ . \tag{6.2}$$

Equation (6.1) under condition (6.2) describes an equilibrium state of a rectangular elastic plate, lengthwise compressed and fixed by hinges at the edge. Here $\Omega_a = [0, a] \times [0, 1]$, Δ is a Laplace operator, $[\omega, \varphi] = \omega_{xx}\varphi_{yy} + \omega_{yy}\varphi_{xx} - 2\omega_{xy}\varphi_{xy}$, ω is a deflection function, φ is a stress function, λ is a load parameter. Let ω and φ belong to space H^4 of the class of functions $W_2^4(\Omega_a)$, satisfying (6.2).

Equation (6.1) is the Euler-Lagrange equation for the extremals of the functional

$$V(\omega, \varphi, \lambda) = \frac{1}{2}(|\Delta\omega|^2 - \langle[\omega, \omega], \varphi\rangle - \lambda|\omega_x|^2 - |\Delta\varphi|^2) \tag{6.3}$$

on space $H^4 \times H^4$. Here $|\varphi|^2 = \langle\varphi, \varphi\rangle$ is a norm in $H = L_2(\Omega_a)$. Functional (6.3) is usually investigated via its reduction

$$\hat{V}(\omega, \lambda) = \sup_{\varphi \in H^4} V(\omega, \varphi, \lambda) . \tag{6.4}$$

The marginal mapping $H^4 \to H^4 \times H^4$, corresponding to this reduction, is given by the correspondence

$$\omega \longmapsto \left(\omega, -\frac{1}{2}\Delta^{-2}[\omega, \omega]\right) . \tag{6.5}$$

Consequently, for (6.4) we have a representation

$$\hat{V}(\omega, \lambda) = \frac{1}{2}(|\Delta\omega|^2 - \lambda|\omega_x|^2) + \frac{1}{8}|\Delta^{-1}[\omega, \omega]|^2 . \tag{6.6}$$

Here Δ^{-1} means Green's operator $\psi \to \varphi$, where φ is a solution to Poisson equation [41]: $\Delta\varphi = \psi$, $\varphi|_{\partial\Omega_a} = 0$. According to [41] operator Δ acts as an isomorphism from the space of functions of the class $C^{2+\alpha}(\Omega_a)$, satisfying (6.2), onto F_0, a subspace of Hölder functions of the class $F = C^{0+\alpha}(\Omega_a)$, satisfying the condition of agreement:

$$\psi(0, 0) = \psi(a, 0) = \psi(0, 1) = \psi(a, 1) = 0 .$$

Let $E = \{\omega \in H^4 \mid \Delta^2\omega \in F\}$, $||\omega||_E = ||\Delta^2\omega||_F$.

Note that it is via function (6.6) that in the theory of elastic hulls the stability of equilibrium states is estimated: equilibrium is stable if it realizes a point of a local minimum of functional (6.6) (in energy space). The loss of stability of a rectangular form of equilibrium ($\omega = 0$) occurs when λ goes through a critical value (the upper critical load):

$$\lambda^*(a) = \sup\left\{\bar{\lambda} \,\middle|\, \frac{\partial^2 \hat{V}}{\partial \omega^2}(0,\lambda)(h,h) > 0, \quad \lambda < \bar{\lambda}\right\}.$$

The upper critical load continuously and piece-wise smoothly depends on the length of plate a. The discontinuity of a derivative $\lambda^*(a)$ occurs at points $a_m = \sqrt{m(m+1)}$, $m = 1, 2 \ldots$. Let $\lambda_m = \lambda^*(a_m)$. It is known that for $a = a_m$ and $\lambda = \lambda_m$ the Euler-Lagrange equation of extremals $\hat{V}$ has a two-dimensional degeneration at zero (the loss of stability with respect to two modes $e_1 = \omega_{m,i}$, $e_2 = \omega_{m+1,1}$, $\omega_{m,n} = 2\sin(\frac{m\pi x}{a})\sin(n\pi y)$). The analysis of supercritical equilibrium states is reduced to the study of critical points of a marginal function [26]:

$$W(\xi,\lambda,a) = \inf_{u \in O^{\infty-2}} \hat{V}(\xi_1 e_1 + \xi_2 e_2 + u, \lambda), \quad \xi \in O^2 \tag{6.7}$$

(for (λ, a), sufficiently close to (λ_m, a_m)), where O^2 and $O^{\infty-2}$ are sufficiently small neighbourhoods of zeros in $\mathbb{R}^2$ and $E^{\infty-2} = \{u \in E | \langle u, e_j\rangle = 0,\ j = 1,2\}$. The local analysis of (6.7) is provided by the fact that there takes place the representation

$$W(\xi,\lambda,a) = \frac{1}{2}(\alpha_1(\lambda,a)\xi_1^2 + \alpha_2(\lambda,a)\xi_2^2) + \sum h_{i,j}\xi_i^2\xi_j^2 + \cdots \tag{6.8}$$

where α_1, α_2 are eigenvalues, corresponding to eigenvectors e_1, e_2 for operator $\Delta^2 + \lambda(\frac{\partial}{\partial x})^2$,

$$h_{j,j} = \frac{1}{8}|\Delta^{-1}[e_j, e_j]|^2, \quad h_{1,2} = h_{2,1} = \frac{1}{8}(2|\Delta^{-1}[e_1,e_2]|^2 + \langle \Delta^{-1}[e_1,e_1], \Delta^{-1}[e_2,e_2]\rangle).$$

From the results of the calculations, published in [32], it follows that for several initial values m holds the inequality

$$h_{1,2} > \sqrt{h_{1,1}h_{2,2}} \tag{6.9}$$

on the basis of which it is possible to make a complete analysis of bifurcations of equilibrium states (see [26]). The question of property of having fixed sign of the form $\sum h_{i,j}\xi_i^2\xi_j^2$ for an arbitrary m is still open.

We can have the same result for Kármán equation on the basis of marginal functions, obtained via different bundles of space E.

7. Elastically Strengthened Kirchhoff's Rod

The problem of the equilibrium of a space rod in Kirchhoff's model [42,43] was studied on the basis of a kinetic analogy with the problem of the motion of a heavy rigid body around a fixed point. The situation corresponding to the cases of a complete integrability of equations of the motion of a rigid body were analyzed in detail [42–44]. In [45–48] the Kirchhoff's rod (including a multi-dimensional generalization) was investigated by the methods of the theory of singularities of smooth functions without the assumption of integrability. The loss of stability of the Kirchhoff's rod under the lengthwise compression is one-mode, though the greater critical loads can be double.

Let us consider a rectangular Kirchhoff's rod strengthened by an elastic force, acting on the rod's end in the plane of two main axes of inertia of the rod's cross section. Let an external load be axial and the rod's ends rigidly fixed. And let the strengthening elastic force have a potential $\frac{\mu}{2}(\int_0^1 (r_2, g_3(s))ds)^2$. Here $r_2 = (0,1,0)^T$, $g_3(s)$ is a basis vector, tangent to the mean line of the rod at the point corresponding to parameter (of length) s, $0 \leq s \leq 1$, μ is a parameter of the force of elastic strengthening (reacting to the deviation of the rod's end in the direction r_2). Assume that $\mu > 4\pi^2$. Let $g_1(s)$ and $g_2(s)$ be basic vectors directed along the main axes of inertia of a normal cross section and let $r_1 = (1,0,0)^T$, $r_3 = (0,0,1)^T$. The equation of an equilibrium state of the rod in this situation is written in the following form

$$-A\frac{dx}{ds} + [Ax, x] + \lambda[r_3, g^{-1}r_3] + \mu\langle r_3, g^{-1}r_2\rangle[r_3, g^{-1}r_2] = 0 \ . \qquad (7.1)$$

Here $\langle \varphi, \varphi\rangle = \int_0^1 (\varphi(s), \psi(s))ds$, λ is a parameter of the force of axial compression, $A = \operatorname{diag}(A_1, A_2, A_3)$ is a tensor of elasticity in cross section ($A_j > 0$), for which the E. L. Nikolai condition is fulfilled: $1/(1+\nu)A_3 > (A_1 + A_2)/2A_1A_a$ (ν is Poisson coefficient), $x(s)$ is an angular velocity of motion of the cross section depending on s, written in the coordinates with respect to a triple $g_1(s), g_2(s), g_3(s)$. The matrix composed of the columns of vector coordinates $g_1(s), g_2(s), g_3(s)$ in the basis $g_1(0), g_2(0), g_3(0)$ is denoted by $g(s), [Ax, x]$ is a vector product. The boundary value condition

$$g(0) = g(1) = I \qquad (7.2)$$

corresponds to the rigidly fixation at the ends. The functional

$$\frac{1}{2}\langle Ax, x\rangle + \lambda\langle r_3, g^{-1}r_3\rangle + \frac{\mu}{2}\langle r_3, g^{-1}r_2\rangle^2 \tag{7.3}$$

is a functional of Eq. (7.1) under condition (7.2). If vector x is identified with a matrix

$$X = \begin{pmatrix} 0 & -x_3 & x_2 \\ x_3 & 0 & -x_1 \\ -x_2 & x_1 & 0 \end{pmatrix} \tag{7.4}$$

then for $X(s)$ it is possible to write a matrix representation [49]:

$$X(s) = g^{-1}(s)\frac{dg}{ds}(s) \ .$$

Let the functional $V(\varphi, \lambda, \mu, A), \varphi(s) \in \mathbb{R}^3$, be obtained from (7.3) by the change

$$g = \exp(\varphi_3 r_3)\exp(\varphi_2 r_2)\exp(\varphi_1 r_1)$$

where $\exp(x) = e^X$ (see (7.4)). Consider V on a linear Banach space $E = C_0^2([0,1], \mathbb{R}^3)$ of the functions of class C^2, defined on [0,1] and taking values in $\mathbb{R}^3$ under the condition

$$\varphi(0) = \varphi(1) = 0 \ .$$

Put $F = C([0,1], \mathbb{R}^3)$, $H = L_2([0,1], \mathbb{R}^3)$. Using an easily verifiable relation

$$\exp(\varphi r_j) r_k = \sin\varphi [r_j, r_k] + \cos\varphi r_k, \quad k \neq j \ ,$$

we can show that

$$V(\varphi, \lambda, \mu, A) = \text{const} + V^{(2)}_{\lambda,\mu,A}(\varphi) + V^{(3)}_{\lambda,\mu,A}(\varphi) + V^{(4)}_{\lambda,\mu,A}(\varphi) + o(\|\varphi\|^4_E)$$

where

$$V^{(2)}_{\lambda,\mu,A}(\varphi) = \frac{1}{2}\Big(\sum A_j \int_0^1 \dot{\varphi}_j^2 ds - \lambda \int_0^1 (\varphi_1^2 + \varphi_2^2) ds + \mu\Big(\int_0^1 \varphi_1 ds\Big)^2\Big)$$

$$V^{(3)}_{\lambda,\mu,A}(\varphi) = (A_2 - A_3)\int_0^1 \varphi_1 \dot{\varphi}_2 \dot{\varphi}_3 ds - A_1 \int_0^1 \dot{\varphi}_1 \varphi_2 \dot{\varphi}_3 ds$$
$$- \mu \int_0^1 \varphi_1 ds \int_0^1 \varphi_2 \varphi_3 ds$$

$$V^{(4)}_{\lambda,\mu,A}(\varphi) = \frac{(A_3 - A_2)}{2}\int_0^1 \varphi_1^2\dot\varphi_2^2 ds + \frac{(A_2 - A_3)}{2}\int_0^1 \varphi_1^2\dot\varphi_3^2 ds$$
$$+ \frac{(A_1 - A_3)}{2}\int_0^1 \varphi_2^2\dot\varphi_3^2 ds + \frac{\lambda}{24}\int_0^1 (\varphi_1^4 + \varphi_2^4 + 6\varphi_1^2\varphi_2^2) ds$$
$$- \mu\left(\int_0^1 \varphi_1 ds\right)\left(\int_0^1 \left(\frac{\varphi_1^3}{6} - \frac{\varphi_1\varphi_2^2}{2}\right) ds\right).$$

When the parameters $\lambda = 4\pi^2 + \delta$, $A_1 = 1$, $A_2 = 4 + p$, $\mu > 4\pi^2$, are localized,

$$e_1 = \sqrt{2}(\sin 2\pi s) r_1, \quad e_2 = \sqrt{2}(\sin \pi s) r_2$$

are bifurcation modes and $\alpha_1 = -\delta$ and $\alpha_2 = -\delta + p$ are spectral functions. From relation (2.9) for the function

$$W(\xi, \delta, p) = \inf_{\langle\varphi, e_j\rangle = \xi_j} V(\varphi, \lambda, \mu, A), \quad \xi \in \mathbb{R}^2$$

we obtain the representation

$$-\frac{1}{2}(\delta\xi_1^2 + (\delta - p)\xi_2^2) + \sum h_{i,j}\xi_i^2\xi_j^2 + o(|\xi|^4).$$

From E. L. Nikolai inequality it follows that $h_{1,2} < -\sqrt{h_{1,1}h_{2,2}}$ (a rigid loss of stability).

The restriction $\mu > 4\pi^2$ formulated above "closes" the mode

$$e_*(\mu) = d(\mu)(\cos\theta(\mu)\left(s - \frac{1}{2}\right) - \cos\frac{\theta(\mu)}{2}\Big) r_1$$

(here $d(\mu)$ is a norming constant, $\theta(\mu)$ is obtained from equation $\theta^2 = \mu(1 - \frac{2}{\theta}\mathrm{tg}\frac{\theta}{2})$). If $\mu < 4\pi^2$ then the loss of stability occurs by the scheme of a one-dimensional assembly (for $\theta^2(\mu)\int_0^1 e_*^4(\mu)(s)ds \neq 4\mu(\int_0^1 e_*(\mu)(s)ds)$ $(\int_0^1 e_*^3(\mu)(s)ds)$). When $\mu \approx 4\pi^2$ is localized there arises a three-mode bifurcation.

8. Bifurcation of Stationary Rotations of Multi-Dimensional Top

The stability of stationary rotations of a rigid body in a three-dimensional space was studied by many specialists in mechanics. It is well known that the stability of a sleeping top is determined by the inequality $C^2\omega^2 > 4Amgl$ for the case of a dynamic symmetry and by the inequality

$\min\{C-A, C-B\}\omega^2 > mgl$ for an asymmetric top. The bifurcation of the rotation of classical (three-dimensional) tops is studied in detail by V. N. Rubanovskii [50].

The peculiarity of a multi-dimensional case is that the factorization with respect to orbits of the symmetry group of an effective potential does not give a function on a sphere. The reduced effective potential depends also on additional variables, responsible for the arrangement of a symmetry subgroup of the level manifold with respect to the kinetic moment in a group of rotations, preserving a vertical line. If a value with respect to a kinetic moment is regular (the annihilator is Cartan's subalgebra), then this subgroup is a maximal torus in a group of rotations around a vertical line. In this case additional variables can be excluded from the above scheme. Upon the exclusion there arises a function of coordinates on a sphere, whose critical points correspond to stationary rotations.

The motion of n-dimensional body in a constant gravitational field is given by the generalized Euler-Poisson equations on group SO(n) (V. I. Arnold, A. S. Mishchenko, S. V. Vishik, F. V. Dolzhanskii, S. V. Monakov, A. T. Fomenko and others):

$$\begin{gathered} M\dot{\Omega} + \dot{\Omega}M + [\Omega, M\Omega + \Omega M] + \lambda r \wedge \gamma = 0 \\ \dot{\gamma} + \Omega\gamma = 0 \end{gathered} \tag{8.1}$$

where $\Omega = f^{-1}\frac{df}{dt}$ is an angular velocity (in body), $f(t)$ is a function with values in SO(n), $M = \mathrm{diag}(m_1, \dots, m_n)$, λ is a load parameter ($\lambda > 0$), $\mathfrak{r} = f^{-1}e_n$ is Poisson vector, $e_n = (0, \dots, 0, 1)^T$ is a vertical basic vector. By $[x, y]$ the commutator of matrices is denoted and by $r \wedge \mathfrak{r}$ it denotes a skew-symmetric matrix, determined by the relation $(r \wedge \mathfrak{r})x = (r, x)\mathfrak{r} - (\mathfrak{r}, x)r$. The constant vector r denotes the centre of gravity. System (8.1) is reduced to one equation

$$M\dot{\Omega} + \dot{\Omega}M + [\Omega^2, M] + \lambda r \wedge f^{-1}e_n = 0\ . \tag{8.2}$$

By $E(f, \Omega)$ a complete top's energy is denoted: $E(f, \Omega) = K(\Omega, \Omega) + U(f)$, where $K(\Omega, \Omega) = \frac{1}{2}\langle M\Omega + \Omega M, \Omega\rangle$ is a kinetic energy (here $\langle X, Y\rangle = \frac{1}{2}\mathrm{tr}XY^*$ is Cartan-Killing formula), $U(f) = \lambda(r, \mathfrak{r})$ is a potential energy of gravitational field. The mapping with respect to a kinetic moment $p(f, \Omega) = \pi J(f, \Omega)$ where $J(f, \Omega) = f(M\Omega + \Omega M)f^{-1}$, π is an orthoprojector in metrics $K(X, Y)$ from so(n) on $\mathfrak{n} = \mathrm{Ann}(e_n) = \{Y \in \mathrm{so}(n)/Ye_n = 0\}$ is an integral of Eq. (8.2) and the corresponding integral set $F_p = \{(f, \Omega)|p(f, \Omega) =$

$p\}$ is a smooth submanifold of dimension $(n-1)(n+2)/2\ \forall\ p \in \mathfrak{n}$. Besides, F_p is a subbundle of a trivial bundle $\mathrm{SO}(n) \times \mathrm{so}(n) \to \mathrm{SO}(n)$ with a fibre $F_{p,f} = \{\Omega \mid p(f,\Omega) = p\}$, $f \in \mathrm{so}(n)$. It is obvious that F_p is invariant with respect to action $(g,(f,\Omega)) \to (gf,\Omega)$, $g \in G_p$, where $G_p \subset \mathrm{SO}(n-1)$ is a subgroup of isotropy at point p of action $(g,p) \to gpg^{-1}$. Here $\mathrm{SO}(n-1)$ is identified with a subgroup $\{u \in \mathrm{SO}(n) \mid ue_n = e_n\}$.

Stationary rotations are found out as critical orbits of action G_p on F_p with respect to $E(f,\Omega)$ [51,49]. Critical orbits are discovered by the reduction of the reduced potential (the first reduction)

$$U_p(f) = \min_{\Omega \in F_{p,f}} E(f,\Omega)\ . \tag{8.3}$$

Since $E(f,\Omega)$ is a convex function with respect to Ω, then there exists a unique cross section h_p: $\mathrm{SO}(n) \to F_p$, smooth with respect to (f,p), for which

$$U_p(f) = E(f, h_p(f))\ .$$

And $h_p(g^{-1}f) = h_{gpg^{-1}}(f)$, $g \in \mathrm{so}(n-1)$, and, consequently, $U_p(g^{-1}f) = U_{gpg^{-1}}(f)$. Thus $U_p(f)$ admits factorization: $U_p(f) = \tilde{U}_p([f])$, where $[f]$ is a class of an element in factor group $\mathrm{SO}(n)/G_p$. Let $p \in \mathrm{Reg}\,\mathfrak{n}$ (i.e. $\mathrm{Ann}(p)$ is Cartan subalgebra in $\mathfrak{n}$). For the investigation of stationary rotations, bifurcating from the rotation around a vertical line, it is necessary to introduce the coordinates in the neighbourhood of a unit of group $\mathrm{SO}(n)/G_p$. For this purpose it is sufficient to give in $\mathrm{SO}(n)$ any parametrized submanifold S, zero-dimensional and transversely intersecting G_p in unit. Let $H_p = \mathrm{Ann}(p)$ and let $H_p^\perp$ be an orthogonal complement to H_p in $\mathrm{so}(n)$. Give an image of the mapping $(u,\nu) \to \exp(\nu)\exp(u)$, $u \in \mathfrak{n}^\perp$, $\nu \in \mathfrak{n} \cap H_p^\perp$ as S. Then

$$U_p(\exp(\nu)\exp(u)) = U_p(\exp(\omega)\exp(\nu)\exp(u)),\quad \omega \in H^p\ . \tag{8.4}$$

Consequently, the function $V_p(u,\nu) = U_p(\exp(\nu)\exp(u))$ is a factorized reduced potential (the second reduction).

In the assumption of the symmetry of a top ($m_i \neq m_j$ for $i \neq j$) and for $r = e_n$ the annihilator of an angular velocity of a general position, ω is Cartan subgalgebra in $\mathfrak{n}$ and $\mathrm{Ann}(\omega) = \mathrm{Ann}(p)$. For $n = 2l+1$ (in the case $\gamma = r = e_n$ the stationary rotation can be stable only in an odd-dimensional space [52]) for the appropriately enumerated vectors $e_1, \ldots, e_{n+1}$ we have

$$\omega = \sum_{j=1}^{n} \omega_j \varepsilon_{2j-1,2j}, \quad p = \sum p_j \varepsilon_{2j-1,2j}$$

$$\omega_i \neq \omega_j, \quad p_i \neq p_j$$
$$\forall \{i, j | i \neq j\}$$

(here $\varepsilon_{i,j} = e_i \wedge e_j$) and the stability condition is written in the form of the system of inequalities

$$\begin{cases} (m_{2j-1} - m_q)(m_{2j} - m_{\hat{q}})(\omega_j^2 - \omega_k^2)(p_j^2 - p_k^2) > 0 \\ (p_j^2 - p_k^2)(p_j \omega_j (m_r - m_q) + p_k \omega_k (m_{\hat{r}} - m_{\hat{q}})) > 0 \\ \omega_k^2 \min\{(m_{2k-1} - m_n), (m_{2k} - m_n)\} > 0 \end{cases} \tag{8.5}$$

$j < k$, $r \neq \hat{r}$, $q \neq \hat{q}$; $r, \hat{r} \in \{2j-1, 2j\}$; $q, \hat{q} \in \{2k-1, 2k\}$ V giving a positive definiteness of the second differential in zero [52].

When λ passes through λ_*, where

$$\lambda_* = \min_k \{\omega_k^2 \min\{(m_{2k-1} - m_n), (m_{2k} - m_n)\}\}$$

a topological type of the function $V(u, \nu)$ in the neighbourhood of zero is reconstructed. When the first two inequalities from (8.5) are fulfilled for $\lambda = \lambda_*$, the dimension of Hesse matrix of function V in zero is equal to unit [53]. For $\lambda > \lambda_*$ and when $\lambda - \lambda_*$ are sufficiently small the Morse index of a zero critical point of function V is equal to unit. If $(m_j - m_n)/(m_{2k} + m_{2k-1}) > \frac{1}{4}$, $j \in \{2k-1, 2k\}$, then the transition of λ upwards through λ_* results in the generation from zero of a pair of critical points of a zero Morse index. These points can be represented in the form

$$g = \exp(\pm b \sqrt{\lambda - \lambda_*} \varepsilon_{j,n} + o(\lambda - \lambda_*)) \tag{8.6}$$

where $b = 1/\omega_k \sqrt{\frac{2(m_j - m_n)^2}{m_{2k-1} + m_{2k}} + \frac{(m_j - m_n)}{2}}$. In the case $\frac{m_j - m_n}{m_{2k-1} + m_{2k}} < \frac{1}{4}$ the transition of λ through λ_* leads to the disappearance of a pair of critical points of index 1 at zero presentable in the form analogous to (8.6).

The Poisson vector presentable in the form

$$r = (\cos b \sqrt{|\lambda - \lambda_*|}) e_n \pm (\sin b \sqrt{|\lambda - \lambda_*|}) e_j + o(|\lambda - \lambda_*|)$$

corresponds to the stationary rotation. The proof is based on the transition of the function (the third reduction)

$$\tilde{V}(u,\lambda) = \min_{\nu} V(u,\nu,\lambda)$$

and then to the function (the fourth reduction)

$$W(\xi,\lambda) = \min_{y} \tilde{V}(\xi\varepsilon_{j,n} + y,\lambda), \quad y \perp \varepsilon_{j,n} \, .$$

Function W is presented in the form

$$\begin{aligned} W(\xi,\lambda) &= a(\lambda)\xi^2 + b(\lambda)\xi^4 + o(|\xi|^5) + \text{const} \, , \\ a(\lambda) &= \frac{\lambda_* - \lambda}{2}, \quad b(\lambda) = \frac{(m_j - m_n)}{2}\omega_k^2\left(\frac{m_j - m_n}{m_{2k-1} + m_n} - \frac{1}{4}\right) \\ &+ O(\lambda - \lambda_*), \quad j \in \{2k-1, 2k\} \, . \end{aligned}$$

The deduction of this representation makes use of the representation

$$\tilde{V}(\tilde{u},\lambda) = \text{const} + \tilde{V}_\lambda^{(2)}(u) + \tilde{V}_\lambda^{(4)}(u) + o(|u|^5)$$

where

$$\begin{aligned} \tilde{V}_\lambda^{(2)}(u) &= k_{2,0}(u) - \frac{\lambda}{2}|u|^2 \, , \\ k_{2,0}(u) &= \frac{1}{2}\langle Cu, u\rangle, C = \mathrm{ad}_\omega \mu \ \mathrm{ad}_\omega - \frac{1}{2}(\mathrm{ad}_\omega \ \mathrm{ad}_p + \mathrm{ad}_p \ \mathrm{ad}_\omega) \, , \\ \mu(x) &= Mx + xM \, , \\ \tilde{V}_\lambda^{(4)}(u) &= k_{4,0}(u) + \frac{\lambda}{24}|u|^4 \, , \\ k_{4,0}(u) &= \frac{1}{24}\langle [\mathrm{ad}_u, [\mu_{2,0}, \mathrm{ad}]](\omega), \omega\rangle + \frac{1}{2}\langle \mu_{2,0}(\omega), \ \mu^{-1}\mu_{2,0}(\omega)\rangle \, , \\ \mu_{2,0} &= \frac{1}{2}[\mathrm{ad}_u, \mu_{1,0}] = \frac{1}{2}(\mathrm{ad}_u^2 \mu + \mu \ \mathrm{ad}_u^2) - \mathrm{ad}_u \mu \ \mathrm{ad}_u \, , \\ \mu_{1,0} &= [\mathrm{ad}_u, \mu] = \mathrm{ad}_u \mu - \mu \ \mathrm{ad}_u \, . \end{aligned}$$

References

1. J.-P. Aubin and I. Ekeland, *Applied Nonlinear Analysis*, John Wiley & Sons, New York.
2. W. T. Koiter, *Stability and supercritical behaviour of elastic systems*, Mekhanika **5** (1960), 99–110 (in Russian).
3. V. A. Trenogin and N. A. Sidorov, *Investigation of bifurcation points and nontrivial branches of solutions to nonlinear equations*, Irkutsk Univ. Press 1072, Vyp.1, pp. 216–247 (in Russian).
4. M. A. Krasnosel'skii, N. A. Bobylev, and E. M. Mukhamadiev, *On one problem of investigation of degenerate extremals of functionals of variational calculus*, DAN SSSR. **240**, no. 3 (1978), 530–533 (in Russian).
5. J. E. Marsden, *On the Geometry of the Liapunow-Schmidt Procedure*, Lect. Notes in Math. **755** (1979), 77–82.
6. Yu. I. Sapronov, *Multimode bifurcations of elastic equilibriums*, Prikladnaya matematika i mekhanika (Applied Mathematics and Mechanics) **52** (1988) Vyp.6, 997–1006 (in Russian).
7. J. Milnor, *Morse Theory*, Princeton Univ. Press, 1963.
8. B. A. Dubrovin, S. P. Novikov, and A. T. Fomenko, *Moden Geometry. Methods of Homology Theory*, Nauka, Moscow, 1984, p. 344 (in Russian).
9. C. C. Conley and E. Zehnder, *The Birkhoff-Lewis fixed point theorem and a conjecture of V. I. Arnol'd*, Invent. Math. **73** (1983), 33–49.
10. S. V. Bolotin, *Periodic solutions of systems with gyroscopic forces*, Prikladnaya matematika i mekhanika (Applied Mathematics and Mechanics) **51** (1987) Vyp.4, 696–687 (in Russian).
11. A. S. Vol'mir, *Elastic Plates and Hulls*, Gostekhizdat, Moscow, 1956 (in Russian).
12. E. P. Popov, *Nonlinear Problems of Statics of Thin Rods*, OGIZ, Moscow, 1948, p. 170 (in Russian).
13. D. I. Shil'krut and P. M. Vyrlan, *Stability of Nonlinear Hulls*, Shtinitsa, Kishinev, 1977, p. 104 (in Russian).
14. D. Shilkrut, *Stability and vibrations of geometrically nonlinear cylindrically orthotropic circular plates*, J. Appl. Mechanics **51** (1874), 354–360.
15. T. Poston and I. Stewart, *Catastrophe Theory and its Applications*, Pitman, 1978.
16. E. I. Grigolyuk and I. Shalashilin, *Problems of Nonlinear Deformation: Method of Extension with Respect to Parameter in Nonlinear Problems of Mechanics of Rigid Deformable Body*, Nauka, Moscow, 1988, p. 232 (in Russian).
17. A. Love, *Mathematical Theory of Elasticity*, NKTN SSSR, Moscow-Leningrad, 1935, p. 674 (in Russian).
18. J. Eells, *A setting for global analysis*, Bull. Amer. Math. Soc. **72** (1966), 751–807.
19. M. M. Vainberg, *Variational Methods of Investigation of Nonlinear Operators*, Gostekhizdat, Moscow, 1956 (in Russian).

20. R. T. Rockafellar, *Convex Analysis*, Princeton Math. Series **28**, Princeton Univ. Press, 1970.
21. L. C. Young, *Lectures on Calculus of Variations and Optimal Control Theory*, N. B. Saunders Co., Philadelphia, 1969.
22. A. I. Perov, *Variational Methods in the Theory of Nonlinear Oscillations*, Voronezh Univ. Press, 1981, p. 196 (in Russian).
23. Yu. G. Borisovich, V. G. Zvyagin, and Yu. I. Sapronov, *Nonlinear Fredholm mappings and Leray-Schauder theory*, Uspekhi Matem. Nauk **32** (1977), 3–54 (in Russian).
24. E. M. Galeev and V. M. Tikhomirov, *Concise Course of the Theory of Extremal Problems*, Moscow Univ. Press, Moscow, 1989, p. 204 (in Russian).
25. S. Lang, *Introduction to Differentiable Manifolds*, Interscience, New York, 1962.
26. Yu. I. Sapronov, *Two-mode bifurcation of solutions to Kármán equation*, Diff. Uravneniya **25**, no. 6 (1989), 1078–1081 (in Russian).
27. Yu. I. Sapronov, *Semiregular angular singularities of smooth functions*, Matemat. Sbornik **180**, no. 10 (1989), 1299–1310 (in Russian).
28. Yu. I. Sapronov, *Angular singularities and multi-dimensional cusps in nonlinear analysis*, Theory of Operators in Functional Spaces (Survey lectures of the 13th All-Union School), Saratov Univ. Press, Kuibyshev branch, 1989, pp. 192–201.
29. M. A. Kracsnosel'skii, G. M. Vainikko, P. P. Zabreiko, Ya. B. Rutitskii, and V. Ya. Stetsenko, *Approximated Solutions of Operator Equations*, Nauka, Moscow, 1969, p. 456 (in Russian).
30. M. M. Vainberg and V. A. Trenogin, *Theory of Branching of Solutions to Nonlinear Equations*, Nauka, Moscow, 1969, p. 528 (in Russian).
31. L. S. Srubshchik, *Swelling and Post-critical Behaviour of Hulls*, Rostov Univ. Press, 1981, p. 96 (in Russian).
32. E. J. Holder and D. Schaeffer, *Boundary conditions and mode jumping in the Karman equations*, SIAM J. Math. Anal. **15**, no. 3 (1984), 446–457.
33. I. I. Vorovich, *Mathematical Problems of Nonlinear Theory of Slating Hulls*, Nauka, Moscow, 1989, p. 376 (in Russian).
34. J. A. Ize, *Bifurcation theory for Fredholm operators*, Memoirs Amer. Math. Soc. **174** (1976).
35. S. Chow, J. Hale, and J. Mallet-Paret, *Applications of generic bifurcation theory* I, II, Arch. Rational Mech. Anal. **59** (1975), 159–188; **62** (1976), 209–235.
36. B. V. Loginov, *Theory of Branching of Solutions to Nonlinear Equations Under the Conditions of Group Invariance*, Fan, Tashkent, 1985, p. 184 (in Russian).
37. V. I. Arnol'd, A. N. Varchanko, and S. M. Gusein-Zade, *Singularities of differentiable mappings*, Classification of Critical Points of Caustics and Wave Fronts, Nauka, Moscow, 1982, p. 304 (in Russian).

38. I. R. Shafarevich, *Fundamentals of Algebraic Geometry*, Nauka, Moscow, 1972 (in Russian).
39. K. Yosida, *Functional Analysis*, Springer-Verlag, Berlin, 1965.
40. P. G. Ciarlet and P. Rabier, *Les equations de von Kármán*, ed. A. Dold and B. Eckmann. Lecture Notes in Math. **826**, Springer, 1980.
41. E. A. Volkov, *On solution of boundary value problems for Poisson equation in a rectangle*, DAN SSSR **147**, no. 2 (1962), 13–16 (in Russian).
42. E. L. Nikolai, *To the problem of elastic line of double curvature*, Transactions in Mechanics, Gostekhizdat, Moscow, 1955, pp. 45–277 (in Russian).
43. E. P. Popov, *Nonlinear problems of statics of thin rods*, Gostekhizdat, Leningrad, Moscow, 1948, p. 170 (in Russian).
44. A. A. Ilyukhin, *Space Problems of Nonlinear Theory of Elastic Rods*, Nauk Dumka, Kiev, 1979, p. 216 (in Russian).
45. Yu. N. Zavarovskii and Yu. I. Sapronov, *Normal form of a key function in the problem of critical loads of elastic rods*, Voronezh, 1981, p. 28. VINITI. No. 4185–81 (in Russian).
46. Yu. N. Zavarovskii, *Branching of solutions to Kirchhoff symmetric space rod*, VGU, Voronezh, 1981, p. 21. VINITI. No. 4610–81 (in Russian).
47. Yu. N. Zavarovskii, *On Lyapunov-Schmidt method for variational problems with parameter*, VGU, Voronezh, 1981, p. 13. VINITI. No. 478–82 (in Russian).
48. Yu. N. Zavarovskii, *Normal form of a key function of a generalized Kirchhoff equation*, Uspekhi Mat. Nauk **38** (1983), Vyp.3, 177–178. (in Russian).
49. V. I. Arnol'd, *Mathematical Methods of Classical Mechanics*, Nauka, Moscow, 1989, p. 472 (in Russian).
50. V. N. Rubanovskii and V. A. Samsonov, *Stability of Stationary Motions in Examples and Problems*, Nauka, Moscow, 1988, p. 304 (in Russian).
51. S. Smale, *Topology and mechanics*, Invent. Math. **10**, no. 6 (1970), 305–331; **11**, no. 1, 45–64.
52. Yu. I. Sapronov, *Multi-dimensional sleeping tops*, Global Analysis and Mathematical Physics, Voronezh Univ. Press, 1987, pp. 95–109 (in Russian).
53. Yu. I. Sapronov, *Bifurcation of stationary rotations of multi-dimensional asymmetric rigid body from the regime of a sleeping top*, Global Analysis and Nonlinear Equations, Voronezh Univ. Press, 1988, pp. 141–152 (in Russian).
54. Yu. I. Sapronov, *Nonlocal finite-dimensional reductions in variational boundary value problems*, Mate. Zametki. **49** (1991), Vyp.1, 94–103.

Geometry in Partial Differential Equations (pp. 376–388)
eds. A. Pràstaro and Th. M. Rassias

ON THE SCHRÖDINGER EQUATION FOR AN N-ELECTRON ATOM

C. S. SHARMA
Mathematical Research Unit, University of London, Birkbeck College,
43 Gordon Square, London WC1H OPD, England

ABSTRACT

A curious method of finding an approximate solution to the Schrödinger equation for an N-electron atom is described. The approximate solution turns out to be the sum of a thousand *exact* solutions of similar but different second order ordinary differential equations. The geometric concept of orthogonality plays a crucial role in this theory.

1. Introduction

Many years ago the author and his collaborators [10, 12, 13] developed a very efficient way of calculating approximate wave functions and energies of stationary states of an N-electron atom. Though there are good reasons for believing that the method developed is both more efficient (in that it saves both time and effort) and more accurate than those actually used by physicists, very little practical applications were made of the method partly because the author was not particularly interested in engaging in lengthy numerical computations and was unable to find financial resources to employ others to do these computations for him. In this article we describe the main mathematical features of this work. A by-product of the method was a purely formal (by which I mean, an algorithm which has little mathematical justification) method for calculating level shifts of unstable states of many-electron atoms. It seems that the effectiveness of this by-product is now being appreciated in atomic physics [7,8] and various variants of the method are being actually used: since this method seems to produce accurate answers consistently, finding mathematical justification of the method is an interesting open problem. In what follows we describe in outline the main mathematical features of the mathematical method mentioned above: we leave out tedious details which the interested reader can find in references cited and we conclude by describing the aforementioned by-product and some of the interesting open questions it poses.

2. Stationary States of *n*-electron Atoms

Physical states of an *N*-electron atom are described by vectors in the Hilbert space $\mathcal{H} = ((\mathcal{L}^2(\mathbb{R}^3)) \otimes \mathcal{S}) \otimes \overset{n\ copies}{\cdots\cdots} \otimes ((\mathcal{L}^2(\mathbb{R}^3)) \otimes \mathcal{S})$, where $\mathcal{S}$ is a two-dimensional abstract Hilbert space which takes into the account the spin of an electron.. We denote a point in the *i*th copy of $\mathcal{L}^2(\mathbb{R}^3)$ by $\boldsymbol{r}_i$, its norm by r_i and its Cartesian coordinates by (x_{1i}, x_{2i}, x_{3i}) and $\mathbb{R}^3$ is regarded as an Euclidean space, thus

$$r_i = \|\boldsymbol{r}_i\| = (x_{1i}^2 + x_{2i}^2 + x_{3i}^2)^{1/2}. \qquad (2.1)$$

Stationary states of the atom are eigenvectors of the Hamiltonian *H* belonging to point eigenvalues and satisfying certain symmetry requirements: the Hamiltonian *H* is a partial differential operator which is an essentially self-adjoint operator on $\mathcal{H}$ and is given by:

$$H = -\tfrac{1}{2}\sum_{i=1}^{n} \nabla_i^2 - \frac{Z}{r_i} + \sum_{i<j} \frac{1}{r_{ij}}, \qquad (2.2)$$

where

$$r_{ij} = [(x_{1i} - x_{1j})^2 + (x_{2i} - x_{2j})^2 + (x_{3i} - x_{3j})^2]^{1/2}. \qquad (2.3)$$

In this expression *Z* is the number of protons in the nucleus of the atom and a term with the index *i* acts on the *i*th copy of $\mathcal{L}^2(\mathbb{R}^3)$ and on remaining copies and components has the effect of the identity operator: this convention avoids the use of unwieldy expression involving tensor products of many identity operators. From the convention it follows also that the terms with two indices *i* and *j* act on both the *i*th and *j*th copies of $\mathcal{L}^2(\mathbb{R}^3)$ and as the identity operator on all the remaining copies and components. In writing the Hamiltonian in this form we have used natural atomic units in which the reduced mass of the electron (the reduced mass takes into account the fact that it is moving round a nucleus of finite mass), velocity of light and Planck's constant divided by 2π all take unit values. We use a peculiar geometric device by which the unit of length is changed to $1/Z$ natural atomic units while the unit of energy is changed to Z^2 natural atomic units. In these modified units the Hamiltonian takes the form

$$H = -\tfrac{1}{2}\sum_{i=1}^{n}\nabla_i^2 - \frac{1}{r_i} + \sum_{i<j}\frac{1}{Zr_{ij}} \,. \tag{2.4}$$

If we ignore the terms which depend on Z, then all the stationary states are known in terms of the known solutions for the hydrogen atom. This approximation is called the *zero*th order solution and ignores the repulsive forces which electrons exert on each other. One then assumes that the eigenfunctions of the correct H and the corresponding eigenvalues have Taylor's expansions in powers of $1/Z$. A considerable amount of work has been done on the justification of this assumption or rather the precise conditions under which this assumption may be valid and an account of these with appropriate references will be found in [5]. For the purposes of the present work, we assume that the appropriate conditions are satisfied and the assumed power series expansions exist and are valid. We write for a typical stationary state represented by the eigenvector Φ belonging to the eigenvalue E (which actually is the energy in the appropriate units):

$$H\Phi = E\Phi, \tag{2.5}$$

$$\Phi = \sum_{i=0}^{\infty} \Phi_i/Z^i, \tag{2.6}$$

and

$$E = \sum_{i=0}^{\infty} E_i/Z^i. \tag{2.7}$$

Substitution of the second and third of these equations in the first and equating the coefficients of of $(1/Z^i)$ for $i = 0, 1, 2, 3$ gives

$$H_0\Phi_0 = E_0\Phi_0, \tag{2.8}$$

$$(H_0 - E_0)\Phi_1 = (E_1 - H_1)\Phi_0, \tag{2.9}$$

$$(H_0 - E_0)\Phi_2 = (E_1 - H_1)\Phi_1 + E_2\Phi_0, \tag{2.10}$$

$$(H_0 - E_0)\Phi_3 = (E_1 - H_1)\Phi_2 + E_2\Phi_1 + E_3\Phi_0, \tag{2.11}$$

where
$$H_1 = \sum_{i<j} \frac{1}{Zr_{ij}} \tag{2.12}$$

and
$$H_0 = H - H_1. \tag{2.13}$$

If we take Φ to be normalized: $(\Phi,\Phi) = 1$, then a similar equating of coefficients in the expansion of the normalizing expression gives

$$(\Phi_0,\Phi_0) = 1, \tag{2.14}$$

$$(\Phi_1,\Phi_0) + (\Phi_0,\Phi_1) = 0, \tag{2.15}$$

$$(\Phi_2,\Phi_0) + (\Phi_1,\Phi_1) + (\Phi_0,\Phi_2) = 0, \tag{2.16}$$

$$(\Phi_3,\Phi_0) + (\Phi_2,\Phi_1) + (\Phi_1,\Phi_2) + (\Phi_0,\Phi_3) = 0, \tag{2.17}$$

With the help of equations (2.8-2.17) one easily proves that

$$E_1 = (H_1\Phi_0,\Phi_0), \tag{2.18}$$

$$2E_2 = (H_1\Phi_1,\Phi_0) + (H_1\Phi_0,\Phi_1), \tag{2 .19}$$

$$E_3 = ((H_1 - E_1)\Phi_1,\Phi_1). \tag{2 .20}$$

Thus an accurate knowledge of Φ_1 enables one to estimate accurately the eigenvalue up to the third order.

It has already been stated earlier that if H_1 is ignored the complete solution is known in terms of the known solutions of the hydrogen atom, that is, in terms of the eigenfunctions of the operator $-\frac{1}{2}\nabla^2 - \frac{1}{r}$. In the expansions assumed above, the zero-order approximation is the one in which H_1 is completely ignored. It is possible to solve the first-order equation (2.9) exactly for any atom if the solutions of equation (2.9) for all the states of the helium atom were known exactly. So in the first-order the problem reduces to finding the solutions of the two electron atom. These solutions are formally known as sums of integrals over the spectrum of the operator $-\frac{1}{2}\nabla^2 - \frac{1}{r}$. Though a laborious numerical evaluation of these integrals can give estimates of the second- and third-order eigenvalues a similar computation of the eigenfunction is quite a hopeless task. The method developed by this author and co-workers has reduced this task to a level where numerical estimates of the first-order eigenfunction and, therefore, of the eignevalue up to the third-order is feasible. The main feature of this method is to reduce the solution as a double integral over the spectrum to a single integral where the integrand which is therefore another single integral over the spectrum, is the solution of a second-order ordinary differential equation which is easy to solve. By solving this differential equation which depends on a single parameter which varies over the spectrum of the hydrogenic Hamiltonian $-\frac{1}{2}\nabla^2 - \frac{1}{r}$ for numerous valid values of the spectrum an accurate approximation of the first-order wave function can be obtained. Details of the calculation are described elsewhere [12,13], but the problem reduces to solving many differential equations of the family:

$$(L_\ell + \lambda)\Phi = F, \tag{2.21}$$

where

$$L_\ell = \left(-\frac{1}{2}\frac{d^2}{dr} - \frac{1}{r}\frac{d}{dr} - \frac{1}{r} + \frac{\ell(\ell+1)}{2r^2}\right), \tag{2.22}$$

F is a known member of the Hilbert space of normed functions on $[0, \infty[$ with the inner-product defined by

$$(f,g) = \int_0^\infty f(r)g(r)r^2 \; dr, \tag{2.23}$$

and λ is a known real number such that $-\lambda \notin \Lambda(L_\ell)$. Let

$$\nu = (1/2\lambda)^{1/2}. \tag{2.24}$$

The exact analytical solution of the equation (2.21) can be obtained by expanding Φ in terms of functions

$$e^{-r/\nu}\left(\tfrac{2r}{\nu}\right)^{\ell} L_i^{2\ell+1}\left(\tfrac{2r}{\nu}\right), \tag{2.25}$$

where $L_n^k(x)$ is the associated Laguerre polynomial of order k and degree $n - k$ and is explicitly defined in [17]. The task of evaluating the expansion coefficients is extremely easy, because with a simple redefinition of the inner-product, these functions become an orthonormal family. Further details of the methods used in calculating the expansion coefficients and estimates of the second- and third-order terms in the expansion of the eigenvalue in powers of $1/Z$ are fully described in [10,12,17]. Thus the geometric concept of perpendicularity plays a crucial role in this theory. All the estimates can be evaluated even more easily numerically by the remarkably simple method described in [15]. What we have described here is actually a summary of the method which evaluates the contribution to the first-order eigenfunction from that region in the double integral over the spectrum in which one of the integrating parameters belongs to the discrete spectrum: the method for evaluating the contribution from the region where both parameters belong to the continuum is still to be discovered and is an open question. Fortunately, empirical evidence suggests that the latter contribution is relatively small. Furthermore, numerical methods for evaluating rough estimates of this contribution already exist and they too support the empirical evidence: further details will be found in [10]. The most important feature of this method is that just as the zero-order eigenfunction of any stationary state of any atom is known in terms of the eigenfunctions of the hydrogen atom, if the first-order terms in the expansion of eigenfunctions of all the stationary states of the helium atom are worked out, then the first-order terms in the expansion of eigenfunctions of all stationary states of all atoms are easily expressed as weighted sums of the solutions for the helium atom. This leads to a substantial economy of time and effort in obtaining these solutions.

3. An Interesting Open Problem

The theory described above has given rise to an interesting open problem in the area of differential equations whose solution may have considerable impact on the very foundations of quantum theory. Hydrogen atom which has just one electron moving round the nucleus has infinitely many points in the discrete spectrum of its Hamiltonian, each of these points correspond to a stationary state. The Hamiltonian of the hydrogen atom also has a continuous spectrum, energy values belonging to the continuous spectrum correspond to a system of a proton and an electron in which the two particles are not bound to each other. The same remark applies to the spectra of the Hamiltonians of other atoms also but there is one peculiar feature: in the continuous spectrum there are regions or bands which correspond to states in which all the electrons appear to be bound to the nucleus as in a stationary state for a while but ultimately one or more of the electrons part company with the others and move to infinity thus giving rise to the so-called unstable states of the system. The existence of these states have something to do with the phenomenon of spectral concentration [16 and references therein], they have something to do also with the complex eigenvalues of the Hamiltonian though, of course, the Hamiltonian of a quantum system is required to be self-adjoint and self-adjoint operators cannot, under any circumstances, have complex eigenvalues. However, when physicists talk of complex eigenvalues of Hamiltonians, they mean complex eigenvalues of an operator which is obtained by modifying the self-adjoint operator by a procedure which is often ill-defined. A review of some of these aspects will be found in another work of this author [9]. What is clear is that the unstable states of atoms are intermediate between stationary states and states in which at least one electron is not bound to the nucleus. In atomic physics, in calculating energies of stationary states, as a first approximation one assumes that each electron moves in the field of the nucleus and the average of the fields of other electrons, so that the eigenfunction corresponding to the stationary state does not depend on inter-electronic distances: this approximation is known as the Hartree-Fock approximation. Eigenenergies and eigenfunctions in this approximation are easy to calculate. However, it is clear that because electrons repel each other, they will try to avoid each other and the exact eigenfunctions will, therefore, depend on inter-electronic distances. This effect is known as electronic correlation. Corrections for electronic correlation lowers the energy eigenvalue and it is generally true that that the lowering of energy eigenvalues gives a more stable state. One of the strange outcomes of the theory described in the preceding Section, is that if it is applied to bring in correlation to the doubly excited states of the helium atom, then it is seen that it is precisely correlation which causes the system to become unstable: in the Hartree-Fock approximation these states are quite stable.

Thus, as is often the case in quantum theory, what seems intuitively obvious is shown to be false: in doubly excited states of helium electrons by avoiding each other do not produce a system which is more stable but they cause the disintegration of the system. One successful way of calculating the energy of an unstable state decomposes the Hilbert space into two complementary and mutually orthogonal subspaces and if the projection on these subspaces are P and P' respectively, with a choice of P based on physical intuition, PHP, where H is the Hamiltonian of the system, has a point eigenvalue corresponding to the unstable state but the energy of the unstable state belongs to the continuous spectrum of H. In this description the difference between the eigenvalues of H and PHP is called the level shift and the method described in the previous Section is a very effective method of finding the level shift and has actually been used by several workers. We shall consider a typical differential equation which gives the level shift and consider three different methods of evaluating the level shift: the methods look very different but give almost identical results. Proving the equivalence of the three methods is an interesting and useful problem which is still open. We consider a particular second-order ordinary differential equation which arises in the calculation of the level shift of one of the unstable states of the helium atom. The simplest such equation is:

$$(L_1 - \tfrac{1}{4})\Phi = (1/27\sqrt{3})e^{-2r}r(3r + 2) \tag{3.1}$$

where L_1 is as defined in (2.22) with $\ell = 1$. No solution exists in the Hilbert space of functions on which inner-product is defined by (2.23), nevertheless if we denote the right hand side of (3.1) by F, then assuming that the correct solution whose norm is infinite according to (2.23) has been found, one of the terms in the correct contribution to the level shift is given by (Φ,F), the inner-product of Φ and F evaluated according to (2.23). The only definite knowledge we have about Φ, apart from the fact that it satisfies equation (3.1), is that it is zero at the origin and at infinity. The homogeneous equation

$$(L_1 - \tfrac{1}{4})\Phi_0 = 0, \tag{3.2}$$

definitely has such a solution and if Φ is any solution satisfying (3.1) and the boundary conditions, then so is $\Phi + \alpha\Phi_0$, where α is any real number and Φ_0 is the solution of (3.2) satisfying the given boundary conditions. Sharma and

Bowtell [14] proposed that the unique solution required by the problem satisfies the orthogonality condition

$$(\Phi,\Phi_0) = 0, \tag{3.3}$$

some physical justification for this requirement exists. Sharma and Bowtell then solved equations (3.1) and (3.2) numerically by setting $\Phi(0) = \Phi_0(0) = 0$ and $\Phi'(0) = \Phi_0'(0) = 1$ and evaluated α by calculating the asymptotic value of the ratio of the two solutions at the local maxima of Φ, for large enough r this is indeed constant and is taken to be α, the unwanted contribution of Φ_0 in the numerical solution of (3.1) was then removed by using this value of α and (Φ,F) was evaluated using the solution thus corrected. Other terms were calculated by essentially the same method and the final numerical values were in excellent agreement with experimental values.

The second method for evaluation of (Φ,F) of the preceding paragraph is due to Horák and Lewis [3]. They evaluated this inner-product both analytically and numerically without solving the differential equation (3.1). Their work involves quite complicated classical analysis and it is difficult to draw an overall picture of their method – this author has a copy of the handwritten original manuscript of [3] with the message: "*It is difficult to describe what one is doing.*" from the first author. Putting their work in proper perspective is another open problem which will have to be solved before the equivalence of their method with that of the preceding paragraph can be demonstrated. We shall give here only their final result, which is impressive both in form and in substance. They find

$$(\Phi,F) = -2^{-8}\,3^{-18}\,90650177 - 2^{15}\,3^{-18}\,5^2\,S(2,2) \tag{3.4}$$

where

$$S(2,2) = \mathrm{Re}[(1+k)^{-1}F(1, k+1, k+2, z)\,z], \tag{3.5}$$

here F is the hypergeometric function [4], $k = i\sqrt{2}$, $z = exp(i\psi)$, $cos\psi = 17/81$ and $sin\psi = -56\sqrt{2}/81$. Horák and Lewis [3] say that removing the imaginary part from (3.5) is equivalent to the the orthogonalization of the preceding paragraph and removes the contribution of the unwanted solution of the homogeneous equation. The numerical value of (Φ,F) from (3.4) was 4.489117×10^{-4} compared to 4.6898×10^{-4} obtained by the method of the preceding paragraph, the agreement seems poor but for comparison with experimental values only the first figure was significant. Nevertheless, some years later Thomas and Sharma [18] recalculated this value by the method of Sharma and Bowtell [14] using greater precision and their value was 4.48915×10^{-4} which is in excellent agreement with the value of Horák and Lewis [3].

Yet another method for evaluating (Φ,F) has been much more recently developed by Manning and Sanders [7]. The method is based on the complex rotation of the Hamiltonian, the associated theory and the mathematics has been reviewed in a paper by Ho [2]. In this method the rotated Hamiltonian is obtained by a complex scaling of the radial coordinates by the transformation $r\mapsto e^{i\theta}r$, consequently $-\frac{1}{2}\nabla^2 - \frac{1}{r}$ is transformed into $-\frac{1}{2}e^{-i2\theta}\nabla^2 - e^{-i\theta}\frac{1}{r}$. Transformations which were made in Section 2 to convert the partial differential equations into ordinary differential equations are not made but the same perturbation theory is used. The first order equation (2.9) with the present scaling reads:

$$\left(-\tfrac{1}{2}e^{-2i\theta}\nabla_1{}^2 - e^{-i\theta}\frac{1}{r_1} - \tfrac{1}{2}e^{-2i\theta}\nabla_2{}^2 - e^{-i\theta}\frac{1}{r_2} - E_0\right)\Phi_1$$

$$= \left(E_1 - e^{-i\theta}\frac{1}{r_{12}}\right)\Phi_0 \qquad (3.6)$$

The operator P' corresponding to the subspace which causes the instability is then applied to both sides and then the functional is found whose extremum corresponds to the resulting solution. Manning and Saunders [7] then optimize the functional to get an approximation to that part of the first-order function which causes the instability. This enables them to calculate the level shift and the values they obtain are in excellent agreement with those of Horák and Lewis [3].

Here we have three different methods of computing the same quantity. The methods are as different from each other as one can imagine, but they consistently give the same answer. There can be very little doubt that they tell the same truth or in other words the methods are equivalent. Since the primary problem is in partial differential equations, the ultimate explanation of the equivalence of the three methods must lie somewhere in the theory of partial differential equations and since such an explanation will assist in a development of an important and useful physical theory, we believe that this is an important problem. The third method also brings us close to the theory of complex eigenvalues of the Hamiltonian. If the Hamiltonian is scaled as in the preceding paragraph, the scaled operator is obviously non-self-adjoint and can have complex eigenvalues. It has been claimed that under such a transformation, discrete eigenvalues remain unchanged on the real axis but the continuous spectrum is rotated by an angle 2θ into the lower half of the complex energy plane. The unstable states are the discrete complex eigenvalues of the rotated Hamiltonian. For these states the real part of the eigenvalue is the energy and the imaginary part is the so-called half-width which determines the rate at which these states decay. In the words of Manning and Sanders [7] *as increasing θ sweeps the associated continua past a resonance, the wave function for that resonance becomes square-integrable. With the function square-integrable, calculation of the eigenvalues proceeds in the manner of a standard variational calculation for a bound state.*

4. Concluding Remarks

Schrödinger's equation for an atom is one of the best known partial differential equations in both mathematics and physics and more papers have been written on this equation than on probably any other partial differential equation. Since there cannot be any doubt that this equation contains one of the most profound truths about the nature of physical reality, it is not surprising that there has been such intense activity concentrated on this single equation. However, the methods used by mathematicians and physicists to understand and solve this equation have been often radically different and a paper written on some aspect of this problem by an average physicist is often as unintelligible to an average mathematician as a paper written on the same problem by the latter is to the former. In this work, we have brought together work done by physicists and chemists on a certain very important aspect of the problem and we find that three different, apparently unrelated routes, lead to the same truth. We have tried to describe the mathematics behind the methods as far as they are understood and tried to state the associated mathematical problem in a way which might make sense to a mathematician, in doing so we have tried to remove the physical jargon as far as practicable because the physical jargon is often what makes such work so unintelligible to mathematicians. Yet we have tried to state briefly the physical problem which lies behind all this. We give references to a relatively large number of works for a small paper, where the interested reader will find further information. Before concluding, we would like to remind the reader that the theory of bound states of atoms is completely well-defined in a mathematical sense and is in complete conformity with the mathematical foundations of the subject as built by von Neumann and Birkhoff [1], Mackey [6] and others [11,19]. On the other hand, the theory of unstable states of atoms is somewhat vague and large parts of the theory are completely alien in the mathematical territory in which the mathematical foundations of quantum theory reside. Some of the difficulties in applying rigorous quantum theory to unstable states have been described by Sharma [9] and Sharma and Ogunsulire [16]. Anyone who takes a decisive step in bringing the naturalization of the theory of unstable states as a resident of the territory of the preceding sentence could be a good candidate for the award of both the Fields Medal (in mathematics) and the Nobel Prize in physics. Thus the underlying fundamental problem is a big one, but there are also a number of small problems to which this work draws attention in the hope that interested readers will bring their solution nearer.

Acknowledgement

It is acknowledged with great pleasure that this work was written at the suggestion of Professor Themistocles M. Rassias.

References

1. G. Birkhoff and J. von Neumann, *The logic of quantum mechanics,* Ann. Math. **37** (1936) 823-843.

2. Y. K. Ho, *The method of complex coordinate rotation and its applications to atomic collision processes,* Phys. Rep. **99**(1983), 1-68.

3. Z. J. Horák and M. N. Lewis, *Z-dependent theory of many electron atoms. II. Level shifts of autoionizing states,* Czech. J. Phys. **26** (1976), 1088-1094.

4. H. Jeffreys and B. Swirles (Lady Jeffreys), *Methods of mathematical physics,* 3rd ed., Cambridge University Press, Cambridge, 1962.

5. T. Kato, *Perturbation theory for linear operators,* Springer, Berlin, 1966.

6. G. W. Mackey, *Mathematical foundations of quantum mechanics,* Benjamin, New York, 1963.

7. L. W. Manning and F. C. Saunders, *Lowest-order contribution to the width and shift of autoionizing states of two-electron atoms,* Phys. Rev. A, **43** (1991), 272-281.

8. J. M. Seminario and F. C. Saunders, *Application of Z-dependent perturbation theory to autoionizing states of heliumlike atoms: Feshbach projection method,* Phys. Rev. A, **42** (1990), 2562-2572.

9. C. S. Sharma, *A critical look at conjectures in the theory of autoionizing states of atoms,* Quantum chemistry: the state of the arty (edited by V. R. Saunders and J. Brown), Science Research council, Didcot, 1975.

10. C. S. Sharma, *Correlation energies in atoms,* Physics Reports, **26C** (1976), 1-67.

11. C. S. Sharma, *Mackey's eighth axiom and quantum logics,* Phys. Lett. **80A** (1980) 135-139.

12. C. S. Sharma and G. Bowtell, *A new approach to correlation energies in atoms,* Nature Phys. Sc. **233** (1971), 34-35.

13. C. S. Sharma and G. Bowtell, *The exact evaluation of certain partial sums of the second order energies of atoms I. The ground and singly excited states,* J. Phys. B (London), **5** (1972), 1062-1070.

14. C. S. Sharma and G. Bowtell, *A new method for the calculation of level shifts of autoionizing states of atoms with application to the 2s2p $^{1,3}P$ autoionizing states of two-electron atoms,* Canad. J. Phys. **51** (1973), 1637-1645.

14. C. S. Sharma and G. Bowtell, *Numerical solution of first-order correlation equations for atoms,* Quantum chemistry: the state of the arty (edited by V. R. Saunders and J. Brown), Science Research council, Didcot, 1975.

15. C. S. Sharma and E. O. Ogunsulire, *Axiomatic quantum theory and resonances,* J. Phys. A (London), **11** (1978), 2159-2177.

16. C. S. Sharma and R. G. Wilson, *A perturbation treatment of the Hartree-Fock equations for 1s2p ^3P and ^1P states of helium like ions,* J. Phys. B (London), **1** (1968), 1-12.

17. R. K. Thomas and C. S. Sharma, *Numerical calculation of the level shift of the autoionizing states of atoms,* Phys. Rev. A, **27** (1983), 1187-1188.

18. V. Varadarajan, *Geometry of quantum theory, Vols. I & II,* Van Nostrand, Princeton, 1968.

Geometry in Partial Differential Equations (pp. 389–422)
eds. A. Pràstaro and Th. M. Rassias

HIGHER SYMMETRIES AND CONSERVATION LAWS OF EULER-DARBOUX EQUATIONS

VALERY E. SHEMARULIN
All-Russian Research of Experimental Physics
Arzamas-16 607200, Russia

ABSTRACT

Generating functions are found for all local symmetries and conservation laws of Euler-Darboux equation $\mathcal{Y}_n$ with the parameter $n \notin \mathbb{Z}$. It is proved that in the symmetries algebra the basis of subalgebra, which is the direct complement to the infinite-dimensional Abelian ideal consisting of smooth solutions of the equation $\mathcal{Y}_n$, is generated by three recursion operators from zero-order trivial symmetry. Explicit differential form is given for conservation laws corresponding to the generating functions found.

1. Introduction

The paper considers solving the problem of description of higher (local) symmetries and conservation laws spaces for the Euler-Darboux equation

$$\mathcal{Y}_n \;=\; \{F_n(u) \equiv u_{\xi\eta} - \frac{n}{\xi+\eta} \cdot (u_\xi + u_\eta) = 0\}, \qquad n \in \mathbb{R}, \tag{1}$$

belonging to one of the classes of linear hyperbolic second-order equations with two independent variables admitting most broad algebra of point symmetries[1]. The urgency of this problem is determined by the role Eq. 1, denoted further as $\mathcal{Y}_n$, plays in the theory of partial differential equations and in gas dynamics. Algebra of classical (point) symmetries of the equation $\mathcal{Y}_n$ is well known[1]. In the present paper $\mathrm{Sym}\,\mathcal{Y}_n$ algebra of higher symmetries and local conservation laws group for the equation $\mathcal{Y}_n$ with a parameter $n \notin \mathbb{Z}$ are found. It is proved that in $\mathrm{Sym}\,\mathcal{Y}_n$ algebra the basis of subalgebra, which is the direct complement to the infinite-dimensional Abelian ideal consisting of smooth solutions of the equation $\mathcal{Y}_n$, is generated by three recursion operators from zero-order trivial symmetry. Explicit differential form is given for conservation laws corresponding to principal generating functions. It should be noted that the results obtained here can be easily adapted to the equation for velocity potentials of polytropic gas one-dimensional plane isentropic flows

$$\varphi_{tt} \;+\; 2\varphi_x \cdot \varphi_{xt} \;+\; [(k-1)\varphi_t + \frac{k+1}{2}\varphi_x^2] \cdot \varphi_{xx} \;=\; 0, \qquad k = \mathrm{const}, \tag{2}$$

which reduces to the equation $\mathcal{Y}_n$ by Legendre transformation with consequent transition to characteristic variables[2,3]. Besides, they allow to solve the problem of description of local conservation laws group and higher variation symmetries algebra for the system of two equations describing the same gas dynamic flows as Eq. 2. The solving of this problem will be considered in a special paper.

Principal results concerning description of the higher symmetries algebra of the equation $\mathcal{Y}_n$ and the group of its local conservation laws obtained here are announced by the author in[3]. The paper employs results and methods of modern differential equations geometry and, in particular, of higher symmetries and conservation laws theory developed in[4,5].

2. Internal coordinates on $\mathcal{Y}_{n,\infty}$ and some formulas for total differentiation operators

According to the geometrical theory, the equation $\mathcal{Y}_n$ is a submanifold in the manifold $J^2(\pi) = J^2(2,1)$ of 2-jets of local sections of the trivial bundle $\pi : \mathbb{R}^2 \times \mathbb{R} \to \mathbb{R}^2$. Independent variables ξ and η are the coordinates in the base $\mathbb{R}^2$, while the dependent variable u is a coordinate in the $\mathbb{R}$ fibre. This trivial bundle is the single one further encountered. Therefore we shall use the symbols J^l and J^∞ instead of $J^l(\pi) = J^l(2,1)$ and $J^\infty(\pi) = J^\infty(2,1)$ denoting l-jets and infinite jets manifolds of trivial bundle π local sections (smooth functions of variables ξ and η) throughout the paper for brevity. A special system of coordinates exists on J^∞:

$$\xi, \eta, u;\ u_{m\xi,l\eta};\ m \geq 0, l \geq 0 \quad (u_{0\xi,0\eta} \equiv u) \tag{3}$$

such that

$$j_\infty(\varphi)^*(u_{m\xi,l\eta}) = \frac{\partial^{m+l}\varphi(\xi,\eta)}{\partial\xi^m\partial\eta^l}; \qquad m \geq 0,\ l \geq 0 \tag{4}$$

for any function $\varphi(\xi,\eta)$. The same coordinates Eq. 3 subject to the condition $m+l \leq q$ will be used on J^q with $q \geq 0$. Let D_ξ and D_η be the operators of total differentiation with respect to ξ and η:

$$\begin{aligned} D_\xi &= \frac{\partial}{\partial\xi} + \sum_{m,l\geq 0} u_{(m+1)\xi,l\eta} \cdot \frac{\partial}{\partial u_{m\xi,l\eta}}, \\ D_\eta &= \frac{\partial}{\partial\eta} + \sum_{m,l\geq 0} u_{m\xi,(l+1)\eta} \cdot \frac{\partial}{\partial u_{m\xi,l\eta}}, \end{aligned} \tag{5}$$

and $\mathcal{Y}_{n,\infty}$ - an infinite prolongation of the equation $\mathcal{Y}_n$.$\mathcal{Y}_{n,\infty}$ is a submanifold in J^∞ defined by an infinite system of equations obtained from the equation $\mathcal{Y}_n$ by attaching all of its differential consequences derived with the operators Eq. 5:

$$\mathcal{Y}_{n,\infty} = \{D_\xi^m D_\eta^l(F_n) = 0; \qquad m, l = 0, 1, 2, \ldots\}. \tag{6}$$

According to Eqs. 6, all variables $u_{m\xi,l\eta}$; $m > 0, l > 0$ (corresponding to mixed derivatives of the dependent variable u) are represented on the manifold $\mathcal{Y}_{n,\infty}$ in terms of variables $\xi, \eta, u_{m\xi}, u_{m\eta}$; $m \geq 1$, which may take arbitrary values on $\mathcal{Y}_{n,\infty}$ together with u. Therefore we take the functions

$$\xi, \eta, u, u_\xi, u_\eta, \ldots, u_{m\xi}, u_{m\eta}, \ldots; \qquad m = 1, 2, 3, \ldots. \tag{7}$$

as internal coordinates on $\mathcal{Y}_{n,\infty}$. The equation $\mathcal{Y}_n$ symmetries are identified with functions φ (generating functions) of the variables Eq. 7 which are solutions of the equation

$$\bar{l}_{F_n}(\varphi) = 0; \qquad l_{F_n} = D_\xi \circ D_\eta - \frac{n}{\xi + \eta} \cdot (D_\xi + D_\eta), \tag{8}$$

where $\bar{l}_{F_n}$ is the restriction on the equation $\mathcal{Y}_{n,\infty}$ of the universal linearization operator l_{F_n}. Since $u_{\xi\eta} = \frac{n}{\xi+\eta} \cdot (u_\xi + u_\eta)$ on the manifold $\mathcal{Y}_n$, then the restrictions $\bar{D}_\xi$ and $\bar{D}_\eta$ on the equation $\mathcal{Y}_{n,\infty}$ of the operators D_ξ and D_η have the form:

$$\begin{aligned} \bar{D}_\xi &= \frac{\partial}{\partial \xi} + \sum_{m=0}^{\infty} u_{(m+1)\xi} \cdot \frac{\partial}{\partial u_{m\xi}} + \sum_{m=1}^{\infty} n \cdot \bar{D}_\eta^{m-1}\left(\frac{u_\xi + u_\eta}{\xi + \eta}\right) \cdot \frac{\partial}{\partial u_{m\eta}}; \\ \bar{D}_\eta &= \frac{\partial}{\partial \eta} + \sum_{m=1}^{\infty} n \cdot \bar{D}_\xi^{m-1}\left(\frac{u_\xi + u_\eta}{\xi + \eta}\right) \cdot \frac{\partial}{\partial u_{m\xi}} + \sum_{m=0}^{\infty} u_{(m+1)\eta} \cdot \frac{\partial}{\partial u_{m\eta}}. \end{aligned} \tag{9}$$

It can be easily seen from direct calculations that relations

$$\begin{aligned} \bar{D}_\xi\left(\frac{u_\xi + u_\eta}{\xi + \eta}\right) &= (n-1) \cdot \frac{u_\xi + u_\eta}{(\xi + \eta)^2} + \frac{u_{2\xi}}{\xi + \eta}; \\ \bar{D}_\xi^2\left(\frac{u_\xi + u_\eta}{\xi + \eta}\right) &= (n-1)(n-2) \cdot \frac{u_\xi + u_\eta}{(\xi + \eta)^3} + (n-2)\frac{u_{2\xi}}{(\xi + \eta)^2} + \frac{u_{3\xi}}{\xi + \eta} \end{aligned} \tag{10}$$

are true. Furthermore, the following general formula is easily proved by induction:

$$\begin{aligned} \bar{D}_\xi^r\left(\frac{u_\xi + u_\eta}{\xi + \eta}\right) &= (n-1)(n-2)(n-3)\ldots(n-r) \cdot \frac{u_\xi + u_\eta}{(\xi + \eta)^{r+1}} \\ &+ \sum_{i=2}^{r} (n-i)(n-(i+1))\ldots(n-r) \cdot \frac{u_{i\xi}}{(\xi + \eta)^{r-i+2}} + \frac{u_{(r+1)\xi}}{\xi + \eta}; r \geq 3. \end{aligned} \tag{11}$$

Analogous formulas for $\bar{D}_\eta^r\left(\frac{u_\xi + u_\eta}{\xi + \eta}\right)$ with $r \geq 1$ are obtained from Eq. 10 and Eq. 11 by formal substitution: $u_{i\xi} \to u_{i\eta}$; $\quad i = 2, 3, \ldots, r+1$.

3. The equation $\mathcal{Y}_n$ contact symmetries algebra and first order recursion operators, n $\neq 0, -1$

We shall assume $n \neq 0, -1$ throughout this section. The generating functions of contact symmetries have the form

$$\varphi \equiv \varphi(\xi, \eta, u, u_\xi, u_\eta). \tag{12}$$

According to the formulas (9)-(11) and those for $\bar{D}_\eta^r\left(\frac{u_\xi+u_\eta}{\xi+\eta}\right)$ analogous to the formulas Eq. 10-Eq. 11 it is easily seen that the left-hand side of Eq. 8 for the function Eq. 12 is a second-degree polynomial with respect to the variables $u_{2\xi}$ and $u_{2\eta}$ of the form $\bar{l}_F(\varphi) \equiv \Lambda_1 \cdot u_{2\xi} \cdot u_{2\eta} + \Lambda_2 \cdot u_{2\xi} + \Lambda_3 \cdot u_{2\eta} + \Lambda_4$, where Λ_i; $i = 1, 2, 3, 4$ are some known functions of the variables $\xi, \eta, u, u_\xi, u_\eta$ and various first and second order φ partial derivatives. By setting coefficients Λ_1, Λ_2 and Λ_3 to 0, we obtain the following equations, correspondingly:

$$\begin{aligned}
&\frac{\partial^2\varphi}{\partial u_\xi \partial u_\eta} = 0,\\
&\frac{\partial^2\varphi}{\partial\eta\partial u_\xi} + u_\eta \cdot \frac{\partial^2\varphi}{\partial u \partial u_\xi} + n \cdot \frac{u_\xi + u_\eta}{\xi+\eta} \cdot \frac{\partial^2\varphi}{\partial u_\xi^2} = 0,\\
&\frac{\partial^2\varphi}{\partial\xi\partial u_\eta} + u_\xi \cdot \frac{\partial^2\varphi}{\partial u \partial u_\eta} + n \cdot \frac{u_\xi + u_\eta}{\xi+\eta} \cdot \frac{\partial^2\varphi}{\partial u_\eta^2} = 0.
\end{aligned} \tag{13}$$

The first of the equations Eq. 13 immediately yields

$$\varphi \equiv \varphi_1(\xi, \eta, u, u_\xi) + \varphi_2(\xi, \eta, u, u_\eta).$$

By differentiating the second of the equations Eq. 13 with respect to u_η we obtain

$$\frac{\partial^2\varphi_1}{\partial u \partial u_\xi} + \frac{n}{\xi+\eta} \cdot \frac{\partial^2\varphi_1}{\partial u_\xi^2} = 0. \tag{14}$$

First integrating the final equation with respect to u_ξ and then using the method of characteristics, we get $\varphi_1 \equiv \delta_1(\xi, \eta, u) + \rho(\xi, \eta, u_\xi - \frac{n}{\xi+\eta} \cdot u)$. The second of the equations Eq. 13 and Eq. 14 give the following equation :

$$\frac{\partial^2\rho}{\partial\eta\partial u_\xi} + \frac{n}{\xi+\eta} \cdot u_\xi \cdot \frac{\partial^2\rho}{\partial u_\xi^2} = 0,$$

which takes the form

$$\frac{\partial^2\rho}{\partial\mu\partial\eta} + \frac{n}{\xi+\eta} \cdot \mu \cdot \frac{\partial^2\rho}{\partial\mu^2} + \frac{n(n+1)}{(\xi+\eta)^2} \cdot u \cdot \frac{\partial^2\rho}{\partial\mu^2} = 0; \qquad \rho \equiv \rho(\xi, \eta, \mu), \tag{15}$$

after introducing a new variable $\mu \equiv u_\xi - \frac{n}{\xi+\eta} \cdot u$. For $n \neq 0, -1$, the variable u in Eq. 15 is free. Therefore

$$\frac{\partial^2 \rho}{\partial \mu^2} \equiv \frac{\partial^2 \rho}{\partial \mu \partial \eta} \equiv 0$$

and $\rho = \rho_1(\xi,\eta) + B(\xi)\cdot\mu \equiv \rho_1(\xi,\eta) + B(\xi)\cdot(u_\xi - \frac{n}{\xi+\eta}\cdot u)$. Thus $\varphi_1 \equiv \delta(\xi,\eta,u) + B(\xi)\cdot u_\xi$. Analogously, we find $\varphi_2 = \lambda(\xi,\eta,u) + C(\eta)\cdot u_\eta$ from the last of the equations Eq. 13 (however this immediately follows from the symmetry considerations). So we obtain for φ the expression

$$\varphi \equiv \tilde{\varphi}(\xi,\eta,u) + C(\eta)\cdot u_\eta + B(\xi)\cdot u_\xi, \tag{16}$$

which yields a general solution of system Eq. 13 for arbitrary smooth functions, $\tilde{\varphi}, C, B$. For φ symmetry of the form Eq. 16, $\bar{l}_{F_n}(\varphi) = \Lambda_4 = 0$ with the left-hand side of $\Lambda_4 = 0$ being a second-degree polynomial with respect to u_ξ, u_η. Setting these polynomial coefficients to 0, we obtain the equations:

$$\frac{\partial^2 \tilde{\varphi}}{\partial u^2} = 0; \tag{17}$$

$$\begin{cases} \dfrac{\partial^2 \tilde{\varphi}}{\partial u \partial \eta} + \dfrac{n}{\xi+\eta}\cdot\dfrac{\partial C}{\partial \eta} - \dfrac{n}{(\xi+\eta)^2}(C+B) = 0, \\ \dfrac{\partial^2 \tilde{\varphi}}{\partial u \partial \xi} + \dfrac{n}{\xi+\eta}\cdot\dfrac{\partial C}{\partial \xi} - \dfrac{n}{(\xi+\eta)^2}(C+B) = 0; \end{cases} \tag{18}$$

$$\frac{\partial^2 \tilde{\varphi}}{\partial \xi \partial \eta} - \frac{n}{\xi+\eta}\cdot\left(\frac{\partial \tilde{\varphi}}{\partial \xi} + \frac{\partial \tilde{\varphi}}{\partial \eta}\right) = 0. \tag{19}$$

The general solution of Eq. 17 and Eq. 19 is given by the formula

$$\tilde{\varphi} = A_1(\xi,\eta) + A_2(\xi,\eta)\cdot u,$$

where

$$\begin{aligned} \frac{\partial^2 A_1}{\partial \xi \partial \eta} - \frac{n}{\xi+\eta}\cdot\left(\frac{\partial A_1}{\partial \xi} + \frac{\partial A_1}{\partial \eta}\right) &= 0, \\ \frac{\partial^2 A_2}{\partial \xi \partial \eta} - \frac{n}{\xi+\eta}\cdot\left(\frac{\partial A_2}{\partial \xi} + \frac{\partial A_2}{\partial \eta}\right) &= 0. \end{aligned} \tag{20}$$

Eqs. (18) result in additional relations for $A_2(\xi,\eta)$:

$$\begin{aligned} \frac{\partial A_2}{\partial \xi} &= \frac{n}{(\xi+\eta)^2}(C+B) - \frac{n}{\xi+\eta}\cdot\frac{\partial B}{\partial \xi}, \\ \frac{\partial A_2}{\partial \eta} &= \frac{n}{(\xi+\eta)^2}(C+B) - \frac{n}{\xi+\eta}\cdot\frac{\partial C}{\partial \eta}. \end{aligned} \tag{21}$$

It is easily seen that the compatibility condition of Eqs. 21, $\frac{\partial}{\partial\eta}\left(\frac{\partial A_2}{\partial\xi}\right) = \frac{\partial}{\partial\xi}\left(\frac{\partial A_2}{\partial\eta}\right)$, is met identically. Eqs. 20 and Eqs. 21 yield

$$2n(n+1)\cdot[C(\eta)+B(\xi)] \;=\; n(n+1)(\xi+\eta)\cdot\left(\frac{\partial C}{\partial\eta}+\frac{\partial B}{\partial\xi}\right),$$

which gives

$$(\xi+\eta)\cdot\frac{\partial B}{\partial\xi} - 2\cdot B(\xi) \;=\; -(\xi+\eta)\cdot\frac{\partial C}{\partial\eta} + 2\cdot C(\eta), \tag{22}$$

since $n(n+1)\neq 0$. Differentiating Eq. 22 twice with respect to ξ, we have $(\xi+\eta)\cdot B'''(\xi)=0$, that is

$$B(\xi) \;\equiv\; p\xi^2+q\xi+r; \qquad p,q,r=\text{const.} \tag{23}$$

Analogously, differentiating Eq. 22 twice with respect to η, we obtain

$$C(\eta) \;\equiv\; a\eta^2+b\eta+c; \qquad a,b,c=\text{const.} \tag{24}$$

By substituting the expressions Eq. 23 and Eq. 24 into Eq. 22, we determine the relation between C and B polynomials: $B(\xi) \equiv -C(-\xi) \equiv -a\xi^2+b\xi-c$. Now it can be easily seen from Eqs. 21 that $\frac{\partial A_2}{\partial\xi}=na$, $\frac{\partial A_2}{\partial\eta}=-na$. Therefore $A_2(\xi,\eta)\equiv na\cdot(\xi-\eta)+d$, $d=\text{const.}$ Thus the following theorem is proved.

Theorem 1. *Contact symmetries $\varphi=\varphi(\xi,\eta,u,u_\xi,u_\eta)\in C^\infty(J^1)$ of the equation $\mathcal{Y}_n$ with $n\neq 0,-1$ are point symmetries and have the form*

$$\varphi \;=\; A_1(\xi,\eta) \;+\; [na\cdot(\xi-\eta)+d]\cdot u \;+\; (a\eta^2+b\eta+c)\cdot u_\eta \;+\; (-a\xi^2+b\xi-c)\cdot u_\xi,$$

where $a,b,c,d=\text{const}$, $A_1(\xi,\eta)$ is an arbitrary solution of the equation $\mathcal{Y}_n$. $C^\infty(J^l)$ denotes as usually smooth functions algebra on J^l.

According to the general theory[6] each symmetry φ of the linear equation $\mathcal{Y}_n$ linear with respect to the variables $u, u_{i\xi}, u_{i\eta}$; $i=1,2,3,\ldots$, defines the linear recursion operator l_φ for the equation $\mathcal{Y}_n$ which has the following property:

$$l_{F_n}\circ l_\varphi \;=\; \tilde{R}\circ l_{F_n}, \qquad \tilde{R} \;=\; \sum_\sigma a_\sigma(\xi,\eta)\cdot D_\sigma, \tag{25}$$

so that the operator l_φ converts each solution (symmetry) of the equation $\mathcal{Y}_n$ to a solution (symmetry) of the same equation. Particularly, Theorem 1 immediately yields

Corollary. *The operators*

$$\Box = D_\xi - D_\eta; \qquad \sigma_1 = \xi\cdot D_\xi + \eta\cdot D_\eta - n\cdot I; \qquad \tau = \xi^2\cdot D_\xi - \eta^2\cdot D_\eta + n\cdot(\eta-\xi)\cdot I$$

are recursion operators for the equation $\mathcal{Y}_n$ *and the equation* $\bar{l}_{F_n}(\varphi) = 0$.

Hereafter I will be an identical operator. The following commutation relations are easily derived by direct calculations:

$$[\Box, \sigma_1] = \Box, \quad [\Box, \tau] = 2\sigma_1, \quad [\sigma_1, \tau] = \tau, \tag{26}$$

so that the algebra $\{\Box, \sigma_1, \tau\}$ generated by $\Box, \sigma_1, \tau$ operators is isomorphic to Lie algebra sl_2^0, which is opposite (anti-isomorphic) to the classical algebra sl_2 of zero-trace 2×2-matrices:

$$\{\Box, \sigma_1, \tau\} \cong sl_2^0. \tag{27}$$

Denote the contact symmetries algebra of the equation $\mathcal{Y}_n$ by Symc$\mathcal{Y}_n$. Let A_n^∞ be a linear space of smooth solutions of the equation $\mathcal{Y}_n$ and

$$\mathrm{NSymc}\mathcal{Y}_n = \{u, \Box(u), \sigma_1(u), \tau(u)\}$$

(symbol $\{S\}$ denotes the linear span of the family S of functions). Then it can be easily seen that A_n^∞ and NSymc$\mathcal{Y}_n$ are subalgebras in Symc$\mathcal{Y}_n$ and there is Lie algebras isomorphism $\{\Box(u), \sigma_1(u), \tau(u)\} \cong sl_2$. Owing to it Theorem 1 yields

Theorem 2. *The contact symmetries algebra of the equation* $\mathcal{Y}_n$ *with* $n \neq 0, -1$ *is a semidirect sum of the infinite-dimensional Abelian ideal* A_n^∞ *and subalgebra* NSymc$\mathcal{Y}_n$*:*

$$\mathrm{Symc}\mathcal{Y}_n = A_n^\infty \oplus \mathrm{NSymc}\mathcal{Y}_n; \qquad \mathrm{NSymc}\mathcal{Y}_n \cong \{u\} \oplus sl_2. \tag{28}$$

For all real $n, m \neq 0, -1$ there is Lie algebras isomorphism NSymc$\mathcal{Y}_n \cong$ NSymc$\mathcal{Y}_m$.

4. Higher - order symmetries $\varphi \in C^\infty(J^l), l \geq 2$ of the equation $\mathcal{Y}_n$; $n \neq 0, \pm 1, \pm 2, \ldots, \pm(l-1), -l$

The following theorem is proved by analogous but more complicated calculations.

Theorem 3. *The symmetries* $\varphi = \varphi(\xi, \eta, u, u_\xi, u_\eta, u_{2\xi}, u_{2\eta}) \in C^\infty(J^2)$ *of the equation* $\mathcal{Y}_n$ *with* $n \neq 0, \pm 1, -2$ *have the form*

$$\varphi = A_1(\xi, \eta) + A_2(\xi, \eta) \cdot u + \{N(\xi) - \frac{2n}{\xi + \eta} \cdot B(\xi)\} \cdot u_\xi +$$

$$\{L(\eta) - \frac{2n}{\xi + \eta} \cdot C(\eta)\} \cdot u_\eta + B(\xi) \cdot u_{2\xi} + C(\eta) \cdot u_{2\eta},$$

where

$A_1(\xi, \eta)$ *is an arbitrary solution of the equation* $\mathcal{Y}_n$,

$$B(\xi) = a + b\xi + c\xi^2 + d\xi^3 + e\xi^4; \quad a, b, c, d, e = \text{const}; \quad C(\eta) = B(-\eta);$$

$$N(\xi) = \alpha + \beta\xi + \gamma\xi^2 + 2e\xi^3; \quad \alpha, \beta, \gamma = \text{const}; \quad L(\eta) = -N(-\eta);$$

$$A_2(\xi, \eta) = n(n+1)(\xi - \eta) \cdot [d + e \cdot (\xi - \eta)] - \\ n \cdot [\beta + \gamma \cdot (\xi - \eta) + 2e \cdot (\xi^2 - \xi\eta + \eta^2)] + \delta; \quad \delta = \text{const}.$$

4.1 Linearity of generating functions of symmetries with respect to variables $u, u_{i\xi}, u_{i\eta}$; $i = 1, 2, \ldots, l$

Now consider the general case when symmetry of the equation $\mathcal{Y}_n$ is

$$\varphi = \varphi(\xi, \eta, u, u_\xi, u_\eta, \ldots, u_{l\xi}, u_{l\eta}) \in C^\infty(J^l), l \geq 3. \tag{29}$$

If follows from the formulas Eq. 9-Eq. 11 and those for $\bar{D}_\eta^r\left(\frac{u_\xi + u_\eta}{\xi + \eta}\right)$ analogous to the formulas Eq. 10-Eq. 11 that the left-hand side of Eq. 8 for function φ of the form Eq. 29 is a second-degree polynomial with respect to the variables $u_{(l+1)\xi}$ and $u_{(l+1)\eta}$, that is

$$\bar{l}_{F_n}(\varphi) \equiv \Lambda_1 \cdot u_{(l+1)\xi} \cdot u_{(l+1)\eta} + \Lambda_2 \cdot u_{(l+1)\xi} + \Lambda_3 \cdot u_{(l+1)\eta} + \Lambda_4,$$

where $\Lambda_i; i = 1, 2, 3, 4$ are some known functions of the variables

$$\xi, \eta, u, u_\xi, u_\eta, \ldots, u_{l\xi}, u_{l\eta}$$

and of various first and second order partial derivatives of φ. Since Λ_i coefficients do not contain the variables $u_{(l+1)\xi}$ and $u_{(l+1)\eta}$, it follows from $\bar{l}_{F_n}(\varphi) = 0$ that $\Lambda_i = 0; i = 1, 2, 3, 4$. Particularly,

$$\Lambda_1 \equiv \frac{\partial^2 \varphi}{\partial u_{l\xi} \partial u_{l\eta}} = 0,$$

which immediately yields

$$\varphi = \varphi_1(\xi, \eta, u, u_\xi, u_\eta, \ldots, u_{(l-1)\xi}, u_{(l-1)\eta}; u_{l\xi}) + \\ \varphi_2(\xi, \eta, u, u_\xi, u_\eta, \ldots, u_{(l-1)\xi}, u_{(l-1)\eta}; u_{l\eta}). \tag{30}$$

By substituting the expression Eq. 30 into the equality $\Lambda_2 = 0$, we obtain the following equation for φ_1:

$$\frac{\partial^2 \varphi_1}{\partial \eta \partial u_{l\xi}} + \sum_{j=0}^{l-1} u_{(j+1)\eta} \cdot \frac{\partial^2 \varphi_1}{\partial u_{j\eta} \partial u_{l\xi}} + \sum_{j=1}^{l} n \cdot \bar{D}_\xi^{j-1} \left(\frac{u_\xi + u_\eta}{\xi + \eta} \right) \cdot \frac{\partial^2 \varphi_1}{\partial u_{j\xi} \partial u_{l\xi}} = 0. \tag{31}$$

Solving Eq. 31 containes a cumbersome and complicated inductive procedure making an essential use of the conditions

$$n \neq 0, \pm 1, \pm 2, \ldots, \pm(l-1), -l. \tag{32}$$

As a result for such restrictions on n we obtain

$$\varphi_1 \equiv \tilde{\varphi}_1(\xi, \eta, u, u_\xi, u_\eta, \ldots, u_{(l-1)\xi}, u_{(l-1)\eta}) + B_l(\xi) \cdot u_{l\xi}.$$

Since the variables $\xi, u, u_\xi, \ldots, u_{l\xi}$ and $\eta, u, u_\eta, \ldots, u_{l\eta}$ are completely "symmetric" in the equation $\bar{l}_{F_n}(\varphi) = 0$, it follows analogously from the equality $\Lambda_3 = 0$, for the same restrictions on a parameter n:

$$\varphi_2 \equiv \tilde{\varphi}_2(\xi, \eta, u, u_\xi, u_\eta, \ldots, u_{(l-1)\xi}, u_{(l-1)\eta}) + C_l(\eta) \cdot u_{l\eta}$$

.

The above consideration together with the expression Eq. 30 yield

Proposition 1. *Let $\varphi \in C^\infty(J^l)$ be the symmetry of the equation $\mathcal{Y}_n, l \geq 3$ and $n \neq 0, \pm 1, \pm 2, \ldots, \pm(l-1), -l$. Then*

$$\varphi \equiv A(\xi, \eta, u, u_\xi, u_\eta, \ldots, u_{(l-1)\xi}, u_{(l-1)\eta}) + B_l(\xi) \cdot u_{l\xi} + C_l(\eta) \cdot u_{l\eta}. \tag{33}$$

As above it can be easily seen that the expression $\bar{l}_{F_n}(\varphi)$ for the function φ of the form Eq. 33 is a polynomial with respect to the variables

$$u_{l\xi}, u_{l\eta}, u_{(l+1)\xi}, u_{(l+1)\eta},$$

with Q_k^1 coefficients depending on the variables

$$\xi, \eta, u, u_\xi, u_\eta, \ldots, u_{(l-1)\xi}, u_{(l-1)\eta}.$$

Therefore the coefficients $Q_k^1 \equiv 0$ for the symmetry φ, and we obtain equations for the functions A, B_l, C_l. Particularly, $u_{l\xi} \cdot u_{l\eta}$-coefficient Q_k^1 has the form

$$\frac{\partial^2 \varphi}{\partial u_{(l-1)\xi} \partial u_{(l-1)\eta}} + \frac{n^2}{(\xi + \eta)^2} \cdot \frac{\partial^2 \varphi}{\partial u_{l\xi} \partial u_{l\eta}} = 0. \tag{34}$$

Hence, $\frac{\partial^2 A}{\partial u_{(l-1)\xi}\partial u_{(l-1)\eta}} \equiv 0$ and

$$A \equiv A_1(\xi,\eta,u,u_\xi,u_\eta,\ldots,u_{(l-2)\xi},u_{(l-2)\eta};u_{(l-1)\xi})+ A_2(\xi,\eta,u,u_\xi,u_\eta,\ldots,u_{(l-2)\xi},u_{(l-2)\eta};u_{(l-1)\eta}).$$

The $u_{l\xi}$-coefficient Q_k^1 with account of Eq. 34 leads to the equation

$$-\frac{nl}{(\xi+\eta)^2}\cdot B_l(\xi)+\frac{\partial^2 A_1}{\partial\eta\partial u_{(l-1)\xi}}+\sum_{j=o}^{l-2}u_{(j+1)\eta}\cdot\frac{\partial^2 A_1}{\partial u_{j\eta}\partial u_{(l-1)\xi}}+ +\sum_{j=1}^{l-1}n\cdot\bar{D}_\xi^{j-1}\left(\frac{u_\xi+u_\eta}{\xi+\eta}\right)\cdot\frac{\partial^2 A_1}{\partial u_{j\xi}\partial u_{(l-1)\xi}}=0. \tag{35}$$

Eq. 35 for A_1 is obtained from Eq. 31 for φ_1 by replacing l with $(l-1)$ and introducing an addend containing $B_l(\xi)$. When solving Eq. 31 we use the conditions Eq. 32 which are considered satisfied here. Analogous conditions for Eq. 35 have the form $n\neq 0,\pm1,\pm2,\ldots,\pm(l'-1),-l';l'=l-1$ and are evidently met according to Eq. 32. The solving procedure of Eq. 35 almost completely coincides with that for Eq. 31. Only the final step differs since the addend $-\frac{nl}{(\xi+\eta)^2}\cdot B_l(\xi)$ needs to be accounted. The general solution of Eq. 35 has the form

$$A_1\equiv\tilde{A}_1(\xi,\eta,u,u_\xi,u_\eta,\ldots,u_{(l-2)\xi},u_{(l-2)\eta})+\{N(\xi)-\frac{nl}{\xi+\eta}\cdot B_l(\xi)\}\cdot u_{(l-1)\xi},$$

where $\tilde{A}_1$ and N are arbitrary functions. It is clear from symmetry considerations that after examining $u_{l\xi}$-coefficient Q_k^1 in the equation $\bar{l}_{F_n}(\varphi)=0$ we obtain analogous representation for A_2. This proves

Proposition 2. *Let $\varphi\in C^\infty(J^l)$ be the symmetry of the equation $\mathcal{Y}_n, l\geq 3$ and $n\neq 0,\pm1,\pm2,\ldots,\pm(l-1),-l$. Then*

$$\varphi \equiv A(\xi,\eta,u,u_\xi,u_\eta,\ldots,u_{(l-2)\xi},u_{(l-2)\eta})+\{N(\xi)-\frac{nl}{\xi+\eta}\cdot B_l(\xi)\}\cdot u_{(l-1)\xi}+ \{L(\eta)-\frac{nl}{\xi+\eta}\cdot C_l(\eta)\}\cdot u_{(l-1)\eta}+B_l(\xi)\cdot u_{l\xi}+C_l(\eta)\cdot u_{l\eta}.$$

Suppose (an induction assumption), φ is proved to be linear with respect to $u_{j\xi},u_{j\eta}; j=l-(i-1),l-i+2,\ldots,l;l-i\geq 1$ and

$$\varphi \equiv A(\xi,\eta,u,u_\xi,u_\eta,\ldots,u_{(l-i)\xi},u_{(l-i)\eta})+B_{l-(i-1)}(\xi,\eta)\cdot u_{(l-(i-1))\xi}+ C_{l-(i-1)}(\xi,\eta)\cdot u_{(l-(i-1))\eta}+\ldots+B_{l-1}(\xi,\eta)\cdot u_{(l-1)\xi}+ C_{l-1}(\xi,\eta)\cdot u_{(l-1)\eta}+B_l(\xi)\cdot u_{l\xi}+C_l(\eta)\cdot u_{l\eta}. \tag{36}$$

For simplicity, we further assume

$$B_{l-(i-1)}(\xi,\eta) \equiv B_{l-i+1}(\xi,\eta);\, C_{l-(i-1)}(\xi,\eta) \equiv C_{l-i+1}(\xi,\eta).$$

Let us prove φ to be also linear with respect to the variables $u_{(l-i)\xi}$ and $u_{(l-i)\eta}$. For the function φ of the form Eq. 36, the expression $\bar{l}_{F_n}(\varphi)$ is a second-degree polynomial with respect to the variables $u_{j\xi}, u_{j\eta}; j = l-i+1, l-i+2, \ldots, l, l+1$ with the coefficients Q_k^i, depending on the variables $\xi, \eta, u, u_\xi, u_\eta, \ldots, u_{(l-i)\xi}, u_{(l-i)\eta}$. For φ symmetry, the identity $\bar{l}_{F_n}(\varphi) \equiv 0$ is met, so that the coefficients $Q_k^i \equiv 0$ and we obtain equations for the functions A, B_j, C_j. Particularly, the $u_{(l-i+1)\xi} \cdot u_{(l-i+1)\eta}$-coefficient Q_k^i will be

$$\frac{\partial^2 \varphi}{\partial u_{(l-i)\xi} \partial u_{(l-i)\eta}} \equiv \frac{\partial^2 A}{\partial u_{(l-i)\xi} \partial u_{(l-i)\eta}} = 0.$$

Hence,

$$\begin{aligned} A \equiv\ & A_1(\xi,\eta,u,u_\xi,u_\eta,\ldots,u_{(l-i-1)\xi},u_{(l-i-1)\eta};u_{(l-i)\xi})+ \\ & A_2(\xi,\eta,u,u_\xi,u_\eta,\ldots,u_{(l-i-1)\xi},u_{(l-i-1)\eta};u_{(l-i)\eta}). \end{aligned}$$

From $u_{(l-i+1)\xi}$-coefficient Q_k^i being zero-equal, in the case of $l-i \geq 1$ we obtain the following equation for A_1:

$$\begin{aligned} & \beta(\xi,\eta) + \frac{\partial^2 A_1}{\partial \eta \partial u_{(l-i)\xi}} + \sum_{j=0}^{l-i-1} u_{(j+1)\eta} \cdot \frac{\partial^2 A_1}{\partial u_{j\eta} \partial u_{(l-i)\xi}} + \\ & \sum_{j=0}^{l-i} n \cdot \bar{D}_\xi^{j-1}\left(\frac{u_\xi + u_\eta}{\xi+\eta}\right) \cdot \frac{\partial^2 A_1}{\partial u_{j\xi} \partial u_{(l-i)\xi}} \quad = \quad 0, \end{aligned} \tag{37}$$

where

$$\begin{aligned} \beta(\xi,\eta) \equiv\ & \frac{\partial^2 B_{l-i+1}(\xi,\eta)}{\partial \xi \partial \eta} - \frac{n}{\xi+\eta} \cdot \frac{\partial B_{l-i+1}(\xi,\eta)}{\partial \eta} - \frac{n(l-i+1)}{(\xi+\eta)^2} \cdot B_{l-i+1}(\xi,\eta) - \\ & \sum_{j=0}^{i-2} \frac{n(n-(l-i+1))\ldots(n-(l-(j+1)))\cdot(l-j)}{(\xi+\eta)^{i+1-j}} \cdot B_{l-j}(\xi,\eta) + \\ & \sum_{j=0}^{i-2} \frac{n(n-(l-i+1))\ldots(n-(l-(j+1)))}{(\xi+\eta)^{i-j}} \cdot \frac{\partial B_{l-j}(\xi,\eta)}{\partial \xi}. \end{aligned} \tag{38}$$

Eq. 37 is solved analogously to Eq. 31 and Eq. 35; its general solution is given by the formulas

$$A_1 \equiv \tilde{A}_1(\xi,\eta,u,u_\xi,u_\eta,\ldots,u_{(l-i-1)\xi},u_{(l-i-1)\eta}) + B_{l-i}(\xi,\eta)\cdot u_{(l-i)\xi};$$

$$B_{l-i}(\xi,\eta) \equiv -\int \beta(\xi,\eta)d\eta + B_{l-i}(\xi), \quad i \leq l-1, \tag{39}$$

where $\beta(\xi,\eta)$ is specified by the identity Eq. 38, $B_{l-i}(\xi)$ is an arbitrary function. Hereafter the functions $B_{l-i}(\xi,\eta)$ and $B_{l-i}(\xi)$ will be accounted different: $B_{l-i}(\xi,\eta) \not\equiv B_{l-i}(\xi), i \neq 0$. It is clear from symmetry considerations that the function A_2 is linear with respect to $u_{(l-i)\eta}$. Thus, the following proposition is proved by induction.

Proposition 3. *Let $\varphi \in C^\infty(J^l)$ be the symmetry of the equation $\mathcal{Y}_n, l \geq 3$ and $n \neq 0, \pm 1, \pm 2, \ldots, \pm(l-1)$, -l. Then φ is linear with respect to the variables $u_{i\xi}, u_{i\eta}; i = 1, 2, \ldots, l$:*

$$\varphi \equiv A(\xi,\eta,u) + \sum_{i=1}^{l-1} B_i(\xi,\eta)\cdot u_{i\xi} + \sum_{i=1}^{l-1} C_i(\xi,\eta)\cdot u_{i\eta} + B_l(\xi)\cdot u_{l\xi} + C_l(\eta)\cdot u_{l\eta}. \quad (40)$$

For the function φ of the form Eq. 40, the exression $\bar{l}_{F_n}(\varphi)$ is a polynomial of second degree with respect to u_ξ, u_η and of first degree with respect to $u_{i\xi}, u_{i\eta}; i = 2, 3, \ldots, l, l+1$ with the coefficients depending on ξ, η, u. For φ symmetry, these coefficients vanish. The $u_\xi \cdot u_\eta$-coefficient is

$$\frac{\partial^2\varphi}{\partial u^2} \equiv \frac{\partial^2 A}{\partial u^2} = 0,$$

hence, the representation $A(\xi,\eta,u) \equiv A_1(\xi,\eta) + A_2(\xi,\eta)\cdot u$ is valid for the function A from Eq. 40. This together with Proposition 3, Theorems 1 and 3 give the following general

Theorem 4. *Let $\varphi \equiv \varphi(\xi,\eta,u,u_\xi,u_\eta,\ldots,u_{l\xi},u_{l\eta}) \in C^\infty(J^l)$ be the symmetry of the equation $\mathcal{Y}_n, l \geq 1$ and $n \neq 0, \pm 1, \pm 2, \ldots, \pm(l-1)$, -l. Then φ is linear with respect to $u, u_{i\xi}, u_{i\eta}; i = 1, 2, \ldots, l$:*

$$\varphi \equiv A_1(\xi,\eta) + A_2(\xi,\eta)\cdot u + \sum_{i=1}^{l-1} B_i(\xi,\eta)\cdot u_{i\xi} +$$
$$\sum_{i=1}^{l-1} C_i(\xi,\eta)\cdot u_{i\eta} + B_l(\xi)\cdot u_{l\xi} + C_l(\eta)\cdot u_{l\eta}. \quad (41)$$

For the function φ of the form Eq. 41, the left-hand side of the equation $\bar{l}_{F_n}(\varphi) = 0$, which determines the symmetries, is a first-degree polynomial with respect to $u, u_{i\xi}, u_{i\eta}; i = 1, 2, \ldots, l, l+1$, with the coefficients depending on ξ, η. By setting the coefficients of this polynomial to 0, we obtain the equations for the functions A_1, A_2, B_i, C_i from the representation Eq. 41 of the φ symmetry. In particular, the coefficient at u variable is

$$\frac{\partial^2 A_2(\xi,\eta)}{\partial\xi\partial\eta} - \frac{n}{\xi+\eta}\cdot\left(\frac{\partial A_2}{\partial\xi} + \frac{\partial A_2}{\partial\eta}\right) = 0, \quad (42)$$

and the free term (not containing the variables $u, u_{i\xi}, u_{i\eta}$) is

$$\frac{\partial^2 A_1(\xi,\eta)}{\partial\xi\partial\eta} - \frac{n}{\xi+\eta}\cdot\left(\frac{\partial A_1}{\partial\xi} + \frac{\partial A_1}{\partial\eta}\right) = 0. \tag{43}$$

After simple transformations of the u_ξ- and u_η-coefficients we come to the following equations, correspondingly:

$$\begin{aligned} &\frac{\partial A_2}{\partial\eta} + \frac{\partial^2 B_1(\xi,\eta)}{\partial\xi\partial\eta} - \frac{n}{\xi+\eta}\cdot\left(\frac{\partial B_1}{\partial\xi} + \frac{\partial B_1}{\partial\eta}\right) + \\ &\quad \sum_{i=1}^{l}\left[\frac{n(n-1)\dots(n-(i-1))}{(\xi+\eta)^i}\cdot B_i(\xi,\eta)\right]_\xi + \\ &\quad \sum_{i=1}^{l}\left[\frac{n(n-1)\dots(n-(i-1))}{(\xi+\eta)^i}\cdot C_i(\xi,\eta)\right]_\eta = 0, \end{aligned} \tag{44}$$

$$\begin{aligned} &\frac{\partial A_2}{\partial\xi} + \frac{\partial^2 C_1(\xi,\eta)}{\partial\xi\partial\eta} - \frac{n}{\xi+\eta}\cdot\left(\frac{\partial C_1}{\partial\eta} + \frac{\partial C_1}{\partial\xi}\right) + \\ &\quad \sum_{i=1}^{l}\left[\frac{n(n-1)\dots(n-(i-1))}{(\xi+\eta)^i}\cdot C_i(\xi,\eta)\right]_\eta + \\ &\quad \sum_{i=1}^{l}\left[\frac{n(n-1)\dots(n-(i-1))}{(\xi+\eta)^i}\cdot B_i(\xi,\eta)\right]_\xi = 0, \end{aligned} \tag{45}$$

hereafter the subscripts ξ and η denote the differentiation with respect to ξ, η and the following notation is adopted:

$$n(n-1)\dots(n-(i-1)) \equiv \prod_{j=0}^{i-1}(n-j); \quad i = 1,2,\dots,l.$$

After a number of simple transformations the compatibility condition of Eq. 44 and Eq. 45, $\frac{\partial}{\partial\xi}\left(\frac{\partial A_2}{\partial\eta}\right) = \frac{\partial}{\partial\eta}\left(\frac{\partial A_2}{\partial\xi}\right)$ takes the form

$$\frac{\partial^3 B_1(\xi,\eta)}{\partial\xi\partial\eta\partial\xi} - \frac{2n}{\xi+\eta}\cdot\frac{\partial^2 B_1(\xi,\eta)}{\partial\xi\partial\eta} + \frac{2n}{(\xi+\eta)^2}\cdot\frac{\partial B_1(\xi,\eta)}{\partial\eta} +$$

$$\sum_{i=2}^{l}\Big[\frac{n(n-1)\dots(n-(i-1))}{(\xi+\eta)^i}\cdot\Big(\frac{\partial B_i(\xi,\eta)}{\partial\xi} - \frac{\partial B_i(\xi,\eta)}{\partial\eta}\Big)\Big]_\xi \qquad =$$

$$\frac{\partial^3 C_1(\xi,\eta)}{\partial\eta\partial\xi\partial\eta} - \frac{2n}{\xi+\eta}\cdot\frac{\partial^2 C_1(\xi,\eta)}{\partial\eta\partial\xi} + \frac{2n}{(\xi+\eta)^2}\cdot\frac{\partial C_1(\xi,\eta)}{\partial\xi} +$$

$$\sum_{i=2}^{l}\Big[\frac{n(n-1)\dots(n-(i-1))}{(\xi+\eta)^i}\cdot\Big(\frac{\partial C_i(\xi,\eta)}{\partial\eta} - \frac{\partial C_i(\xi,\eta)}{\partial\xi}\Big)\Big]_\eta \tag{46}$$

The compatibility condition Eq. 46 for even $l = 2m$, $m > 1$, will be represented further as a nontrivial differential equation for the functions $B_l(\xi)$ and $C_l(\eta)$ containing some polynomials of the variables ξ and η. If $B_i(\xi,\eta)$ and $C_i(\xi,\eta); i = 1,2,\dots,l$, are already known and the compatibility condition Eq. 46 is met, then the function $A_2(\xi,\eta)$ is defined by a curvilinear integral

$$A_2(\xi,\eta) = \int\limits_C \frac{\partial A_2}{\partial\xi}\cdot d\xi + \frac{\partial A_2}{\partial\eta}\cdot d\eta, \tag{47}$$

where C is an arbitrary contour on the plane $\mathbb{R}^2 = \{(\xi,\eta)\}$ connecting the fixed point (ξ_0,η_0) with the current one (ξ,η), while $\frac{\partial A_2}{\partial\xi}$ and $\frac{\partial A_2}{\partial\eta}$ are represented in terms of $B_i(\xi,\eta)$ and $C_i(\xi,\eta)$ according to the formulas Eq. 44 and Eq. 45.

4.2 Linear operators L_l in Sym$\mathcal{Y}_n$ algebra generated by contact symmetries

According to Theorem 1, the functions

$$\varphi_{-1} = u_\xi - u_\eta,\quad \varphi_0 = \xi\cdot u_\xi + \eta\cdot u_\eta,\quad \varphi_{+1} = -\xi^2\cdot u_\xi + \eta^2\cdot u_\eta + n\cdot(\xi-\eta)\cdot u \tag{48}$$

are the symmetries of the equation $\mathcal{Y}_n$. Each φ symmetry of the equation $\mathcal{Y}_n$ determines the linear operator L_φ : Sym$\mathcal{Y}_n \to$ Sym$\mathcal{Y}_n$ in Sym$\mathcal{Y}_n$ algebra operating in accordance with the rule $L_\varphi(f) = \{\varphi, f\}; f \in$Sym$\mathcal{Y}_n$, where $\{\varphi, f\}$ is the higher Jacobi bracket of φ and f functions. For brevity, the operators in Sym$\mathcal{Y}_n$ corresponding to φ_{-1},φ_0 and φ_{+1} functions will be denoted by the symbols L_{-1}, L_0 and L_{+1}. Having defined the operator L_l : Sym$\mathcal{Y}_n \to$ Sym$\mathcal{Y}_n$ for $l \geq 1$ by the equality

$$L_l = L_{-1}\circ L_{+1} + L_0^2 - L_0 - l\cdot(l+1)\cdot I$$

and performed some simple but very cumbersome transformations we obtain the following general formula for L_l operation on the arbitrary symmetry $\varphi \in C^\infty(J^l)$ of the equation $\mathcal{Y}_n$ which is writtenas Eq. 41:

$$\begin{aligned}
L_l(\varphi) \; = \; & \{[2n - l(l+1)] \cdot A_1 + n \cdot (\xi - \eta) \cdot (A_{1,\xi} - A_{1,\eta}) + (\xi+\eta)^2 \cdot A_{1,\xi\eta}\} + \\
& u \cdot \{-l(l+1) \cdot A_2 + (\xi+\eta)^2 \cdot A_{2,\xi\eta} + n \cdot (B_{1,\eta} - B_{1,\xi}) + n \cdot (-C_{1,\eta} + C_{1,\xi})\} + \\
& \sum_{i=1}^{l-1} u_{i\xi} \cdot \{[i(i+1) - l(l+1)] \cdot B_i - 2i(\xi+\eta) \cdot B_{i,\eta} + (\xi+\eta)^2 \cdot B_{i,\xi\eta} + \\
& (i+1)(n-i) \cdot (B_{(i+1),\eta} - B_{(i+1),\xi})\} + \\
& \sum_{i=1}^{l-1} u_{i\eta} \cdot \{[i(i+1) - l(l+1)] \cdot C_i - 2i(\xi+\eta) \cdot C_{i,\xi} + (\xi+\eta)^2 \cdot C_{i,\xi\eta} + \\
& (i+1)(i-n) \cdot (C_{(i+1),\eta} - C_{(i+1),\xi})\}, \qquad (49)
\end{aligned}$$

here $A_{1,\xi} \equiv \frac{\partial A_1}{\partial \xi}$; $A_{1,\xi\eta} \equiv \frac{\partial^2 A_1}{\partial \xi \partial \eta}$, etc. It is evident from Eq. 49 that

$$\varphi \in \mathrm{Sym}\mathcal{Y}_n \bigcap C^\infty(J^l) \quad \Longrightarrow \quad L_l(\varphi) \in \mathrm{Sym}\mathcal{Y}_n \bigcap C^\infty(J^{l-1}).$$

Besides, it is easily seen that $L_l(u) = -l(l+1) \cdot u$, hence $L_l \neq 0$. The operator L_l will be further used in the inductive proof of $B_l(\xi)$ and $C_l(\eta)$ polynomiality (more precisely, in the inductive proof of Theorem 6).

4.3 Structure formulas for the coefficients $B_i(\xi,\eta)$ and $C_i(\xi,\eta)$ of $\varphi \in C^\infty(J^l)$ symmetry

Theorem 5. *Let $\varphi \in C^\infty(J^l)$ be the symmetry of the equation $\mathcal{Y}_n, l \geq 3$ and $n \neq 0, \pm 1, \pm 2, \ldots, \pm(l-1)$, -l. Then the following identities*

$$\begin{aligned}
B_{l-k}(\xi,\eta) \; & \equiv B_{l-k}(\xi) + \sum_{j=1}^{k} \sum_{m=0}^{j-1} \frac{\delta_{j,m}^{(k)}}{(\xi+\eta)^{j-m}} \cdot B_{l-k+j}^{(m)}(\xi); \\
C_{l-k}(\xi,\eta) \; & \equiv C_{l-k}(\eta) + \sum_{j=1}^{k} \sum_{m=0}^{j-1} \frac{\delta_{j,m}^{(k)}}{(\xi+\eta)^{j-m}} \cdot C_{l-k+j}^{(m)}(\eta); \\
& k = 0, 1, 2, \ldots, l-1; \quad \delta_{j,m}^{(k)} = \mathrm{const}(j,m,k), \qquad (50)
\end{aligned}$$

are valid for the coefficients $B_{l-k}(\xi,\eta)$ and $C_{l-k}(\xi,\eta)$ from the representation Eq. 41 *of φ symmetry, here $\delta_{j,m}^{(k)}$ does not depend on $B_\alpha(\xi)$ and $C_\alpha(\eta)$,* $B_\alpha^{(m)}(\xi) \equiv \frac{d^m B_\alpha(\xi)}{d\xi^m}$, $C_\alpha^{(m)}(\eta) \equiv \frac{d^m C_\alpha(\eta)}{d\eta^m}$.

Note, that in fact $\delta_{j,m}^{(k)}$ are polynomial functions of n and l parameters. However, since n and l are fixed here, $\delta_{j,m}^{(k)}$ is not shown explicitly to depend on them.

Theorem 5 is easily proved by induction. For $k = 0$ and $k = 1$, it follows from Proposition 2 that the formulas Eq. 50 are true. Now suppose that the formulas Eq. 50 for $B_{l-k}(\xi,\eta)$, are found to be true for each k satisfying the restrictions $1 \leq k \leq i-1, i \leq l-1$, and prove Eq. 50 for k=i. By substituting the expansions Eq. 50 for $B_{l-k}(\xi,\eta); 1 \leq k \leq i-1$ into the expression Eq. 38 for $\beta(\xi,\eta)$ and then calculating $B_{l-i}(\xi,\eta)$ according to the formula Eq. 39, we obtain the expression for $B_{l-i}(\xi,\eta)$ coinciding with Eq 50 after simple but cumbersome transformations. Thus, the identities Eq. 50 for $B_{l-k}(\xi,\eta)$ are proved to be valid for each $k \leq l-1$. Now it follows immediately from ordinary symmetry considerations that the formulas Eq. 50 are valid for $C_{l-k}(\xi,\eta)$. Theorem 5 is proved.

The following theorem is formulated for the functions $B_{s-i}(\xi)$, however it is evident that analogous theorem takes place for the functions $C_{s-i}(\eta)$.

Theorem 6. *Let $l \geq 3; n \neq 0, \pm1, \pm2, \dots, \pm(l-1), -l$ and the corresponding functions $B_{s-i}(\xi)$ from the expansions* Eq. 50 *for all natural $s \leq l-1$ and for all symmetries $\varphi \in C^\infty(J^s)$ of the equation $\mathcal{Y}_n$ have the following properties:*

$$\begin{aligned}
&B_{s-i}(\xi) \text{ is a polynomial, } \deg \leq B_{s-i}(\xi) \leq 2\cdot s - i;\\
&B_{s-i}(\xi) \equiv \nu_{s,i}\cdot B_s^{(i)}(\xi) + \pi_{2s-i-1}^{s,i}(\xi); \quad \nu_{s,i} = \text{const}(s,i),\\
&B_s^{(i)}(\xi) = \frac{d^i B_s(\xi)}{d\xi^i},\\
&\pi_{2s-i-1}^{s,i}(\xi) \text{ is a polynomial of degree } \leq 2s-i-1;\\
&\nu_{s,i} \text{ does not depend on } B_s(\xi); \quad 0 \leq i \leq s-1, s \leq l-1.
\end{aligned} \tag{51}$$

Then, the following indentity is valid for any symmetry $\varphi \in C^\infty(J^l)$:

$$\begin{aligned}
&B_{l-j}(\xi) \equiv \nu_{l,j}\cdot B_l^{(j)}(\xi) + \pi_{2l-j-1}^{l,j}(\xi); \quad \nu_{l,j} = \text{const}(l,j),\\
&B_l^{(j)}(\xi) = \frac{d^j B_l(\xi)}{d\xi^j},\\
&\pi_{2l-j-1}^{l,j}(\xi) \text{ is a polynomial of degree } \leq 2l-j-1;\\
&\nu_{l,j} \text{ does not depend on } B_l(\xi); \quad 0 \leq j \leq l-1.
\end{aligned} \tag{52}$$

The proof is carried out by induction over j index. For $j = 0$, the identity Eq. 52 becomes the tautology $B_{l-0}(\xi) \equiv B_l(\xi)$. In case $j = 1$, applying the operator L_l to the symmetry $\varphi \in C^\infty(J^l)$ we obtain $\tilde{B}_{l-1}(\xi) \equiv 2l \cdot \{-B_{l-1}(\xi) + \frac{l-1}{2}B_l^{(1)}(\xi)\}$ according to the formula Eq. 49 for the coefficient $\tilde{B}_{l-1}(\xi)$ at the variable $u_{(l-1)\xi}$ in the expression $L_l(\varphi)$. Under the condition of the theorem, $\tilde{B}_{l-1}(\xi)$ is a polynomial of degree $\leq 2\cdot(l-1)$. It follows immediately from this remark and the previous identity that Eq. 52 is true for $j = 1$. Now suppose that the relation Eq. 52 has been proved for each j such that $0 \le j \le i \le l-2, i \ge 1$, and show this to make Eq. 52 true also for $j = i+1$. Using the representations Eq. 50 for $B_{l-i}(\xi,\eta)$ and the assumption made, we find

$$B_{l-i}(\xi,\eta) \equiv \nu_{l,i} \cdot B_l^{(i)}(\xi) + \sum_{j_1=1}^{i} \frac{\delta_{j_1}}{(\xi+\eta)^{j_1}} \cdot B^{(i-j_1)}(\xi) + \sum_{j_1=0}^{i-1} \frac{1}{(\xi+\eta)^{j_1}} \cdot \pi_{2l-i-1+j_1}(\xi); \quad \delta_{j_1} = \mathrm{const}(j_1), \tag{53}$$

hereafter $\pi_\alpha(\xi)$ is a polynomial of degree $\le \alpha$, and all the constants are independent on $B_l(\xi)$. Analogously, we obtain

$$B_{l-(i+1)}(\xi,\eta) \equiv B_{l-(i+1)}(\xi) + \sum_{j_1=1}^{i+1} \frac{\delta_{j_1}^{(1)}}{(\xi+\eta)^{j_1}} \cdot B_l^{(i+1-j_1)}(\xi) + \sum_{j=1}^{i} \frac{1}{(\xi+\eta)^{j_1}} \cdot \pi_{2l-i-2+j_1}(\xi); \quad \delta_{j_1}^{(1)} = const(j_1). \tag{54}$$

Let $\tilde{B}_{(l-1)-i}(\xi,\eta)$ be the coefficient at the variable $u_{((l-1)-i)\xi} \equiv u_{(l-(i+1))\xi}$ in the symmetry $L_l(\varphi)$. Then the formulas Eq. 49, Eq. 53 and Eq. 54 yield

$$\tilde{B}_{(l-1)-i}(\xi,\eta) \equiv \{[l-(i+1)]\cdot(l-i) - l(l+1)\} \cdot B_{l-(i+1)}(\xi) + \sum_{j_1=0}^{i+1} \frac{\tilde{\delta}_{j_1}}{(\xi+\eta)_1^j} \cdot B_l^{(i+1-j_1)}(\xi) + \sum_{j_1=0}^{i} \frac{\tilde{\gamma}_{j_1}}{(\xi+\eta)_1^j} \cdot \pi_{2l-i-2+j_1}(\xi); \quad \tilde{\delta}_{j_1}, \tilde{\gamma}_{j_1} = \mathrm{const}. \tag{55}$$

On the other hand, we find from the representation Eq. 50 for $\tilde{B}_{(l-1)-i}(\xi,\eta)$ and from the relations Eq. 51 for $\tilde{B}_{(l-1)-i}(\xi)$:

$$\tilde{B}_{(l-1)-i}(\xi,\eta) \equiv \nu_{l-1,i} \cdot \tilde{B}_{(l-1)}^{(i)}(\xi) + \tilde{\pi}_{2(l-1)-i-1}^{l-1,i}(\xi) + \sum_{j_1=1}^{i} \frac{\delta_{j_1}^{(2)}}{(\xi+\eta)^{j_1}} \cdot \tilde{B}_{(l-1)}^{(i-j_1)}(\xi) + \sum_{j_1=1}^{i-1} \frac{1}{(\xi+\eta)^{j_1}} \cdot \tilde{\pi}_{2(l-1)-i-1+j_1}. \tag{56}$$

By comparing Eq. 55 and Eq. 56 we conclude that

$$\{[l-(i+1)]\cdot(l-i) - l(l+1)\} \cdot B_{l-(i+1)}(\xi) + \tilde{\delta}_0 \cdot B_l^{(i+1)}(\xi) + \tilde{\gamma}_0 \cdot \pi_{2l-i-2}(\xi) \equiv \gamma_{l-1,i} \cdot \tilde{B}_{(l-1)}^{(i)}(\xi) + \tilde{\pi}_{2l-i-3}(\xi). \tag{57}$$

According to the above assumption, $i \ge 1$ and $l-(i+1) \ge 1$. Hence, the coefficient at $B_{l-(i+1)}(\xi)$ in Eq. 57 is nonzero. Besides, according to Eq. 51 $\tilde{B}_{(l-1)}(\xi)$ is a polynomial of degree $\le 2\cdot(l-1)$. Therefore, Eq. 57 can be rewritten as

$$B_{l-(i+1)}(\xi) \equiv \nu_{l,i+1} \cdot B_l^{(i+1)}(\xi) + \pi_{2l-2-i}^{l,i+1}(\xi); \quad \nu_{l,i+1} = \mathrm{const}(l, i+1),$$

where $\nu_{l,i+1}$ does not depend on $B_l(\xi)$. Thus, the identity Eq. 52 and Theorem 6 are proved.

Theorem 7. *Let $l \geq 3; n \neq 0, \pm 1, \pm 2, \ldots, \pm(l-1), -l$ and corresponding functions $B_{s-i}(\xi)$ from the expansion* Eq. 50 *satisfy the relations* Eq. 51 *and the functions $C_{s-i}(\eta)$ from the same expansion satisfy the relations obtained from* Eq. 51 *by formally replacing B_{s-i} with C_{s-i} and ξ with η for all natural $s \leq l-1$ and for all symmetries $\varphi \in C^\infty(J^s)$ of the equation $\mathcal{Y}_n$. Then the following identities are true for the coefficients $B_i(\xi,\eta)$ and $C_i(\xi,\eta)$ from the representation* Eq. 41 *of an arbitrary symmetry $\varphi \in C^\infty(J^l)$ of the equation $\mathcal{Y}_n$:*

$$B_i(\xi,\eta) \equiv \sum_{j=0}^{l-i} \frac{\delta_j^{(i)}}{(\xi+\eta)^j} \cdot B_l^{(l-i-j)}(\xi) + \sum_{j=0}^{l-i-1} \frac{1}{(\xi+\eta)^j} \cdot \pi_{l+i+j-1}^{j,i}(\xi);$$

$$C_i(\xi,\eta) \equiv \sum_{j=0}^{l-i} \frac{\delta_j^{(i)}}{(\xi+\eta)^j} \cdot C_l^{(l-i-j)}(\eta) + \sum_{j=0}^{l-i-1} \frac{1}{(\xi+\eta)^j} \cdot \pi_{l+i+j-1}^{j,i}(\eta);$$

$$1 \leq i \leq l; \qquad \delta_j^{(i)} = \mathrm{const}(i,j), \tag{58}$$

where $\pi_\alpha^{j,i}(\cdot)$ is a polynomial of degree $\leq \alpha$, $\delta_j^{(i)}$ does not depend on $B_l(\xi)$ and $C_l(\eta)$.

The representation Eq. 58 for $B_i(\xi,\eta)$ according to Theorem 6 follows from Eq. 52 and Eq. 50, while the representation for $C_i(\xi,\eta)$ is directly obtained from the representation for $B_i(\xi,\eta)$ and symmetry considerations.

4.4 Explicit formulas for one infinite series of higher symmetries of the equation $\mathcal{Y}_n$

Now derive the formulas for the symmetries $\varphi_{2s} \equiv (D_\xi - D_\eta)^{2s}(u)|_{\mathcal{Y}_\infty}; s = 1,2,3,\ldots$ of the equation $\mathcal{Y}_n$ which will be further required (starting from fhis point, we shall write $\mathcal{Y}_\infty$ instead of $\mathcal{Y}_{n,\infty}$ for brevity). Since $\Box = D_\xi - D_\eta$ is a recursion operator for the equation $\bar{l}_{F_n}(\varphi) = 0$, then really $\varphi_l \in \mathrm{Sym}\, \mathcal{Y}_n$. As the operators D_ξ and D_η commutate, the Newton binomial formula is valid: $(D_\xi - D_\eta)^m = \sum_{i=0}^{m} (-1)^i \cdot \binom{m}{i} \cdot D_\eta^i \circ D_\xi^{m-i}$; $m = 1,2,3,\ldots$; $\binom{m}{i}$ is a binomial coefficient. The following representation for φ_{2s} can be easily obtained using this formula:

$$\varphi_{2s} = (D_\xi - D_\eta)^{2s}(u)|_{\mathcal{Y}_\infty} = \sum_{q=0}^{s-1} (-1)^q \cdot \binom{2s}{q} \cdot (D_\xi^{2s-2q} + D_\eta^{2s-2q}) \circ$$

$$(D_\xi^q \circ D_\eta^q)(u)|_{\mathcal{Y}_\infty} + (-1)^s \cdot \binom{2s}{s} \cdot D_\xi^s D_\eta^s(u)|_{\mathcal{Y}_\infty}; \quad s = 1,2,3,\ldots \tag{59}$$

Hence, to find φ_{2s} function expression in terms of internal coordinates on $\mathcal{Y}_\infty$ it is sufficient to obtain corresponding formulas for each addend in Eq. 59. These formulas are given by the following theorems.

Theorem 8. *The identity*

$$D^l_\xi \circ D^l_\eta(u)|_{\mathcal{Y}_\infty} \equiv \sum_{i=1}^{l} \frac{\Gamma^l_{l-i}}{(\xi+\eta)^{2l-i}} \cdot (u_{i\xi} + u_{i\eta}); l = 1, 2, 3, \ldots, \tag{60}$$

is valid, with the coefficients Γ^l_i *given by the following formulas*

$$\Gamma^l_i \equiv \binom{l-1+i}{i} n(n-1)(n-2)\ldots(n-(l-i-1)) \cdot (n-(l-i))^2 \cdot$$
$$(n-(l-i+1))^2 \ldots (n-(l-1))^2, \quad 1 \le i \le l-1;$$

$$\Gamma^l_0 \equiv n(n-1)(n-2)\ldots(n-(l-1)).$$

Theorem 9. *The identity*

$$(D^m_\xi + D^m_\eta) \circ (D^l_\xi \circ D^l_\eta)(u)|_{\mathcal{Y}_\infty} \equiv \sum_{i=1}^{l+m} \frac{\Gamma^{l,m}_i}{(\xi+\eta)^{2l+m-i}} \cdot (u_{i\xi} + u_{i\eta});$$
$$m, l = 1, 2, 3, \ldots, \tag{61}$$

is valid, with the coefficients $\Gamma^{l,m}_i$ *given by the following formulas*

$$\Gamma^{l,m}_i = \lambda^{l,m}_i \cdot n(n-1)(n-2)\ldots(n-(i-1)) \cdot (n-i)^2 \ldots (n-(l-1))^2 \cdot$$
$$(n-l)(n-(l+1))\ldots(n-(l+m-1)), \quad 1 \le i \le l-1;$$
$$\Gamma^{l,m}_\delta = \lambda^{l,m}_\delta \cdot n(n-1)(n-2)\ldots(n-(l-1)) \cdot$$
$$(n-\delta)(n-(\delta+1))\ldots(n-(l+m-1)), \quad l \le \delta \le l+m-1;$$
$$\Gamma^{l,m}_{l+m} = n(n-1)(n-2)\ldots(n-(l-1));$$
$$\lambda^{l,m}_i = \binom{2l+m-i-1}{l-1} + \binom{2l+m-i-1}{l+m-1}, \quad 1 \le i \le l+m.$$

Theorems 8 and 9 are proved by induction, with Theorem 9 being proved by induction over m exponent. These proofs are omitted because they are highly cumbersome and complicated. It follows from Eq. 59-Eq. 61 that

$$\varphi_{2s} = (D_\xi - D_\eta)^{2s}(u)|_{\mathcal{Y}_\infty} \equiv \sum_{i=1}^{2s} \frac{A_{i,s}}{(\xi+\eta)^{2s-i}} \cdot (u_{i\xi} + u_{i\eta});$$
$$A_{i,s} \equiv \text{const}(i, s, n); \qquad A_{2s,s} = 1; s = 1, 2, 3, \ldots. \tag{62}$$

Concrete expressions for $A_{i,s}$ with $i \neq 1, 2s$ are not given here, because we shall not need them. The coefficient $K(s) \equiv A_{1,s}$ plays the key role when calculating the

Sym$\mathcal{Y}_n$ algebra. A compact representation of $K(s)$, which will be derived below, is used in this calculation. It is easily seen that $\varphi_2 = u_{2\xi} + u_{2\eta} - \frac{2n}{\xi+\eta} \cdot (u_\xi + u_\eta)$. Therefore

$$K(1) = -2n. \tag{63}$$

For $s \geq 2$, we obtain from the formulas Eq. 59-Eq. 62:

$$K(s) \equiv \sum_{l=1}^{s-1} (-1)^l \cdot \binom{2s}{l} \cdot \Gamma_1^{l,2s-2l} + (-1)^s \cdot \binom{2s}{s} \cdot \Gamma_{s-1}^s.$$

By substituting the above expressions for $\Gamma_1^{l,2s-2l}$ and Γ_{s-1}^s into this identity we find

$$K(s) \equiv 2n(n-1)(n-2)\dots(n-(s-1)) \cdot \{-\binom{2s}{1} \cdot (n-s)\dots(n-(2s-2))+$$
$$\sum_{l=2}^{s-1} (-1)^l \binom{2s}{l} \binom{2s-2}{l-1} (n-1)(n-2)\dots(n-(l-1)) \cdot (n-s)\dots(n-(2s-$$
$$(l+1))) + (-1)^s \cdot \frac{1}{2} \cdot \binom{2s}{s} \binom{2s-2}{s-1} \cdot (n-1)(n-2)\dots(n-(s-1))\};$$
$$s = 2, 3, 4, \dots \tag{64}$$

This immediately yields

Lemma 1. *For $s \geq 2, K(s)$ is a polynomial with respect to the variable n, with*

$$K(s) \equiv 2n(n-1)(n-2)\dots(n-(s-1)) \cdot \mathcal{P}_{s-1}(n);$$

$\mathcal{P}_{s-1}(n)$ is a polynomial with respect to n; $\deg_n \mathcal{P}_{s-1}(n) \leq s-1$; $\deg_n K(s) \leq 2s-1$.

The symbol $\deg_n \pi$ denotes as usually the π polynomial degree with respect to the variable n. For $s = 2, 3$ we obtain simply from Eq. 64

$$K(2) = 4n(n-1)(n+1); \quad K(3) = -12n(n-1)(n-2)(n+1)(n+2).$$

These equalities and Lemma 1 allow to suggest that

$$K(s) = C(s) \cdot n(n-1)(n-2)\dots(n-(s-1)) \cdot (n+1)(n+2)\dots(n+(s-1));$$
$$C(s) = \text{const}(s); s = 2, 3, 4, \dots. \tag{65}$$

The validity of the formula Eq. 65 will be proved below.

Proposition 4. *If the representation Eq. 65 takes place for $K(s)$, then*

$$C(s) = (-1)^s \cdot 2 \cdot \binom{2s-2}{s-1}; \qquad s = 2, 3, 4, \dots \tag{66}$$

Proof. If the formula Eq. 65 is true, it evidently yields

$$\frac{K(s)}{2n(n-1)(n-2)\dots(n-(s-1))}\Big|_{n=0} = \frac{1}{2}\cdot C(s)\cdot(s-1)!. \qquad (67)$$

On the other hand, after simple transformations we obtain from Eq. 64

$$\frac{K(s)}{2n(n-1)(n-2)\dots(n-(s-1))}\Big|_{n=0} = (-1)^s\frac{(2s-2)!}{(s-1)!}\cdot A(s), \qquad (68)$$

where $A(s) = \sum_{l=1}^{s-1}(-1)^{l-1}\binom{2s}{l} + (-1)^{s-1}\cdot\frac{1}{2}\cdot\binom{2s}{s}$. For $s \geq 2$, it is easily shown that $A(s) = 1$ using the trivial equality $0 = -(1-1)^{2s} = \sum_{l=0}^{2s}(-1)^{l-1}\binom{2s}{l}$). Now by comparing Eq. 67 and Eq. 68 we obtain Eq. 66.

It is also easy to see that the following two lemmas are valid.

Lemma 2. *The identity*

$$\sum_{l=1}^{s}(-1)^{l-1}\cdot\gamma_l\cdot\binom{2s-2}{l-1} \equiv 0; \quad \gamma_l = \begin{cases}1, & l < s;\\ \frac{1}{2}, & l = s; s \geq 2\end{cases}$$

is valid.

Lemma 3. *The identity*

$$\sum_{\mu=0}^{r}(-1)^{\mu}\cdot\gamma_\mu\cdot\binom{2r}{\mu} \equiv 0; \quad \gamma_\mu = \begin{cases}1, & \mu < r;\\ \frac{1}{2}, & \mu = r; r \geq 1\end{cases}$$

is valid.

Eq. 64 and Lemma 2 yield

Lemma 4. *For $s \geq 2$, the equality*

$$\frac{K(s)}{2n(n-1)(n-2)\dots(n-(s-1))}\Big|_{n=-1} = 0$$

takes place.

Analogously, Eq. 64 and Lemma 3 yield

Lemma 5. *For $s \geq 3$, the equality*

$$\frac{K(s)}{2n(n-1)(n-2)\dots(n-(s-1))}\Big|_{n=-2} = 0$$

takes place.

Using simple considerations, we prove

Lemma 6. *The identities*

$$\sum_{l=1}^{s}(-1)^{l-1}\cdot\gamma_l\cdot\binom{2s-2}{l-1}\cdot(l-1)(2s-2-(l-1)) \equiv 0,$$

$$\sum_{l=1}^{s}(-1)^{l-1}\cdot\gamma_l\cdot\binom{2s-2}{l-1}\cdot l\cdot(2s-l) \equiv 0; \qquad \gamma_l=\begin{cases}1, & l<s;\\ \frac{1}{2}, & l=s; s\geq 3\end{cases}$$

are valid.

Proposition 5. *The identities*

$$B_{s,m} \equiv \sum_{l=1}^{s}(-1)^{l-1}\cdot\gamma_l\cdot\binom{2s}{l}\binom{2s-2}{l-1}(m+1)\dots(m+l-1)$$
$$(m+s)\cdot\ldots\cdot(m+2s-(l+1)) \equiv 0, \qquad 1\leq m\leq s-1, s\geq 3;$$

$$\sum_{l=1}^{s}(-1)^{l-1}\cdot\gamma_l\cdot\binom{2s-2}{l-1}^2\cdot(m+1)\dots(m+l-1)(m+s)$$
$$\dots(m+2s-(l+1)) \equiv 0, \qquad 1\leq m\leq s-2, s\geq 3; \qquad \gamma_l=\begin{cases}1, & l<s,\\ \frac{1}{2}, & l=s\end{cases}$$

are valid for all integer s and m, satisfying the given restrictions.

Proposition 5 is proved by induction over s parameter, while the induction over m parameter is accomplished for each fixed s; Lemma 6 is used.

Proposition 6.

$$\left.\frac{K(s)}{2n(n-1)(n-2)\dots(n-(s-1))}\right|_{n=-m} = 0$$

for all integer s and m satisfying the inequalities $1\leq m\leq s-1; s\geq 2$.

For $s=2$, the equality $K(2)=4n(n-1)(n+1)$ shows Proposition 6 to be obviously valid. In a general case for $s\geq 3$, we obtain by taking $n=-m$ in Eq. 64:

$$\left.\frac{K(s)}{2n(n-1)(n-2)\dots(n-(s-1))}\right|_{n=-m} \equiv (-1)^s\cdot B_{s,m},$$

where $B_{s,m}$ is the same as in the Proposition 5. According to Proposition 5, this indicates that Proposition 6 is also true for $s\geq 3$.

The formula Eq. 63 for $K(1)$, Lemma 1 and Propositions 6 and 4 give

Theorem 10. *The general formula*

$$K(s) \equiv (-1)^s \cdot 2 \cdot \binom{2s-2}{s-1} \cdot n(n-1)(n-2)\dots(n-(s-1))\cdot$$
$$(n+1)(n+2)\dots(n+(s-1)); \qquad s = 1,2,3,4,5,\dots \quad (69)$$

is true.

Corollary. *For $n \neq 0, \pm1, \pm2, \dots, \pm(s-1)$, the coefficient $K(s) \neq 0$.*

4.5. Polynomiality of functions $B_l(\xi)$ and $C_l(\eta)$

Theorem 11. *Let $\varphi \in C^\infty(J^l)$ be the symmetry of the equation $\mathcal{Y}_n; l \geq 1$ and $n \neq 0, \pm1, \dots, \pm(l-1)$, -l. Then for any integer $i : 0 \leq i \leq l-1$, the functions $B_{l-i}(\xi)$ from the expansion Eq. 50 of the coefficients $B_{l-i}(\xi, \eta)$ of φ symmetry are polynomials possessing the following properties:*

$$\deg B_{l-i}(\xi) \leq 2l - i, \qquad 0 \leq i \leq l-1;$$
$$B_{l-i}(\xi) \equiv \nu_{l,i} \cdot B_l^{(i)}(\xi) + \pi^{l,i}_{2l-i-1}(\xi), \qquad \nu_{l,i} = \text{const}(l,i,n), 0 \leq i \leq l-1, \quad (70)$$

where $B_l^{(i)} \equiv \frac{d^i B_l(\xi)}{d\xi^i}, \pi^{l,i}_\alpha(\xi)$ is a polynomial of degree $\leq \alpha$, with constants $\nu_{l,i}$ being independent on $B_l(\xi)$. The functions $C_{l-i}(\eta)$ possess analogous properties.

Proof. For $l = 1$ and $l = 2$, the theorem statements follow immediately from Theorems 1 and 3. Now suppose that $l \geq 3$ and Theorem 11 statements are valid for $\varphi \in C^\infty(J^s)$ symmetries of the equation $\mathcal{Y}_n$ for each integer $s : 1 \leq s \leq l-1$, that is the properties Eq. 51 and analogous properties for $C_{s-i}(\eta)$ take place. Let us prove the validity of Theorem 11 statements also for $\varphi \in C^\infty(J^l)$ hence it follows from. Note at once, that according to Theorem 7 the identities Eq. 58 are met due to Eq. 51 and analogous relations for $C_{s-i}(\eta)$. First, derive some identities required further and related to the symmetries $\varphi_{2s+1} \equiv (D_\xi - D_\eta)^{2s+1}(u)|_{\mathcal{Y}_\infty}; s = 1,2,3,\dots$ After a number of transformations we obtain, using the formula Eq. 62 for φ_{2s} :

$$\varphi_{2s+1} \equiv (D_\xi - D_\eta)(\varphi_{2s})|_{\mathcal{Y}_\infty} \equiv \sum_{i=2}^{l} \frac{B_{i,l}}{(\xi+\eta)^{l-i}}(u_{i\xi} - u_{i\eta}),$$
$$B_{i,l} = \text{const}(i,l,n); \qquad B_{l,l} = 1; \; l = 2s+1; \; s = 1,2,3,\dots \quad (71)$$

By comparing the representations Eq. 58 with the representations Eq. 62 and Eq. 71 we conclude that

$$\delta^{(i)}_{l-i} = \begin{cases} K(s), \; l = 2s; \\ 0, \; l = 2s+1; \quad s = 1,2,3,\dots \end{cases} \quad (72)$$

since $\delta_j^{(i)}$ constants do not depend on $B_l(\xi)$ and $C_l(\eta)$. Since $B_l(\xi) \equiv B_{l,l} = 1$ evidently for φ_l with $l = 2s+1$, according to the formulas Eq. 58 we can rewrite Eq. 71 as

$$\varphi_l = \sum_{i=1}^{l} \frac{\delta_{l-i}^{(i)}}{(\xi+\eta)^{l-i}} \cdot (u_{i\xi} - u_{i\eta}); \quad \delta_{l-1}^{(1)} = 0; \; l = 2s+1; \; s = 1,2,3,\ldots \tag{73}$$

It is easy to derive the relation

$$\varphi_{l+1} \equiv (D_\xi - D_\eta)(\varphi_l)|_{\mathcal{Y}_\infty} \equiv \sum_{i=1}^{l} -2n(n-1)(n-2)\ldots(n-(i-1)) \cdot \delta_{l-i}^{(i)} \cdot \frac{u_\xi + u_\eta}{(\xi+\eta)^l} + \{ \text{ addends containing variables } u_{i\xi} \text{ and } u_{i\eta} \text{ with } i \geq 2\}$$

from Eq. 73. On the other hand, according to the formula Eq. 62 with $l = 2s+1$

$$\varphi_{l+1} \equiv \varphi_{2(s+1)} \equiv \sum_{i=1}^{2(s+1)} \frac{A_{i,s+1}}{(\xi+\eta)^{2(s+1)-i}} \cdot (u_{i\xi} + u_{i\eta}); \quad A_{i,s+1} = K(s+1).$$

These expressions for φ_{l+1} give an equality which is important further:

$$\sum_{i=2}^{l} n(n-1)(n-2)\ldots(n-(i-1)) \cdot \delta_{l-i}^{(i)} = -\frac{1}{2} \cdot A_{i,s+1} = -\frac{1}{2} \cdot K(s+1); \quad l = 2s+1; s = 1,2,3,\ldots \tag{74}$$

Under the assumptions made it is also easy to derive the following recurrence relation

$$\delta_i^{(l-i)} = -\frac{1}{i} \cdot \{[nl - (2n+i)(i-1)] \cdot \delta_{i-1}^{(l-i+1)} + \sum_{j=0}^{i-2} n(n-(l-i+1))\ldots(n-(l-j-1)) \cdot l \cdot \delta_j^{(l-j)}\}.$$

However this relation is not used further.

Resume the proof of the theorem. By substituting the expressions Eq. 58 for $B_i(\xi,\eta)$ and $C_i(\xi,\eta)$ into compatibility condition Eq. 46, we come to the following differential equation for $B_l(\xi)$ and $C_l(\eta)$:

$$\lambda_{l+2} \cdot B_l(\xi) + \sum_{i=2}^{l+1} r_i(\xi+\eta)^{l+2-i} \cdot B_l^{(l+2-i)}(\xi) + \sum_{i=2}^{l+1} (\xi+\eta)^{l+2-i} \cdot \pi_{l+i-3}^{*,i}(\xi) \equiv \lambda_{l+2} \cdot C_l(\eta) + \sum_{i=2}^{l+1} \sigma_i(\xi+\eta)^{l+2-i} \cdot C_l^{(l+2-i)}(\eta) + \sum_{i=2}^{l+1} (\xi+\eta)^{l+2-i} \cdot \pi_{l+i-3}^{\Delta,i}(\eta);$$
$$r_i, \sigma_i = \text{const}(i,l), \quad \lambda_{l+2} = -\delta_{l-1}^{(1)} \cdot (l-1)(l+1)(l+2n), \tag{75}$$

here as usually π_α^i denotes a polynomial of degree $\leq \alpha$; the concrete form of π_α^i is not needed. If $l = 2m$ with integer $m \geq 2$, then according to (72) $\lambda_{l+2} = -2 \cdot K(m) \cdot (2m-1)(2m+1)(m+n)$. Under the restrictions on n given by the theorem, $K(m) \neq 0$ according to formula Eq. 69, hence $\lambda_{2m+2} \neq 0$. Thus, with even $l = 2m, m \geq 2$, Eq. 75 actually contains the functions $B_l(\xi)$ and $C_l(\eta)$. For this reason, we shall speak of Eq. 75 being nontrivial for $l = 2m$. In case of odd $l = 2m+1, m \geq 1$, Eq. 75 and Eq. 72 yield $\lambda_{l+2} = 0$. Since we have no information on r_i and σ_i coefficients, we cannot conclude whether Eq. 75 is nontrivial. Moreover, as was noted above in the contact case, when $l = 1$, the corresponding compatibility condition of Eqs. 21 is met identically. It can be also shown, that the functions $B_l(\xi), C_l(\eta)$ disappear from Eq. 75 together with their derivatives for $l = 3$ while the equation itselt turns out to be a relation connecting the polynomials $\pi_\alpha^{*,i}(\xi)$ and $\pi_\alpha^{\Delta,i}(\eta)$ only. Both for $l = 1$ and $l = 3$, the nontrivial differential equation for $B_l(\xi)$ and $C_l(\eta)$ allowing to complete calculating symmetries on J^1 and J^3 is obtained from Eq. 42 for $A_2(\xi, \eta)$. In a general case for arbitrary odd $l \geq 3$ the nontrivial equation for $B_l(\xi)$ and $C_l(\eta)$ is also obtained from Eq. 42. In fact, by substituting the expressions Eq. 44 and Eq. 45 for $\frac{\partial A_2}{\partial \xi}$ and $\frac{\partial A_2}{\partial \eta}$ into Eq. 42 we obtain a differential equation containing only functions $B_i(\xi, \eta)$ and $C_i(\xi, \eta)$.Wu replacing $B_i(\xi, \eta)$ and $C_i(\xi, \eta)$ with their expressions Eq. 58 in this equation, we come to the following equation for $B_l(\xi)$ and $C_l(\eta)$ analogous to Eq. 75:

$$K_l \cdot B_l(\xi) + \sum_{i=1}^{l+1} \rho_i (\xi + \eta)^{l+2-i} \cdot B_l^{(l+2-i)}(\xi) + \sum_{i=1}^{l+1} (\xi + \eta)^{l+2-i} \cdot \tilde{\pi}_{l+i-3}^{*,i}(\xi) \equiv$$

$$- K_l \cdot C_l(\eta) + \sum_{i=1}^{l+1} \tau_i (\xi + \eta)^{l+2-i} \cdot C_l^{(l+2-i)}(\eta) + \sum_{i=1}^{l+1} (\xi + \eta)^{l+2-i} \cdot \tilde{\pi}_{l+i-3}^{\Delta,i}(\eta);$$

$$\rho_i, \tau_i = \text{const}(i, l), K_l = \text{const}(l), \tag{76}$$

where $\tilde{\pi}_\alpha^{*,i}(\xi)$ and $\tilde{\pi}_\alpha^{\Delta,i}(\eta)$ are the polynomials of degree $\leq \alpha$. The coefficient K_l for odd l is given by the formula

$$K_l = l(l+1+2n) \cdot \sum_{i=2}^{l} n(n-1)\dots(n-(i-1)) \cdot \delta_{l-i}^{(i)}; \; l = 2m+1; \; m = 1, 2, 3, \dots$$

By comparing this expression with Eq. 74, we find $K_l = -\frac{1}{2} l(l+1+2n) \cdot K(m+1)$ for $l = 2m+1$. Under the restrictions on the parameter n given by the theorem, $l+1+2n \neq 0$ and $K(m+1) \neq 0$. Hence, $K_l \neq 0$ with $l = 2m+1; m = 1, 2, 3, \dots$ and Eq. 76 is nontrivial for odd l.

Formulate some simple propositions of general nature before going on with the proof of the theorem. By applying the recursion operators $\Box, \sigma_1, \tau$ to the trivial symmetry $\varphi_0 \equiv u$ of the equation $\mathcal{Y}_n$ we easily prove

Proposition 7. *For each integer $l \geq 1$, there exists the symmetry*

$$\varphi \quad \equiv \quad A + B_l(\xi) \cdot u_{l\xi} + C_l(\eta) \cdot u_{l\eta}; \qquad A \in C^\infty(J^{l-1}) \tag{77}$$

of the equation $\mathcal{Y}_n$, *such that* $B_l(\xi)$ *is any given polynomial of degree* $\le 2l$.

If $B_l(\xi) \equiv \sum_{i=0}^{m} a_i\xi^i, a_m \neq 0$, for the symmetry φ of the form Eq.77, then we obtain by commutating φ and $\tau(u)$:

$$\tilde{\varphi} \equiv \{\varphi, \tau(u)\} \equiv H + \tilde{B}_l(\xi) \cdot u_{l\xi} + \tilde{C}_l(\eta) \cdot u_{l\eta}; \ H \in C^\infty(J^{l-1}),$$

$$\tilde{B}_l(\xi) \equiv \xi^2 \cdot B_l'(\xi) - 2l\xi \cdot B_l(\xi) \equiv (m - 2l) \cdot a_m\xi^{m+1} + \pi(\xi),$$

where $\pi(\xi)$ is a polynomial of degree $\le m$. From this immediately follows

Proposition 8. *If the equation* $\mathcal{Y}_n$ *has the symmetry* $\varphi \in C^\infty(J^l)$ *of the form* Eq. 77 *with the highest coefficient* $B_l(\xi)$ *being a polynomial of degree* $m \neq 2l$, *this equation has the symmetry* $\tilde{\varphi}$ *of analogous form with the highest coefficient* $\tilde{B}_l(\xi)$ *of degree* $\tilde{m} = m + 1$. *Particularly, if* $m \ge 2l + 1$, *then for any integer* $i > m$ *the equation* $\mathcal{Y}_n$ *has the symmetry of the form* Eq. 77 *with* $B_l(\xi)$ *being a polynomial of degree* i .

It is easily verified that the following identity for the higher Jacobi bracket is valid for any function $\varphi \in C^\infty(J^\infty)$:

$$\{\varphi, u_\xi - u_\eta\} \quad \equiv \quad \frac{\partial\varphi}{\partial\xi} - \frac{\partial\varphi}{\partial\eta},$$

which evidently leads to

Proposition 9. *If* φ *is the symmetry of the equation* $\mathcal{Y}_n$, *then* $\tilde{\varphi} \equiv \frac{\partial\varphi}{\partial\xi} - \frac{\partial\varphi}{\partial\eta}$ *is also its symmetry. Particularly, if* φ *has the form* Eq. 77 *with* $B_l(\xi)$ *being a polynomial of degree* m, *then for any integer* $i : 0 \le i \le m$ *there exists the symmetry* $\tilde{\varphi}$ *of the equation* $\mathcal{Y}_n$ *having the form* Eq. 77 *where* $B_l(\xi)$ *is a polynomial of degree* i.

Let the symbol S_l denote the linear space over the field $\mathbb{R}$ formed by coefficients $B_l(\xi)$ at the highest variables $u_{l\xi}$ in the symmetries of the form Eq. 77 of the equation $\mathcal{Y}_n$. Denote the subspace in S_l consisting of polynomial coefficients $B_l(\xi)$ by $S\mathcal{P}_l$. According to Proposition 7, the space $S\mathcal{P}_l$ is not empty. Proposition 8 trivially yields

Proposition 10. *Let the polynomial degrees of the space* $S\mathcal{P}_l$ *have in common restricted, that is there exists* $N > 0$ *such that* $\deg B_l(\xi) \le N$ *for any* $B_l(\xi) \in S\mathcal{P}_l$. *Then the inequality* $\deg B_l(\xi) \le 2l$ *is met for any* $B_l(\xi) \in S\mathcal{P}_l$.

Resume the proof of Theorem 11. If l is even, then by differentiating Eq. 75 $2l$ times with respect to ξ, we obtain

$$\sum_{i=0}^{3l+1-m} \nu_i \cdot (\xi + \eta)^{j(i)} \cdot B_l^{(i)}(\xi) + r_m \cdot (\xi + \eta)^{l+2-m} \cdot B_l^{(3l+2-m)}(\xi) \ = \ 0,$$

$$\nu_i,\ j(i) = \mathrm{const}(i),\ 0 \le j(i) \le l+1-m;\ r_m \ne 0,\ 2 \le m \le l+2. \qquad (78)$$

According to Proposition 9, $\tilde{\varphi} \equiv \frac{\partial\varphi}{\partial\xi} - \frac{\partial\varphi}{\partial\eta} \equiv H_1 + B_l'(\xi)\cdot u_{l\xi} - C_l'(\eta)\cdot u_{l\eta} \in \mathrm{Sym}\,\mathcal{Y}_n$, here $H_1 \in C^\infty(J^{l-1})$. Hence, $B_l'(\xi)$ satisfies Eq. 78:

$$\sum_{i=0}^{3l+1-m} \nu_i\cdot(\xi+\eta)^{j(i)}\cdot(B_l'(\xi))^{(i)} + r_m\cdot(\xi+\eta)^{l+2-m}\cdot(B_l'(\xi))^{(3l+2-m)} = 0. \qquad (79)$$

Differentiating Eq. 78 with respect to ξ and subtracting Eq. 79 from the resulting equation, we have

$$\sum_{i=0}^{3l+1-m} \nu_i\cdot j(i)\cdot(\xi+\eta)^{j(i)-1}\cdot B_l^{(i)}(\xi) + (l+2-m)r_m\cdot(\xi+\eta)^{l+1-m}\cdot B_l^{(3l+2-m)}(\xi) = 0.$$

Evidently, the previous equation is formally obtained from Eq. 78 if the coefficients at $B_l^{(i)}(\xi)$ in Eq. 78 are differentiated with respect to ξ. Differentiating the coefficients of Eq. 78 $(l+2-m)$ times with respect to ξ, we obtain the following relation: $(l+2-m)!\cdot r_m\cdot B_l^{(3l+2-m)}(\xi) = 0$. Since $r_m \ne 0$ and $2 \le m \le l+2, B_l^{(3l)}(\xi) = 0$. In case of odd l we find $B_l^{(3l+1)}(\xi) = 0$ quite analogously from Eq. 76. Differentiating Eq. 75 and Eq. 76 $2l$ times with respect to η we obtain in the same way $C_l^{(3l)}(\eta) = 0$ for even l and $C_l^{(3l+1)}(\eta) = 0$ for odd l. Thus $B_l(\xi)$ and $C_l(\eta)$ are polynomials and inequalities $\deg B_l(\xi) \le 3\cdot l$, $\deg C_l(\eta) \le 3\cdot l$ being true for their degrees. In the above notation, this means that $S_l = S\mathcal{P}_l$ and the degrees of polynomial $B_l(\xi)$ and $C_l(\eta)$ are in common restricted by the constant $3l$. According to Proposition 10, this yields $\deg B_l(\xi) \le 2\cdot l$ and $\deg C_l(\eta) \le 2\cdot l$. Proposition 7 shows that previous inequalities cannot be intensified. According to the inductive assumption on the validity of the identities Eq. 51, Theorem 6 results in the validity of the identities Eq. 52: $B_{l-i}(\xi) \equiv \nu_{l,i}\cdot B_l^{(i)}(\xi) + \pi_{2l-i-1}^{l,i}(\xi); 0 \le i \le l-1$, where $\nu_{l,i}$ are the constants independent on $B_l(\xi)$; $\pi_\alpha^{l,i}(\xi)$ is a polynomial of degree $\le \alpha$. As $B_l(\xi)$ is a polynomial of degree $\le 2l$, it follows that $B_{l-i}(\xi)$ are polynomials and the relations Eq. 70 are met. Evidently, analogous relations take place also for $C_{l-i}(\eta)$. Theorem 11 is proved.

Since $B_l(\xi)$ and $C_l(\eta)$ are polynomials, the variables ξ and η in Eq. 75 and Eq. 76 can take arbitrary values. We obtain the important identity $C_l(\eta) \equiv (-1)^l\cdot B_l(-\eta); l = 3,4,5,\ldots$, by assuming $\xi = -\eta$ in these equations and noting that $\lambda_{l+2} \ne 0$ for even l and $K_l \ne 0$ for odd l. Theorems 1, 3, 4, equality Eq. 43 and the previous identity directly lead to the following theorem which is one of the major theorems in this paper.

Theorem 12. *Let $\varphi \equiv \varphi(\xi,\eta,u,u_\xi,u_\eta,\ldots,u_{l\xi},u_{l\eta}) \in C^\infty(J^l)$ be the symmetry of the equation $\mathcal{Y}_n, l \ge 1$ and $n \ne 0,\pm1,\pm2,\ldots,\pm(l-1),-l$. Then φ is linear with respect to the variables $u, u_{i\xi}, u_{i\eta}$; $i = 1,2,\ldots,l$:*

$$\varphi \equiv A_1(\xi,\eta) + A_2(\xi,\eta)\cdot u + \sum_{i=1}^{l-1} B_i(\xi,\eta)\cdot u_{i\xi} + \sum_{i=1}^{l-1} C_i(\xi,\eta)\cdot u_{i\eta} + B_l(\xi)\cdot u_{l\xi} + C_l(\eta)\cdot u_{l\eta}, \tag{80}$$

with $B_l(\xi)$ and $C_l(\eta)$ being polynomials of degree $\leq 2\cdot l$,

$$C_l(\eta) \equiv (-1)^l \cdot B_l(-\eta), \tag{81}$$

and $A_1(\xi,\eta)$ being an arbitrary solution of the equation $\mathcal{Y}_n$.

5. Description of the Sym$\mathcal{Y}_n$ algebra of higher symmetries of the equation $\mathcal{Y}_n$; $n \notin \mathbb{Z}$

As shown above, the functions $\varphi_0 = u_\xi - u_\eta, \varphi_1 = \xi^2 u_\xi - \eta^2 u_\eta + n\cdot(\eta - \xi)u \equiv \tau(u), \varphi_2 = \xi\cdot u_{2\xi} - \eta\cdot u_{2\eta} + \frac{2n}{\xi+\eta}(-\xi\cdot u_\xi + \eta\cdot u_\eta)$ are the symmetries of the equation $\mathcal{Y}_n$. If $\varphi \in$Sym $\mathcal{Y}_n$ and $\varphi \equiv H_{l-1} + u_{l\xi} + (-1)^l u_{l\eta}$; $H_{l-1} \in C^\infty(J^{l-1})$, then it easily follows from the coordinate definition of the higher Jacobi bracket that

$$\{\varphi,\varphi_2\} = H_l - l\cdot u_{(l+1)\xi} + (-1)^l\cdot l\cdot u_{(l+1)\eta};\ H_l \in C^\infty(J^l).$$

Thus commutating to φ_2 moves the symmetry from J^l to J^{l+1}, keeping the coefficients at the highest variables constancy. Analogously (see Proposition 8), strengthed with φ_1 increases by one the degree of the highest coefficients $B_l(\xi)$ and $C_l(\eta)$ of the symmetry φ represented in the form (80) for the case when deg$B_l(\xi) < 2l$ while the l symmetry order retains . We obtain the function

$$\varphi_l^m \equiv \{\ldots\{\underbrace{\{\ldots\{\varphi_0,\varphi_2\}\ldots\varphi_2\}}_{l-1}\underbrace{\varphi_1\}\ldots\varphi_1\}}_{m} \in \mathrm{Sym}\mathcal{Y}_n, \tag{82}$$

by commutating $\varphi_0(l-1)$ times to φ_2 and m times to φ_1. According to the remarks made

$$\varphi_l^m \equiv H_{l-1} + B_l^m(\xi)\cdot u_{l\xi} + C_l^m(\eta)\cdot u_{l\eta}; H_{l-1} \in C^\infty(J^{l-1}),$$

with $B_l^m(\xi)$ and $C_l^m(\eta)$ being polynomials of degree m, if $m \leq 2l$.

Theorem 13. *The* Sym $\mathcal{Y}_n$ *algebra of the higher symmetries of the equation $\mathcal{Y}_n$ with $n \notin \mathbb{Z}$ is a semidirect sum of the Abelian infinite-dimensional ideal A_n^∞ and subalgebra* NSym $\mathcal{Y}_n$*:*

$$\mathrm{Sym}\mathcal{Y}_n = A_n^\infty \oplus \mathrm{NSym}\mathcal{Y}_n. \tag{83}$$

As a vector space, NSym$\mathcal{Y}_n$ *is generated by the elements*

$$u;\ \varphi_l^m,\ 0 \le m \le 2l,\ l = 1, 2, 3, \ldots, \tag{84}$$

and as Lie algebra, by the elements $\varphi_0, \varphi_1, \varphi_2, u$.

Proof. The representation Eq. 83 follows directly from Theorem 12 and the mentioned property of φ_l^m functions. The theorem statements that A_n^∞ is the Abelian ideal and NSym $\mathcal{Y}_n$ is a subalgebra simply follow from the coordinate definition of the higher Jacobi bracket.

Let $\Box^i$ be the i-degree of the operator $\Box$,

$$\Box_0^i = \Box^i, \qquad \Box_j^i = \underbrace{[\ldots[\Box^i, \tau], \ldots, \tau]}_{j}; \qquad i, j = 0, 1, 2, \ldots,$$

and $\bar{\Box}_j^i(u)$ be $\Box_j^i(u)$ restriction on $\mathcal{Y}_\infty$. Using commutation relations Eq. 26 the following formulas are proved by induction:

$$\Box^i \circ \sigma_1 = \sigma_1 \circ \Box^i + i \cdot \Box^i, \qquad \Box^i \circ \tau = \tau \circ \Box^i + 2i \cdot \sigma_1 \circ \Box^{i-1} + i(i-1) \cdot \Box^{i-1},$$

$$\sigma_1^i \circ \tau = \tau \circ (\sigma_1 + I)^i = \sum_{j=0}^{i} \binom{i}{j} \tau \circ \sigma_1^j, \qquad \sigma_1 \circ \tau^i = \tau^i \circ (\sigma_1 + i \cdot I),$$

$$\Box \circ \tau^i = \tau^i \circ \Box + 2i \cdot \tau^{i-1} \circ \sigma_1 + i(i-1) \cdot \tau^{i-1}; \qquad i \ge 0.$$

Simple calculations using these formulas and formulas Eq. 26 result in important commutation relations

$$\begin{aligned} &[\Box_j^i, \Box] = -j(2i - (j-1)) \cdot \Box_{j-1}^i, \qquad 1 \le j \le 2i; \\ &[\Box_j^i, \sigma_1] = (i - j) \cdot \Box_j^i, \qquad 0 \le j \le 2i; \\ &[\Box_j^i, \tau] = \Box_{j+1}^i, \qquad 0 \le j \le 2i; i \ge 1. \\ &[\Box_1^i, \Box^j] = -2ij \cdot \Box^{i+j-1}; \qquad i, j \ge 0. \end{aligned} \tag{85$'$}$$

For φ and ψ linear with respect to variables u_σ, there is a known equality $l_{\varphi,\psi} = [l_\psi, l_\varphi]$ which is easy to verify. According to this, Eq. 85$'$ immediately yields the following relations for the higher Jacobi bracket

$$\begin{aligned} &\{\bar{\Box}_j^i(u), \bar{\Box}(u)\} = j(2i - (j-1)) \cdot \bar{\Box}_{j-1}^i(u), \qquad 1 \le j \le 2i; \\ &\{\bar{\Box}_j^i(u), \bar{\sigma}_1(u)\} = (j - i) \cdot \bar{\Box}_j^i(u), \qquad 0 \le j \le 2i; \\ &\{\bar{\Box}_j^i(u), \bar{\tau}(u)\} = -\bar{\Box}_{j+1}^i(u), \qquad 0 \le j \le 2i; i = 1, 2, 3, \ldots \end{aligned} \tag{85}$$

The relations Eq. 85 and NSym$\mathcal{Y}_n$ algebra definition (see Theorem 13) easily yield

Proposition 11. *The functions* $u, \bar{\Box}_j^i(u), 0 \le j \le 2i, i = 1, 2, 3, \ldots$ *are linearly independent and form a linear basis of* NSym$\mathcal{Y}_n$ *algebra.*

Corollary. *The functions* Eq. 84 *are linearly independent and also form a linear basis in* NSym$\mathcal{Y}_n$.

6. Description of a group of local conservation laws for the equation $\mathcal{Y}_n; \quad n \notin \mathbb{Z}$

6.1. Lagrangian for the equation $\mathcal{Y}_n$. Description of the space of conservation laws generating functions

Proposition 12. *The equation $b^{-1} \cdot F_n(u) = 0$, where $b \equiv (\xi + \eta)^{2n}, F_n(u)$ is the left-hand side of the equation $\mathcal{Y}_n$, has a Lagrangian which may be taken $L = \frac{1}{2} b^{-1} \cdot u \cdot F_n(u)$.*

Proving this is that the variational derivative $\frac{\delta L}{\delta u}$ is calculated in a standard manner. Since the Lagrangian existence condition is the self-adjointness of the corresponding operator of universal linearization (here the operator[6] $l_b - 1_{F_n}$), Proposition 12 directly leads to

Corollary. *The operator equality*

$$l^*_{F_n} \circ b^{-1} = b^{-1} \circ l_{F_n}, b \equiv (\xi + \eta)^{2n} \tag{86}$$

is valid.

Now start calculating generating functions g of higher (local) conservation laws for the equation $\mathcal{Y}_n$. The linear space created by these functions will be denoted by $\mathbb{Z}_p(\mathcal{Y}_n)$. We shall assume that generating functions g are already restricted on $\mathcal{Y}_\infty$ and, hence, are the functions of the variables Eq. 7. It is known[4,6], that the functions $g \in \mathbb{Z}_p(\mathcal{Y}_n)$ are the solutions of the equation $\bar{l}^*_{F_n}(g) = 0$. This and Eq. 86 yield

Lemma 7. *If $g \in \mathbb{Z}_p(\mathcal{Y}_n)$, then $g = b^{-1} \cdot \varphi, \varphi \in$Sym$\,\mathcal{Y}_n$.* (87) (87)

Hereafter b is assumed to be defined by the formula Eq. 86. Since the equation $\mathcal{Y}_n$ is evidently normal and can be obtained from the variational principle, there is a bijective correspondence between $\mathbb{Z}_p(\mathcal{Y}_n)$ and the space of Noetherian (variational) symmetries of the equation $\mathcal{Y}_n$. However instead of studying whether the symmetries are Noetherian, another approach is used here (also applicable to equations that are not variational) based on the criterion from[4]. According to this criterion, generating functions g of conservation laws for the equation $\mathcal{Y}_n$ are those and only those solutions of the equation $\bar{l}^*_{F_n}(\bar{g}) = 0$, for which the operator equality

$$\bar{l}_g + \bar{\Delta}^* = \bar{B} \circ \bar{l}_{F_n}, \tag{88}$$

is met, where $\Delta(F_n) \equiv l^*_{F_n}(g), B^* = B; \Delta, B$- are $\mathcal{C}$-differential operators[4]. Thus the $\mathbb{Z}_p(\mathcal{Y}_n)$ space description problem is reduced to selection from set of the functions g of the form Eq. 87 such that the equality Eq. 88 is met. If $g \equiv g(\xi,\eta) \in \mathbb{Z}_p(\mathcal{Y}_n)$, then, according to Lemma 7 and Theorem 13, $g \equiv b^{-1} \cdot a$, where $a \equiv a(\xi,\eta) \in A_n^\infty$ is some solution of the equation $\mathcal{Y}_n$.A trivial verification of the condition Eq. 88 for the function $g \equiv b^{-1} \cdot a$ with an arbitrary function $a \in A_n^\infty$ leads to

Proposition 13. *The function $g \equiv g(\xi,\eta) \in \mathbb{Z}_p(\mathcal{Y}_n)$ if and only if $g \equiv b^{-1} \cdot a$ where $a \equiv a(\xi,\eta) \in A_n^\infty$ is an arbitrary solution of the equation $\mathcal{Y}_n$.*

Simple verification of the condition Eq. 88 proves also

Proposition 14. *The function $g_l = b^{-1} \cdot \square^l(u) \in \mathbb{Z}_p(\mathcal{Y}_n)$ if and only if $l = 2m+1, m = 0,1,2,3,\ldots$.*

There is a canonical method of deriving generating functions of conservation laws from the found ones[4]. Namelly, if g is a generating function of the conservation law for an equation $F = 0$ and φ is the symmetry of the same equation, then

$$\varphi[g] \quad \equiv \quad l_g(\varphi) + \Delta^*(g), \tag{89}$$

where $\Delta(F) \equiv l_F(\varphi)$, is also the generating function of some conservation law for an equation $F = 0$. By taking $\varphi = \square(u)$ and then $\varphi = \tau(u)$ in Eq. 89, we come after simple transformations to

Lemma 8. *Let $g \in \mathbb{Z}_p(\mathcal{Y}_n), \varphi_0 = \square(u) = u_\xi - u_\eta$ and $\varphi_1 = \tau(u)$. Then $g \equiv b^{-1} \cdot f, f \in$Sym $\mathcal{Y}_n$ and the functions*

$$\varphi_0[g] = \{\varphi_0, g\} = b^{-1} \cdot \{\varphi_0, f\} = b^{-1} \cdot \left(\frac{\partial f}{\partial \eta} - \frac{\partial f}{\partial \xi}\right),$$

$$\varphi_1[g] = \{\varphi_1, g\} + 2n(\eta - \xi) \cdot g = b^{-1} \cdot \{\varphi_1, f\}$$

are also generating functions of some conservation laws for the equation $\mathcal{Y}_n$ (here as usually $\{\varphi_i, g\}$ is the Jacobi bracket of φ_i and g functions).

By repeatedly applying Lemma 8 to the functions g_{2m+1} from Proposition 14 we prove

Proposition 15. *The functions*

$$g^0_{2m+1} \quad \equiv \quad b^{-1} \cdot \square^{2m+1}(u),$$

$$g^l_{2m+1} \quad \equiv \quad b^{-1} \cdot \{\underbrace{\varphi_1\{\varphi_1 \ldots \{\varphi_1}_{l}, \square^{2m+1}(u)\} \ldots\}\},$$

$$m = 0,1,2,3,\ldots, \quad 1 \le l \le 2 \cdot (2m+1), \tag{90}$$

where $\varphi_1 = \tau(u)$, *are generating functions of some conservation laws for the equation* $\mathcal{Y}_n$.

Proposition 16. *If* $g \equiv b^{-1} \cdot f, f \in \mathrm{Sym}\, \mathcal{Y}_n$ *and*

$$f \equiv f(\xi, \eta, u, u_\xi, u_\eta, \ldots, u_{2m\xi}, u_{2m\eta}) \in C^\infty(J^{2m}); m = 0, 1, 2, 3, \ldots$$

with $\frac{\partial f}{\partial u_{2m\xi}} \not\equiv 0$ *and* $\frac{\partial f}{\partial u_{2m\eta}} \not\equiv 0$, *then* $g \notin \mathbb{Z}_p(\mathcal{Y}_n)$.

This is easily proved by contradiction using the relations Eq. 85 and Lemma 8.

Lemma 7, Theorem 13 and Propositions 13-16 yield

Theorem 14. *The linear space* $\mathbb{Z}_p(\mathcal{Y}_n)$ *of generating functions of local conservation laws for the equation* $\mathcal{Y}_n$ *with* $n \notin \mathbb{Z}$ *is a direct sum*

$$\mathbb{Z}_p(\mathcal{Y}_n) \quad = \quad b^{-1} A_n^\infty \oplus \mathbb{Z}_p^N(\mathcal{Y}_n),$$

where $b = (\xi + \eta)^{2n}, b^{-1} A_n^\infty = \{b^{-1} \cdot a(\xi, \eta) : a(\xi, \eta) \in A_n^\infty\}, \mathbb{Z}_p^N(\mathcal{Y}_n)$ *is a vector space with one of its bases formed by the functions* $g_{2m+1}^l; m = 0, 1, 2, 3, \ldots, 0 \le l \le 2 \cdot (2m + 1)$, *which are defined by the formulas* Eq. 90.

6.2. *Differential form of conservation laws*

Let $F(u) = 0$ be an arbitrary linear differential equation. Write the Green identity for the operator $l_F = F$:

$$g \cdot l_F(u) - u \cdot l_F^*(g) \quad \equiv \quad \mathrm{Div} \bar{P}_g, \tag{91}$$

here $\bar{P}_g$ is some vector-function. If $\bar{l}_F^*(\bar{g}) = 0$ ($\bar{l}_F$ and $\bar{g}$ are the restrictions of l_F and g on prolongation of the equation $F = 0$), then Eq. 91 evidently defines some conservation law for the equation $F(u) = 0$. We shall denote it by $Gr_F(g)$:

$$Gr_F(g) \quad = \quad \{\mathrm{Div} \bar{P}_g = 0\}.$$

Proposition 17. *Let* g *be the generating function of some conservation law* θ_g *for the linear equation* $\mathcal{Y} = \{F = 0\}$. *Then the equality*

$$\bar{g} \quad \equiv \quad g + \sum_\sigma \frac{\partial g}{\partial u_\sigma} \cdot u_\sigma$$

is true for the generating function $\bar{g}$ *of the conservation law* $Gr_F(g)$ *for the equation* $\mathcal{Y}$.

Indeed, since g is the generating function of the conservation law, then $l_F^*(g) + l_g^*(F) \equiv 0$ (see[6]). Hence, $g \cdot l_F(u) - u \cdot l_F^*(g) \equiv (g \cdot I + u \cdot l_g^*)(F)$ and we obtain[4,6] for the generating function $\bar{g}$ of the conservation law $Gr_F(g)$: $\bar{g} = \overline{(g \cdot I + u \cdot l_g^*)^*(1)} = g + l_g(u)$, which coincides with the expression given in the proposition.

It is easily seen that $g_{2m+1}^l \equiv \sum_\sigma a_\sigma \cdot u_\sigma, a_\sigma \equiv a_\sigma(\xi, \eta)$. Therefore, Theorem 14 and Proposition 17 immediately result in

Theorem 15. *The linear space $\mathbb{Z}(\mathcal{Y}_n)$ of local conservation laws for the equation $\mathcal{Y}_n$ with $n \notin \mathbb{Z}$ is a direct sum*

$$\mathbb{Z}(\mathcal{Y}_n) \quad = \quad \mathbb{Z}^{-1}(\mathcal{Y}_n) \oplus \mathbb{Z}^N(\mathcal{Y}_n),$$

where $\mathbb{Z}^{-1}(\mathcal{Y}_n) = \{Gr_{F_n}(g_{-1}) : g_{-1} \equiv g_{-1}(\xi, \eta) \equiv b^{-1} \cdot a(\xi, \eta), a(\xi, \eta) \in A_n^\infty\}, \mathbb{Z}^N(\mathcal{Y}_n)$ is a vector space with one of its bases formed by the conservation laws $Gr_{F_n}(g_{2m+1}^l)$; $m = 0, 1, 2, 3, \ldots, 0 \leq l \leq 2 \cdot (2m+1)$. The function $\bar{g} = g_{-1}(\xi, \eta)$ is the generation function of the conservation law $Gr_{F_n}(g_{-1}(\xi, \eta))$ and the function $\bar{g} = 2 \cdot g_{2m+1}^l$ is that of the conservation law $Gr_{F_n}(g_{2m+1}^l)$.

In particular, the conservation law for the equation $\mathcal{Y}_n$ with the generating function $g = g_{-1}(\xi, \eta)$ or $g \in \mathbb{Z}_p^N(\mathcal{Y}_n)$ has the form $Gr_{F_n}(g)$ and is obtained from the general Green formula[7] for the operator l_{F_n}:

$$D_\xi\{-\frac{n}{\xi+\eta} u \cdot g + \frac{1}{2} u_\eta \cdot g - \frac{1}{2} D_\eta(g) \cdot u\} + D_\eta\{-\frac{n}{\xi+\eta} u \cdot g + \frac{1}{2} u_\xi \cdot g - \frac{1}{2} D_\xi(g) \cdot u\} = 0.$$

7. On symmetries and conservation laws for the equation $\mathcal{Y}_n$ with $n \in \mathbb{Z}$

Now consider the case of integer parameter n of the equation $\mathcal{Y}_n$. Theorem 12 shows that symmetries $\varphi \in C^\infty(J^l)$ of the equation $\mathcal{Y}_n$ for $l < n+1$ with $n \geq 0$ and for $l < -n$ with $n < 0$, as in general case $n \notin \mathbb{Z}$, are linear combinations of the form:

$$\varphi = A(\xi, \eta) + \sum_{0 \leq \lambda \leq l} \sum_{0 \leq m \leq 2l} \alpha_m^\lambda \cdot \varphi_\lambda^m; \qquad \alpha_m^\lambda = \mathrm{const}_m^\lambda, \tag{92}$$

where $A(\xi, \eta) \in A_n^\infty$ and φ_λ^m are defined by the formulas Eq. 82. For $l = n+1$ in case $n \geq 0$ and for $l = -n$ in case $n < 0$, more symmetries may occur in addition to Eq. 92 as follows from the solving procedure of the equation $\bar{l}_{F_n}(\bar{\varphi}) = 0$. This is due to the fact that one of the equations defining the Sym$\mathcal{Y}_n$ algebra contains the factor $(n+l) \cdot (n - (l-1))$ and, hence, is met identically. The case of wave equation ($n = 0$ or $n = -1$) indicates Sym$\mathcal{Y}_n$ algebra expansion being really possible. Since the function $u(\xi, \eta)$ is a solution of the equation $\mathcal{Y}_n$ if and only if $v \equiv (\xi + \eta)^{-(2n+1)} \cdot u(\xi, \eta)$ is a solution of the equation $\mathcal{Y}_{n_1}$ with $n_1 = -(n+1)$, then the following lemma is valid.

Lemma 9. *For any real n the equations* $\mathcal{Y}_n$ *and* $\mathcal{Y}_{n_1}, n_1 = -(n+1)$ *are equivalent with respect to the point transformation*

$$\mu : (\xi, \eta, u) \to (\xi, \eta, \ v), \quad v = (\xi + \eta)^{-(2n+1)} \cdot u.$$

According to Lemma 9 it is sufficient to consider the case $n \geq 0$ when calculating symmetries and conservation laws for the equation $\mathcal{Y}_n$. If n is integer and $n > 0$, then the equation $\mathcal{Y}_n$ is reduced to the form

$$[w_\xi \cdot (\xi + \eta)^{-2n}]_\eta \quad = \quad 0, \tag{93}$$

by repeating Laplace η-transformation n times and then using the equivalence transformation with respect to the function[1]. Finally, we prove

Theorem 16. *For* $n \in \mathbb{Z}; n \neq 0, -1$ *the equation* $\mathcal{Y}_n$ *is equivalent to* Eq. 93 *and for* $n = -1$ *to the wave equation* $w_{\xi\eta} = 0$.

Since higher symmetries of one-dimensional wave equation are well known[8], then according to Theorem 16, the calculation of higher symmetries and conservation laws for the equation $\mathcal{Y}_n$ with $n \in \mathbb{Z}$ reduces to solving simpler analogous problem for Eq. 93 with integer $n > 0$ and to calculating generating functions of higher conservation laws for the equation $w_{\xi\eta} = 0$. Generating functions and corresponding conservation laws for the equation $w_{\xi\eta} = 0$ are calculated in[9]. Eq. 93 for $n > 0$ will not be considered in this paper.

References

1. L. V. Ovsyannikov, *Group Analysis of Differential Equations* (Nauka, Moscow, 1978, in Russian. English translation: Academic Press, 1982).
2. L. V. Ovsyannikov, *Lectures on Foundations of Gas Dynamics* (Nauka, Moscow, 1981, in Russian).
3. V. E. Shemarulin, *Matem. Zametki.* **47** (1990) 138 (in Russian).
4. A. M. Vinogradov, *Acta Appl.Math.* **2** (1984) 21.
5. A. M. Vinogradov, I. S. Krasilshchik, V. V. Lychagin, *Introduction to Geometry of Nonlinear Differential Equations* (Nauka, Moscow, 1986, in Russian).
6. Peter J. Olver, *Applications of Lie Groups to Differential Equations* (Springer-Verlag, New York, 1986. Russian translation: Mir, Moscow, 1989).
7. V. V. Zharinov, *Docl. Acad. Nauk SSSR* **246** (1987) 785 (in Russian).
8. N. N. Ibragimov, *Groups of Transformations in the Mathematical Physics* (Nauka, Moscow, 1983, in Russian. English translation: Reidel, 1985).
9. V. E. Shemarulin, *VANT, ser.: Mat. Mod. Fiz. Proc.* **1** (1992) 11 (in Russian).

Geometry in Partial Differential Equations (pp. 423–445)
eds. A. Pràstaro and Th. M. Rassias

STRINGS AND MEMBRANES

K. S. STELLE
The Blackett Laboratory
Imperial College
Prince Consort Road
London SW7 2BZ, U.K.

ABSTRACT

We review the relationship between string theory and theories of higher-dimensional extended objects. Considered in particular are the coupling to Yang-Mills background fields in the presence of gauge anomalies and the relationship between string and membrane κ fermionic gauge symmetries.

1. Introduction

The constraints imposed by relativistic covariance on extended-object theories have both been the source of much of the difficulty of the subject and at the same time give much of the hope for eventual success in achieving a satisfactory physical interpretation for them. The tight constraints on the interactions in these theories give rise to the hope that at least the general structure of quantised extended-object theories might be determined purely by mathematical principles. Over the past decade, much progress has been made in exploring the properties of string theories, and the subject has progressed through several different phases. The original insistence on maintaining worldsheet conformal invariance in a string theory formulated in its critical dimension has yielded to the realisation that non-critical strings may reveal reveal more of their non-perturbative secrets. There is also a recognition, inspired by conformal field theory, that the specific Lagrangian realisation of at least the compactified sector of a theory may be less important than its algebraic characterisation by conformal eigenvalues. Moreover, the varieties of worldsheet symmetry that may prove to be relevant in string theory are now seen in a larger context, with a recognition of the possible rôles to be played by Kač-Moody algebras and W algebras.

Yet another aspect of the emerging unifying theory is the proposal that there is a duality between strings and membranes[1], according to which a "solitonic" sector of string theory would best be described by a heterotic super five-brane theory. Attempts[2] to make sense out of quantised extended objects of spatial dimensionalities $p > 1$ have been stalled by the apparent possibility that the mass spectra of such objects may be continuous, owing to the infinite possibilities for zero-area "spike" configurations leading to an infinity of flat "valleys" in the effective potential[3]. This undesirable possibility is based on analysis of a regularised version the $p = 2$ supermembrane, and may turn out not to hold in the exact non-perturbative theory, or might fail for extended objects of higher dimensionalities. If the proposal of string–five-brane duality proves to be correct, then there must be a way to find a discrete mass spectrum at least for the heterotic case with $p = 5$.

While the details of the quantisation problem still remain a delicate subject for future study, some aspects of the relationship between strings and membranes can be stated already at a classical or semiclassical level. One of the unsolved problems of the subject is the precise structure of the heterotic 5-brane Lagrangian. It is in particular the details of the fermion couplings in this chiral theory that remain unclear. In this article, we consider two issues that may help to shed light upon these problems. If one integrates out the fermions, one obtains an effective bosonic extended-object theory in which the traces of the integrated-out fermions are seen in the chiral gauge anomaly that they leave behind. This anomaly is seen in the effective theory via the appearance of a Wess-Zumino term in the worldsheet Lagrangian. The cancellation of the anomaly requires a Green-Schwarz type mechanism[4,5], in which an antisymmetric-tensor background field acquires an anomalous gauge-symmetry transformation. In the presence of the anomaly, the gauge algebra is modified by the inclusion of a $U(1)$ extension term that may be identified with the antisymmetric tensor's own gauge symmetry, after a pull-back to the world-volume. In the case of the string, this extension is in fact a central extension, but for higher-dimensional extended objects, the modification to the gauge algebra that is allowed by the Jacobi identities necessarily involves the presence of the Yang-Mills gauge field in "soft structure constants".

In section two, we shall study the coupling of Yang-Mills fields to bosonic p-branes with Wess-Zumino terms of this type, and shall construct a $U(1)$-extended covariant derivative for the modified gauge algebra.

The addition of fermions to extended-object theories is consistent with relativistic covariance thanks to the existence of a local fermionic worldsheet symmetry known as κ-symmetry. While in principle this might be a clue to the construction of unknown Lagrangians such as that for the heterotic 5-brane, the nonlinearity of the presently-known realisations of κ-symmetry makes this a difficult task. Nonetheless, one may be able to learn about the structure of higher-dimensional extended objects by comparison with known Lagrangians of lower dimension. Lower-dimensional super p-branes may be obtained from higher-dimensional ones *via* a process of "double dimensional reduction", *i.e.* of simultaneously reducing both the worldsheet and the target spacetime dimensions[6]. Conversely, it may be hoped that features of higher-dimensional theories may be learned from lower dimensions through a process of "dimensional oxidation", *i.e.,* looking for theories that would be consistent with the known ones in lower dimensions upon dimensional reduction. Study of the reduction process has already revealed that the string's characteristic world-sheet Weyl symmetry is obtained upon dimensional reduction out of the "internal" part of the higher-dimensional diffeomorphism invariance[7].

In the section three, we review the procedure of double dimensional reduction and show how the lower-dimensional κ symmetry is in the end obtained by a consistent dimensional reduction from the higher super p-brane theory. In the conclusion, we discuss some open problems.

2. Strings and p-branes in a Yang-Mills background*

A characteristic feature of the coupling to gauge-field backgrounds in extended-object theories with chiral fermions is that the transformations of abelian and non-abelian background fields become linked. In the effective field theory describing the low-energy behaviour of such models, the graviton supermultiplet contains a $(p+1)$-form gauge field that couples naturally to the $(p+1)$-dimensional world-volume of the extended object. Two examples of interest are a superstring coupled to $d = 10$ supergravity in the standard formulation involving a two-form abelian gauge field, and a super fivebrane coupled to the dual formulation of $d = 10$ supergravity, which involves a six-form abelian gauge field. The gauge anomalies induced when quantising the fermions in the superstring theory coupled to a background $d = 10$ Yang-Mills supermultiplet may be cancelled by an anomalous variation of the two-form gauge field of the supergravity multiplet[9]. These modified transformations are precisely those needed for invariance of the classical effective field theory action of $d = 10$ supergravity coupled to a $d = 10$ Yang-Mills supermultiplet[10].

In the six-form formulation of the effective classical action for $d = 10$ supergravity coupled to super Yang-Mills theory, there is no anomalous variation of the six-form gauge field. Upon quantisation of this effective field theory, however, anomaly cancellation once again does require such a variation[11]. In both formulations of the effective field theory, one thus finds that the Yang-Mills gauge algebra is modified by the presence of an abelian antisymmetric tensor gauge transformation in the commutator of two Yang-Mills gauge transformations. One may describe the consequences of this modification by its effect upon the covariant derivative that would appear in the functional Schrödinger equation that would govern the evolution of the quantum wavefunctional for the p-brane. This equation, analogous to the Wheeler-De Witt equation in quantum gravity, will not itself be the focus of the present discussion. Instead, we shall concentrate on the way in which functional derivatives must be modified in order to maintain gauge covariance. The p-brane wavefunctional is a function on p-loop space, *i.e.,* the space of maps from the p-brane to spacetime. The

* This section reviews work done in collaboration with E. Bergshoeff, R. Percacci, E. Sezgin and P.K. Townsend[8].

$(p+1)$-form gauge transformation in spacetime is equivalent to a $U(1)$ transformation of an associated one-form on p-loop space[12]. We expect, accordingly, a $U(1)$ extension of the gauge algebra.

The action for a bosonic extended object coupled to a general gravitational and Yang-Mills background describes a p-brane propagating on a background $\mathcal{M} \times \mathcal{G}$, where $\mathcal{M}$ is spacetime and $\mathcal{G}$ is the group manifold of the Yang-Mills group under consideration. For $p = 1$, this action reduces to the bosonic sector of the heterotic string[13]. For odd values of p, an action was proposed in a highly condensed notation in Ref.[5]; we shall give a more explicitly-written version of this below[8]. From the action for the $p = 1$ (string) case, one may deduce the anomaly-modified set of Yang-Mills generators acting on $p = 1$ (string) wavefunctionals and one may then calculate their commutator algebra, which turns out[14] to be an affine Kač-Moody algebra, whose central-charge operator may be interpreted as a $U(1)$ extension associated to the antisymmetric tensor[15].

Strings: $p = 1$

To be specific, let us begin with the case of a closed bosonic string. For an action describing motion on a background $\mathcal{M} \times \mathcal{G}$, we let x^μ and y^m be the coordinates on $\mathcal{M}$ and on $\mathcal{G}$, respectively. The background gauge and antisymmetric-tensor fields are $A_\mu(x) = A^a_\mu T_a$ and $B_{\mu\nu}(x)$. Given the group manifold $\mathcal{G}$, there will be left-invariant vector fields $L_a = L_a{}^m(y)\partial_m$ satisfying the group's Lie algebra $[L_a, L_b] = L_c f^c{}_{ab}$ where $f^c{}_{ab}$ are the structure constants. The gravitational background is described by a Riemannian metric $g_{\mu\nu}$; on the group manifold $\mathcal{G}$, there is also an invariant metric $g_{mn} = L_m{}^a L_n{}^b d_{ab}$, where $d_{ab} = \mathrm{tr}(T_a T_b)$ is a certain multiple of the Cartan-Killing metric for $\mathcal{G}$. We shall also need an antisymmetric potential b_{mn} on $\mathcal{G}$, chosen so as to satisfy $3\partial_{[m}b_{np]} = f_{abc}L_m{}^a L_n{}^b L_p{}^c$, where raising and lowering of indices is done with d_{ab} and its inverse, *i.e.* $f_{abc} = d_{ad}f^d{}_{bc}$.

Letting the worldsheet coordinates be denoted by $\xi^i = (\tau, \sigma)$, and letting the independent worldsheet metric be denoted by $\gamma_{ij}(\xi)$, the string action on $\mathcal{M} \times \mathcal{G}$ can be written

$$S = \int d^2\xi \Big[-\tfrac{1}{2}\sqrt{-\gamma}\gamma^{ij}\Big(\partial_i x^\mu \partial_j x^\nu g_{\mu\nu} + (D_i y)^m (D_j y)^n g_{mn}\Big) + \epsilon^{ij}\left(\tfrac{1}{2}B_{ij} - k_1 L^a_i A_{ja} - \tfrac{1}{2}k_1 b_{ij}\right)\Big], \tag{2.1}$$

where B_{ij}, b_{ij}, L^a_i and A^a_i are the pullbacks to the worldsheet of $B_{\mu\nu}$, b_{mn}, L^a_m and A^a_μ, respectively, *i.e.* $B_{ij} = \partial_i x^\mu \partial_j x^\nu B_{\mu\nu}$, $b_{ij} = \partial_i y^m \partial_j y^n b_{mn}$, *etc.* Also, $\gamma = \det \gamma_{ij}$, and $(D_i y)^m \equiv \partial_i y^m - \partial_i x^\mu A^a_\mu L^m_a$ is a Yang-Mills covariant derivative. The coefficient k_1 in (2.1) is a normalisation constant needed to make the exponentiated Wess-Zumino term $\exp(-\frac{1}{2}ik_1 \int \epsilon^{ij} b_{ij})$ be globally well-defined; the value of k_1 is an integral multiple of a numerical factor whose precise value will not concern us here. The last two terms in (2.1) are not Yang-Mills invariant by themselves, but their variation can be cancelled, up to a total derivative by an anomalous variation of $B_{\mu\nu}$. Specifically, under Yang-Mills and antisymmetric-tensor gauge transformations, the fields transform as

$$\begin{aligned} \delta x^{\mu} &= 0 \ , \\ \delta y^{m} &= \epsilon^{a}(x) L_{a}{}^{m}(y) \ , \\ \delta A_{\mu}^{a} &= \partial_{\mu}\epsilon^{a} + f^{a}{}_{bc} A_{\mu}^{b}\epsilon^{c} = (D_{\mu}\epsilon)^{a} \ , \\ \delta B_{\mu\nu} &= -\ 2k_1 A_{[\mu}^{a}\partial_{\nu]}\epsilon_{a} + 2\partial\lambda_{[\mu}\Lambda_{\nu]} \ . \end{aligned} \tag{2.2}$$

The variation of the action (2.1) under these transformations is

$$\begin{aligned} \delta S &= \int d^2\xi\, \varepsilon^{ij}\partial_i\big[k_1\epsilon^a\partial_j y^m L_m{}^b(d_{ab} - b_{ab}) + \partial_j x^{\mu}\Lambda_{\mu}\big] \\ &= \int d\tau\partial_\tau \oint d\sigma \big[k_1\epsilon^a\partial_\sigma y^m L_m{}^b(d_{ab} - b_{ab}) + \partial_\sigma x^{\mu}\Lambda_{\mu}\big] \ , \end{aligned} \tag{2.3}$$

where the second line follows from Stokes' theorem for a closed string. We note that the inclusion of the Wess-Zumino term in the action is not obligatory from a worldsheet point of view, but it is required in order for the background gauge field transformations that are needed for invariance of the action to coincide with those known from D-dimensional supergravity/Yang-Mills theory for the particular (D=10) case of most interest.

Now consider a string wavefunctional $\Psi(x,y)$. When making gauge transformations, we shall be varying the background fields A_μ and $B_{\mu\nu}$ as well as the string variables. Since the string wavefunctional will also respond to the changes in these backgrounds, we may denote the arguments of this functional as $\Psi(x,y;A,B)$. In order to determine the structure of the covariant derivative that should operate on such a functional, for example in a functional Schrödinger equation, we shall take a specific example that is as close as possible to an *invariant* functional[8], *i.e.* it will be invariant up to a phase. This procedure is logically analogous to discovering the structure of the ordinary covariant derivative for the spatial rotation group $SO(3)$ by investigating the properties of invariant tensor functions like $f(x^2)$, $x^i f(x^2)$, *etc.* In the present case, we pick the "ground-state" wave-functional

$$\Psi[x,y;A,B] = \int^{x,y} [dX][dY] e^{iS[X,Y;A,B]} \ , \tag{2.4}$$

where the arguments $\big(x^\mu(\sigma), y^m(\sigma)\big)$ are the boundary values of the dummy integration variables $\big(X^\mu(\tau,\sigma), Y^m(\tau,\sigma)\big)$. The world-sheet of the string in this case is considered to be a two-manifold with a single S^1 boundary coinciding with the configuration of the string at the final time. The absence of an initial boundary condition may be viewed in analogy to the 'no-boundary' treatment of the wavefunction of the universe in quantum cosmology. Upon variation, assuming an invariant path-integral measure and keeping only first-order infinitesimal terms, we have

$$\begin{aligned}\Psi[x, y+\delta y; A+\delta A, B+\delta B] &= \int^{x,y+\delta y} [dX][dY] e^{iS[X,Y;A+\delta A,B+\delta B]} \\ &= \int^{x,y} [dX][dY] e^{iS[X,Y+\delta Y;A+\delta A,B+\delta B]} \\ &= \int^{x,y} [dX][dY] e^{i(S+\delta S)} \\ &= (1+i\delta S)\Psi[x,y;A,B] ,\end{aligned} \tag{2.5}$$

where the last line follows from the assumption that δS is a surface term and therefore is independent of the integration variables. The assumption that the functional integral measure is invariant is justified for the bosonic string case considered here. If we were dealing with the true heterotic string, in which the Yang-Mills fields couple to the worldsheet *via* chiral fermions, then the surface term would arise not from the variation of the classical action but from the non-invariance of the measure. The final result would be the same, however. Expanding the left-hand-side of (2.5) to first order in variations, one deduces that Ψ satisfies the functional differential equation

$$\begin{aligned}\Big\{ \int d^D x' \Big[\delta A^a_\mu(x') \frac{\delta}{\delta A^a_\mu(x')} + \delta B_{\mu\nu}(x') \frac{\delta}{\delta B_{\mu\nu}(x')} \Big] \\ + \oint d\sigma \delta y^m(\sigma) \frac{\delta}{\delta y^m(\sigma)} - i\, \delta S(x,y) \Big\} \Psi[x,y;A,B] = 0 .\end{aligned} \tag{2.6}$$

where D is the dimension of $\mathcal{M}$. The functional derivatives here are taken to be densities; *e.g.*, $\delta y^n(\sigma')/\delta y^m(\sigma) = \delta^n_m \delta(\sigma - \sigma')$, where the delta function is a density. Substituting the particular variations of equations (2.2) into (2.6) and taking into account the independence of ϵ^a and Λ, we find the invariance conditions for Ψ:

$$G_\epsilon \Psi(x,y;A,B) = 0 \qquad G_\Lambda \Psi(x,y;A,B) = 0 , \tag{2.7}$$

where

$$\begin{aligned}G_\epsilon = \int d^D x' \Big\{ \big(D_\mu \epsilon(x')\big)^a \frac{\delta}{\delta A^a_\mu(x')} - 2k_1 \big(\partial_\nu \epsilon(x')\big)_a A^a_\mu(x') \frac{\delta}{\delta B_{\mu\nu}(x')} \Big\} \\ + \oint d\sigma \epsilon^a\big(x(\sigma)\big) D_a(\sigma)\end{aligned} \tag{2.8a}$$

$$G_\Lambda = \int d^D x' \, 2\partial_\mu \Lambda_\nu(x') \frac{\delta}{\delta B_{\mu\nu}(x')} - i \oint d\sigma \, \partial_\sigma x^\mu(\sigma) \Lambda_\mu\big(x(\sigma)\big) \tag{2.8b}$$

and $D_a(\sigma)$ is given by

$$D_a(\sigma) = L_a{}^m \frac{\delta}{\delta y^m(\sigma)} - i k_1 \partial_\sigma y^m L_m{}^b (d_{ab} - b_{ab}) . \tag{2.9}$$

Equations (2.7) encode the fact that the wavefunctional Ψ (2.4) is invariant under the transformations (2.2) up to a phase. Although this result was found for a particular invariant wavefunctional, the structure of the covariant derivative so found remains valid in general, as is clearly to be required for the gauge-invariance of physical quantities.

Given the generators (2.8) of Yang-Mills and antisymmetric-tensor gauge transformations, one may compute their algebra to see how the anomalous phase term in (2.5) affects things. One may easily find the structure of the corrected algebra by confining attention initially to just the background fields, which transform among themselves. In this fashion, one obtains

$$[G_{\epsilon_1}, G_{\epsilon_2}] = G_{[\epsilon_1,\epsilon_2]} - G_\Lambda \tag{2.10}$$

where

$$\Lambda = k_1 \omega_1^2(A, \epsilon_1, \epsilon_2) \tag{2.11}$$

is a 1–form 2-cocycle of the algebra of Yang-Mills gauge transformations and k_1 is the normalisation constant referred to above. Specifically, we have

$$\omega_1^2(A, \epsilon_1, \epsilon_2) = -2\mathrm{tr}\, \epsilon_1 d\epsilon_2 \ . \tag{2.12}$$

A special feature of the string is that in this case the cocycle (2.11) is background-field independent, so the algebra defined by (2.10) is a Lie algebra. For $p > 1$ the structure 'constants' of the algebra will be background-field dependent.

Of course, verifying the consistency of the algebra (2.10) requires verifying that the variations of all the string variables yield the same algebra. In the case of the string, verifying this algebra directly on the string variables is not too difficult because the operators $D_a(\sigma)$ appearing in the generators (2.8) are background-field independent. As a result, one finds that all commutators of the complete generators vanish except for (2.10), with $[\epsilon_1, \epsilon_2] \equiv f^a{}_{bc}\epsilon_1^b \epsilon_2^c T_a$ and $\Lambda_\mu = k_1 \epsilon_2^a \overleftrightarrow{\partial}_\mu \epsilon_{1a}$.

General odd p

Now let us consider the case of a p-brane for general odd p, propagating as above on $\mathcal{M} \times \mathcal{G}$. The corresponding background fields in the spacetime $\mathcal{M}$ are the metric $g_{\mu\nu}$, the Yang-Mills field A_μ and the $(p+1)$-form gauge field $B_{\mu_1\ldots\mu_{p+1}}(x)$. On the 'vertical' part $\mathcal{G}$ of the target space, as before, the backgrounds are the left-invariant metric g_{mn} and the potential $b_{m_1\ldots m_{p+1}}(y)$, which is defined to satisfy

$$\partial_{[m_1} b_{m_2\ldots m_{p+2}]} = -c_p(p+1)!\mathrm{tr} L_{[m_1} \ldots L_{m_{p+2}]} \ , \tag{2.13}$$

where now we write $L_m = L_m{}^a T_a$, and $c_p = (-1)^{\frac{p+1}{2}} \left(\frac{p+3}{2}\right) \Gamma(\frac{p+3}{2})^2 / \Gamma(p+3)$.

We may now write the odd-p-brane action, for unit p-volume tension, as[5,8]

$$S = \int d^{p+1}\xi \Big\{ -\tfrac{1}{2}\sqrt{-\gamma}\gamma^{ij}\left(\partial_i x^\mu \partial_j x^\nu g_{\mu\nu} + D_i y^m D_j y^n g_{mn}\right) + \frac{(p-1)}{2}\sqrt{-\gamma} + \frac{1}{(p+1)!}\varepsilon^{i_1\ldots i_{p+1}}\left[B_{i_1\ldots i_{p+1}} + k_p C_{i_1\ldots i_{p+1}} - k_p b_{i_1\ldots i_{p+1}}\right]\Big\} , \tag{2.14}$$

where k_p is a normalisation constant, and $B_{i_1\ldots i_{p+1}}$ and $b_{i_1\ldots i_{p+1}}$ are the pullbacks of the corresponding antisymmetric tensors on $\mathcal{M}$ and $\mathcal{G}$ to the worldvolume. The expression $C_{i_1\ldots i_{p+1}}$ denotes the components of the pullback of a $(p+1)$-form C_{p+1} on $\mathcal{M}\times\mathcal{G}$ that is constructed as follows. Define $A_t = tA + (1-t)L$ and $F_t = dA_t + A_t^2$, where $A = dx^\mu A_\mu$ and $L = dy^m L_m$. Defining the operator

$$\ell_t = dt(A^a - L^a)\frac{\partial}{\partial F_t^a} , \tag{2.15}$$

one has

$$C_{p+1}(A,F,L) = \int_0^1 \ell_t\, \omega^0_{p+2}(A_t,F_t) , \tag{2.16}$$

where ω^0_{p+2} is the Chern-Simons form defined by the relation

$$d\omega^0_{p+2} = \mathrm{tr}F^{\frac{p+3}{2}}. \tag{2.17}$$

The cases of most interest are $p = 1, 3, 5$, for which

$$\begin{aligned} \omega_3^0(A,F) &= \mathrm{tr}\Big(FA - \frac{1}{3}A^3\Big) \\ \omega_5^0(A,F) &= \mathrm{tr}\Big(F^2A - \tfrac{1}{2}FA^3 + \frac{1}{10}A^5\Big) \\ \omega_7^0(A,F) &= \mathrm{tr}\Big(F^3A - \frac{2}{5}F^2A^3 - \frac{1}{5}FAFA^2 + \frac{1}{5}FA^5 - \frac{1}{35}A^7\Big) ; \end{aligned} \tag{2.18}$$

substituting these expressions into (2.16), one finds that

$$\begin{aligned} C_2 =& \,\mathrm{tr}(AL) \\ C_4 =& \frac{1}{4}\mathrm{tr}\left[2(FA+AF-A^3)L + ALAL - 2AL^3\right] \\ C_6 =& \frac{1}{30}\mathrm{tr}\big[(10F^2A + 10FAF + 10AF^2 - 8FA^3 - 8A^3F - 4AFA^2 - 4A^2FA + 6A^5)L \\ & + 2F(A^2L^2 - L^2A^2 + 3ALAL - 3LALA) - 6A^3LAL \\ & + 3F(LAL^2 - L^2AL + 2L^3A - 2AL^3) + 6A^3L^3 \\ & - 3L^2A^2LA + 3A^2L^2AL + 2ALALAL + 6L^3ALA + 6AL^5\big] . \end{aligned} \tag{2.19}$$

The Yang-Mills gauge variation of the Chern–Simons forms defines the $(p+1)$-form ω^1_{p+1}:

$$\delta_\epsilon \omega^0_{p+2}(A,F) = d\omega^1_{p+1}(A,F,\epsilon) \ . \tag{2.20}$$

Moreover, the form ω^1_{p+1} can be written as[16]

$$\omega^1_{p+1}(A,F,\epsilon) = \mathrm{tr}\, d\epsilon\, \phi_p(A,F) \tag{2.21}$$

where the p-form $\phi_p = \phi^a_p T_a$ is a Lie algebra-valued polynomial in A and F, given, for $p = 1,3,5$, by

$$\begin{aligned}
\phi_1 &= -A \\
\phi_3 &= -\tfrac{1}{2}(FA + AF - A^3) \\
\phi_5 &= -\tfrac{1}{3}\left[(F^2A + FAF + AF^2) - \tfrac{4}{5}(A^3F + FA^3) - \tfrac{2}{5}(A^2FA + AFA^2) + \tfrac{3}{5}A^5\right] \ .
\end{aligned} \tag{2.22}$$

The components of the form ω^1_{p+1} appear in the Yang-Mills transformation of the antisymmetric tensor field; the full Yang-Mills and antisymmetric tensor gauge transformations are then

$$\begin{aligned}
\delta x^\mu &= 0 \\
\delta y^m &= \epsilon^a(x) L^m_a(y) \\
\delta A^a_\mu &= \partial_\mu \epsilon^a + f^a{}_{bc} A^b_\mu \epsilon^c \\
\delta B_{\mu_1 \dots \mu_{p+1}} &= -\, k_p \omega^1_{\mu_1 \dots \mu_{p+1}}(A,F,\epsilon) + (p+1)\partial_{[\mu_1} \Lambda_{\mu_2 \dots \mu_{p+1}]}
\end{aligned} \tag{2.23}$$

The variation of C_{p+1} is determined as follows. Note first that ℓ_t commutes with δ_ϵ, since $A - L$ and F both transform homogeneously. Then, from the definition (2.16) of C_{p+1}, and (2.20) it follows that

$$\delta_\epsilon C_{p+1} = \int_0^1 \ell_t \; d\omega^1_{p+1}(A_t, F_t, \epsilon) \ . \tag{2.24}$$

Next, apply the homotopy formula[16]

$$d_t \equiv dt \frac{d}{dt} = \ell_t d - d\ell_t \ ; \tag{2.25}$$

this leads to the formula

$$\delta_\epsilon C_{p+1} = \omega^1_{p+1}(A,F,\epsilon) - \omega^1_{p+1}(L,\epsilon) + d\int_0^1 \ell_t \mathrm{tr}\, d\epsilon \phi_p(A_t, F_t) \ , \tag{2.26}$$

where $\omega^1_{p+1}(L,\epsilon) = \omega^1_{p+1}(L,0,\epsilon)$, and the last term has been rewritten using (2.21). Finally, defining a Lie algebra-valued $(p-1)$-form $\chi_{p-1} = \chi^a_{p-1} T_a$ by

$$\chi_{p-1} = \int_0^1 \ell_t \; \phi_p(A_t, F_t) \ , \tag{2.27}$$

one obtains

$$\delta_\epsilon k_p C_{p+1} = k_p\Big(\omega^1_{p+1}(A,F,\epsilon) - \omega^1_{p+1}(L,\epsilon) + d\big(\mathrm{tr}\,[\epsilon d\chi_{p-1}]\big)\Big)\,. \tag{2.28}$$

The variation of the last term in (2.14) is given by

$$\begin{aligned} &-\frac{k_p}{p!}\epsilon^{i_1\ldots i_{p+1}}\partial_{i_1}\Big\{\epsilon^a L_a^{m_1}\partial_{i_2}y^{m_2}\ldots\partial_{i_{p+1}}y^{m_{p+1}}b_{m_1m_2\ldots m_{p+1}}\Big\}\\ &-\frac{(p+2)k_p}{(p+1)!}\epsilon^{i_1\ldots i_{p+1}}\epsilon^a L_a^n\partial_{i_1}y^{m_1}\ldots\partial_{i_{p+1}}y^{m_{p+1}}\partial_{[n}b_{m_1\ldots m_{p+1}]}\,. \end{aligned} \tag{2.29}$$

The first term in (2.29) is a surface term; using (2.13), the second term is seen to be equal to

$$k_pc_p(p+2)\mathrm{tr}\;\epsilon L^{p+1} = k_p\mathrm{tr}\;\epsilon d\phi_p(L) = k_p\Big(\omega^1_{p+1}(L,\epsilon) - d(\mathrm{tr}\;\epsilon\phi_p(L))\Big)\,, \tag{2.30}$$

where the fact that $\phi_p(L)\equiv\phi_p(L,0) = c_p(p+2)L^p$ and (2.21) have been used. One can now easily prove the invariance of the action up to a surface term. The terms in the first line of (2.14) are manifestly invariant. The variation of B cancels the first term on the r.h.s. of (2.28). The first term on the r.h.s. of (2.30) cancels the second term on the r.h.s. of (2.28).

All that remains now are the first term in (2.29), the third term on the r.h.s. of (2.28) and the second term on the r.h.s. of (2.30). Taking into account the surface terms coming from the tensor gauge transformations as well, one finds

$$\begin{aligned} \delta S = \int d\tau\partial_\tau\oint d^p\sigma\frac{1}{p!}\epsilon^{r_1\ldots r_p}\Big\{&\Lambda_{r_1\ldots r_p}\\ &-k_p\epsilon^a\Big[L_a{}^m b_{mr_1\ldots r_p} + d_{ab}\big(\phi^b_p(L)\big)_{r_1\ldots r_p} - pd_{ab}\partial_{[r_1}\big(\chi^b_{p-1}\big)_{r_2\ldots r_p]}\Big]\Big\}\,, \end{aligned} \tag{2.31}$$

where the i-index has been decomposed into a time and a space part according to $i = (0, r_1\ldots r_p)$. In this formula, appropriate pull-backs with $\partial_r x^\mu$ and $\partial_r y^m$ are understood. (For comparison with the string formulas given earlier, note here $\chi^a_0 = 0$ for the string and for the 3-brane and the 5-brane, one has respectively $\chi^a_2 = -\frac{1}{2}d^a{}_{bc}A^bL^c$ and $\chi^a_4 = d^a{}_{bcd}L^bA^c[F^d - \frac{1}{10}f^d{}_{fg}(3A^fA^g + 3L^fL^g - 2L^fA^g)]$, where $d_{abcd} = \mathrm{tr}(T_{(a}T_bT_cT_{d)})$).

From (2.31), one deduces the generators of the (anomaly-modified) Yang-Mills gauge transformations acting on p-brane wavefunctionals:

$$\begin{aligned} G_\epsilon = \int d^Dx'\Big\{\big(D_\mu\epsilon^a(x')\big)\frac{\delta}{\delta A^a_\mu(x')} - (p+1)k_p\big(\partial_{\mu_1}\epsilon^a(x')\big)d_{ab}(\phi^b_p)_{\mu_2\ldots\mu_{p+1}}(A,F)\frac{\delta}{\delta B_{\mu_1\ldots\mu_{p+1}}(x')}\\ +\oint d^p\sigma\;\epsilon^a D_a(\boldsymbol{\sigma}) \end{aligned} \tag{2.32a}$$

$$G_\Lambda = \int d^D x' \left\{ (p+1)\partial_{\mu_1}\Lambda_{\mu_2\ldots\mu_{p+1}}(x') \frac{\delta}{\delta B_{\mu_1\ldots\mu_{p+1}}(x')} \right\}$$
$$- i \oint d^p\sigma \, \varepsilon^{r_1\ldots r_p}\partial_{r_1}x^{\mu_1}\cdots\partial_{r_p}x^{\mu_p}\Lambda_{\mu_1\ldots\mu_p} , \tag{2.32b}$$

where $D_a(\boldsymbol{\sigma})$ is given by

$$D_a(\boldsymbol{\sigma}) = L_a{}^m \Bigg[\frac{\delta}{\delta y^m(\boldsymbol{\sigma})}$$
$$+ i\frac{k_p}{p!}\varepsilon^{r_1\ldots r_p}\Big\{ b_{mr_1\ldots r_p} + L_{mb}\Big[\big(\phi_p^b(L)\big)_{r_1\ldots r_p} - p\partial_{[r_1}\big(\chi_{p-1}^b\big)_{r_2\ldots r_p]}\Big]\Big\}\Bigg] \tag{2.33}$$

The algebra of these generators closes to form an algebra of the form (2.10), but now with

$$\Lambda = k_p\omega_p^2(A, \epsilon_1, \epsilon_2) , \tag{2.34}$$

and in the specific cases with $p = 3, 5$ one has, respectively,

$$\begin{aligned} \omega_3^2(A, \epsilon_1, \epsilon_2) &= -\operatorname{tr}\{d\epsilon_1, d\epsilon_2\}A \\ \omega_5^2(A, \epsilon_1, \epsilon_2) &= \frac{1}{15}\operatorname{tr}(5F - 3A^2)\left[2A\{d\epsilon_1, d\epsilon_2\} - d\epsilon_1 A d\epsilon_2 + d\epsilon_2 A d\epsilon_1\right] . \end{aligned} \tag{2.35}$$

These forms of the extensions of the gauge algebra may also be derived directly from the Jacobi identities simply upon consideration of the transformations of the background fields A_μ^a and $B_{\mu_1\ldots\mu_{p+1}}$ as given in (2.23). Of course, such considerations can only establish the possible consistent forms of the modified gauge algebras, but cannot directly verify full closure of the particular modified transformations of all the fields including y^m. Once this closure has been verified explicitly[17], however, the specific form (2.10, 2.34) of the modified algebra is then determined entirely by consistency with the Jacobi identities.

Given the Yang-Mills generators (2.32,2.33) for a p-brane wavefunctional, one can construct corresponding covariant derivatives. For this, it is convenient first to rewrite the action (2.14) in canonical form:

$$S = \int d\tau \oint d^p\sigma \left[\dot{x}^\mu p_\mu + \dot{y}^m p_m - \ell\,\mathcal{H}_0 - s^r\,\mathcal{H}_r \right] . \tag{2.36}$$

In this form of the action, we have introduced independent momenta $p_\mu(\tau, \boldsymbol{\sigma})$ and $p_m(\tau, \boldsymbol{\sigma})$ conjugate to the worldvolume fields x^μ and y^m, and also Lagrange multipliers $\ell(\tau, \boldsymbol{\sigma})$ and $s^r(\tau, \boldsymbol{\sigma})$ for the time and space reparametrisation constraints. The constraint functions are

$$\begin{aligned} \mathcal{H}_0 &= g^{\mu\nu}\mathcal{P}_\mu\mathcal{P}_\nu + d^{ab}\mathcal{P}_a\mathcal{P}_b + \det\left(\partial_r x^\mu\partial_s x^\nu g_{\mu\nu} + (D_r y)^m (D_s y)^n g_{mn}\right) \\ \mathcal{H}_r &= \partial_r x^\mu\mathcal{P}_\mu + (D_r y)^m L_m{}^a\mathcal{P}_a \end{aligned} \tag{2.37}$$

where

$$\mathcal{P}_\mu = p_\mu - \mathbf{B}_\mu + A^a_\mu \mathcal{P}_a$$
$$\mathcal{P}_a = L_a{}^m (p_m + k_p \mathbf{b}_m) \ . \tag{2.38}$$

The terms $\mathbf{B}_\mu$ and $\mathbf{b}_m$ in (2.38) are taken from the coupling to the antisymmetric tensor $B_{\mu\nu}$ and from the Wess-Zumino term in (2.14). These terms are linear in τ derivatives and so can be expressed as

$$\int d\tau \oint d^p\sigma \left[\dot{x}^\mu \mathbf{B}_\mu - k_p \dot{y}^m \mathbf{b}_m\right] , \tag{2.39}$$

where the overdot indicates differentiation with respect to τ. Explicitly, we have

$$\mathbf{B}_\mu = \frac{1}{p!}\varepsilon^{r_1 \ldots r_p} \partial_{r_1} x^{\nu_1} \cdots \partial_{r_p} x^{\nu_p} B_{\mu\nu_1\ldots\nu_p} + k_p A^a_\mu \frac{\partial C}{\partial A^a_0} + k_p F^a_{\mu\nu} \partial_r x^\nu \frac{\partial C}{\partial F^a_{0r}}$$
$$\mathbf{b}_m = \frac{1}{p!}\varepsilon^{r_1 \ldots r_p} \partial_{r_1} y^{n_1} \cdots \partial_{r_p} y^{n_p} b_{mn_1\ldots n_p} - L_m{}^a \frac{\partial C}{\partial L_0{}^a} \ . \tag{2.40}$$

The couplings (2.39) are invariant as written under the gauge transformations (2.23), up to a surface term. In order for the remainder of the action (2.36) to be invariant, it is necessary to supplement the transformations given in (2.23) by appropriate transformations of p_μ and p_m. Observing that the action (2.36) can be rewritten in the form

$$S = \int d\tau \oint d^p\sigma \left[\dot{x}^\mu \mathcal{P}_\mu + (D_0 y)^m L_m{}^a \mathcal{P}_a - \ell\, \mathcal{H}_0 - s^r\, \mathcal{H}_r\right]$$
$$+ \int d\tau \oint d^p\sigma \left[\dot{x}^\mu \mathbf{B}_\mu - k_p \dot{y}^m \mathbf{b}_m\right] , \tag{2.41}$$

it follows that invariance of the action will follow if the transformations of p_μ and p_m are chosen in such a way as to make the functions $\mathcal{P}_\mu$ and $\mathcal{P}_a$ covariant, *i.e.*

$$\delta \mathcal{P}_\mu = 0 \qquad\qquad \delta \mathcal{P}_a = -\mathcal{P}_c f^c{}_{ab} \epsilon^b \ . \tag{2.42}$$

Upon quantisation, one makes the replacements

$$p_\mu(\boldsymbol{\sigma}) \to \hat{p}_\mu(\boldsymbol{\sigma}) = -i \frac{\delta}{\delta x^\mu(\boldsymbol{\sigma})} \qquad p_m(\boldsymbol{\sigma}) \to \hat{p}_m(\boldsymbol{\sigma}) = -i \frac{\delta}{\delta y^m(\boldsymbol{\sigma})} , \tag{2.43}$$

and thus one obtains the corresponding p-loop-space covariant derivatives

$$\mathcal{D}_\mu \equiv i \hat{\mathcal{P}}_\mu \tag{2.44a}$$
$$\mathcal{D}_a \equiv i \hat{\mathcal{P}}_a \ . \tag{2.44b}$$

Corresponding to the classical result (2.42), they satisfy the algebra

$$\begin{aligned} [\mathcal{D}_\mu, G_\Lambda] &= 0 \qquad [\mathcal{D}_\mu, G_\epsilon] = 0 \\ [\mathcal{D}_a, G_\Lambda] &= 0 \qquad [\mathcal{D}_a, G_\epsilon] = \epsilon^b f^d{}_{ab}\mathcal{D}_d \ . \end{aligned} \tag{2.45}$$

The horizontal covariant derivative $\mathcal{D}_\mu$ is given explicitly by[8]

$$\begin{aligned} \mathcal{D}_\mu = & \frac{\delta}{\delta x^\mu(\boldsymbol{\sigma})} - \frac{i}{p!}\varepsilon^{r_1\ldots r_p}\partial_{r_1}x^{\nu_1}\cdots\partial_{r_p}x^{\nu_p}B_{\mu\nu_1\ldots\nu_p} - ik_pF^a_{\mu\nu}\partial_r x^\nu\frac{\partial C}{\partial F^a_{0r}} \\ & + A^a_\mu L_a{}^m\Big[\frac{\delta}{\delta y^m(\boldsymbol{\sigma})} + \frac{ik_p}{p!}\varepsilon^{r_1\ldots r_p}\partial_{r_1}y^{n_1}\cdots\partial_{r_p}y^{n_p}b_{mn_1\ldots n_p} - k_pL_m{}^b\Big(\frac{\partial C}{\partial L_0{}^b} + \frac{\partial C}{\partial A^b_0}\Big)\Big]. \end{aligned} \tag{2.46}$$

This can be simplified by use of the relation[17]

$$\frac{\partial C}{\partial A^a_0} + \frac{\partial C}{\partial L_0{}^a} = d_{ab}\Big(\phi^b_p(A,F) - \phi^b_p(L) + d\chi^b_{p-1}\Big) \ , \tag{2.47}$$

from which it follows that

$$\mathcal{D}_\mu = \frac{\delta}{\delta x^\mu(\boldsymbol{\sigma})} - i\mathcal{C}_\mu + A^a_\mu D_a(\boldsymbol{\sigma}) \ , \tag{2.48}$$

where $D_a(\boldsymbol{\sigma})$ is the p–loop space differential operator given above in (2.33) and

$$\mathcal{C}_\mu = \frac{1}{p!}\varepsilon^{r_1\ldots r_p}\partial_{r_1}x^{\nu_1}\cdots\partial_{r_p}x^{\nu_p}B_{\mu\nu_1\ldots\nu_p} + k_pA^a_\mu\phi^b_p(A,F)d_{ab} + k_pF^a_{\mu\nu}\partial_r x^\nu\frac{\partial C}{\partial F^a_{0r}} \ . \tag{2.49}$$

The expression $\mathcal{C}_\mu$ may be identified as a $U(1)$ gauge potential. In the case of the string, $p=1$, the result (2.48) reproduces the results of Refs[14,15].

3. String and membrane κ symmetries*

Covariant derivatives such as (2.48) become particularly relevant in the context of κ-symmetric super p-brane theories[19], because their algebra, when taken together with the principle of integrability along null super-planes[20,14], is a powerful tool for establishing the effective equations of motion for the background fields. From them, one can construct the gauge-symmetry constraints of the theory, as we have seen; from this, one may eventually be able to determine the action of the theory in as-yet unknown cases such as that of the heterotic 5-brane[1]. Much remains to be done in understanding the structure of the higher-dimensional supersymmetric extended-object theories. Much of the difficulty lies in the need for a fermionic "κ" symmetry[21]

* This section presents work done in collaboration with A. Achucarro and P. Kapusta[18].

in order to ensure equality of the numbers of bosonic and fermionic degrees of freedom on the worldsheet while also maintaining full Lorentz covariance. The κ symmetries are very nonlinear and, in most formulations to date, suffer from the characteristic supersymmetric problem of forming a closed algebra only modulo equations of motion. This makes them specific to a particular action, and thus difficult to generalise from one case to another.

In this section, we shall have a more modest aim than that of understanding the implementation of κ symmetries in general supersymmetric p-branes. Instead, we shall concentrate upon the relationship between two known cases: the superstring ($p = 1$) and the supermembrane ($p = 2$). This discussion shall follow the pattern of the known "double dimensional reduction" relationship between the bosonic membrane and the bosonic string[6], which has also been used to establish a Kaluza-Klein origin to the conformal invariance of the string[7]. As in the last section, much depends on the structure of Wess-Zumino terms in the coupling of extended objects. In the present discussion, we shall limit ourselves to the case where the spacetime is flat and the antisymmetric tensor background is absent. Nonetheless, the Wess-Zumino term continues to play an important role owing to the presence of fermionic torsion.

We shall be concerned here also with the consistency of the double dimensional reduction procedure, according to which one simultaneously shrinks away a target spacetime and a worldsheet dimension by considering membrane solutions wrapped around the compact spacetime dimension to be reduced. Consistency in this context essentially means commutativity between the dimensional-reduction procedure and variation of the action to generate field equations. This notion of consistency is the one familiar from Kaluza-Klein theories, and amounts to a decoupling of the massive Kaluza-Klein fields at linear order from currents built from the massless fields desired to be retained after dimensional reduction, which corresponds to taking the Kaluza-Klein masses to infinity.

We shall be concerned here with the double dimensional reduction of the $D = 11$ supermembrane to the $D = 10$ type IIA superstring[6]. The $D = 11$ supermembrane action is[19]

$$I = -\frac{1}{2}\int d^3\hat{\xi}\Big[(\sqrt{-\hat{\gamma}}\hat{\gamma}^{\hat{\imath}\hat{\jmath}}\hat{\Pi}_{\hat{\imath}}^{\hat{\mu}}\hat{\Pi}_{\hat{\jmath}}^{\hat{\nu}}\eta_{\hat{\mu}\hat{\nu}} - \sqrt{-\hat{\gamma}} + i\varepsilon^{\hat{\imath}\hat{\jmath}\hat{k}}\hat{\bar{\theta}}\hat{\Gamma}_{\hat{\mu}\hat{\nu}}\partial_{\hat{\imath}}\hat{\theta}\left(\hat{\Pi}_{\hat{\jmath}}^{\hat{\mu}}\hat{\Pi}_{\hat{k}}^{\hat{\nu}} + i\hat{\Pi}_{\hat{\jmath}}^{\hat{\mu}}\hat{\bar{\theta}}\hat{\Gamma}^{\hat{\nu}}\partial_{\hat{k}}\hat{\theta} - \frac{1}{3}\hat{\bar{\theta}}\hat{\Gamma}^{\hat{\mu}}\partial_{\hat{\jmath}}\hat{\theta}\hat{\bar{\theta}}\hat{\Gamma}^{\hat{\nu}}\partial_{\hat{k}}\hat{\theta}\right)\Big]. \tag{3.1}$$

The target superspace coordinates in (3.1) are $(\hat{x}^{\hat{\mu}}, \hat{\theta}^{\hat{\alpha}})$ where $\hat{\mu} = 0, \ldots, 10$ and $\hat{\alpha} = 1, \ldots, 32$. Here, $\xi^{\hat{\imath}}(\hat{\imath} = 0, 1, 2)$ are the membrane worldvolume coordinates τ, σ, ρ; the $d = 11$ supersymmetric derivative $\hat{\Pi}_{\hat{\imath}}$ is defined by $\hat{\Pi}_{\hat{\imath}}^{\hat{\mu}} = \partial_{\hat{\imath}}\hat{x}^{\hat{\mu}} - i\hat{\bar{\theta}}\hat{\Gamma}^{\hat{\mu}}\partial_{\hat{\imath}}\hat{\theta}$. The worldvolume metric $\hat{\gamma}_{\hat{\imath}\hat{\jmath}}$ will initially be varied independently of the coordinates $x^{\hat{\mu}}$; $\hat{g} = \det(\hat{\gamma}_{\hat{\imath}\hat{\jmath}})$. Throughout, $(11, 3)$-dimensional quantities will always be represented by fields and indices with hats; quantities without hats will be $(10, 2)$-dimensional.

Varying the action (3.1) with respect to the fields $x^{\hat{\mu}}$, $\hat{\theta}^{\hat{\alpha}}$ and $\hat{\gamma}^{\hat{\imath}\hat{\jmath}}$ respectively yields the field equations

$$\frac{\delta I}{\delta x_{\hat{\mu}}} = \partial_{\hat{\imath}}\left(\sqrt{-\hat{\gamma}}\hat{\gamma}^{\hat{\imath}\hat{\jmath}}\hat{\Pi}_{\hat{\jmath}}^{\hat{\mu}}\right) - i\varepsilon^{\hat{\imath}\hat{\jmath}\hat{k}}\hat{\Pi}_{\hat{\imath}}^{\hat{\nu}}\partial_{\hat{\jmath}}\hat{\bar{\theta}}\hat{\Gamma}_{\hat{\nu}}^{\hat{\mu}}\partial_{\hat{k}}\hat{\theta} = 0 \tag{3.2}$$

$$\frac{\delta I}{\delta\hat{\bar{\theta}}} = 2i\sqrt{-\hat{\gamma}}\hat{\gamma}^{\hat{\imath}\hat{\jmath}}\hat{\Gamma}_{\hat{\mu}}\Pi_{\hat{\imath}}^{\hat{\mu}}\partial_{\hat{\jmath}}\hat{\theta} - i\varepsilon^{\hat{\imath}\hat{\jmath}\hat{k}}\hat{\Gamma}_{\hat{\mu}\hat{\nu}}\hat{\Pi}_{\hat{\imath}}^{\hat{\mu}}\hat{\Pi}_{\hat{\jmath}}^{\hat{\nu}}\partial_{\hat{k}}\hat{\theta} = 0 \tag{3.3}$$

$$\frac{\delta I}{\delta\gamma^{\hat{\imath}\hat{\jmath}}} = -\frac{1}{2}\sqrt{-\hat{\gamma}}\left(\hat{\Pi}_{\hat{\imath}}^{\hat{\mu}}\hat{\Pi}_{\hat{\mu}\hat{\jmath}} - \frac{1}{2}\hat{\gamma}_{\hat{\imath}\hat{\jmath}}\hat{\gamma}^{\hat{k}\hat{l}}\hat{\Pi}_{\hat{k}}^{\hat{\nu}}\hat{\Pi}_{\hat{\nu}\hat{l}} + \frac{1}{2}\hat{\gamma}_{\hat{\imath}\hat{\jmath}}\right) = 0\ . \tag{3.4}$$

Now consider the procedure of double dimensional reduction[6]. We make an appropriate split in the target-space and worldvolume coordinates:

$$\begin{aligned} x^{\hat{\mu}} &= (x^{\mu}, y) \quad \mu = 0, 1, \ldots, 9 \\ \hat{\theta}^{\hat{\alpha}} &= \theta^{\alpha} \quad \hat{\alpha}, \alpha = 0, \ldots, 32 \\ \xi^{\hat{\imath}} &= (\xi^{i}, \rho) \quad i = 0, 1\ . \end{aligned} \tag{3.5}$$

and we impose the following dimensional-reduction ansatz:

$$\partial_{\rho}x^{\mu} = 0 \tag{3.6}$$

$$\partial_{\rho}y = 1 \tag{3.7}$$

$$\partial_{\rho}\hat{\theta} = 0 \tag{3.8}$$

$$\partial_{\rho}\hat{\gamma}_{\hat{\imath}\hat{\jmath}} = 0 \tag{3.9}$$

Equation (3.7) may be integrated to give

$$y = \rho + f(\sigma, \tau), \tag{3.10}$$

where $f(\sigma,\tau)$ is an arbitrary function of the worldsheet coordinates. To obtain in the end the known superstring action, one must also choose a suitable parametrisation of the worldvolume metric. A choice of parametrisation is known[7] that yields upon reduction the correct *bosonic* symmetries of the superstring; we shall adopt the same choice here and shall enquire whether one can obtain a similarly consistent reduction of the κ symmetry. Accordingly, we choose the parametrisation

$$\hat{\gamma}_{\hat{\imath}\hat{\jmath}} = \phi^{-\frac{2}{3}}\begin{pmatrix} e^{-f(\sigma,\tau)}\gamma_{ij} + \phi^2 A_i A_j & \phi^2 A_j \\ \phi^2 A_i & \phi^2 \end{pmatrix}\ . \tag{3.11}$$

From equation (3.11), it follows that

$$\sqrt{-\det(\hat{\gamma}_{\hat{\imath}\hat{\jmath}})} = e^{-f}\sqrt{-\det(\gamma_{ij})}\ . \tag{3.12}$$

From equation (3.12), it is clear that the factor $e^{-f(\sigma,\tau)}$ in (3.11,3.12) plays the rôle of the conformal factor of the worldsheet metric in string theory. Detailled analysis of the behaviour of the diffeomorphisms in the compactified $i = 3$ worldsheet dimension shows them to be the Kaluza-Klein origin of the string's Weyl symmetry[7]. In the following, the e^{-f} factor will remain undetermined. Substitution of the metric ansatz (3.11) into the supermembrane action (3.1) while simultaneously imposing conditions (3.6-3.9) gives an action that contains the superstring as well as additional terms that may be eliminated using algebraic equations of motion. The result is

$$\begin{aligned} I_{\text{red}} = -\frac{1}{2}\int d^2\xi \Big[& \phi^{\frac{2}{3}}\sqrt{-\gamma}\gamma^{ij}\{\Pi_i^\mu\Pi_{\mu j} \\ & + \left(\Pi_i^{11} - A_i\right)\left(\Pi_j^{11} - A_j\right)\} + e^{-f}\sqrt{-\gamma}\left(\frac{1}{\phi^{\frac{4}{3}}} - 1\right) \\ & + 2i\varepsilon^{ij}\bar{\theta}\Gamma_\mu\Gamma_{11}\partial_i\theta\Pi_j^\mu - \varepsilon^{ij}\bar{\theta}\Gamma_\mu\Gamma_{11}\partial_i\theta\theta\Gamma^\mu\partial_k\theta\Big] \ . \end{aligned} \tag{3.13}$$

We have used here the convention that $\varepsilon^{ij3} = \varepsilon^{ij}$ and $\Pi_i^{11} = \partial_i Y - i\bar{\theta}\Gamma^{11}\partial_i\theta$. Equation (3.13) will provide us with the algebraic equations of motion for the A_i, ϕ and f fields, which will need to be employed throughout the reduction procedure. These equations furnish us with auxiliary conditions that we will require to be preserved by the symmetries of the reduced theory. These conditions are found to be:

$$\frac{\delta I}{\delta A_i} = -\frac{1}{2}\phi^{\frac{2}{3}}\sqrt{-\gamma}\gamma^{ij}\left(\Pi_j^{11} - A_j\right) = 0 \tag{3.14}$$

$$\frac{\delta I}{\delta \phi} = -\frac{2\sqrt{-\gamma}}{3\phi^{\frac{1}{3}}}\left[\gamma^{ij}\left(\Pi_i^\mu\Pi_{\mu j} + \left(\Pi_i^{11} - A_i\right)\left(\Pi_j^{11} - A_j\right)\right) - \frac{2e^{-f}}{\phi^2}\right] = 0 \tag{3.15}$$

$$\frac{\delta I}{\delta f} = -\frac{1}{2}e^{-f}\sqrt{-\gamma}\left(1 - \frac{1}{\phi^{\frac{4}{3}}}\right) = 0 \ . \tag{3.16}$$

Next, we solve the equations (3.14-3.16) and substitute the results into the action (3.13). By decomposing the 32-component Majorana spinor into two 16-component Majorana-Weyl spinors $\theta^- = \frac{1}{2}(1 - \Gamma_{11})\theta$ and $\theta^+ = \frac{1}{2}(1 + \Gamma_{11})\theta$, we obtain the standard type IIA superstring action in $D = 10$ dimensions[22]:

$$\begin{aligned} I_{\text{string}} = \int d^3\xi \Big[& -\frac{1}{2}\sqrt{-\gamma}\gamma^{ij}\Pi_i^\mu\Pi_{\mu j} \\ & - i\varepsilon^{ij}\partial_j x^\mu\left(\bar{\theta}^+\Gamma_\mu\partial_i\theta^+ - \bar{\theta}^-\Gamma_\mu\partial_i\theta^-\right) - \varepsilon^{ij}\bar{\theta}^+\Gamma^\mu\partial_i\theta^+\bar{\theta}^-\Gamma_\mu\partial_j\theta^-\Big] \ . \end{aligned} \tag{3.17}$$

It remains to be checked whether substitution of (3.6-3.9) and (3.11) into the supermembrane equations of motion (3.2-3.4) does in fact yield the desired type IIA string equations of motion. This is is done straightforwardly, giving

$$\left[\frac{\delta I}{\delta x_\mu}\right]_{\rm red} = \partial_i\left(\sqrt{-\gamma}\phi^{\frac{2}{3}}\gamma^{ij}\Pi^\mu_j\right) - i\varepsilon^{ij}\partial_i\bar\theta\Gamma_\mu\Gamma_{11}\partial_j\theta = 0 \tag{3.18}$$

$$\left[\frac{\delta I}{\delta\bar\theta}\right]_{\rm red} = 2i\sqrt{-\gamma}\phi^{\frac{2}{3}}\gamma^{ij}\Gamma_\mu\Pi^\mu_i\partial_j\theta - 2i\varepsilon^{ij}\Gamma_\mu\Gamma_{11}\Pi^\mu_i\partial_j\theta = 0 \tag{3.19}$$

$$\begin{aligned}\left[\frac{\delta I}{\delta\hat\gamma^{ij}}\right]_{\rm red} &= -\frac{1}{2}e^{-f}\sqrt{-\gamma}\Big((\Pi^\mu_i\Pi_{\mu j} - \frac{1}{2}\gamma_{ij}\gamma^{kl}\Pi^\nu_k\Pi_{\nu l}\\ &-\frac{1}{2}\left(\gamma_{ij}+\phi^2e^fA_iA_j\right)\gamma^{kl}\left(\Pi^{11}_k - A_k\right)\left(\Pi^{11}_l - A_l\right)\\ &+\frac{1}{2\phi^{\frac{2}{3}}}\left(1-\frac{1}{\phi^{\frac{4}{3}}}\right)\left(e^{-f}\gamma_{ij}+\phi^2A_iA_j\right)+\frac{1}{2}\phi^{\frac{4}{3}}A_iA_j\left(2-\phi^{\frac{2}{3}}e^f\gamma^{kl}\Pi^\nu_k\Pi_{\nu l}\right)\\ &+\left(\Pi^{11}_i - A_i\right)\Pi^{11}_j + A_i\left(\Pi^{11}_j - A_j\right)\Big)\,.\end{aligned} \tag{3.20}$$

The terms in equation (3.20) are grouped so as to show that substitution of solutions to (3.6-3.9) does in fact yield the correct superstring equations of motion. In particular, one may choose a solution with $\phi = 1$ in order to satisfy (3.16).

Fermionic symmetries and consistency of the reduction

The supermembrane action (3.1) is invariant under world-volume reparametrisations and super-Poincaré transformations. Our immediate concern here, however, is the local fermionic κ symmetry that it posesses. To analyse this symmetry, we shall employ the 1.5 order formalism, following Ref[23]. For the transformation of the $x^{\hat\mu}$ and $\theta^{\hat\alpha}$ fields, we take

$$\delta_{\hat\kappa}\hat\theta = \left(1+\hat\Gamma\right)\kappa \tag{3.21}$$

$$\delta_{\hat\kappa}\hat x^{\hat\mu} = i\hat{\bar\theta}\Gamma^{\hat\mu}\left(1+\hat\Gamma\right)\kappa \tag{3.22}$$

where

$$\hat\Gamma = \frac{1}{6\sqrt{-\hat\gamma}}\varepsilon^{\hat\imath\hat\jmath\hat k}\Pi^{\hat\mu}_{\hat\imath}\Pi^{\hat\nu}_{\hat\jmath}\Pi^{\hat\lambda}_{\hat k}\Gamma_{\hat\mu\hat\nu\hat\lambda} \tag{3.23}$$

and $\Gamma_{\hat\mu\hat\nu\hat\lambda}$ is totally antisymmetrised with "strength one" over the three spacetime indices, *i.e.* $\Gamma_{\hat\mu\hat\nu\hat\lambda} = \Gamma_{[\hat\mu}\Gamma_{\hat\nu}\Gamma_{\hat\lambda]} = \frac{1}{6}\Gamma_{\hat\mu}\Gamma_{\hat\nu}\Gamma_{\hat\lambda} + 5$ terms, and $\hat\kappa^{\hat\alpha}$ is a 32-component Majorana spinor in $D = 11$. Henceforth, we shall omit explicit spinorial indices unless necessary. The form of the supermembrane action (3.1) as well as the proposed forms for the κ-symmetry field transformations (3.21-3.22) should determine the κ transformation for the metric $\hat\gamma_{\hat\imath\hat\jmath}$. Now, on calculating the variation of (3.1) under (3.21-3.22), one finds terms that cancel only upon use of the embedding equation

$$\hat\gamma_{\hat\imath\hat\jmath} = \hat\Pi^{\hat\mu}_{\hat\imath}\Pi_{\hat\mu\hat\jmath}\ , \tag{3.24}$$

which is the solution to the metric field equation (3.4). This problem is overcome by absorbing corresponding equation of motion terms into the variation of the metric, which is yet to be chosen. More specifically, suppose that one has a variation of the supermembrane action that gives the following expression

$$\delta_{\hat{\kappa}} I = \hat{X}^{\hat{\imath}\hat{\jmath}} \left(\hat{\gamma}_{\hat{\imath}\hat{\jmath}} - \hat{\Pi}_{\hat{\imath}}^{\hat{\mu}} \Pi_{\hat{\mu}\hat{\jmath}} \right) + \frac{\delta I}{\delta \gamma^{\hat{\imath}\hat{\jmath}}} \delta_{\hat{\kappa}} \gamma^{\hat{\imath}\hat{\jmath}} , \tag{3.25}$$

where $\hat{X}^{\hat{\imath}\hat{\jmath}}$ is some matrix. By considering the form of equation (3.4), notice that the first two terms can be absorbed by the third if one adds to the transformation for the inverse metric the term

$$\delta_{\hat{\kappa}} \hat{\gamma}^{\hat{\imath}\hat{\jmath}} = \frac{2}{\sqrt{-\hat{\gamma}}} \left(\hat{X}^{\hat{\imath}\hat{\jmath}} - \hat{\gamma}^{\hat{\imath}\hat{\jmath}} \hat{X}_{\hat{k}}^{\hat{k}} \right) . \tag{3.26}$$

Equation (3.26) has the effect that the transformation $\delta_{\hat{\kappa}} \left(\sqrt{-\hat{\gamma}} \hat{\gamma}^{\hat{\imath}\hat{\jmath}} \right)$ becomes simply $2\hat{X}^{\hat{\imath}\hat{\jmath}}$. It is useful for the following to write out the variation of the action explicitly:

$$\begin{aligned} \delta_{\hat{\kappa}} I = & \frac{\delta I}{\delta \hat{\gamma}^{\hat{\imath}\hat{\jmath}}} \delta_{\hat{\kappa}} \hat{\gamma}^{\hat{\imath}\hat{\jmath}} - 2i\sqrt{-\hat{\gamma}} \partial_{\hat{\imath}} \bar{\hat{\theta}} \left[\Pi_{\hat{\jmath}}^{\hat{\mu}} \Gamma_{\hat{\mu}} \hat{\gamma}^{\hat{\imath}\hat{\jmath}} \left(1 - \hat{\Gamma}^2 \right) \hat{\kappa} \right] \\ & + i\partial_{\hat{\imath}} \bar{\hat{\theta}} \left[\hat{\gamma}^{\hat{\imath}(\hat{\jmath}} \varepsilon^{\hat{k})\hat{l}\hat{m}} \Pi_{\hat{l}}^{\hat{\rho}} \Pi_{\hat{m}}^{\hat{\tau}} \Gamma_{\hat{\rho}\hat{\tau}} \left(\hat{\gamma}_{\hat{k}\hat{\jmath}} - \hat{\Pi}_{\hat{\jmath}}^{\hat{\mu}} \Pi_{\hat{\mu}\hat{k}} \right) \left(1 + \hat{\Gamma} \right) \hat{\kappa} \right] . \end{aligned} \tag{3.27}$$

From (3.23), one also has

$$\hat{\Gamma}^2 = \frac{1}{6\left(\sqrt{-\hat{\gamma}}\right)^2} \varepsilon^{\hat{\imath}\hat{\jmath}\hat{k}} \varepsilon^{\hat{l}\hat{m}\hat{n}} \eta_{\hat{\lambda}\hat{\sigma}} \eta_{\hat{\nu}\hat{\rho}} \eta_{\hat{\mu}\hat{\tau}} \Pi_{\hat{\imath}}^{\hat{\mu}} \Pi_{\hat{\jmath}}^{\hat{\nu}} \Pi_{\hat{k}}^{\hat{\lambda}} \Pi_{\hat{l}}^{\hat{\sigma}} \Pi_{\hat{m}}^{\hat{\rho}} \Pi_{\hat{n}}^{\hat{\tau}} . \tag{3.28}$$

Using the 1.5 order formalism, the result for the κ transformation of the metric is[23]

$$\begin{aligned} \delta_{\hat{\kappa}} \left(\sqrt{-\hat{\gamma}} \hat{\gamma}^{\hat{\imath}\hat{\jmath}} \right) = & -2i\partial_{\hat{k}} \bar{\hat{\theta}} \hat{\gamma}^{\hat{k}(\hat{\imath}} \varepsilon^{\hat{\jmath})\hat{l}\hat{m}} \Pi_{\hat{l}}^{\hat{\rho}} \Pi_{\hat{m}}^{\hat{\tau}} \Gamma_{\hat{\rho}\hat{\tau}} \left(1 + \hat{\Gamma} \right) \hat{\kappa} \\ & + \frac{2i}{3\sqrt{-\hat{\gamma}}} \partial_{\hat{p}} \bar{\hat{\theta}} \Pi_{\hat{q}}^{\hat{\mu}} \Gamma_{\hat{\mu}} \hat{\gamma}^{\hat{p}\hat{q}} \varepsilon^{\hat{k}\hat{l}(\hat{\imath}} \varepsilon^{\hat{\jmath})\hat{m}\hat{n}} \hat{\kappa} \left(\Pi_{\hat{l}}^{\hat{\lambda}} \Pi_{\hat{\lambda}\hat{m}} \Pi_{\hat{k}}^{\hat{\nu}} \Pi_{\hat{\nu}\hat{n}} \right. \\ & \left. + \Pi_{\hat{l}}^{\hat{\lambda}} \Pi_{\hat{\lambda}\hat{m}} \hat{\gamma}_{\hat{k}\hat{n}} + \hat{\gamma}_{\hat{l}\hat{m}} \hat{\gamma}_{\hat{k}\hat{n}} \right) . \end{aligned} \tag{3.29}$$

Normally, for the type IIA superstring one does not have to go through the procedure of explicitly absorbing equation of motion terms into the transformation of the world-sheet metric. This is because the κ transformation in that case can also be formulated in terms of spinorial parameters that also carry an additional world-sheet index[22]. Variation of the action (3.17) with respect to the x^μ, θ^+ ,θ^- fields yields an expression proportional to $\Pi_i^\mu \Pi_{\mu j}$, so the required transformation law for $\left(\sqrt{-\gamma}\gamma^{ij}\right)$ can simply be read off. However, the presence of the second "cosmological" term in the supermembrane action (3.1) necessitates the present formulation of the κ symmetry, in which it is appropriate to use the 1.5 order formalism.

For the purposes of the reduction, we shall for the moment simply apply (3.6-3.9) and (3.11) to the transformations (3.21-3.22). Thus, we obtain the following transformations:

$$\delta_\kappa \theta = (1+\Gamma)\,\kappa \tag{3.30}$$

$$\delta_\kappa x^\mu = i\bar{\theta}\Gamma^\mu (1+\Gamma)\,\kappa \tag{3.31}$$

where now $\hat{\Gamma}$ is modified to

$$\Gamma = \frac{e^f}{2\sqrt{-\gamma}}\varepsilon^{ij}\Pi_i^\mu \Pi_j^\nu \Gamma_{\mu\nu}\Gamma_{11}\;. \tag{3.32}$$

Using the reduction procedure on equation (3.29) one now gets

$$\begin{aligned}\delta_\kappa\left(\sqrt{-\gamma}\phi^{\frac{2}{3}}\gamma^{ij}\right) &= -\left[4ie^f\phi^{\frac{2}{3}}\partial_k\bar{\theta}\gamma^{k(i}\varepsilon^{j)m}\Pi_m^\mu\Gamma_\mu\Gamma_{11}(1+\Gamma)\,\kappa\right]\\ &+\left[\frac{2i}{\sqrt{-\gamma}}e^{2f}\phi^{\frac{2}{3}}\partial_p\bar{\theta}\Pi_q^\mu\Gamma_\mu\gamma^{pq}\varepsilon^{l(i}\varepsilon^{j)m}\kappa\left(\Pi_l^\lambda\Pi_{\lambda m}+e^{-f}\phi^{\frac{2}{3}}\gamma_{lm}\right)\right]\end{aligned} \tag{3.33}$$

Let the r.h.s. of equation (3.33) be denoted by $2X^{ij}$. From equation (3.26) we have that

$$\begin{aligned}(\delta_{\hat{\kappa}}\hat{\gamma}_{33})_{\rm red} &= \delta_\kappa\left(\phi^{\frac{4}{3}}\right) = \frac{2}{\sqrt{-\gamma}}\phi^{\frac{2}{3}}\gamma_{ij}X^{ij}\\ &= -4i\partial_k\bar{\theta}e^f\phi^{\frac{4}{3}}\varepsilon^{kl}\Pi_l^\mu\Gamma_\mu\Gamma_{11}(1+\Gamma)\,\kappa\\ &\quad+\frac{2i}{\sqrt{-\gamma}}\partial_p\bar{\theta}\Pi_q^\mu\Gamma_\mu\gamma^{pq}e^{2f}\phi^{\frac{4}{3}}\gamma^{lm}\kappa\left(\Pi_l^\lambda\Pi_{\lambda m}+e^{-f}\phi^{\frac{2}{3}}\gamma_{lm}\right)\end{aligned} \tag{3.34}$$

As stated above, we require our reduced κ-transformations to preserve the auxiliary reduction conditions (3.14–3.16). First of all, however, note that the Kaluza-Klein conditions (3.6–3.9) together with (3.33) constrain κ to satisfy

$$\partial_\rho\kappa = 0 \tag{3.35}$$

which is to be expected for the reduced string theory.

Proceeding now to attempt the calculation of a reduced κ transformation, we use the first auxiliary equation (3.14) together with (3.11), (3.26) and (3.34) and find that

$$\begin{aligned}\delta_{\hat{\kappa}}\hat{\gamma}_{i3} &= -\frac{1}{\sqrt{-\hat{\gamma}}}2\left(\hat{X}^{i3}-\hat{\gamma}^{i3}\hat{X}_{\hat{k}}^{\hat{k}}\right)\\ &= \delta_\kappa\left(\phi^{\frac{4}{3}}A_i\right)\end{aligned} \tag{3.36}$$

and thus,

$$\delta_\kappa A_i = -\frac{2\phi^{\frac{-2}{3}}\gamma_{ij}}{\sqrt{-\gamma}}\left(\hat{X}^{jk}A_k + \hat{X}^{j3}\right) \tag{3.37}$$

which yields

$$\begin{aligned}\delta_\kappa\left(A_i - \Pi_i^{11}\right) = &\left[\partial_i\bar{\theta}e^f\phi^{\frac{4}{3}}\varepsilon^{kl}\Pi_k^\mu\Gamma_\mu\Gamma_{11}\left(1+\Gamma\right)\kappa\right]\left(\Pi_l - A_l\right)\\ &- 2i\partial_i\theta\Gamma^{11}\left(1-\Gamma^2\right)\kappa\,.\end{aligned} \tag{3.38}$$

Equations (3.34) and (3.38) imply that one cannot consistently impose (3.14–3.16) using this reduction scheme. However, the form of equation (3.34), which also holds for more general cases, suggests that this might be possible if one could reformulate the κ transformation so as to yield a traceless X^{ij}. This motivates a change over to fermionic parameters carrying a worldvolume index. This turns out to be posible because the supermembrane action (3.1) actually has an enlarged symmetry with a triplet of parameters $\kappa_{\hat{\imath}}$. Explicitly, we write

$$\begin{aligned}\delta_{\hat{\kappa}}\hat{\theta} = &\, i\left(\hat{\gamma}^{\hat{\imath}\hat{\jmath}}\hat{\Pi}_{\hat{\imath}}^\mu\Gamma_\mu + \frac{1}{2\sqrt{\hat{\gamma}}}\varepsilon^{\hat{\imath}\hat{\jmath}\hat{k}}\hat{\Pi}_{\hat{\imath}}^{\hat{\mu}}\hat{\Pi}_{\hat{\jmath}}^{\hat{\nu}}\Gamma_{\hat{\mu}\hat{\nu}}\right)\hat{\kappa}_{\hat{k}}\\ &- i\left(\hat{\gamma}^{\hat{p}(\hat{\imath}}\hat{\gamma}^{\hat{\jmath})\hat{q}} - \frac{1}{2}\hat{\gamma}^{\hat{\imath}\hat{\jmath}}\hat{\gamma}^{\hat{p}\hat{q}}\right)\left(\hat{\Pi}_{\hat{\imath}}^{\hat{\nu}}\hat{\Pi}_{\hat{\nu}\hat{\jmath}} - \hat{\gamma}_{\hat{\imath}\hat{\jmath}}\right)\hat{\Pi}_{\hat{p}}^{\hat{\lambda}}\Gamma_{\hat{\lambda}}\hat{\kappa}_{\hat{q}}\end{aligned} \tag{3.39}$$

The presence in the transformation (3.39) of an equation of motion term is crucial for the consistency of the ensuing reduction. We shall explain this presently, but first we give the form of the new κ transformation on the membrane metric. We again use the 1.5 order formalism. The result obtained is

$$\begin{aligned}\delta_{\hat{\kappa}}\left(\sqrt{-\hat{\gamma}}\hat{\gamma}^{\hat{\imath}\hat{\jmath}}\right) = &\, 4\sqrt{-\hat{\gamma}}\partial_{\hat{m}}\hat{\bar{\theta}}\left(\hat{\gamma}^{\hat{\imath}\hat{m}} - \frac{1}{\sqrt{-\hat{\gamma}}}\varepsilon^{\hat{\imath}\hat{m}\hat{k}}\hat{\Pi}_{\hat{k}}^{\hat{\nu}}\Gamma_{\hat{\nu}}\right)\left(\hat{\gamma}^{\hat{\jmath}\hat{n}} + \frac{1}{\sqrt{-\hat{\gamma}}}\varepsilon^{\hat{\jmath}\hat{l}\hat{n}}\hat{\Pi}^{\hat{\lambda}}\Gamma_{\hat{\lambda}}\right)\hat{\kappa}_{\hat{n}}\\ &- \frac{2}{\sqrt{-\hat{\gamma}}}\partial_{\hat{m}}\hat{\bar{\theta}}\varepsilon^{\hat{m}\hat{k}(\hat{\imath}}\varepsilon^{\hat{\jmath})\hat{l}\hat{n}}\left(\hat{\gamma}_{\hat{k}\hat{l}} - \hat{\Pi}_{\hat{k}}^{\hat{\nu}}\hat{\Pi}_{\hat{\nu}\hat{l}}\right)\\ &+ 4\sqrt{-\hat{\gamma}}\frac{\delta I}{\delta\hat{\bar{\theta}}}\left(\hat{\gamma}^{\hat{p}(\hat{\imath}}\hat{\gamma}^{\hat{\jmath})\hat{q}} - \frac{1}{2}\hat{\gamma}^{\hat{\imath}\hat{\jmath}}\hat{\gamma}^{\hat{p}\hat{q}}\right)\hat{\Pi}_{\hat{p}}^{\hat{\lambda}}\Gamma_{\hat{\lambda}}\hat{\kappa}_{\hat{q}}\end{aligned} \tag{3.40}$$

The last line of (3.40) results directly from the presence of the metric equation of motion in (3.39). We can now choose a simple reduction scheme, supplementing (3.6–3.9) with

$$\hat{\kappa}_i = \kappa_i. \tag{3.41}$$

$$\hat{\kappa}_3 = 0 \tag{3.42}$$

Upon employing these conditions, one finds that both ϕ and $\left(\Pi_i^{11} - A_i\right)$ transform into auxiliary conditions, as required. However, if the last line of equation (3.40) were not present one would find upon reduction that

$$\delta_\kappa\left(\Pi_i^{11} - A_i\right) \sim (\theta \text{ equation of motion}) \tag{3.43}$$

The fully reduced transformations following from equations (3.39) and (3.40) are

$$
\begin{aligned}
\delta_\kappa \theta^+ &= 2i\Pi_i^\mu \Gamma_\mu P_+^{ij}\kappa_j^- - e^f P_+^{ip} P_+^{jq} \Pi_p^\lambda \Gamma_\lambda \kappa_q^- \left(\Pi_i^\mu \Pi_{\mu j}\right) && (3.44)\\
\delta_\kappa \theta^- &= 2i\Pi_i^\mu \Gamma_\mu P_-^{ij}\kappa_j^+ - e^f P_-^{ip} P_-^{jq} \Pi_p^\lambda \Gamma_\lambda \kappa_q^+ \left(\Pi_i^\mu \Pi_{\mu j}\right) && (3.45)\\
\delta_\kappa X^\mu &= i\bar{\theta}^+ \Gamma^\mu \delta_\kappa \theta^+ + i\bar{\theta}^- \Gamma^\mu \delta_\kappa \theta^- && (3.46)\\
\delta_\kappa \left(\sqrt{-\gamma}\gamma^{ij}\right) &= 16\sqrt{-\gamma}\left(\partial_k \bar{\theta}^+ P_+^{ik} P_+^{jl} \kappa_l^- + \partial_k \bar{\theta}^- P_-^{ik} P_-^{jl} \kappa_l^+\right)\\
&\quad - 8e^f \sqrt{-\gamma}\left(\partial_m \bar{\theta}^+ P_+^{ip} P_+^{jq} P_-^{ml} \Pi_l^\nu \Gamma_\nu \Pi_p^\lambda \Gamma_\lambda \kappa_q^-\right.\\
&\quad \left. + \partial_m \bar{\theta}^- P_-^{ip} P_-^{jq} P_+^{ml} \Pi_l^\nu \Gamma_\nu \Pi_p^\lambda \Gamma_\lambda \kappa_q^+\right) . && (3.47)
\end{aligned}
$$

Here, we have decomposed θ and κ_i into positive and negative chirality components using $\frac{1}{2}(1+\Gamma_{11})$ and $\frac{1}{2}(1-\Gamma_{11})$ and the bosonic projectors

$$
\begin{aligned}
P_+^{ik} &= \frac{1}{2}\left(\gamma^{ik} + \frac{\varepsilon^{ik}}{\sqrt{-\gamma}}\right)\\
P_-^{ik} &= \frac{1}{2}\left(\gamma^{ik} - \frac{\varepsilon^{ik}}{\sqrt{-\gamma}}\right)
\end{aligned} \qquad (3.48)
$$

The form of these equations should be compared with the standard Green-Schwarz transformations[22]. Again, note the presence of equations of motion.

Finally, we must check the remaining auxiliary equation, (3.15). Acting on the equation $\left(2e^{-f} - \phi^2\gamma^{ij}\Pi_i^\mu \Pi_{\mu j}\right) = 0$ with our reduced transformations for $\theta^\pm, X^\mu$ and ϕ and picking out the coefficients of $\Pi_i^\mu \Pi_{\mu j}$, one derives the following expression for the κ-symmetry transformation of the worldsheet metric, modulo auxiliary conditions:

$$
\begin{aligned}
\delta_\kappa \gamma^{ij} &= 16\left(\partial_k \bar{\theta}^+ P_+^{ik} P_+^{jl} \kappa_l^- + \partial_k \bar{\theta}^- P_-^{ik} P_-^{jl} \kappa_l^+\right)\\
&\quad - 8e^f \sqrt{-\gamma}\left(\partial_m \bar{\theta}^+ P_+^{ip} P_+^{jq} P_-^{ml} \Pi_l^\nu \Gamma_\nu \Pi_p^\lambda \Gamma_\lambda \kappa_q^-\right. && (3.49)\\
&\quad \left. + \partial_m \bar{\theta}^- P_-^{ip} P_-^{jq} P_+^{ml} \Pi_l^\nu \Gamma_\nu \Pi_p^\lambda \Gamma_\lambda \kappa_q^+\right) + F(\sigma,\tau)\gamma^{ij}
\end{aligned}
$$

Here, $F(\sigma,\tau)$ is an unimportant field-dependent function. Its specific form is unimportant for one now see by inspection that (3.49) is compatible with (3.47). Thus, (3.44–3.47) give the correct κ symmetry transformations of the dimensionaly-reduced type IIA string theory.

4. Conclusion

We have discussed in this article two aspects of the relationship between strings and higher-dimensional extended objects. Both the analysis of the Yang-Mills anomaly structure and the treatment of κ symmetry under dimensional reduction/oxidation can be used to shed light on the main unsolved problems of the subject. The principle unsolved problem is the construction of the action for the most interesting higher extended object, the heterotic 5-brane[1]. This construction of the effective action for this solitonic sector of the heterotic string has proved difficult precisely because of the interplay between anomalous Yang-Mills transformations and the action-specific κ symmetry transformations. It may be hoped that the developments reviewed in this article may be useful in overcoming these difficulties.

The type IIA string κ-symmetry transformations derived *via* dimensional reduction in equations (3.44–3.47) is not the conventional one[22]. However, it is easy to show that the two types of transformation are equivalent on shell[6]. Starting from the transformation

$$\delta_{\hat{\kappa}}\hat{\theta} = \left(1 + \hat{\Gamma}\right)\kappa \tag{4.1}$$

and choosing

$$\hat{\kappa} = \frac{1}{2}i\left(\hat{\gamma}^{\hat{\imath}\hat{\jmath}}\hat{\Pi}^{\mu}_{\hat{\imath}}\Gamma_{\mu} + \frac{1}{2\sqrt{\hat{\gamma}}}\varepsilon^{\hat{\imath}\hat{\jmath}\hat{k}}\hat{\Pi}^{\hat{\mu}}_{\hat{\imath}}\hat{\Pi}^{\hat{\nu}}_{\hat{\jmath}}\Gamma_{\hat{\mu}\hat{\nu}}\right)\hat{\kappa}_{\hat{k}} \tag{4.2}$$

one obtains equation (3.39) up to equations of motion.

A number of interesting questions remain concerning the relation of string and p-brane κ symmetries. In the algebra of the string κ symmetry arises a Weyl transformation and also a local bosonic "λ" symmetry. It is already understood how the Weyl transformation arises from the third diffeomorphism invariance of the supermembrane prior to dimensional reduction[7]. It is to be expected that the full κ-symmetry algebra of the type IIA string theory may similarly be obtained by reduction from that of the supermembrane. All of these considerations are in their turn likely to have a bearing upon the relationship between string and membrane anomalies.

Acknowledgments

I would like to thank all of my collaborators on the work reviewed in this article.

References

1. A. Strominger, *Nucl. Phys.* **B343** (1990) 167;
   M.J. Duff and J.X. Lu, *Phys. Rev. Lett.* **66** (1991) 1402;
   Nucl. Phys. **B357** (1991) 534.
2. For reviews, see
   K.S. Stelle and P.K. Townsend in *Proc. CAP-NSERC Summer Institute in Theoretical Physics*, eds. G. Kunstatter et al. (World Scientific, Singapore, 1988);
   C.N. Pope and K.S. Stelle, *Int. J. Mod. Phys.* **A4** (1989) 4037.
3. B. de Wit, M. Lüscher and H. Nicolai, *Nucl. Phys.* **B320** (1989) 135.
4. C.M. Hull and E. Witten, *Phys. Lett.* **160B** (1985) 398.

5. J.A. Dixon, M.J. Duff and E. Sezgin, *Phys. Lett.* *279B* (1992) 265.
6. M.J. Duff, P.S. Howe, T.Inami and K.S. Stelle, *Phys. Lett.* **191B** (1987) 70.
7. A. Achucarro, P. Kapusta and K.S. Stelle, *Phys. Lett.* **232B** (1989) 302.
8. E. Bergshoeff, R. Percacci,E. Sezgin, K.S. Stelle and P.K. Townsend, "$U(1)$-Extended Gauge Algebras in p-Loop Space," *Nucl. Phys.* **B** (in press).
9. M.B. Green and J.H. Schwarz, *Phys. Lett.* *149B* (1984) 117.
10. E. bergshoeff, B. de Wit, M. de Roo and P. van Nieuwenhuizen, *Nucl. Phys.* **B195** (1982) 97.
11. S.J. Gates and H. Nishino, *Phys. Lett.* **157B** (1985) 157; A. Salam and E. Sezgin, *Phys. Scripta* **32** (1985) 283.
12. R. Nepomechie and P.G.O. Freund, *Nucl. Phys.* **B199** (1982) 482; P.A. Marchetti and R. Percacci, *Lett. Math. Phys.* **6** (1982) 405; H. Zainuddin, *J. Math. Phys.* **31** (1990) 2225; E. Bergshoeff, P.S. Howe, C.N. Pope, E. Sezgin and E. Sokatchev, *Nucl. Phys.* **B354** (1991) 113; J.A. de Azcárraga, J.M. Izquierdo and P.K. Townsend, *Phys. Lett.* **267B** (1991) 366.
13. N.K. Nielsen, *Nucl. Phys.* **B167** (1980) 249; M.J. Duff, B.E.W. Nilsson and C.N. Pope, *Phys. Lett.* **163B** (1985) 343; M.J. Duff, B.E.W. Nilsson, C.N. Pope and N.P. Warner, *Phys. Lett.* **171B** (1986) 170; R. Nepomechie, *Phys. Lett.* **171B** (1986) 195; *Phys. Rev.* **D33** (1986) 3670; R. Kallosh, *Phys. Scripta* **T15** (1987) 118; *Phys. Lett.* **176B** (1986) 50.
14. E. Bergshoeff, F. Delduc and E. Sokatchev, *Phys. Lett.* **262B** (1991) 444.
15. P.S. Howe, *Phys. Lett.* **273B** (1991) 90.
16. J. Mañes, R. Stora and B. Zumino, *Comm. Math. Phys.* **102** (1985) 157.
17. R. Percacci and E. Sezgin, "Symmetries of p-Branes" preprint, CTP-TAMU-32/92.
18. A. Achucarro, P. Kapusta and K.S. Stelle, unpublished.
19. E. Bergshoeff, E. Sezgin and P.K. Townsend *Phys. Lett.* **189B** (1987) 75.
20. E. Witten, *Comm. Math. Phys.* **92** (1984) 455; E. Bergshoeff, P.S. Howe, C.N. Pope, E. Sezgin and E. Sokatchev, *Nucl. Phys.* **B354** (1991) 113.
21. W. Siegel *Phys. Lett.* **128B** (1983) 397.
22. M.B. Green, J.H. Schwarz *Phys. Lett.* **136B** (1984) 367.
23. E. Bergshoeff, E. Sezgin and P.K. Townsend *Ann. Phys.* **185** (1988) 330.

Geometry in Partial Differential Equations (pp. 446–464)
eds. A. Pràstaro and Th. M. Rassias

METHODS FOR SOLVING TWO-DIMENSIONAL NONSTATIONARY MHD EQUATIONS AT SMALL ALFVEN-MACH NUMBERS

VYACHESLAV S. TITOV
Program Systems Institute of the Russian Academy of Sciences
Pereslavl-Zalessky 152140, Russia

ABSTRACT

The two-dimensional ideal magnetohydrodynamic (MHD) equations derived in the limit of small Alfven-Mach numbers are considered.

The magnetic field in this situation is generally described by complex analytical functions having several branch points. We discuss the methods for finding such functions from some boundary conditions.

If the evolution of magnetic field is found, the MHD flows may be described at small and large Mach numbers (hyposonic and hypersonic cases, respectively) by a system of ordinary differential equations.

In hypersonic case there exist the flows with frozen-in potential of magnetic field. They are determined by a linear diffusion-like equation. Two classes of solutions to this equation are found analytically.

1. Introduction

The ideal magnetohydrodynamic (MHD) equations due to nonlinearity are too intricate to solve them analytically. They have been solved only in a few special cases[1]. However, one can hope for success, dealing with the approximate MHD equations derived from the primary ones in important limiting cases, since passing to a limit usually simplifies equations. We review here the recent results obtained in this way for the two-dimensional ideal MHD equations derived in the limits when the Alfven-Mach number is small, while the Mach number may be small or large (hyposonic and hypersonic limits, respectively).

Such equations are relatively simple but rather useful for applications to space and laboratory plasmas[2]. In particular, they may be used to study the plasma flows in stellar atmospheres[3] and to understand some important aspects[4] of the magnetic reconnection process. In spite of their numerous applications, the effective methods for solving these equations appeared only for the last time. Our aim is to review the results obtained in[5–8] and to discuss the new ones.

In Section 2 we revise the approximation of a strong magnetic field to derive the equations of hyposonic and hypersonic MHD flows. The physical meaning of the approximate equations is also discussed there. Then, in Section 3, the general scheme for solving the equations at zero order of approximation is considered. Such equations in hypersonic case have a wide class of solutions representing the flows with frozen-in potential of magnetic field (Section 4). These flows may be described in terms of a linear diffusion-like equation for which two subclasses of solutions are determined analytically. Section 5 summarizes all the results.

2. Derivation of Approximate Equations

2.1 Approximation of Strong Magnetic Field

The MHD flow is said to be *two-dimensional* if it occurs in the same plane where the magnetic field lines lie. In this case all the physical quantities depend on the radius-vector $\mathbf{r} = (x, y)$ and on the time t.

Remind that the magnetic field $\mathbf{B}$ may be written then (see[1,2], for instance) in terms of the flux function $A(\mathbf{r}, t)$:

$$\mathbf{B} = (B_x, B_y) = (\partial_y A, -\partial_x A)\ . \tag{1}$$

The difference of the function values at two points defines a magnetic flux which crosses an arc connecting these points. The contours of $A(\mathbf{r}, t)$ at a fixed moment represent the magnetic field lines.

In the sequel, it is convenient to write the equation of motion in the following dimensionless form:

$$\varepsilon_1 \rho (\partial_t \mathbf{v} + (\mathbf{v} \cdot \nabla)\mathbf{v}) = -\varepsilon_1 \varepsilon_2 \frac{\nabla p}{\gamma} - \Delta A \, \nabla A\ , \tag{2}$$

where $\mathbf{v} = (v_x, v_y)$ is the velocity of plasma, ρ and p are its density and pressure, respectively. The typical values of length l_*, velocity v_*, time l_*/v_*, density ρ_*, pressure p_*, and strength of magnetic field B_* are the corresponding scale units in the considered equation. The dimensionless parameters have the form

$$\varepsilon_1 = v_*^2 / V_A^2\ , \tag{3}$$

$$\varepsilon_2 = c_s^2 / v_*^2\ . \tag{4}$$

Here $V_A = B_*/\sqrt{4\pi\rho_*}$ and $c_s = \sqrt{\gamma p_*/\rho_*}$ are typical values of the Alfven and sound speeds, respectively; the factor $\gamma \sim 1$ is the ratio of specific heats. So the parameter ε_1 equals the Alfven-Mach number squared and the other parameter ε_2 coincides with the inverse Mach number squared.

Since the normal and tangential directions relative to the magnetic field are physically different, it is reasonable to project on them Eq. 2. The normal projection is

$$\varepsilon_1 \rho \nabla A \cdot (\partial_t \mathbf{v} + (\mathbf{v} \cdot \nabla)\mathbf{v}) = -\frac{\varepsilon_1 \varepsilon_2}{\gamma} \nabla A \cdot \nabla p - \Delta A \, (\nabla A)^2\ . \tag{5}$$

The tangential projection can be represented by the vector product

$$\rho\nabla A \times (\partial_t \mathbf{v} + (\mathbf{v}\cdot\nabla)\mathbf{v}) = -\frac{\varepsilon_2}{\gamma}\nabla A \times \nabla p \tag{6}$$

or, speaking exactly, by its single nonzero component, normal to the plane of flow. The parameter ε_1 drops out of Eq. 6 because the magnetic force cannot accelerate plasma along the field lines. As it will be seen in Section 2.3, this fact is very important for deriving approximate equations.

The remaining equations are the induction equation for frozen-in flux

$$\partial_t A + \mathbf{v}\cdot\nabla A = 0\ , \tag{7}$$

the continuity equation

$$\partial_t \rho + \nabla\cdot(\rho\mathbf{v}) = 0\ , \tag{8}$$

and the adiabatic energy equation

$$\partial_t(p/\rho^\gamma) + \mathbf{v}\cdot\nabla(p/\rho^\gamma) = 0\ . \tag{9}$$

Eq. 5–9 constitute the full system of ideal MHD equations.

The approximation of strong magnetic fields corresponds to the case when

$$V_A \gg c_s\ , \tag{10}$$

$$V_A \gg v_*\ , \tag{11}$$

so that

$$\varepsilon_1 \ll 1\ , \tag{12}$$

$$\varepsilon_1\varepsilon_2 \ll 1\ . \tag{13}$$

Under such conditions Eq. 5 yields

$$\Delta A = 0\ . \tag{14}$$

Thus, the strong magnetic field evolves through a sequence of current-free or potential configurations. The frozen-in flux condition (Eq. 7) prohibits plasma particles to leave the contours of the harmonic flux function $A(\mathbf{r}, t)$. So the contours play twofold role: they represent the field lines, as mentioned earlier, and form infinitesimally thin tubes that channel the plasma flow. The shape of the tubes and their length change in course of time, following the field lines, and this changing is independent of motion of particles.

2.2. The Limit of Hyposonic Flows

In the case of hyposonic flows when

$$c_s \gg v_* \tag{15}$$

such that

$$\varepsilon_2 \gg 1 \tag{16}$$

Eq. 6 gives[4]

$$\nabla A \times \nabla p = 0 , \tag{17}$$

which implies that the pressure is constant along the magnetic lines, i.e.

$$p = \tilde{p}(A(\mathbf{r},t),t) , \tag{18}$$

where $\tilde{p}(A,t)$ is generally an arbitrary function. It is determined in each concrete problem by the corresponding boundary conditions, every time in ones own way.

Substituting Eq. 18 into Eq. 9 and using Eq. 1, 7 and 8, one can obtain

$$\nabla \cdot \mathbf{v} = -\frac{1}{\gamma p}\left[\partial_t p - \frac{\partial_t A}{\mathbf{B}^2}(\nabla A \cdot \nabla p)\right] , \tag{19}$$

which should be considered as an equation for $\mathbf{v}$. On account of Eq. 19 we may rewrite Eq. 8 as

$$\partial_t \rho + \mathbf{v} \cdot \nabla \rho = \frac{\rho}{\gamma p}\left[\partial_t p - \frac{\partial_t A}{\mathbf{B}^2}(\nabla A \cdot \nabla p)\right] . \tag{20}$$

Thus, Eq. 7, 14, and Eq. 18–20 describe the hyposonic flows of plasma in strong magnetic fields.

2.3. The Limit of Hypersonic Flows

In the case of hypersonic flows when

$$v_* \gg c_s \tag{21}$$

such that

$$\varepsilon_2 \ll 1 \tag{22}$$

Eq. 6 gives[4]

$$\nabla A \times (\partial_t \mathbf{v} + (\mathbf{v} \cdot \nabla)\mathbf{v}) = 0 . \tag{23}$$

It requires that the acceleration of plasma must be normal to the field lines in hypersonic flows. Notice that if one wanted to tend ε_1 and ε_2 to zero in Eq. 2 directly, Eq. 23 would be lost. It is due to dropping ε_1 out of Eq. 6, as mentioned above.

The set of Eq. 7–9, 14 and 23 describes the hypersonic MHD flows of plasma in strong magnetic fields at zero order of the approximation. It is also possible to derive the equations of higher order[8] by expanding first all the unknown values in powers of ε_1 and ε_2 up to the required order N. For example, the corresponding expansion of the flux function is

$$A(\mathbf{r},t,\varepsilon_1,\varepsilon_2) = \sum_{n=0}^{N}\sum_{k=0}^{n} A^{k,n-k}(\mathbf{r},t)\ \varepsilon_1^k \varepsilon_2^{n-k} . \tag{24}$$

Then it is necessary to substitute these expansions in the primary MHD equations and to collect all the coefficients of the same monomials $\varepsilon_1^k \varepsilon_2^m$ $(k+m \leq N)$ in separate expressions. Equating such expressions with zero, one can obtain the sought equations.

A similar approach was proposed first in[2]. But it has one defect: the authors expand the values in power series of one parameter, assuming the second parameter to be a fixed power of the first. Such an assumption reduces the two-parametric family of solutions to one-parametric and leads to an additional relation between the scale units. All the dimensionless parameters, however, are generally independent and hence the expansion has to be made as shown above.

Further on, we shall need the equations describing the flux function at first order of the approximation. They are obtained from Eq. 5 in the above manner and have the form

$$\Delta A^{10} = -\rho^{00} \nabla A^{00} \cdot (\partial_t \mathbf{v}^{00} + (\mathbf{v}^{00} \cdot \nabla) \mathbf{v}^{00}) / (\nabla A^{00})^2 \,, \tag{25}$$

$$\Delta A^{01} = 0 \,. \tag{26}$$

The values superscripted here by two zeros are determined from Eq. 7, 8, 14, and 23 in which zero superscripts are omitted for brevity.

Let us discuss now what physical properties the hypersonic flows manifest at zero order. As mentioned above, the magnetic field evolves in these flows through a sequence of current-free configurations. Such configurations are also force-free, i.e. the magnetic force is equal to zero everywhere in volume; other forces are absent as well. Nevertheless, the plasma may be accelerated, since Eq. 23 admits an acceleration normal to the magnetic field lines. So we have the paradox: the plasma is accelerated in the absence of forces.

In fact, the appropriate force exists, of course, but it is found at the next order of the approximation. On account of Eq. 26 the first order current density is

$$j = -\varepsilon_1 \Delta A^{10} \,, \tag{27}$$

where ΔA^{10} is determined by Eq. 25. The current passes perpendicularly to the plane of flow and interacts with the potential field of zero order. Such an interaction produces the appropriate magnetic force accelerating the plasma—this just settles the apparent paradox. It is interesting that by reference to individual particles the magnetic tubes behave themselves like stiff strings relative to the threaded light beads (Section 4.1).

3. Scheme of Solving the Equations

From the analytical point of view both systems of equations discussed above may in principle be solved as follows. The first two steps are the same in hyposonic and hypersonic cases: A is found from Eq. 14, then one can find from Eq. 7 the velocity component normal to the magnetic lines. The tangential component is determined by Eq. 19 and 23 in hyposonic and hypersonic cases, respectively. In the first case, however, before determining this component one needs to define the pressure $\tilde{p}(A,t)$. Then, the continuity equations (Eq. 8 and Eq. 20) yield us the density ρ. Finally, the pressure in hypersonic flows is found from Eq. 9.

Moving step by step in accord with this scheme, one should use in each step the information obtained in the previous ones. The calculating of magnetic field and velocity is rather subtle here, so it requires a special consideration.

3.1 Method for Calculating the Magnetic Field

The frozen-in flux law (Eq. 7) not only imposes constraints on the velocity field $\mathbf{v}$ but also requires the preservation of field line topology through the evolution of magnetic configuration. Moffat[9] explained this fact in quite clear way. Such a condition leads to the overdetermining[10] of the known classical boundary value problems (BVP) for Eq. 14, since it requires that the values of $A(\mathbf{r},t)$ at the separatrice field lines passing through the points of zero magnetic field (so called 'X-type neutral points') must be constant in course of time.

One can satisfy, however, this requirement and do the BVP to be well posed by introducing tangential discontinuity arcs (the current sheets or briefly CS) instead of the neutral points existing in the initial configuration. The CS in thus modified BVP play the role of free boundaries such that fitting their forms, positions, and sizes makes the BVP to be well posed. This idea were clearly formulated in[11] for the coronal CS formation caused by redistribution of magnetic flux due to horizontal displacements of field line footpoints at the photosphere. Here the chromosphere and corona is an upper half-plane $y > 0$, and the photospheric line is an x-axes.

One can consider a more general case when such a process may be also caused by an emergence of magnetic flux from below the photosphere. For this purpose it is sufficient to regard to be given not the displacements but the normal field component $B_y(x,0,t)$, as it was done for the problem with one CS in[12].

Since the boundary value $A(x,0,t)$ may be obtained from $B_y(x,0,t)$ by integrating it over x (see Eq. 1), we deal with the following *nonclassical Dirichlet BVP*: from tangential derivative $\partial A/\partial x$, given at each moment t on the boundary $y = 0$, find in the half plane $y \geq 0$ a time sequence of the harmonic flux functions $A(x,y,t)$ with *an initially fixed topology* of contours.

Thus, it is necessary to solve Eq. 14 at $y > 0$ for each t and to take into account the field topology constraints due to Eq. 7 by introducing the CS instead of the X-points so that the values of the flux function at the corresponding separatrices would be constant in course of time. A possible formation of CS near the points of contact of separatrices with photosphere[13,14] is not considered here.

Let the distribution of B_y on the boundary be such that the corresponding purely potential field has a dipole character of vanishing at infinity ($\sim 1/z^2$) and contains N neutral points at $y > 0$ and no ones at $y = 0$ (the case of $N = 1$ was considered in[6]. From these boundary conditions we want to find a potential configuration which contains CS, instead of neutral points. There is a general expression for the analytic function

$$B(z,t) \equiv B_y(x,y,t) + \mathrm{i}\, B_x(x,y,t) \tag{28}$$

of complex variable $z = x + iy$ describing such configurations without any connection with boundary conditions. Following to[11] we may write it in the form

$$B(z,t,\mathbf{r}_e) = P(z,t,\mathbf{r}_e)\, Q(z,\mathbf{r}_e)\,, \tag{29}$$

$$Q(z,\mathbf{r}_e) = \prod_{k=1}^{N} \sqrt{(z - z_{1k})(z - z_{2k})}\,. \tag{30}$$

The function $P(z,t,\mathbf{r}_e)$ here is holomorphic by variable z in the whole domain under study[11] and does not contain any nulls there. The complex values $z_{1k} = x_{1k} + i\, y_{1k}$ and $z_{2k} = x_{2k} + i\, y_{2k}$ are coordinates of the branch points, which are end points of the CS as well. The set of $4N$ real components x_{1k}, x_{2k}, y_{1k} and y_{2k}, $k = 1, \ldots, N$, are denoted here for brevity by the vector $\mathbf{r}_e$.

Eq. 29 was found in[11] for configurations with prescribed form of CS by regarding CS as boundaries on which $P(z,t,\mathbf{r}_e)$ must satisfy some boundary conditions. Following to this idea one can find the magnetic field for symmetric configurations when the form of CS may be guessed[12]. In more general cases a similar problem seems to be unsolvable.

However, suppose for a moment that $P(z,t,\mathbf{r}_e)$ is known and consider CS as corresponding cuts of z-plane for the multivalued function $B(z,t,\mathbf{r}_e)$. The shape of CS depends on the form of $P(z,t,\mathbf{r}_e)$ and coincides with the shape of the field line passing through the corresponding branch points. In particular cases such an approach happened to be rather fruitful[15]. We shall follow therefore this approach as well. Thus, in our case the problem reduces to finding the function $P(z,t,\mathbf{r}_e)$ from the normal field component $B_y(x,0,t)$ given at the boundary $y = 0$.

To solve this problem, introduce in the low half plane $y < 0$ N images of CS about x-axis. The end points of images have the conjugate coordinates z^*_{1k} and z^*_{2k}. In analogy with the end points of the real CS, these ones are also to be branch points. Therefore, we must introduce the corresponding additional factors in the square root of Eq. 30. This results in

$$Q_{hp}(z,\mathbf{r}_e) = \prod_{k=1}^{N} \sqrt{[(z - x_{1k})^2 + y_{1k}^2][(z - x_{2k})^2 + y_{2k}^2]}\,, \tag{31}$$

so that the magnetic field in our boundary problem is represented by

$$B(z,t,\mathbf{r}_e) = P(z,t,\mathbf{r}_e)\, Q_{hp}(z,\mathbf{r}_e)\,. \tag{32}$$

Owing to this trick the expression for $\Re\, P(z,t,\mathbf{r}_e)$ at $y = 0$ includes only *one normal* component $B_y(x,0,t)$:

$$\Re\, P(x,t,\mathbf{r}_e) = B_y(x,0,t)/Q(x,\mathbf{r}_e)\,. \tag{33}$$

Then the holomorphic function $P(z,t,\mathbf{r}_e)$ may be found in terms of Schwarz integral[16]. Taking account of this fact and Eq. 32, we obtain the main result[7]

$$B(z,t,\mathbf{r}_e) = -\frac{\mathrm{i}}{\pi} Q_{hp}(z,\mathbf{r}_e) \int_{-\infty}^{+\infty} \frac{B_y(\xi,0,t)\,d\xi}{(\xi - z)\,Q_{hp}(\xi,\mathbf{r}_e)} \,. \tag{34}$$

One can show that when $z_{1k} = z_{2k} \equiv z_k$, $k = 1,\ldots,N$, this integral is transformed to the classical one

$$B_{cl}(z,t) = -\frac{\mathrm{i}}{\pi} \int_{-\infty}^{+\infty} \frac{B_y(\xi,0,t)\,d\xi}{\xi - z} \tag{35}$$

for a purely potential field.

Notice that, due to the presence of the root, the function in Eq. 34 is doublevalued. Yet its two values at each point differ in signs only. So the tangent at the magnetic field line is defined in each point by Eq. 34 unambiguously. In other words, it defines magnetic lines up to the arrows indicating the field direction. We may always restore the right direction at a point by drawing through it the corresponding field line up to the boundary where unknown direction is determined by a known sign of $B_y(x,0,t)$. This is one of the way to separate the physical branch of the function defined by Eq. 34. When we refer to (34), we shall actually imply this branch further.

The positions of the end points of CS are not arbitrary—they are determined from certain equations. A part of them can be derived from the necessary asymptotic condition at infinity

$$B(z,t,\mathbf{r}_e) \sim 1/z^2 \,. \tag{36}$$

Here we take into account that an appearance of CS together with its 'negative' images of finite length is obviously incapable to change the initial asymptotic behaviour of magnetic field. When applied to Eq. 34, this condition yields[7] the following $2N+1$ equations for $\mathbf{r}_e$:

$$\int_{-\infty}^{+\infty} \frac{\xi^m B_y(\xi,0,t)}{Q_{hp}(\xi,\mathbf{r}_e)} d\xi = 0, \quad m = 0,\ldots,2N \,. \tag{37}$$

In addition to the right asymptotic behaviour (Eq. 36), the obtained equations guarantee an absence of nulls for the function $P(z,t,\mathbf{r}_e)$ in the top half plane. This important fact may be proved in rather general form by using the argument principle.

The frozen-in flux law gives the rest $2N-1$ equations for $\mathbf{r}_e$. The first N equations

$$\Re \int_{-\infty}^{z_{1k}} B(\zeta,t,\mathbf{r}_e)d\zeta - \Re \int_{-\infty}^{z_k} B_{cl}(\zeta,0)d\zeta = 0 \,, \quad k = 1,\ldots,N \,, \tag{38}$$

require an invariability of the flux function $A(x, y, t, \mathbf{r}_e)$ in course of time at one of the end points of each CS. The other $N-1$ equations

$$\Re \int_{z_{1k}}^{z_{2k}} B(\zeta, t, \mathbf{r}_e) d\zeta = 0 \ , \quad k = 1, \ldots, N-1 \ , \tag{39}$$

separate the end points belonging to the same CS. The N-th pair is separated automatically as a last one. We need such equations because without them the branch points are indistinguishable (see Eq. 31). Note also that the above equations do not depend on the contour of integrating. The latter is chosen from the reasons of convenience.

Thus, the solution of the considered problem is given by the integral

$$A(x, y, t, \mathbf{r}_e) = -\Re \int_{z_{11}}^{z} B(\zeta, t, \mathbf{r}_e) d\zeta - \Re \int_{-\infty}^{z_1} B_{cl}(\zeta, 0) d\zeta \ , \tag{40}$$

where the field $B(z, t, \mathbf{r}_e)$ is represented by Eq. 34, and the end points of the current sheets are determined by Eq. 37, 38, and 39.

The proposed method may be easily generalized to the case of the infinitely long CS. In this case it is necessary only to eliminate from Eq. 31 the factors containing x_{2k} and y_{2k} for such CS and to change the asymptotic conditions (Eq. 37) so that the field vanishes at infinity like $1/z$.

The first use of our method shows its high efficiency in calculating of magnetic configurations and investigating of their existence. This is due to that the initially complicated BVP with several free boundaries has reduced to solving some standard problems.

3.2 Method for Calculating the Flows

If the evolution of magnetic field is known, we can determine the velocity component normal to the field lines from the frozen-in flux condition (Eq. 7). So it is reasonable to represent the velocity as a sum of the normal and tangential components:

$$\mathbf{v} = -(\partial_t A \nabla A + \sigma \nabla \Phi)/\mathbf{B}^2 \ , \tag{41}$$

where the function $\Phi(\mathbf{r}, t)$ is a potential of magnetic field related with the flux function $A(\mathbf{r}, t)$ by the Cauchy-Riemann equations

$$\partial_x \Phi = -\partial_y A \ , \tag{42}$$

$$\partial_y \Phi = \partial_x A \ . \tag{43}$$

The new unknown $\sigma(\mathbf{r}, t)$ in Eq. 41 defines a relationship between the tangential component of velocity and magnetic field. The normal component in Eq. 41 is taken so that Eq. 7 would be satisfied automatically. Thus, in order to determine the velocity field, after that the evolution of magnetic field is already known, we need only to find $\sigma(\mathbf{r}, t)$.

In the hyposonic case this value is found from Eq. 19—after substituting Eq. 41 into Eq. 19 and some transformations using Eq. 7, 42, and 43 we obtain

$$\mathbf{B} \cdot \nabla(\sigma/\mathbf{B}^2) = \nabla A \cdot \nabla(\partial_t A/\mathbf{B}^2) + \frac{1}{\gamma p}[\partial_t p - (\nabla A \cdot \nabla p)\, \partial_t A/\mathbf{B}^2] \,. \qquad (44)$$

If we determine in this equation the pressure according to Eq. 18 (in addition to the evolution of magnetic field), it may be solved by the method of characteristics. The time enters in Eq. 44 as a parameter and the corresponding characteristics here are magnetic field lines.

In the hypersonic case analogous manipulations with Eq. 23 yield the following equation for σ:

$$\partial_t \sigma + \mathbf{v} \cdot \nabla \sigma = \mathbf{v} \cdot [\partial_t \mathbf{B} + (\mathbf{v} \cdot \nabla)\mathbf{B}] \,. \qquad (45)$$

Together with Eq. 41 this relation defines an equation for σ which may be solved by the method of characteristics. The characteristics here coincide with the trajectories of plasma particles. As the velocity is proportional to σ, one may expect that the initial single-valued distribution of σ has to turnover after a finite time. We suppose that at the places where this process develops the initial assumption (Eq. 21) breaks down and the pressure gradient becomes important, so in reality the shock waves must form there.

4. Harmonic 'Hypo-Alfvenic' HyperSonic ($\mathrm{H_S^A}$) Flows

4.1 The Representation of Hypersonic Flows in Conformal Coordinates

It is useful to study the MHD flows satisfied to Eq. 10 and 23 in a special system of the Eulerian coordinates the choice of which is based on the two physical properties of these flows[8]. First, the flux function $A(\mathbf{r}, t)$ is frozen into the moving medium (see Eq. 7) and, second, this function is harmonic.

The first property means the following: being at $t = 0$ in a point $\mathbf{r}_0$ such that $A(\mathbf{r}_0, 0)$ equals some a, the particle at any other time t is on the contour of $A(\mathbf{r}, t)$ with *the same* value a. Therefore, it is reasonable to take the value a as one of the new coordinates for particles.

The second property implies an existence of the conjugate harmonic function $\Phi(\mathbf{r}, t)$ related with the flux function $A(\mathbf{r}, t)$ by Eq. 42 and 43. At fixed t the contours of $\Phi(\mathbf{r}, t)$ and $A(\mathbf{r}, t)$ are *orthogonal* to each other. It is convenient therefore to put the value φ of the potential $\Phi(\mathbf{r}, t)$ as a second new coordinate for particles.

Thus, for each particle position in the x, y-plane there is a position in the a, φ-plane and vice versa. The functions $A(\mathbf{r}, t)$ and $\Phi(\mathbf{r}, t)$ define a mapping from the x, y- to the a, φ-plane, while some functions $X(a, \varphi, t)$ and $Y(a, \varphi, t)$ define the reverse mapping. Both these mappings are obviously conformal, so the latter has to satisfy also the Cauchy-Riemann equations

$$\partial_a X = \partial_\varphi Y \; , \tag{46}$$

$$\partial_\varphi X = -\partial_a Y \; . \tag{47}$$

The study of flows in the a, φ-plane is simpler than that in the real plane. Here the flow is *one-dimensional*, since each particle moves in the a, φ-plane along a straight line $a =$ const. Denote by $\theta(a, \varphi, t)$ the velocity of particle along this line in a point (a, φ) at a time t. It defines the rate of changing the potential $\Phi(\mathbf{r}, t)$ in a moving particle with the radius-vector

$$\mathbf{R}(a, \varphi, t) = (X(a, \varphi, t), \, Y(a, \varphi, t))$$

in the real plane. So the real velocity of particle $\mathbf{v}$ consists of the two parts:

$$\mathbf{v} = \partial_t \mathbf{R} + \theta \partial_\varphi \mathbf{R} \; . \tag{48}$$

Here $\partial_t \mathbf{R}$ represents the velocity of the point of the conformal network formed by the contours of flux function and potential. The vector $\theta \, \partial_\varphi \mathbf{R}$ is a particle velocity determined by reference to this point.

Taking account of Eq. 7, one can show that the relationship

$$(\partial_\varphi \mathbf{R})^2 (\partial_t \theta + \theta \, \partial_\varphi \theta) + (\partial_\varphi \mathbf{R} \cdot \partial_\varphi^2 \mathbf{R}) \theta^2 + 2(\partial_\varphi \mathbf{R} \cdot \partial_\varphi \partial_t \mathbf{R}) \theta + (\partial_\varphi \mathbf{R} \cdot \partial_t^2 \mathbf{R}) = 0 \tag{49}$$

represents Eq. 23 in the a, φ-plane[8]. This relationship defines the evolution of $\theta(a, \varphi, t)$ and has a clear meaning: it prohibits particles to be accelerated along the magnetic field lines—a constraint equivalent to that the primary Eq. 23 imposes, namely, that the acceleration must be perpendicular to the field lines. Indeed, Eq. 49 may be rewritten as

$$\partial_\varphi \mathbf{R} \cdot D_t^2 \mathbf{R} = 0 \; . \tag{50}$$

Here the value $D_t^2 \mathbf{R}$ is an acceleration of particle in the real plane, since the vector $\mathbf{R}(a, \varphi, t)$ defines its real position and the operator $\partial_t + \theta \partial_\varphi$ is a material derivative in the a, φ-coordinates. Now, taking into account that the vector $\partial_\varphi \mathbf{R}$ is tangential to the field lines, we come to the above statement.

The obtained Eq. 49 might be integrated by the method of characteristics after that the function $\mathbf{R}(a, \varphi, t)$ is already found from Eq. 46 and 47. All the solutions of Eq. 49 are 'woven' from the integral trajectories of the following system of ordinary differential equations:

$$\frac{d\varphi}{dt} = \theta \; , \tag{51}$$

$$\frac{d\theta}{dt} = -[(\partial_\varphi \mathbf{R} \cdot \partial_\varphi^2 \mathbf{R}) \theta^2 + 2(\partial_\varphi \mathbf{R} \cdot \partial_\varphi \partial_t \mathbf{R}) \theta + (\partial_\varphi \mathbf{R} \cdot \partial_t^2 \mathbf{R})] / (\partial_\varphi \mathbf{R})^2 \; . \tag{52}$$

The variable a here plays the role of parameter. Thus, the whole information about the possible flows is contained in this one-parametric family of the ordinary differential equations.

Write their solutions in the following symbolic form:

$$\varphi = \Phi(a, \varphi_0, \Theta_0(a, \varphi_0), t) , \tag{53}$$

$$\theta = \Theta(a, \varphi_0, \Theta_0(a, \varphi_0), t) , \tag{54}$$

where the values φ_0 and $\Theta_0(a, \varphi_0)$ correspond to the initial φ-coordinate and the velocity of particle in the a, φ-plane.

The set of Eq. 51 and 52 may be reduced to one second order ordinary differential equation

$$(\partial_\varphi \mathbf{R})^2 \frac{d^2\varphi}{dt^2} + (\partial_\varphi \mathbf{R} \cdot \partial_\varphi^2 \mathbf{R}) \left(\frac{d\varphi}{dt}\right)^2 + 2(\partial_\varphi \mathbf{R} \cdot \partial_\varphi \partial_t \mathbf{R}) \frac{d\varphi}{dt} + (\partial_\varphi \mathbf{R} \cdot \partial_t^2 \mathbf{R}) = 0 . \tag{55}$$

Gorbachov and Kelner[23] derived an analogous equation, relying on the Lagrangian approach to the problem. Yet a similarity between their and our equations is not accidental. It is due to that applying the method of characteristics to Eq. 49 is really equivalent to solving the problem in the Lagrangian representation.

They searched for the plasma flow, regarding the magnetic field *to be given* and used coordinates (a', φ') related with ours by some functions $g_1(a) = a'$ and $g_2(a, \varphi) = \varphi'$ and assumed that the field was given in terms of a function $\mathbf{R}(a', \varphi', t)$. However, if we try *to find* the magnetic field as well, we shall deal with the Cauchy-Riemann Eq. 46 and 47 having the simplest form in our system of coordinates. So the latter is more suitable for solving the whole problem.

Eq. 55 coincides with the equation of motion of a light bead sliding without friction along a stiff string. The variable φ here corresponds to some coordinate of the bead on the string. The variable a plays the role of parameter which we can consider simply constant. The string may be deformed in course of time, and the vector-function $\mathbf{R}(a, \varphi, t)$ at fixed a defines the string form at each moment. We would like to emphasize that it is much more than a mere mechanical analogy. This fact should be considered as a *new important heuristic principle* of analysis of the plasma flows in strong magnetic fields.

Write the remaining Eq. 8 and 9 of the basic system in the a, φ-plane

$$\partial_t(\rho J) + \partial_\varphi(\rho J \theta) = 0 , \tag{56}$$

$$\partial_t(p/\rho^\gamma) + \theta\, \partial_\varphi(p/\rho^\gamma) = 0 , \tag{57}$$

where the Jacobian

$$J = \partial_a X\ \partial_\varphi Y - \partial_\varphi X\ \partial_a Y . \tag{58}$$

It is not difficult to find $\rho(a, \varphi, t)$ and $p(a, \varphi, t)$ by the method of characteristics if $\mathbf{R}(a, \varphi, t)$ and $\theta(a, \varphi, t)$ are given. The corresponding expressions are determined by

$$\rho(a, \varphi_0, t) = \rho_0(a, \varphi_0)\, J_0(a, \varphi_0)(\tilde{J}\, \partial_{\varphi_0} \Phi)^{-1} , \tag{59}$$

$$p(a, \varphi_0, t) = p_0(a, \varphi_0)\, J_0^\gamma(a, \varphi_0)(\tilde{J}\, \partial_{\varphi_0} \Phi)^{-\gamma} . \tag{60}$$

Here the initial values are subscripted by zero, the letter $\tilde{J}$ denotes the composite function $J(a, \tilde{\varphi}(a, \varphi_0, t), t)$ with $\tilde{\varphi}(a, \varphi_0, t) = \varphi(a, \varphi_0, \Theta_0(a, \varphi_0), t)$.

Thus, we have the following results[8]. In the a, φ-plane all the values are represented by Eq. 53, 54, 59, and 60. They determine the implicit parametric representation in which φ_0 plays the role of parameter. In addition, one can say about the two-parametric representation of the values in the x, y-plane if we regard a and φ in turn as parameters. Indeed, the distributions of physical quantities on the a, φ-plane and the vector-function $\mathbf{R}(a, \varphi, t)$ define implicitly the corresponding distributions on the x, y-plane whenever we put $\mathbf{r} = \mathbf{R}(a, \varphi, t)$.

4.2 General Properties of H_S^A Flows

It is interesting that there exist the hypersonic MHD flows with the potential $\Phi(\mathbf{r}, t)$ frozen into the moving medium as well as the flux function $A(\mathbf{r}, t)$. In other words, the potential $\Phi(\mathbf{r}, t)$ in such flows is governed by

$$\partial_t \Phi + \mathbf{v} \cdot \nabla \Phi = 0 \,, \tag{61}$$

so that the value of potential is temporary invariable in each individual particle over its motion. Therefore, the particle situated initially in $\mathbf{r}_0$ is always on the equipotential line defined by the condition $\Phi(\mathbf{r}, t) = \Phi(\mathbf{r}_0, 0)$.

A priori, it is not obvious that such flows really exist. Yet it is just so and this fact is especially surprising in view of such flows happen to be many. To show this, note first that Eq. 61 implies that (see Eq. 51)

$$\theta(a, \varphi, t) = 0 \tag{62}$$

in each point of the a, φ-plane. So, instead of Eq. 49, we have

$$\partial_t^2 \mathbf{R} \cdot \partial_\varphi \mathbf{R} = 0 \,. \tag{63}$$

Three Eq. 63, 46, and 47 form an overdetermined system in two unknown functions $X(a, \varphi, t)$ and $Y(a, \varphi, t)$, but nevertheless they do give solutions for the evolution of magnetic configurations with frozen-in potential $\Phi(\mathbf{r}, t)$.

The flow corresponding to such an evolution is trivial in the a, φ-plane: all the particles simply remain at rest. In the x, y-plane they trace therefore the points of the conformal network formed by the contours of the potential and flux function.

Next, the analytical function

$$Z(f, t) = X(a, \varphi, t) + \mathrm{i}\, Y(a, \varphi, t) \tag{64}$$

of the complex variable

$$f = a + \mathrm{i}\, \varphi \tag{65}$$

represents all the solutions of Eq. 46 and 47. The function $Z(f, t)$ at a fixed t is the inverse of the complex magnetic potential

$$F(z, t) = A(\mathbf{r}, t) + \mathrm{i}\, \Phi(\mathbf{r}, t) \tag{66}$$

of the same complex variable z as in Eq. 28.

By using now Eq. 64 and 65, one can transform Eq. 63 to

$$\partial_t^2 Z/\partial_f Z - \overline{\partial_t^2 Z/\partial_f Z} = 0 , \tag{67}$$

where the overbar stands for a complex conjugate. It follows that each term here is an arbitrary real function $\chi(t)$ and so the unknown function $Z(f,t)$ must satisfy[6,8]

$$\partial_t^2 Z = \chi(t)\, \partial_f Z . \tag{68}$$

Thus, the hypersonic MHD flows with frozen-in magnetic potential really exist and Eq. 68 is basic for their determining.

The representation of the flows in terms of the function $Z(f,t)$ has several merits. In particular, knowing $Z(f,t)$, we can easily draw the magnetic field lines at any moment. To manage it, one needs to fix t and a and then, running varies values φ in Eq. 65, to compute the values of $Z(f,t)$. Their real and imaginary parts, respectively, yield the x- and y-coordinates of the points on the contour of $A(\mathbf{r},t) = a$ or, in other words, on the field line.

In addition, with the help of $Z(f,t)$ we can easily find the motion of any individual particle. Let a particle be at the initial moment in some point (x_0, y_0) such that the complex potential in this point

$$F_0(z_0) = F(z_0, 0) , \tag{69}$$

$$z_0 = x_0 + \mathrm{i}\ y_0 . \tag{70}$$

Since the particle is at rest in the a,φ-plane, it is always at the point with $f = F_0(z_0)$. Therefore, in real space the particle is located at the point determined by the composite complex function

$$\mathcal{Z}(z_0, t) = Z(F_0(z_0), t) . \tag{71}$$

Its real and imaginary parts, respectively, define the x- and y-coordinate of the particle at each moment t.

Thus, the flows considered here are the time sequences of the conformal mappings represented by Eq. 71. So the hypersonic MHD flows with the frozen-in magnetic potential $\Phi(\mathbf{r},t)$ possess a standard property of such mappings—they *conserve locally the angle* between any two lines passing formed by moving particles.

The real and imaginary parts of $\partial_t \mathcal{Z}(z_0,t)$ yield the velocity components of particle and so they are conjugate harmonic functions of x_0 and y_0. Note that this property may be used as a starting-point[5] to derive Eq. 68. For brevity, we shall call it *the* H_S^A *equation* and its solutions and the corresponding flows—*the* H_S^A *solutions* and H_S^A *flows*. This notation originates from the *Harmonic 'Hypo-Alfvenic' HyperSonic* flows called so in accordance with the conditions under study.

Note that the $\mathrm{H}_\mathrm{S}^\mathrm{A}$ equation is *linear*, in contrast with Eq. 23 from which it was derived. Also it coincides almost with the form of a spatially inhomogeneous equation of one-dimensional diffusion if one transposes the variables t and f in it. This is a good basis for studying the $\mathrm{H}_{\mathrm{S}}^{\mathrm{A}}$ flows. The study is convenient to do by using the complex z-, z_0-, and f-planes, since the $\mathrm{H}_{\mathrm{S}}^{\mathrm{A}}$ flows are described in terms of the functions $Z(f,t)$ and $\mathcal{Z}(z_0,t)$.

The general expressions of physical values are simplified for the $\mathrm{H}_{\mathrm{S}}^{\mathrm{A}}$ flows. By Eq. 58–60 we have

$$\rho(a,\varphi,t) = \rho_0(a,\varphi)|Z_0'(f)/\partial_f Z(f,t)|^2 \,, \tag{72}$$

$$p(a,\varphi,t) = p_0(a,\varphi)|Z_0'(f)/\partial_f Z(f,t)|^{2\gamma} \,, \tag{73}$$

where the prime stands for the derivative of the initial function

$$Z_0(f) = Z(f,0) \,. \tag{74}$$

It is helpful to have similar expressions for the magnetic and electric fields. Eq. 1 yields for the magnetic field

$$B_y(a,\varphi,t) + \mathrm{i}\, B_x(a,\varphi,t) = -1/\partial_f Z(f,t) \,. \tag{75}$$

The electric field has the one component $E(\mathbf{r},t) = -\partial_t A(\mathbf{r},t)$ perpendicular to the plane of flow. It is written as

$$E(a,\varphi,t) = \Re\left(\partial_t Z(f,t)/\partial_f Z(f,t)\right) . \tag{76}$$

Eq. 75 helps to understand better the constraints imposed on $\mathrm{H}_{\mathrm{S}}^{\mathrm{A}}$ flows. Let us express $\partial_f Z$ from Eq. 75 and substitute it into Eq. 68. Then the imaginary and real parts of the resulting relation, respectively, give

$$B_x\partial_t^2 X + B_y\partial_t^2 Y = 0 \,, \tag{77}$$

$$B_x\partial_t^2 Y - B_y\partial_t^2 X = \chi(t) \,. \tag{78}$$

The first equation shows that the tangential acceleration vanishes—this property is equivalent to the starting Eq. 63. The second equation means that the product of magnetic field and normal acceleration in $\mathrm{H}_{\mathrm{S}}^{\mathrm{A}}$ flows may vary in time but not in space.

The latter constraint leads to a singular behavior of the $\mathrm{H}_{\mathrm{S}}^{\mathrm{A}}$ flows in the vicinity of magnetic neutral point. Near this point the acceleration of plasma tends to infinity if $\chi(t) \neq 0$. The used approximation fails there, because the initial assumption about a strong magnetic field is violated. Yet the tendency of the flow to be accelerated near this point is likely to be valid, since the region where the approximation fails tightens to a point as the parameters ε_1 and ε_2 tend to zero. In order to show that, we estimate the size of this region. Suppose that at dimensionless distance r from the neutral point the magnetic field behaves like $B_* r^n$ with $n > 0$. Then the corresponding local Alfven speed equals $V_A r^n$. Now, taking account of Eq. 3 and 4, we obtain that locally the Alfven speed becomes larger than the sound speed at a distance

$$r > (\varepsilon_1\varepsilon_2)^{1/2n} \,. \tag{79}$$

Thus, as ε_1 and ε_2 tend to zero, the region of inapplicability of our model really decreases. We may hope therefore that, asymptotically, the $\mathrm{H}_\mathrm{S}^\mathrm{A}$ equation describes adequately the motion of the plasma even near the neutral points.

Eq. 78 causes a special evolution of the current density j in $\mathrm{H}_\mathrm{S}^\mathrm{A}$ flows. According to Eq. 27, 74, 75, and 78, its first order value equals

$$j = -\varepsilon_1 \chi(t) \rho_0(a, \varphi) |Z_0'(f)|^2 \, . \tag{80}$$

We see from here that the ratio of current densities at any two particles does not change in time and is determined by the initial magnetic configuration and plasma density.

Summarizing, we can say that there are the hypersonic MHD flows with frozen-in potential of strong magnetic field governed by one diffusion type Eq. 68. They are characterized by two general properties. First, such flows are conformal, and, second, they conserve the ratio of the current densities in any pair of moving particles.

4.3. Acceleration-Free H_S^A Flows

Since there are no restrictions on the function $\chi(t)$, we may put $\chi(t) = 0$ for the simplest case. This condition corresponds to the acceleration-free $\mathrm{H}_\mathrm{S}^\mathrm{A}$ flows described by

$$\partial_t^2 Z = 0 \, . \tag{81}$$

Whence one can easily find

$$Z(f,t) = Z_0(f) + V_0(f)\, t \, . \tag{82}$$

Here the function $Z_0(f)$ defines the initial relation between the z- and f-coordinates of particles; its inverse function is the initial complex potential of magnetic field. The function $V_0(f)$ defines the initial velocity field. Both these functions are assumed to be given.

To obtain the Lagrangian form of solution, one needs to replace f by $F_0(z_0)$ in Eq. 82. As a result we get[5]

$$\mathcal{Z}(z_0, t) = z_0 + \mathcal{V}_0(z_0)\, t \, , \tag{83}$$

$$\mathcal{V}_0(z_0) = V_0(F_0(z_0)) \, . \tag{84}$$

In the considered flows there must appear strong discontinuities. Indeed, any particle in such a flow moves with constant velocity and in fixed direction along a straight line. The velocity value and its direction, however, change from particle to particle. Therefore, several particles with different values f may come together in one point z. If this occurs, the magnetic potential $F(z,t)$ becomes multi-valued. The latter should be interpreted as an evidence of appearance of strong discontinuity[18] in the flow. This point of view is confirmed in our case—as it has been shown in[8], the strong discontinuity appears here as a corresponding cut of the domain of indices for our solution when a physical branch is separated in it.

Using Eq. 82, one can obtain the general expressions for the plasma density

$$\rho(a,\varphi,t) = \rho_0(a,\varphi)|1 + V_0'(f)\,t/Z_0'(f)|^{-2}\,, \tag{85}$$

pressure

$$p(a,\varphi,t) = p_0(a,\varphi)|1 + V_0'(f)\,t/Z_0'(f)|^{-2\gamma}\,, \tag{86}$$

magnetic field

$$B_y(a,\varphi,t) + \mathrm{i}\,B_x(a,\varphi,t) = -(Z_0'(f) + V_0'(f)\,t)^{-1}\,, \tag{87}$$

and electric field

$$E(a,\varphi,t) = \Re\,[V_0(f)(Z_0'(f) + V_0'(f)\,t)^{-1}]\,, \tag{88}$$

respectively. Eq. 82–88 completely describe the acceleration-free $\mathrm{H}_\mathrm{S}^\mathrm{A}$ flows.

We see from here that the found solutions have singularities in the critical points where

$$\partial_f Z(f,t) = Z_0'(f) + V_0'(f)\,t = 0$$

and all the physical values tend to infinity. Near these points the pressure gradient becomes so large that our approximations break down. Such singularities is likely to be smoothed out in solutions of full MHD equations.

4.4 H_S^A Flows with Frozen-in Current Density

Suppose now that $\chi(t) = \text{const} \neq 0$, then the plasma is accelerated in $\mathrm{H}_\mathrm{S}^\mathrm{A}$ flows and Eq. 68 is the form of the diffusion or heat conduction equation. The general solution of this equation may be represented[19] as a power series by variable t

$$Z(f,t) = Z_0(f) + \sum_{k=0}^{\infty}\Big(Z_0^{(k+1)}(f)\frac{(\chi t^2)^{k+1}}{(2k+2)!} + V_0^{(k)}(f)\frac{\chi^k t^{2k+1}}{(2k+1)!}\Big)\,, \tag{89}$$

where the functions $Z_0(f)$ and $V_0(f)$ stand for the same values as in Eq. 82; their $(k+1)$-th and k-th derivatives with respect to f are denoted by $Z_0^{(k+1)}(f)$ and $V_0^{(k)}(f)$, respectively, and $V_0^{(0)}(f) = V_0(f)$. Other parameters of plasma are as before given by Eq. 72–76.

To write the solution in Lagrangian form, one simply replaces f by $F_0(z_0)$ and the derivative d/df by $(F_0'(z_0))^{-1}d/dz_0$ in Eq. 89 . The result is[6,8]

$$\begin{aligned}\mathcal{Z}(z_0,t) = z_0 + \sum_{k=0}^{\infty}\Big[&\Big(\frac{1}{F_0'(z_0)}\frac{d}{dz_0}\Big)^{k+1}(z_0)\frac{(\chi t^2)^{k+1}}{(2k+2)!}\\ &+ \Big(\frac{1}{F_0'(z_0)}\frac{d}{dz_0}\Big)^{k}(\mathcal{V}_0(z_0))\frac{\chi^k t^{2k+1}}{(2k+1)!}\Big]\,,\end{aligned} \tag{90}$$

where the function $\mathcal{V}_0(z_0)$ is defined by Eq. 84. and the resulting H_S^A flows are nontrivial, even if the plasma is initially at rest. By contrast, the acceleration-free H_S^A flows have no such property.

The difference in their behavior is due to different values of current density in these cases. Eq. 80 shows that the first order current density equals zero for acceleration-free H_S^A flows with $\chi = 0$, but not for in the present case of $\chi \neq 0$ where the magnetic force caused by the nonzero current density must accelerate the plasma. So it begins moving, even if its initial velocity is zero.

Note also that the current density is temporary invariable in each moving particle if χ is constant (see Eq. 80). Its value depends only on the initial magnetic field and plasma density. Therefore, the solution represented by Eq. 89 corresponds to the H_S^A flows with *frozen-in current density.*

5. Conclusions

We have considered the two-dimensional nonstationary MHD equations describing the flows of ideal plasma in a strong magnetic field. In two limiting cases of hyposonic and hypersonic flows the procedure for deriving a chain of equations at various orders of the approximation is formulated.

The magnetic field of zero order is determined by complex analytical functions having several branch points, the position of which may be found from the asymptotic and frozen-in flux conditions. It should be done before calculating the flow.

If the evolution of magnetic field is established, one can find the velocity of the flow. Its normal component is explicitly expressed from the frozen-in flux condition, while the tangential component is described in each of the two cases by different partial differential equations of the first order. The latter may be solved by the method of characteristics. The characteristics for each case are also different: in hyposonic flows they coincide with the magnetic field lines, whereas in hypersonic flows—with the trajectories of plasma particles.

It is useful to study the hypersonic flows in a special conformal coordinate system in which the positions of plasma particles at each moment are represented by the values of the flux function and potential of magnetic field. Using this system of coordinates, one can show[8] that there exists a particular class of solutions describing flows with three characteristic properties. First, both the potential and flux function are frozen into the moving plasma. Second, the ratio of current densities of any two moving particles remains constant in course of time at the first order of approximation. Third, each flow is represented by a continuous sequence of conformal mappings of the plane. We use the ligature H_S^A (Harmonic 'Hypo-Alfvenic' HyperSonic) to denote these general class of solutions.

The solutions are described by a linear diffusion-like equation (H_S^A equation) containing an arbitrary function of time. If this function is identically constant, the H_S^A equation is solved analytically. The solutions with zero and nonzero constants are physically different. They represent, respectively, unaccelerated and accelerated flows with zero and nonzero first-order current density.

The found solutions are usually characterized by multi-valued potentials of the magnetic field. Each potential, however, may be split into several single-valued brunches by cutting up its domain of indices. One of the brunches admits a physical interpretation[8]. The obtained results represent an important new step in the nonlinear theory of plasma physics.

References

1. E. R. Priest, *Solar Magnetohydrodynamics* (Reidel, Dordrecht, 1982).
2. B. V. Somov and S. I. Syrovatskii, in *Proc. of the P. N. Lebedev Physical Institute* **74** (Consultants Bureau, New York and London, 1976), p. 13.
3. S. I. Syrovatskii and B. V. Somov, in *Solar and Interplanetary Dynamics*, eds. M. Dryer and E. Tandberg-Hanssen, IAU Symp. **91** (1980), 425.
4. S. I. Syrovatskii, *Sov. Phys. JETP* **33** (1971) 933.
5. V. S. Titov, in *Physics of Solar Plasma*, (Nauka, Moscow, 1989, in Russian), p. 21.
6. V. S. Titov, in *Proc. of the All-Union Meeting II 'Mathematical models of the near cosmos'* (Nucl. Phys. Res. Inst. of the Moscow University, Moscow, 1991, in Russian).
7. V. S. Titov, *Solar Phys.* **139** (1992), 401.
8. V. S. Titov and E. R. Priest, *Geophys. and Astrophys. Fluid Dyn.* (1993), submitted.
9. H. K. Moffat, *J. Fluid Mech.* **159** (1985) 359.
10. B. C. Low, *Astrophys. J.* **323** (1987) 358.
11. J. J. Aly and T. Amari, *Astron. Astrophys.* **221** (1989) 287.
12. T. Amari and J. J. Aly, *Astron. Astrophys.* **227** (1990) 628.
13. N. Seehafer, *Solar Phys.* **145** (1986) 223.
14. B. C. Low *Astrophys. J.* **381** (1991) 295.
15. T. J. Tur and E. R. Priest, *Solar Phys.* **48** (1976) 89.
16. M. Laurentiev and B. Chabat, *Méthode de la théorie des fonctions d'une variable complexe*, (Edn. Mir, Moscou, 1972).
17. V. S. Gorbachev and S. R. Kelner, *Sov. Phys. JETP* **94** (1988) 89, in Russian.
18. V. V. Lychagin, *Acta Appl. Math.* **3** (1985) 135.
19. J. R. Cannon, *The one-dimensional heat equation*, (Addison-Wesley Pub. Com., Menlo Park, California, 1984).

AUTHOR INDEX